The Woodworker's Bible

Macmillan Publishing Company
866 Third Avenue, New York, N.Y. 10022
Collier Macmillan Canada, Inc.

Cover photography by Maggie Murray and Richard Greenhill.
Table (detail right) by John Makepeace.

This book was created and designed by Martensson Books, Ltd, London, and produced by Reference International Ltd.

Library of Congress Cataloging-in-Publication Data

Martensson, Alf.
The woodworker's bible.

Reprint. Originally published: New York : Bobbs-Merrill, 1980.
"A Bobbs-Merrill book."
Includes index.
1. Cabinet-work. 2. Carpentry. 3. Woodworking tools.
4. Woodworking machinery. I. Title.
[TT197.M37 1986 684'.08 86-12690
ISBN 0-02-011940-2

First published in the United States by The Bobbs-Merrill Company, Inc.
Indianapolis/New York 1980

First Macmillan Edition 1986

33 32 31 30

Printed in the United States of America

The Woodworker's Bible

Alf Martensson

A Bobbs-Merrill Book
Macmillan Publishing Company
New York

Important note:

In the interest of clarity, we removed the machine guards for many of the photographs in this book. Proper safety procedures are absolutely essential in woodworking and no machine is safe without all its guards in place.

It would have been completely impossible to compile the amount of diverse information, photographs and drawings in this book without the generous help of many individuals and companies and without the "overtime" efforts of the design team who put the book together.

Zena Flax of Flax and Kingsnorth worked miracles in the design of the book and the splendid drawings by Roger Courthold and Robert Stoneman of Terry Allen Designs give the book its character. My thanks too to Jackie Branch and Anne Jope who made sense of my writing to type the manuscript.

Professor T. A. Oxley of Protimeter Ltd. assisted with technical information about their moisture meter and Philip Cole supplied invaluable information, photographs and data about the Sperberg chain saw mill.

Sergio Mora of the London College of Furniture allowed us to use photographs and data from his research with carbon fiber reinforcement in wood. Mr R. Stevens at the London College of Furniture answered my endless questions and allowed us to take photographs of the facilities at the College.

Mr. L. E. Gilson at Parry and Son in London provided useful photographs, tools for photography and invaluable advice and Professor J. F. Levy of the Botany Department, Imperial College allowed us to use photographs of wood structure from his work. Wadkin Ltd. also provided photographs which proved extremely helpful.

My thanks to the manufacturers who supplied their machinery and tools for photography: Elu Machinery Ltd., DeWalt (McCulloch of Europe), Robin Sorby and Sons Ltd., Stanley Power Tools, Wolf Electric Tools Ltd. and Tinker Engineering Co.

John Makepeace kindly allowed us to spend a day taking photographs at The School for Craftsmen in Wood in Beaminster, Dorset and Maggie Murray took the splendid pictures.

Special thanks are also due to Erland Russell who inspired me with his love for the spindle molder and provided the names of many of the craftsmen whose work appears in the book.

Finally, our thanks to the craftsmen who allowed us to use photographs of their designs and who were patient with us while decisions were made.

Addresses of manufacturers and craftsmen appear on page 288.

Contents

Dictionary of hand tools and devices

ADZ

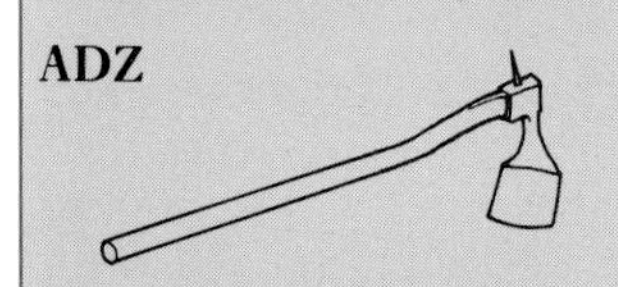

Axe-like tool but with blade at right angle to handle (haft), like the claw of a hammer. Used to dress large timbers by straddling the length and swinging the adz down to scoop up wood with the curved blade.

Carpenter's adz

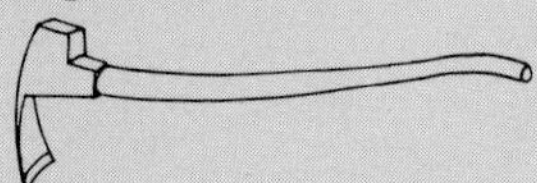

Adz with straight handle and almost flat blade.

Hollowing adz

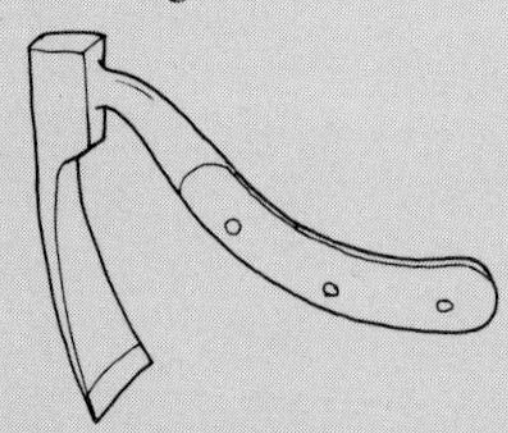

Head tapers back to prevent binding in work. Oval hardwood handle.

ANGLE DIVIDER

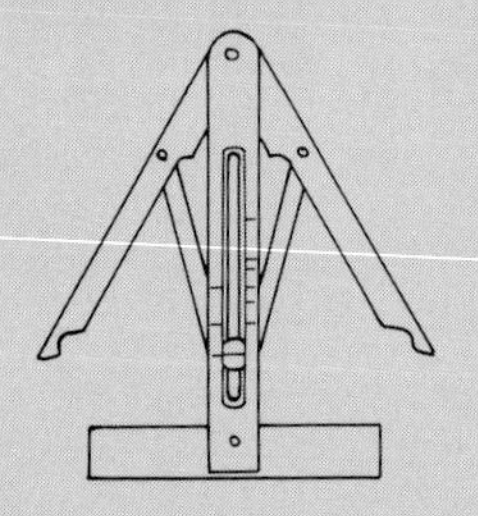

Device used for measuring and bisecting angles, for fitting moldings and for general layout work and laying out of polygons.

AUGER

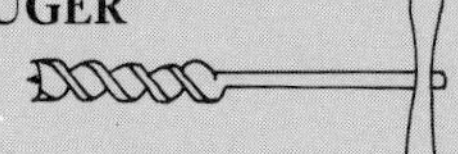

Traditional tool for drilling large and long holes. Handle mounted on shaft up to 24in long, is turned by hand.

Burn auger

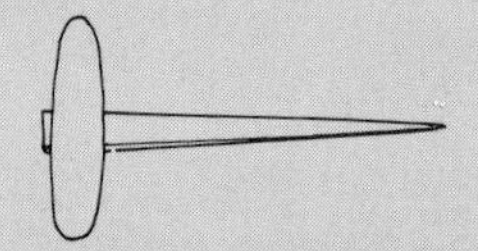

Old-fashioned drill with rectangular section pointed blade used to burn rather than cut holes.

Auger bit see BIT

AWL

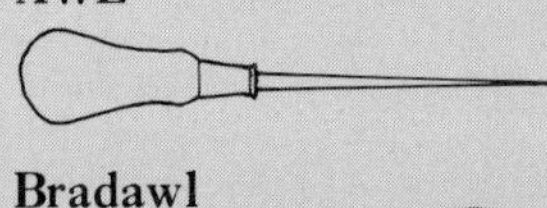

Bradawl

Wood boring tool with sharp, screwdriver-like edge used by hand to start holes for small nails and screws.

Scratch awl

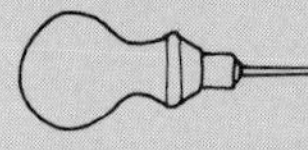

Beech handle with sharp point for marking joints.

AXE

American axe

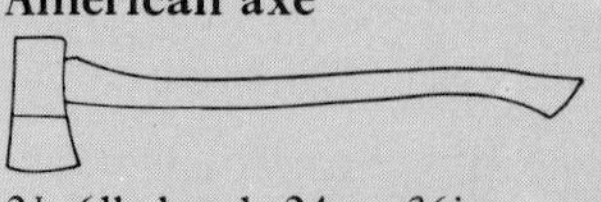

$2\frac{1}{2}$–6lb head, 24 to 36in long hickory handle. Used for general felling work. Head shapes vary regionally.

Broad axe (side axe)

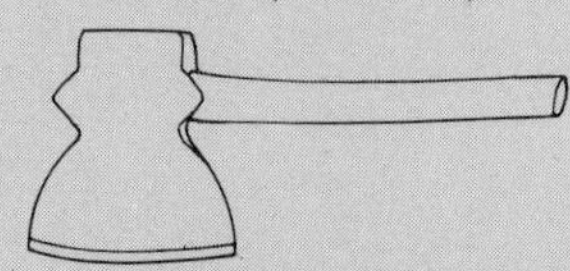

Traditional tool used by carpenters to finish wood. Large head with one side of bit beveled. Small handle angled out to clear side of wood.

Double bit felling axe

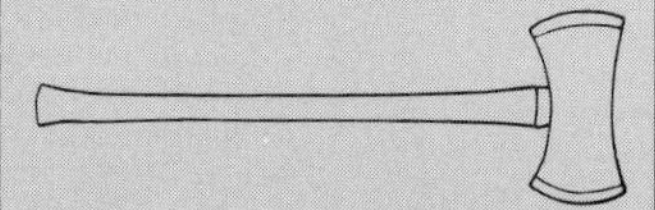

Long handle and hefty $3\frac{1}{2}$lb head, ideal for felling trees. Two cutting edges makes it more versatile and means less sharpening.

Kent axe (broad hatchet)

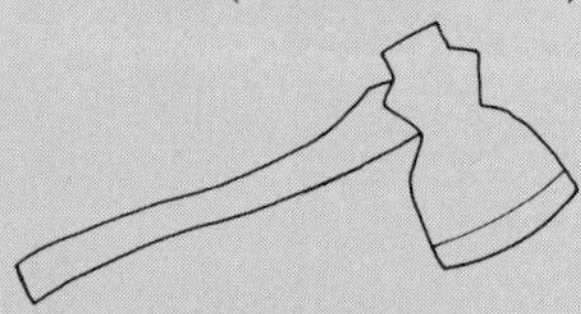

$2\frac{1}{2}$lb head, 18in hickory handle, general purpose axe with symmetrical blade and curved edge.

BEADING TOOL

Used for making molding patterns in places where molding planes cannot be used, e.g. diagonally across work or on irregularly shaped surfaces.

Single handed beader

Beading tool held in one hand and used to cut beads, flutes or reeds.

Universal hand beader

Beader used like a spokeshave for beading or fluting on irregular surfaces. Has square fence or oval fence for curved work. Several different blades for various beads, flutes, reeds etc.

BENCH DOGS

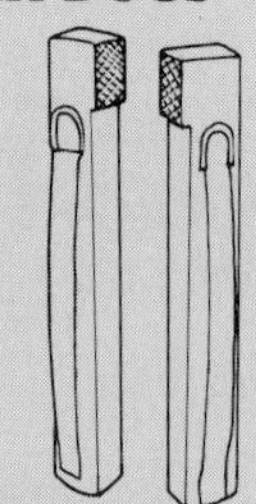

Bench aid like bench stop but all steel. Can be fitted into bench top by mortising rectangular holes at regular intervals along front of bench. 1 × $\frac{5}{8}$in knurled face and 7in long shank. Also available with round shank for fitting into drilled holes.

BENCH HOOK

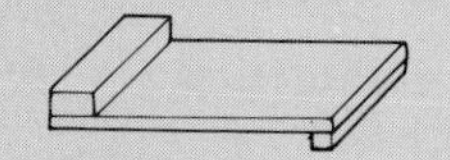

Device usually made of hardwood like beech or maple. 8 × 10in approximately. Used instead of vise on bench top to hold small pieces of wood stationary while sawing with back saw.

BENCH SCREW

Steel screw with handle and collar used to make wooden vises, presses such as a veneering press. Made in various sizes depending on use.

BENCH STOP

Cast iron or hardwood body fixed to underside of bench. Retractable $1\frac{1}{2} \times$ 1in block projects through the surface of workbench. Used as a stop for wood being planed.

BEVEL

Combination bevel

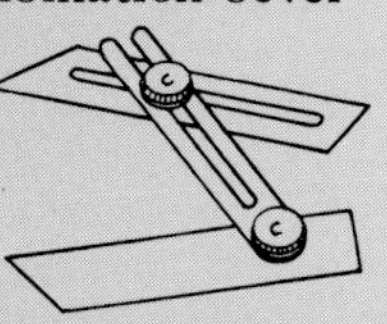

Device used to measure or lay out any desired angle flat on the work. Added blade makes this a versatile bevel.

Sliding bevel (T-bevel)

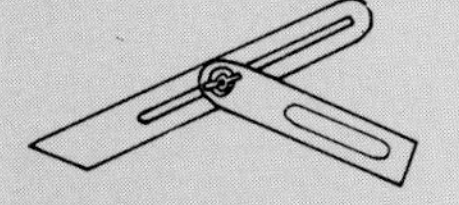

Commonest type of bevel used by woodworkers for checking and marking angles. Consists of flat steel blade that slides and pivots in wood or metal handle with a locking screw. Blades range from 6 to 12 in long. Angle of bevel is set either with a protractor or by fitting the bevel to a piece already beveled.

BIT

Auger bit

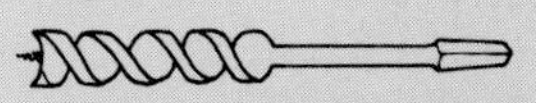

Most commonly used woodboring bit. Tapered square tang is held in jaws of brace. Cutting done by vertical spurs on outside and by horizontal cutters called cutting edges. Available in two types: Jennings pattern, $\frac{1}{4}$ to $1\frac{1}{4}$in diameter, have a double helix twist for faster wood clearance. Solid center type, $\frac{1}{4}$ to $1\frac{1}{4}$in diameter, with single twist is stronger. Alternative type available with single spur for difficult woods.

AUGER BIT GAUGE

Depth stop fixed to auger bit.

Bright spoon bit

Simple, gouge-shaped bit formerly used for drilling holes in end grain.

Center bit

Unlike auger bit, has no twist and can therefore not pull shavings up and out of the hole being bored. A fast, clean cutting bit for drilling shallow holes with brace. Old type has brad point and is therefore frequently used in machine drilling.

Countersink bit

Available in $\frac{3}{8}$, $\frac{1}{2}$ and $\frac{5}{8}$in diameter. Bores cone-shaped hole to sink head of screw flush with surface. Available with square tang for use in brace or straight shanked for hand or electric drill.

Dowel bit (short auger bit)

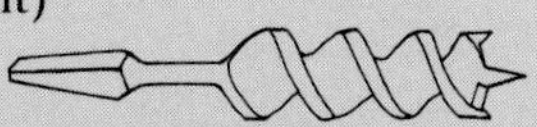

Used in brace to bore holes for dowel pegs.

Expansion bit (expanding bit)

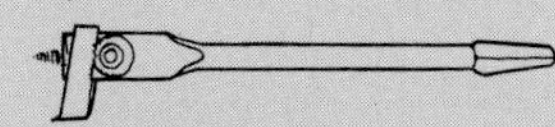

Small, $\frac{1}{2}$ to $1\frac{1}{2}$ or large $\frac{7}{8}$ to 3in diameter. Similar to solid center bit. Cutter can be adjusted to enlarge the diameter of the bit. Square shank to fit brace.

Extension bit holder

Rod-like device used to extend brace bits. Designed to let you drill holes in cramped spaces or to unusual depths.

Forstner bit

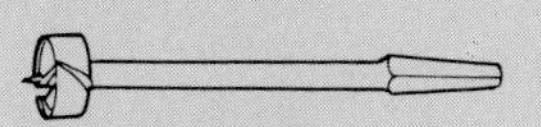

Round or square shank for electric drill or brace. $\frac{3}{8}$ to 2in diameter. Cutting end has deep rim and small pointed center. Therefore cuts a clean hole with flat bottom. Used to drill partway through work where flat bottomed hole is needed and where the spurs or screws of regular auger bit would go through the wood.

Plug cutter

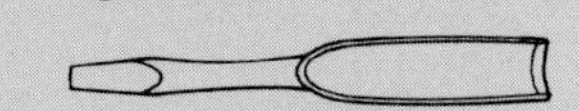

Most plug cutters are used in power drills. This hand tool cuts around a hole to produce a core or plug which can then be inserted in another hole.

BRACE see DRILLS

BRADAWL see AWL

BRAD DRIVER (push pin)

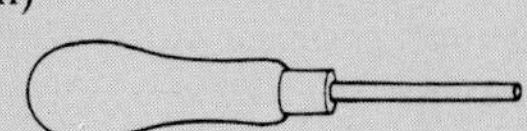

Tool for driving brads and other small nails too small to hold in the hand. Brad is inserted in hollow steel tip and held by magnet. Then handle is depressed and while the steel tip stays on the surface of the work, the interior pin drives the brad in.

BURNISHER

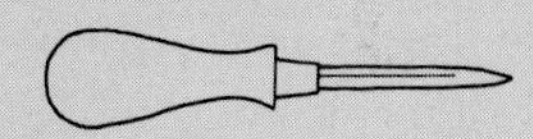

Awl-like tool with tough steel blade for turning the edges on scraper blades. Often homemade from old Swiss files, ground and polished.

Three corner burnisher

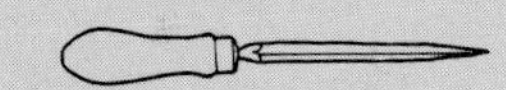

Triangular-shaped burnisher tapering to a point. Tool is 5in long and $\frac{3}{8}$in wide. Also available in oval and round shapes.

Wheel burnisher

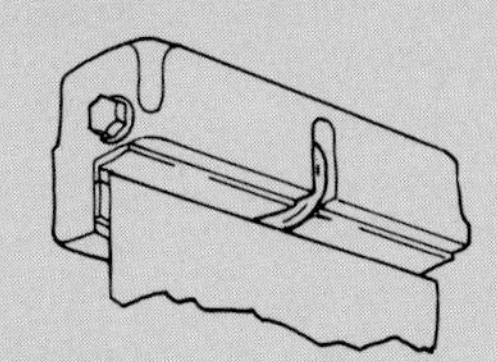

Burnisher used for quick and accurate sharpening of scraper blades after squaring off on oilstone.

BUTT MARKER see BUTT GAUGE

CALIPERS

Instrument for measuring inside and outside dimensions of round, oval, irregular objects where it is difficult to use a rule. Used in general work in thicknessing and turning. Set and measured against a rule.

Double calipers

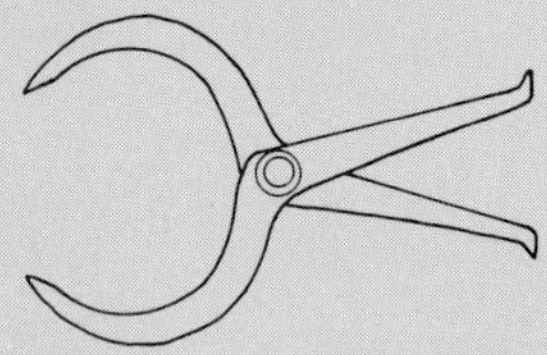

Works as inside and outside caliper and gauge for matching outside to inside diameter.

Dual indicating calipers

Caliper for taking outside and inside measurements up to 6in. Can also be locked for use as a fixed gauge.

Inside calipers

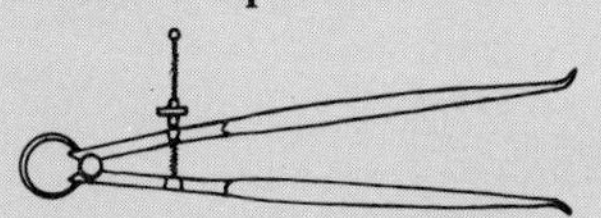

Calipers used to measure inside diameters. Made in a variety of sizes in either plain or spring type.

Odd leg calipers (Jenny or hermaphrodite)

Up to 6in size, used to find center of round or square sections and to scribe lines parallel to edge.

Outside calipers

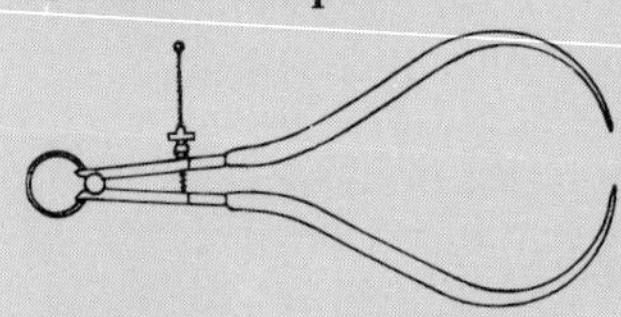

Shaped like compass, but has bowed legs with points turned toward each other. Used for measuring outside diameters.

CARVING TOOLS

CARVING CHISELS

Dog leg chisel

Chisel with bent blade and bevel ground uppermost for finishing flat recesses.

Parting tool (V-tool)

V-shaped blade used to make grooves or square cornered cut-outs. Ground on outside face, made with straight or curved blades.

Skew-cut chisel

Cutting edge is not at right angles to sides. Beveled on two sides.

Spoon bit chisel (short bend or spade chisel)

Designed for deep cutting. Used to finish work after preliminary shaping with spoon bit gouge.

Straight chisel

Chisel with rectangular blade ground on both faces with rounded heels. Either square or skew cutting edge. Edges honed and finely stropped for razor sharpness.

CARVING GOUGES

Gouges are numbered to indicate the shape of the cutting edge and depth of cut. They are lighter than a firmer gouge and available in many sections. All are out-canneled, ground on outside face. Blades are either parallel or taper towards bolster for greater clearance.

Back bent gouge

Curve reversed to form convex cutting edge.

Bent or curved gouge

Has slightly bent tip used for deep work.

Fishtail gouges

Gouges with tapering blades available in larger sizes.

Macaroni tool

Device used as gouge and V-parting tool. Has three working edges ground with both inside and outside bevels.

Spades

Entire shank of gouge curved to form S.

Spoon bit gouge (front bent gouge)

Very sharp curve on front section of blade only for deep recesses.

Straight gouge

Has parallel sides and is available in many different sweeps.

Veiner

Fine, straight bladed gouge with U-shaped cutting edge and parallel sides. Available in widths from $\frac{1}{16}$ to $1\frac{1}{4}$in.

CARVER'S HOOK

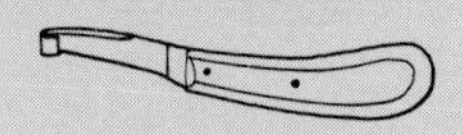

Knife with curved blade, designed to be pushed or pulled, depending on type. Twin-edged so cuts left or right, used for broad strokes and roughing out work. Two sharp edges for shaping, rounding and scooping.

CARVER'S PUNCHES

Steel punches cut with designs used to texture backgrounds.

CARVER'S SCREW

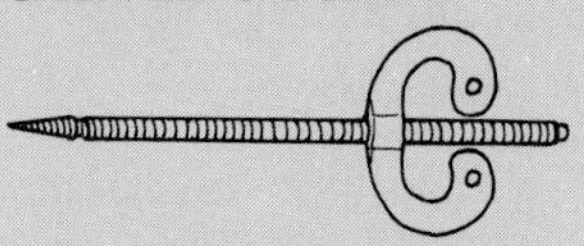

Double-threaded rod holds piece on bench without clamps.

CENTER FINDER

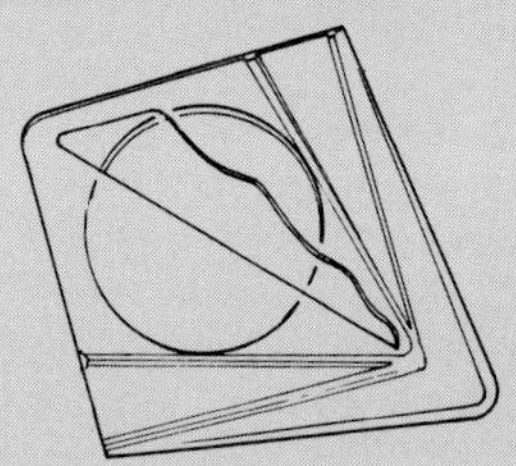

Plastic device for finding center of round or octagonal stock.

CHALK LINE

Used in construction and woodwork for marking a straight line longer than the longest available straightedge by stretching a line coated with chalk and then snapping it, leaving a straight line on the work. End of line has hook for one-person use.

CHAMFER GUIDE

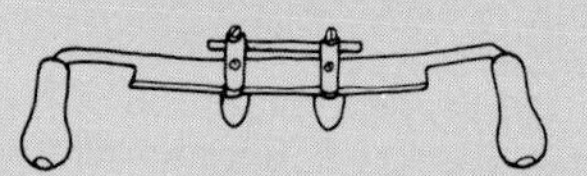

Device fitted to drawknife to make it into chamfer tool.

CHISELS

Butt chisel

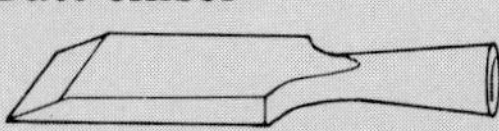

Chisel with blade only $2\frac{1}{4}$ to $3\frac{1}{4}$in long. Suited for putting on small hardware where not much strength is required, e.g. butt hinges.

Corner chisel

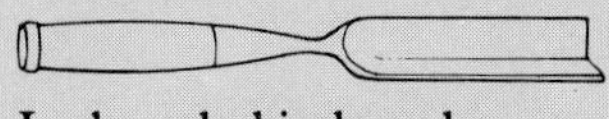

L-shaped chisel used to clear out corners or mortise slots.

Drawer lock chisel
(bolt chisel)

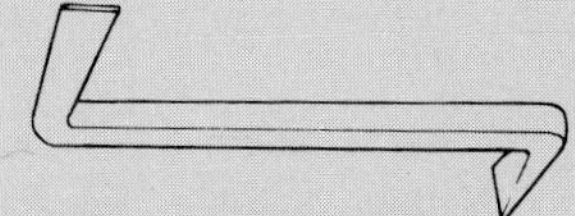

6in long used to cut lock recesses. Square-sectioned steel bar cranked at both ends. Each end tapered and ground to sharp edge. One edge is parallel to long bar of tool, the other is at right angle to it.

Firmer chisel

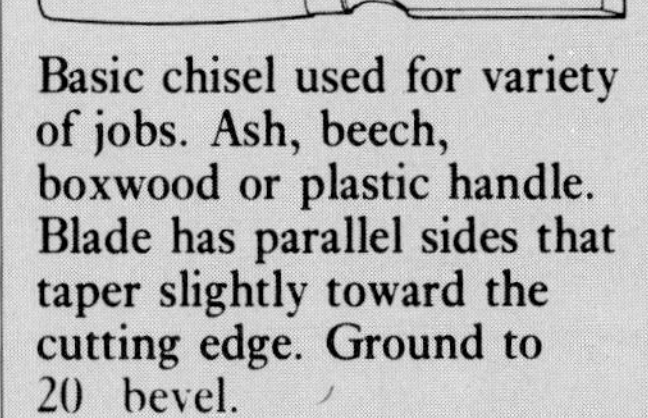

Basic chisel used for variety of jobs. Ash, beech, boxwood or plastic handle. Blade has parallel sides that taper slightly toward the cutting edge. Ground to 20 bevel.

BEVEL EDGE FIRMER CHISEL

Beveled instead of square sides. Good general purpose chisel.

SOCKET FIRMER CHISEL

Socket handle construction allows chisel to withstand heavier blows.

Framing chisel

Heavier version of firmer chisel. Used to cut tenons. Fitted with ferrules on handle to prevent splitting. Flat or with slight curve on bevel side.

Mortise chisel

Socket-type construction for heavy-duty work. Very square and stout, one-purpose tool for chiseling out mortises.

LOCK MORTISE CHISEL (SWAN NECK MORTISE CHISEL)

Curved blade to remove waste from deep mortises. Designed for cutting deep, blind recesses.

SASH MORTISE CHISEL

$\frac{1}{4}$ to $\frac{1}{2}$in wide blade. Used for light work with softwoods.

Paring chisel

Lighter duty tool. Used for shaping and preparing long, planed surfaces such as long housings. Usually worked by hand alone. Smallest angle bevel of all chisels at 15°. Available in bevel edge or firmer socket type.

Ripping chisel

Like a crow bar, but with wider and sharper chisel. Steel bar, either straight or goose-necked and notched at one end for pulling nails.

CHISEL GAUGE

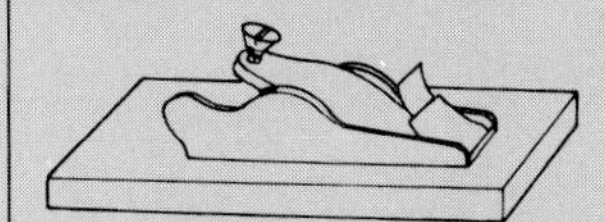

Device used for blind nailing. Gauge is attached to $\frac{1}{4}$in chisel with bevel edge up. Shaving is raised and nail inserted underneath. Shaving is then glued down to hide nail head.

CHISEL GRINDER

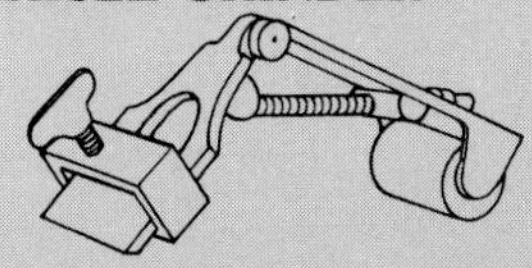

Device for holding a chisel or plane iron at right angle when being sharpened on stone.

CLAMPS see also HOLDFAST

Band clamp

Steel strap with horn, pulled together with a hand screw. Used on round or oval work.

FLEXI-CLAMP

4 to 8ft steel band shaped with angle brackets to required form to clamp irregular shapes such as guitar bodies.

Bar clamp (cabinet clamp)

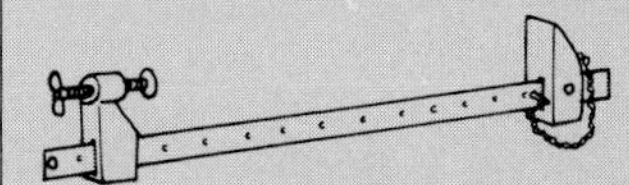

24 to 60in long. Used to hold frameworks or boards together while gluing. Long steel bar with fixed jaw at one end and adjustable jaw with bolt at other end held in place by notches or holes in the bar. Extensions available to increase length.

T-BAR CLAMP

36 to 84in long, heavy-duty clamp with T-section instead of rectangular steel bar.

C-clamp (G-clamp or carriage maker's clamp)

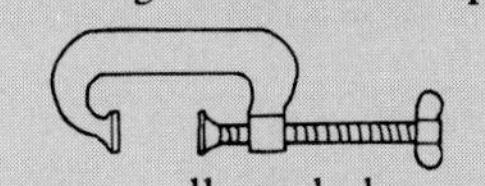

Common all steel clamp, 2 to 12in capacity.

Cam actuated clamp

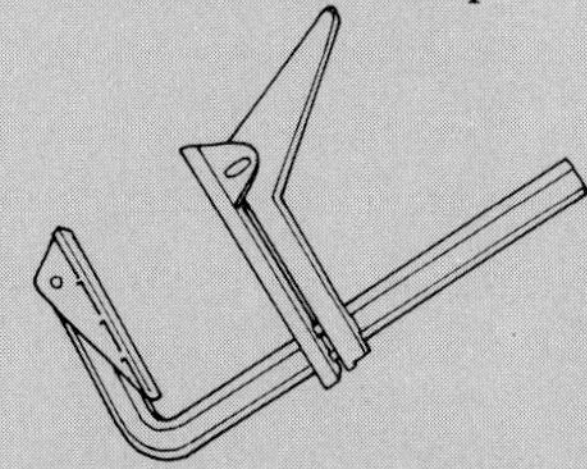

Quick action clamp fitted with plastic jaws. 8 and 18in capacity.

Clamp heads

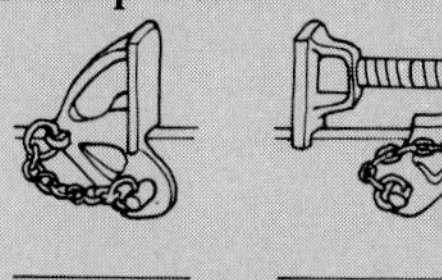

To fit wooden board fixed with $\frac{3}{8}$in steel pins. Used to make bar clamp of any size.

Corner clamp (miter clamp)

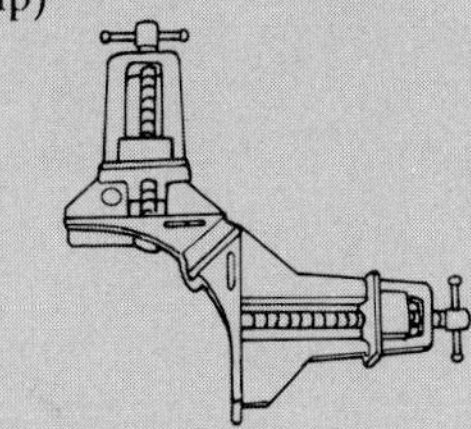

Triangular device used to clamp corners of frames.

Edge clamps

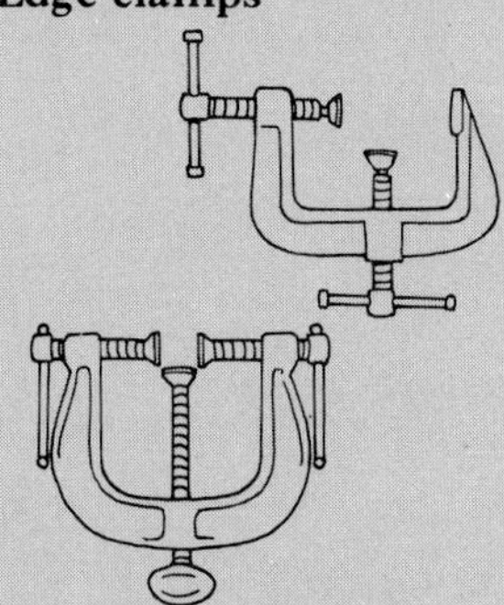

Designed to hold edges onto end of workpiece. Applies right angle pressure to edge or side of work as well as top and bottom. Shaped like C-clamp with additional screw at right angles to first or modified shape with three screw threads for greater flexibility.

Frame clamp

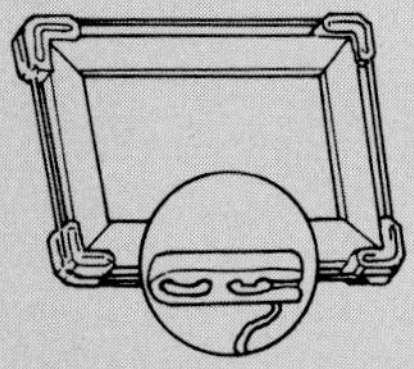

Simple device using plastic or aluminum corner blocks on plastic cord. Used to clamp mitered frames.

Hand screw (parallel clamps)

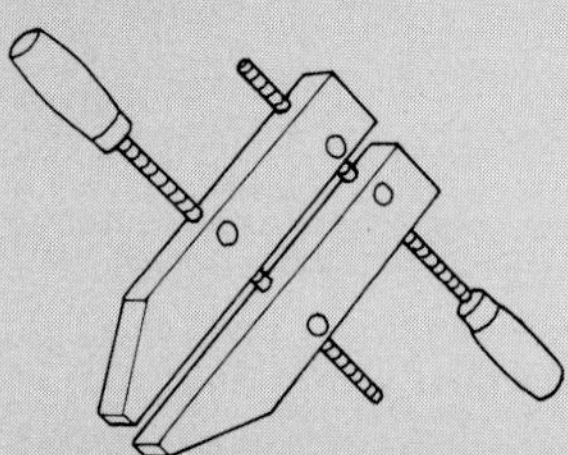

Medium-sized clamp with beech or maple arms, joined by long parallel steel bolts in opposite directions. Bolts are adjusted separately, so clamps are used for parallel or non-parallel work. Traditional hand screws used wood threads.

Jet clamp

Two moveable jaws with slide on steel bar to form clamp of required length. Jaws fitted with swivel brackets and protective pads.

Log dogs

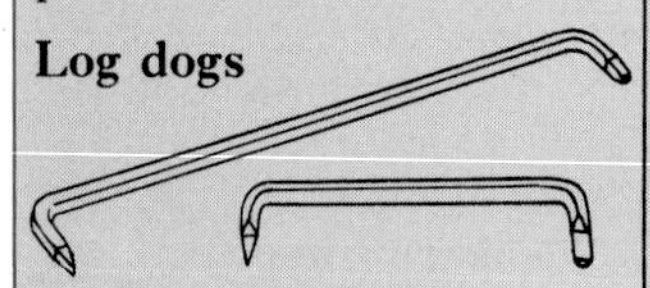

Metal clamps like pinch dogs used for logs.

Long reach clamps

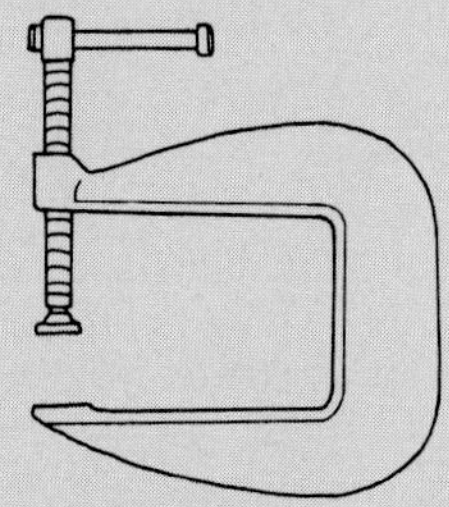

Designed to hold work some distance from edge of bench. Operates like C-clamp.

Miter clamp see corner clamp

Pinch dog (joint clamp)

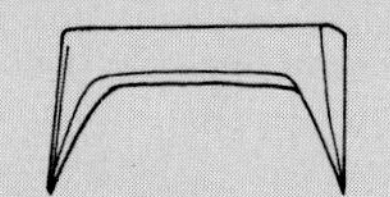

Steel device, $\frac{1}{4}$ to 3in wide used to hold boards together while gluing. Tapered points straddle joint and are driven into end grain to pull boards together.

Pipe clamp

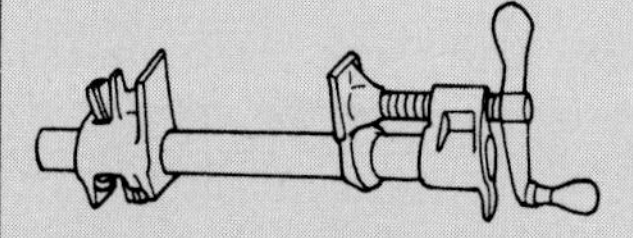

Clamp heads to fit $\frac{1}{2}$ or $\frac{3}{4}$in bore iron pipe. Used to make clamps to any length.

DOUBLE BAR CLAMP

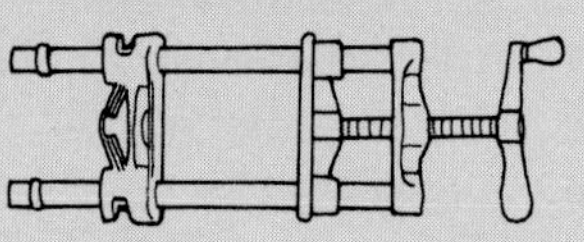

Mounted on pair of $\frac{1}{2}$in pipes.

REVERSIBLE PIPE CLAMP

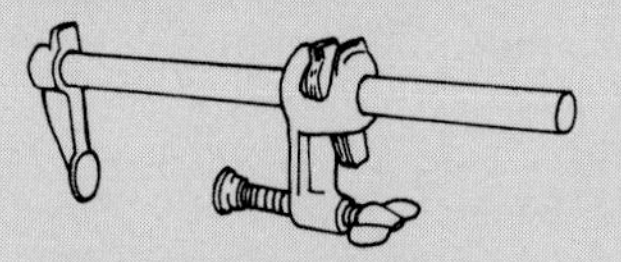

Foot is screwed onto thread of $\frac{3}{4}$in pipe, head slips on pipe and grips with quick action disk clutch.

Quick action clamp
(deep engagement clamp)

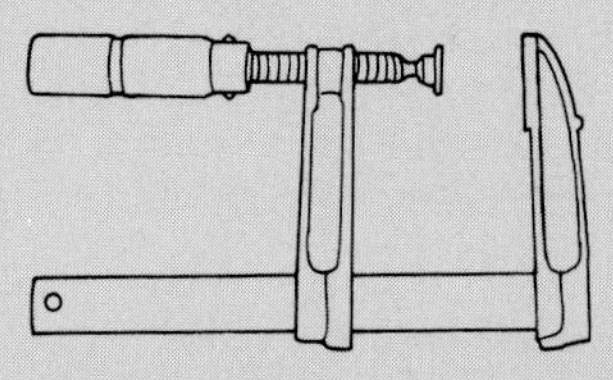

Popular clamp with sliding jaw which moves freely up and down profiled steel bar and is then tightened with one movement. Often fitted with plastic jaw pads. Hole in handle for further leverage.

Spring clamp

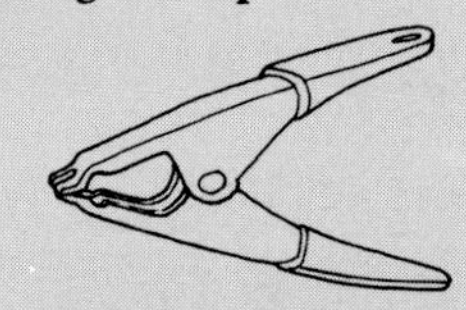

Small, lightweight metal clamp. Like pair of pliers with spring that holds jaws closed.

Spring grip clamp
(spring miter clamp)

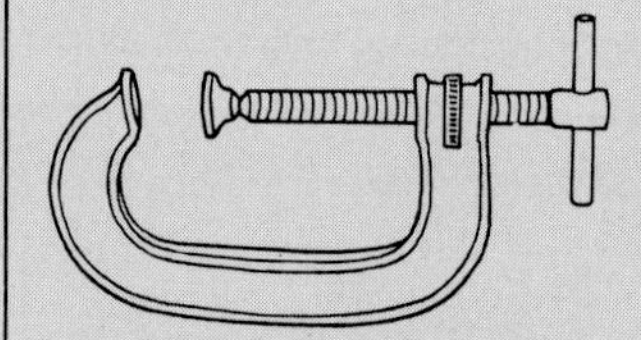

C-clamp operated one handed. Rotate spinner on side to tighten clamp.

Web clamp (strap clamp)

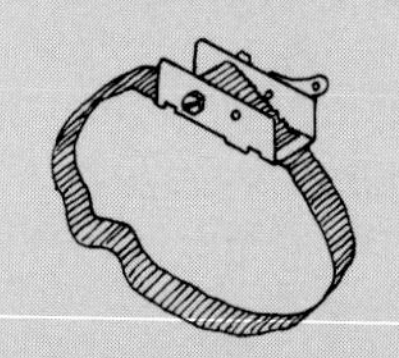

Strap or web with steel web fasteners used to clamp irregular shapes. Tightened with screwdriver or wrench.

COMPASS

Device used for marking circles. One leg holds marking instrument.

CORNERING TOOL

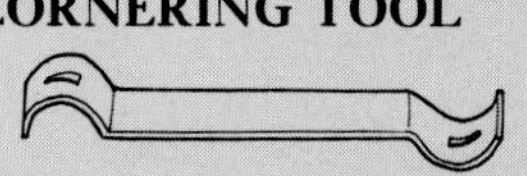

Tool for rounding off sharp edges. Heavy strap of steel with slight hook.

COUNTERSINK

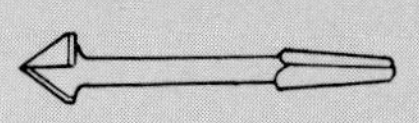

Used to make a conical hole to take the head of a countersunk screw or bolt.

DEPTH GAUGE

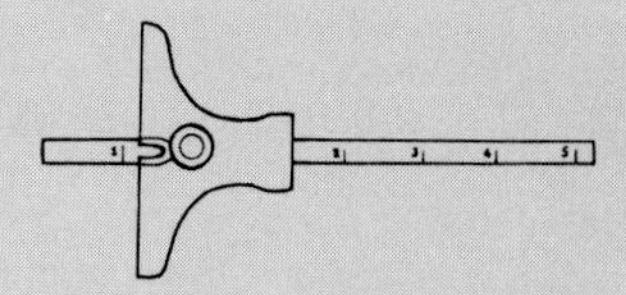

Steel device used to measure depth of holes and mortises up to 15in deep. Graduated rule with sliding head.

DEPTH AND ANGLE GAUGE

Device for measuring depth and angles.

DIVIDERS

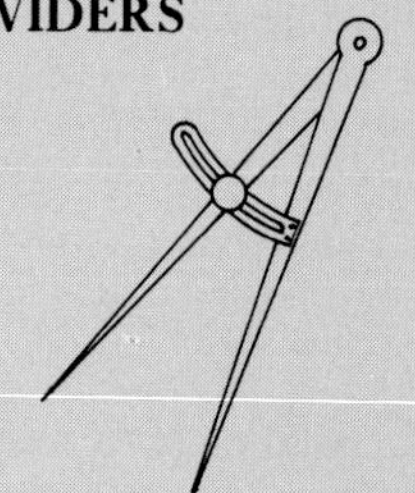

Device like compass but with two steel points used for scribing, dividing, measuring and finding center of a circle.

DOVETAIL MARKER
(dovetail template)

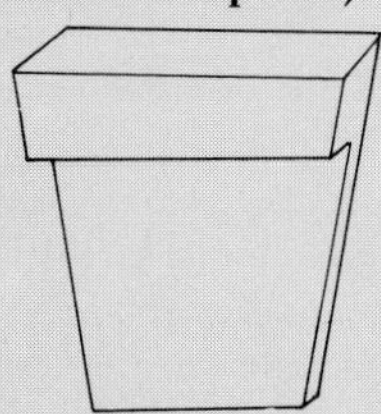

Homemade device made of hardwood. Used to mark out dovetail joints to save repeated setting out.

DOWEL CENTERS
(dowel locators)

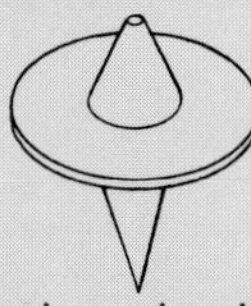

Used for inserting in already bored hole and marking exact centers for opposite holes.

DOWEL FORMER
(dowel sizer)

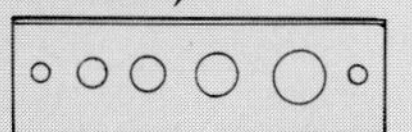

Hardened steel plate to size dowels. Can also cut glue channels in dowels at the same time. Drive oversized dowel through appropriate hole.

DOWELING JIG
(drill guide)

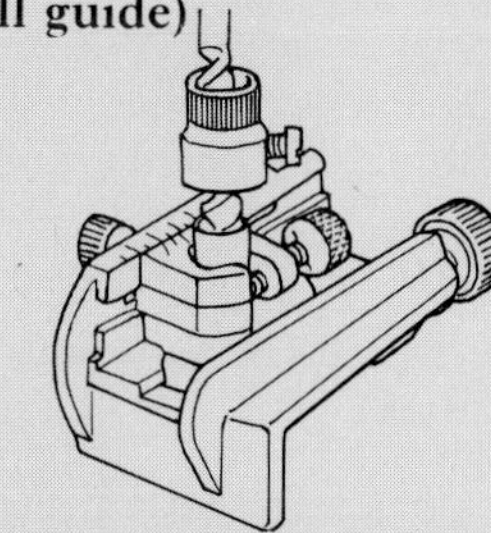

Device to control drilling of holes for dowel joints. Interchangeable for different sized bits. Guides drill bit straight and true. Normally used with electric drill, but can be used with brace and auger bit. Several models available.

DOWEL POINTER

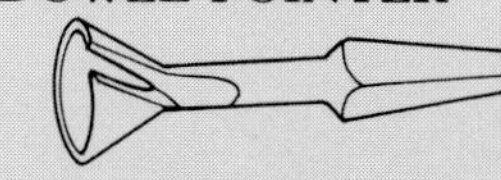

Device much like pocket pencil sharpener used to chamfer the ends of dowels to make them easier to insert. Either handheld or used in brace.

DRAWKNIFE

Used by drawing toward user to rough down and shape wood. Various types for specific applications.

Carpenter's drawknife

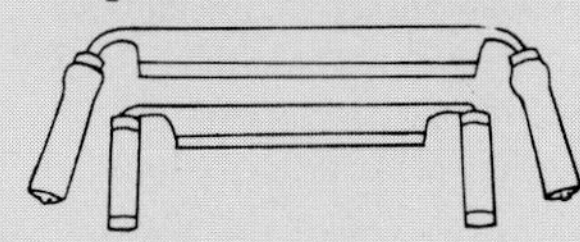

Standard type for general woodworking use.

Chamfer knife

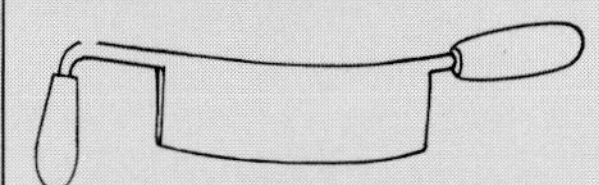

Handles are at right angles to each other. Tool for shaping wood longitudinally. Removes more wood with each stroke than a plane but is more difficult to control.

DRILLS

Archimedian drill

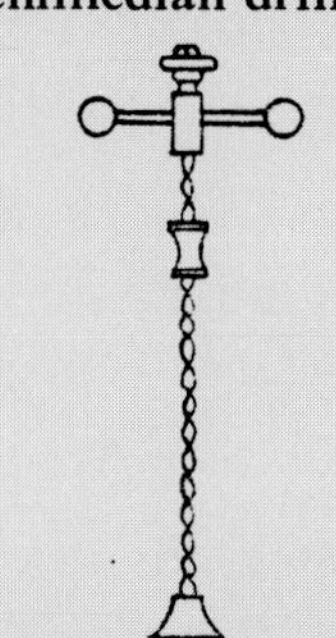

Old-fashioned push drill for small holes. Still available.

Brace

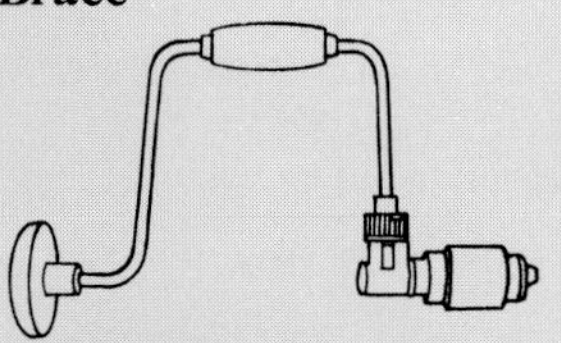

Standard carpenter's brace with sweep of handle between 5 and 14in for boring holes in wood. Accepts bits with square tapered shank. Plain or ratchet type available. Antique braces in all wood or wood with brass reinforcements can still be found.

JOIST BRACE

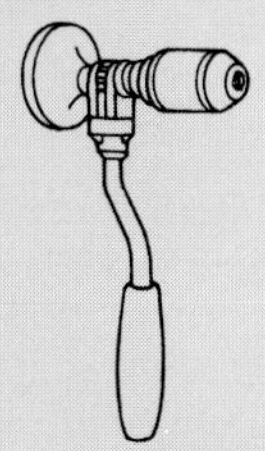

Functions as ordinary brace in restricted space between joists for use by plumbers, electricians etc.

CORNER BIT BRACE

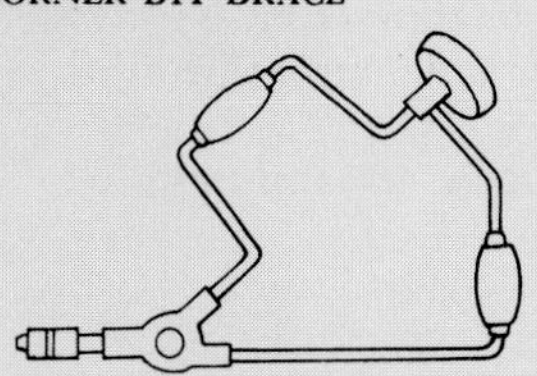

Brace used in restricted spaces along the floor, for example, where rotation of standard brace handle is not possible.

Breast drill

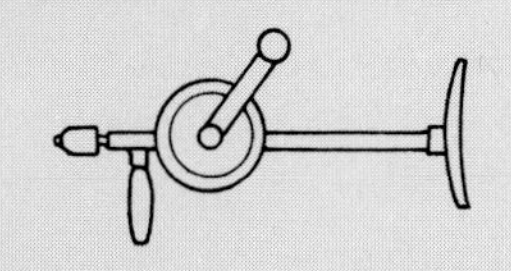

Larger version of hand drill which works on the same principle. Intended for drilling larger holes and therefore requires more pressure. Drill is pressed to chest when in use.

Hand drill (wheel brace)

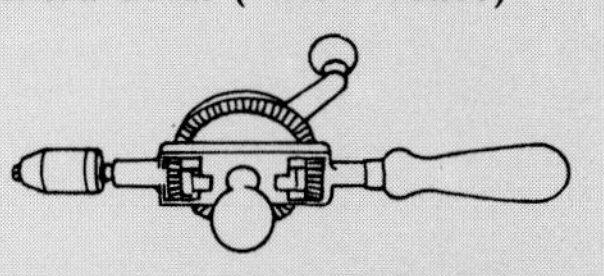

Drill made for smaller bits up to $\frac{5}{16}$in capacity. Some models have mechanism for producing different speeds.

Push drill
(automatic drill)

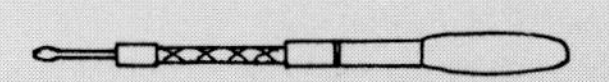

Used to drill small holes by pushing down. Takes special bits called drill points which are single, straight or fluted.

DRILL STOP
(depth stop)

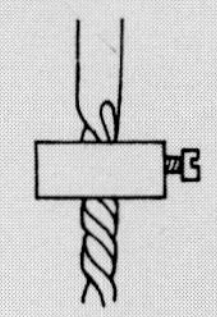

Device for controlling drilling depth.

FILE

Tool for smoothing wood and metal. Hardened steel surface covered with parallel rows of teeth arranged at 65° angle to edge. Classified by type either single or double cut, and by coarseness according to the number of teeth per inch; coarse, middle, common or bastard cut (about 26 teeth per inch), second cut, medium smooth, smooth, dead smooth and double dead smooth. Single cut is primarily for metal. Best cut for wood is bastard double cut.

Auger bit file

Small file designed to grind and sharpen auger bits and other wood boring bits.

Flat file

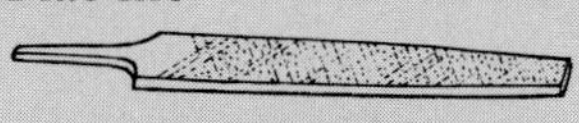

Most common type of file, made in all cuts and variety of coarseness. Usually tapers in width and thickness toward the head. Double cut bastard type most common.

Half round file

All-purpose file for flat and curved surfaces.

Hand file

Flat in cross section but has sides parallel to top. Tapers only in thickness.

Mill file

Used for fine work and tool sharpening. Single cut.

Needle files

34 to 184 teeth to the inch. Usually sold in sets of various cross section. Used to sharpen saws and drill bits.

Piller file

Slimmer version of hand file. Used to file narrow openings.

Round file

Used for enlarging holes and shaping corners. Tapers and is usually single cut bastard type. Small round files called rat tail or mouse tail files.

Saw file

Triangular file, often double ended used to file saw teeth.

FEATHER EDGE SAW FILES

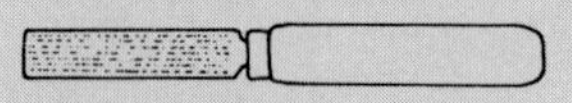

Designed to sharpen fine teeth of Japanese hand saw. saw.

Square file

Used to enlarge rectangular holes. Tapered on all four sides.

Triangular file (three square file)

Longitudinally tapered on all three sides. Used to file acute internal angles and angular stock.

Wood file

File with coarse teeth used after rasp to smooth surface.

Woodcarver's files see RIFFLERS

FILE BRUSH

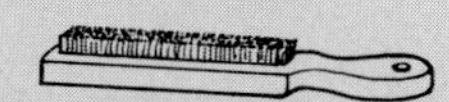

Used to clean clogged file teeth. Some models combine brush, card and scorer for quick and thorough cleaning.

FILE HOLDER

Holding device to allow file to be used for surface smoothing. Can be adjusted for different length files.

FROE (riving axe)

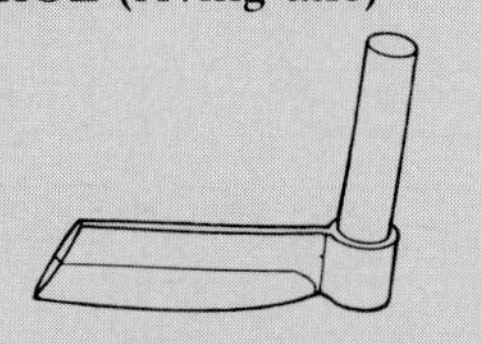

Traditional tool with knife type wedge used to split wood along grain to make shingles, lathe, staves etc.

Froe club see MAUL

GAUGE

Bit gauge

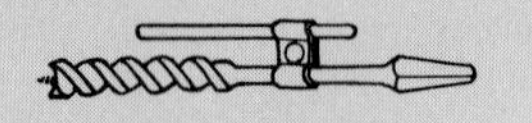

Depth stop clamped to auger bit.

Butt gauge (butt marker)

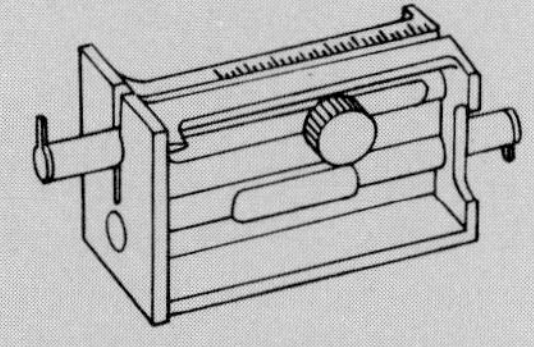

Traditional marking tool used to mark position and thickness of hinge butts. Struck with hammer to mark area to be removed. Also used as marking and mortise gauge.

Cutting gauge

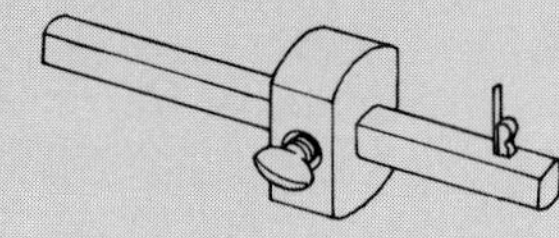

Flat blade used to mark line parallel to edge across grain and can also be used to cut parallel strips of veneer.

Depth gauge

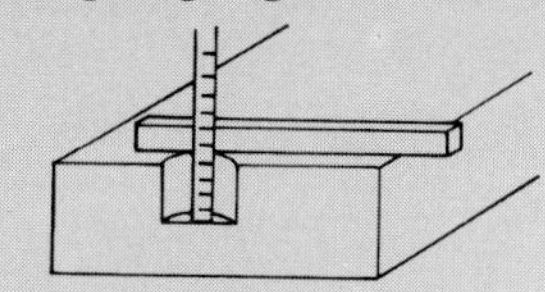

Measures depth of hole.

Drill stop

Depth stop like bit gauge but used on twist drills.

Kebiki

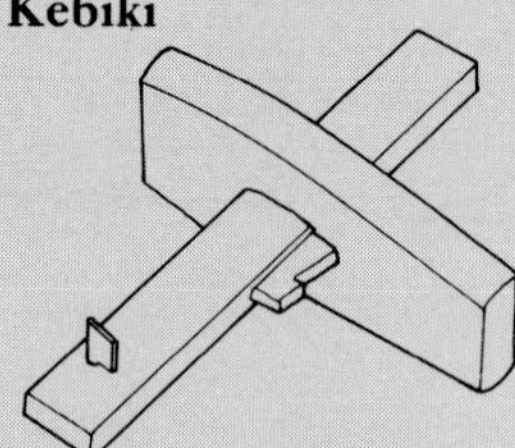

Japanese marking gauge

Marking gauge

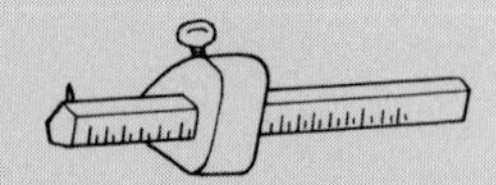

Device fitted with sharp pin for marking line parallel with edge of work along grain.

Mortise gauge

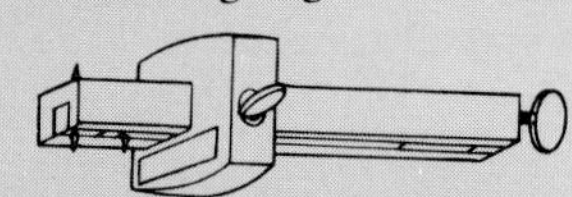

Like marking gauge but with two points instead of one. Second point is adjustable to set points to mortise width.

Panel gauge

Larger marking gauge fitted with pencil. Used for wide panels.

Rabbet gauge

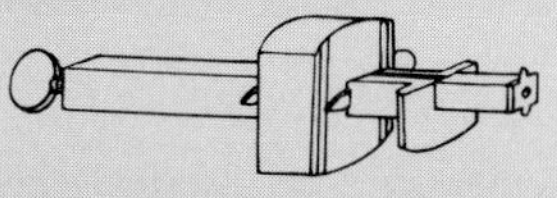

May be used as a butt gauge as well as marking mortises and other joints.

Scratch gauge

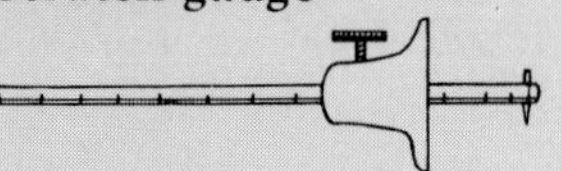

Very exact marking gauge graduated in 64ths of an inch for use on metal or wood.

Slitting gauge (cutting gauge)

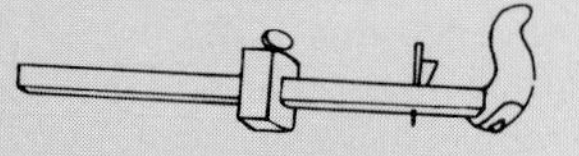

Device for accurately slitting thin boards. Roller set in base of handle allows gauge to move smoothly.

T-gauge (grasshopper gauge)

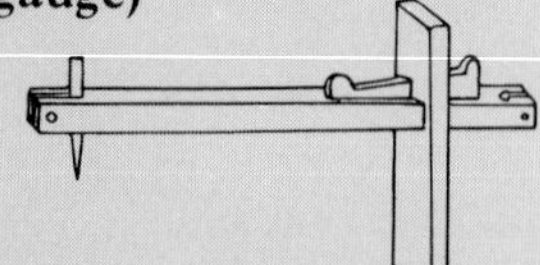

Homemade device with extra long fence for fitting over moldings.

Tenon marking gauge

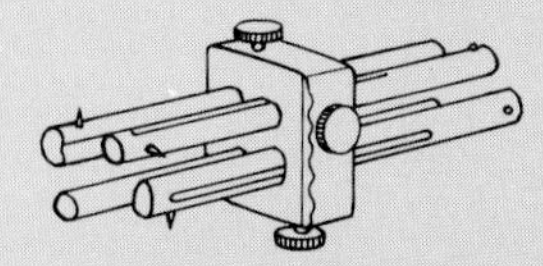

Device for marking tenons. Gauges can be locked in place and each side of fence is numbered so sequence is maintained.

GIMLET

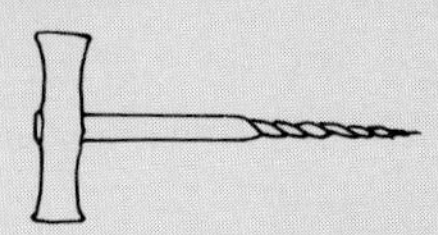

T-shaped tool, like miniature auger. Used to bore small holes such as screw holes in wood. Available with straight or spiral groove.

Japanese gimlet

Straight handled. Used rolled between palms of two hands to make small hole in wood or held and used like screwdriver.

GLUE INJECTOR

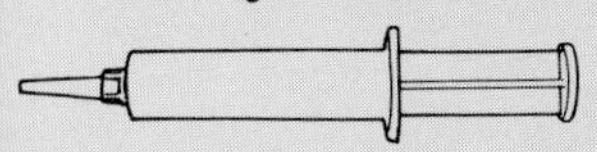

Pressure forces glue through needle nose by pushing down plunger.

GLUE POT

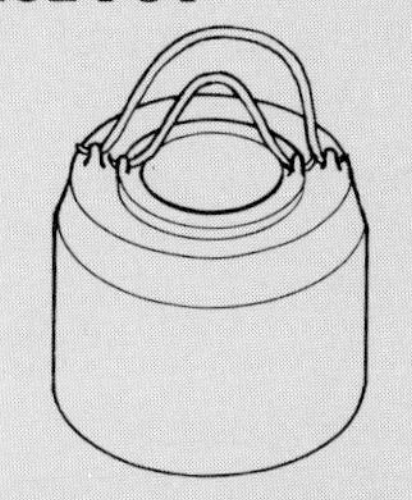

Pot for melting animal glue. Outer pot filled with water and smaller pot suspended inside contains glue.

GLUE SPREADER

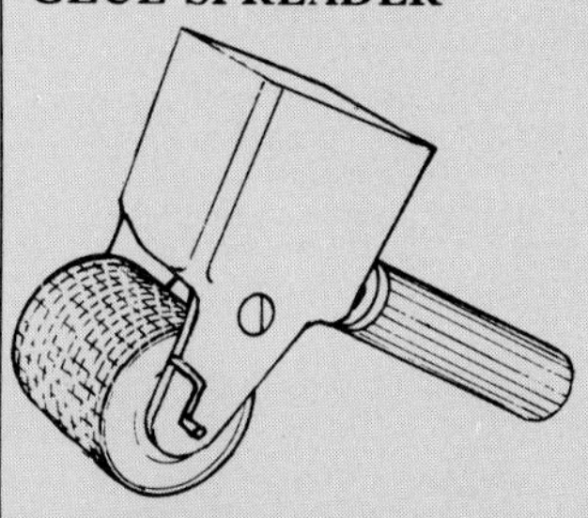

Metal device with hopper to hold glue and rubber roller to spread it evenly and thinly.

GOUGE

Concave chisels for cutting rounded grooves or holes. Manipulated by hand or with aid of mallet or soft-face hammer. Types are carver's, turner's or carpenter's.

Firmer gouge

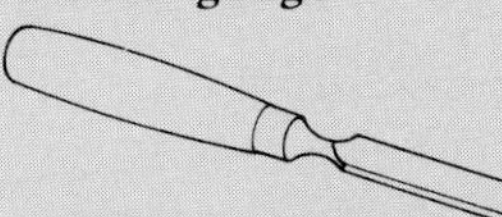

Used for most work. Generally out-canneled but may be ground on inside too. Flat, middle or regular sweep.

SOCKET FIRMER GOUGE

Heavy-duty firmer gouge with socket handle construction.

Paring gouge (scribing gouge)

Made with inside grinding edge and designed for lighter, finer work. Available in regular, flat or middle sweep.

Woodcarver's gouges see CARVING TOOLS

HAMMERS

Claw hammer

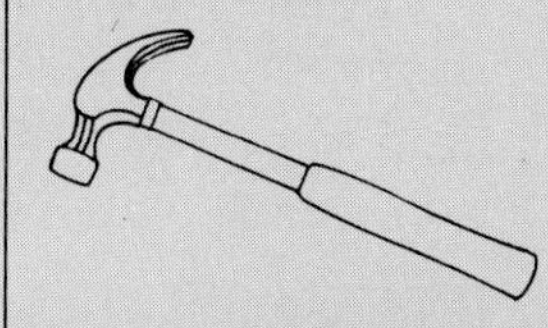

Various head weights. Shaft is wood, steel or fiber glass.

Cross peen hammer

6 to 16oz head. Used for general carpentry work. Tapered peen used to start nails held between fingers.

Framing hammer

Larger nail hammer with ripping claw. Weight at least 20oz. Handle is longer and claw straighter for ripping.

Magnetic tack hammer

Lightweight hammer with end of head opposite face magnetized to hold small tacks.

Soft-face hammer

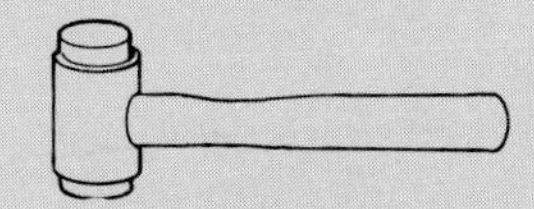

Used when necessary to strike hard blow without marking the work such as in assembly. Made with tough plastic or rubber head that can be replaced. Older types made with rolled rawhide head.

Tack hammer

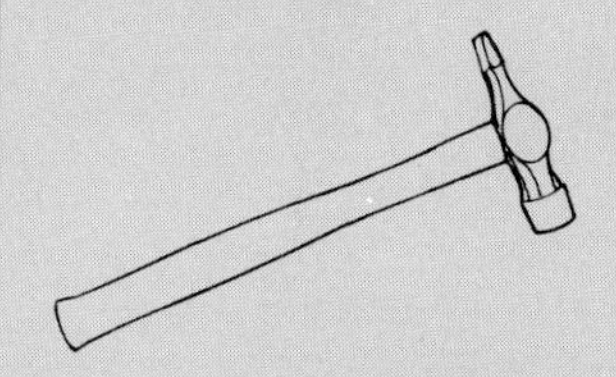

Used to drive pins and tacks. Usually magnetic. Cross peen or ball pattern available.

Sprig hammer
(picture framer's hammer)

Used to drive sprigs or glazing points.

Veneer hammer

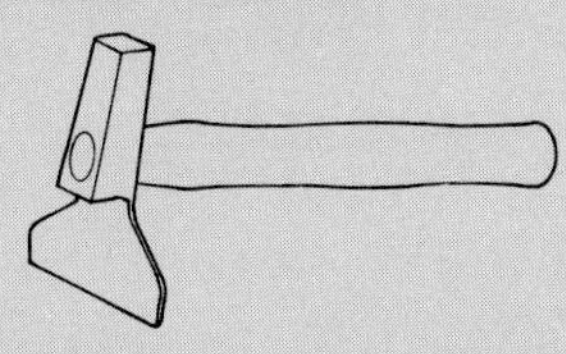

Device used to press down veneer to a glued surface.

Warrington hammer
(joiner's pattern hammer)

Wedge-shaped peen. Used as all-purpose pin hammer.

HAND MITER

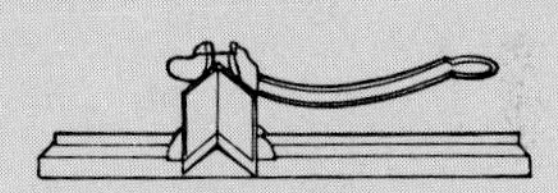

Works like a paper guillotine to cut miters quickly and accurately. Blades are at right angles so that in one motion, the molding for miter joint is cut to form both angles of miter.

HATCHET

Short handled axe usually with hammer head and slot in blade for pulling nails.

Barrel hatchet

Standard hatchet with flat top and poll for use as a hammer.

Broad hatchet
(halving hatchet)

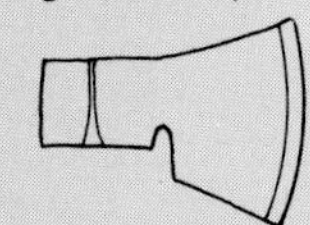

Smaller version of broad axe. Used for shaping and dressing. Beveled on one side only.

Claw hatchet

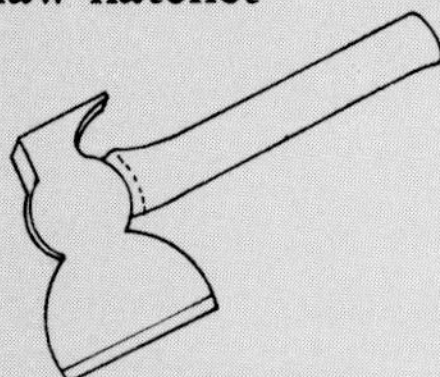

Medium-sized hatchet, similar to broad hatchet with nail pulling claw at side of nail driving poll.

Half hatchet
(shingle hatchet)

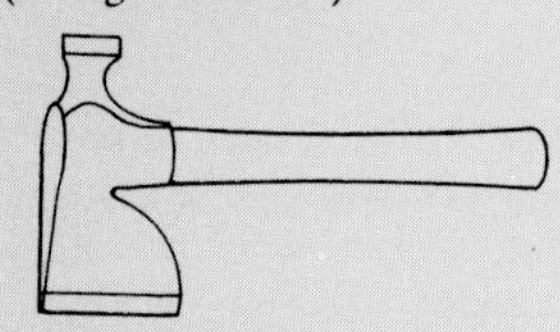

Head with straight front edge. Used to cut and nail shingles.

Hunter's hatchet

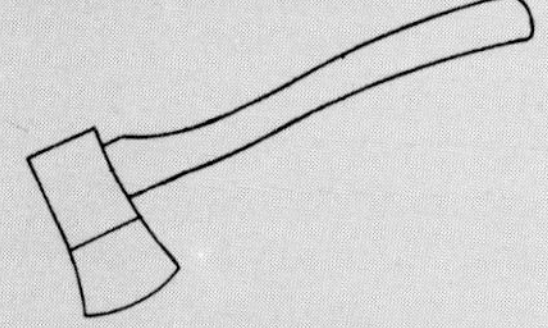

General purpose hatchet used to trim and shape wood.

Japanese hatchet

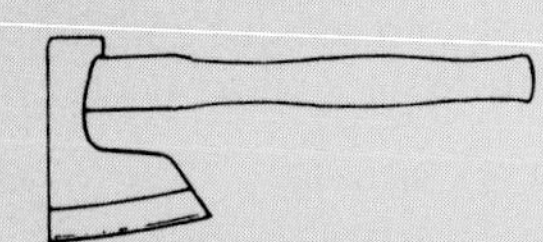

$1\frac{1}{4}$lb head, $4\frac{1}{2}$in cutting edge with hickory handle.

HOLDFAST
(hold down)

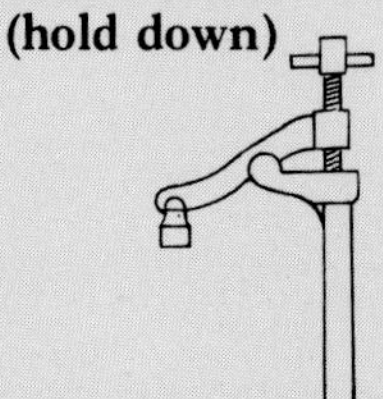

Device used to hold work flat on bench.

Bench holdfast

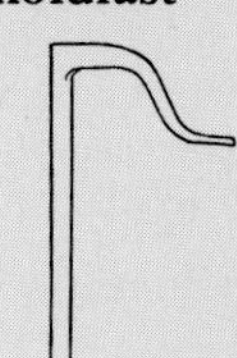

Traditional style holdfast, tightened by hammering head from above and released by tapping back of head.

HONING GUIDE

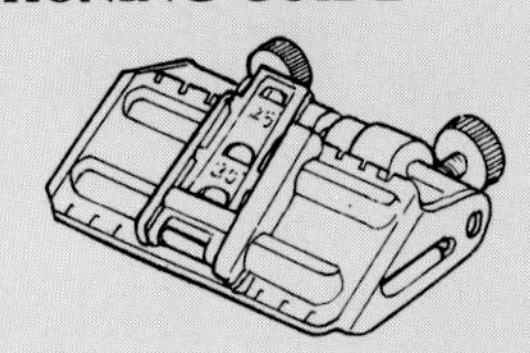

Device used to hold blade of tool being sharpened at correct angle.

IMPACT DRIVER

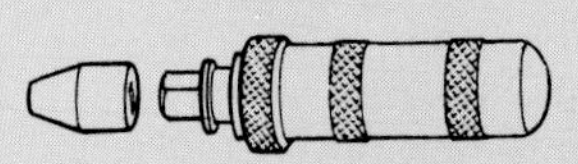

Used to loosen tight screws or nuts. Has replaceable screwdriver bits in different patterns.

INCLINOMETER

Instrument to read angle of work to horizontal.

INSHAVE
(round shave)

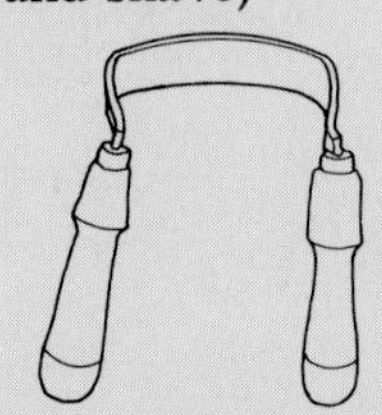

U-shaped cutting tool for hollowing out. Used like a drawknife. Beveled on outside face.

JOINTMASTER
(sawing jig)

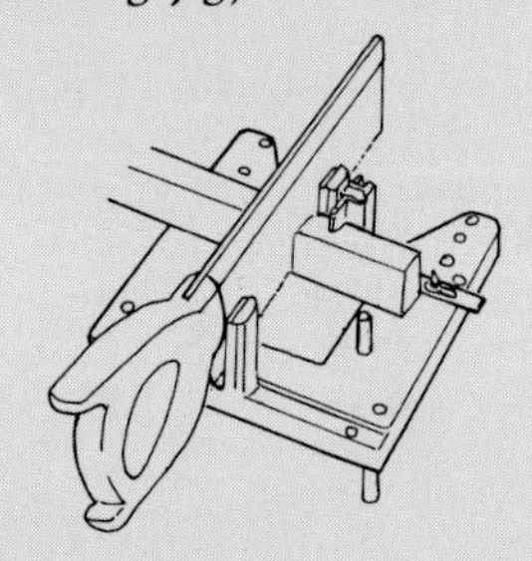

Used as a guide for saw to cut miters, lap, butt, tenon and other joints.

KNIFE

Asian carving knife

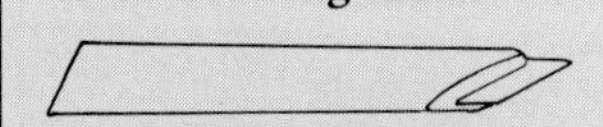

Has opposite bevels for push and pull skew cuts. Handle is extended over cutting edge so that pressure can be applied with thumb. Sold in pairs.

Burn-in knife

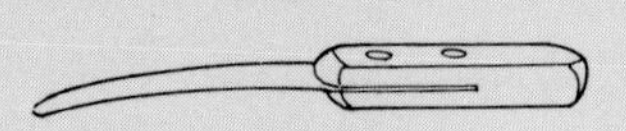

Used for burning and applying shellac sticks to fill dents.

Chip carving knife

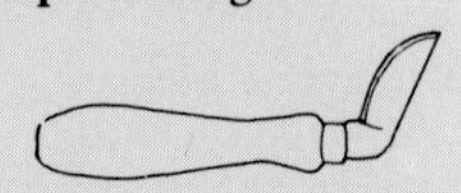

Use to carve low relief decoration.

Crooked knife

Traditional finishing tool used to shape rough cut wood. Used one-handed.

Drawing knife

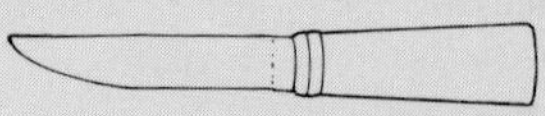

Old-fashioned knife now replaced by sloyd or bench knife. Used to cut groove to guide saw.

Marking knife

Blade beveled one side only so flat face can be run against try square for marking joints.

Razor plow

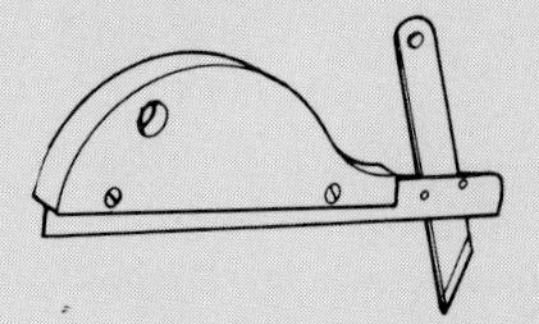

Extremely efficient carving knife. Pushed away from user. Used for carving incised letters and chip carving.

Sloyd knife

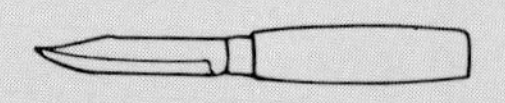

Woodworker's standard bench knife. Used for marking wood and for small cutting jobs.

Striking knife

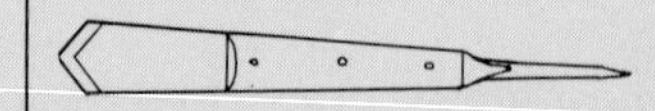

For scribing accurately without pencil. One end is point, the other a blade.

Utility knife

All-purpose knife with replaceable blades that are usually retractable into the handle. Good for marking and slicing veneers.

Veneer knife

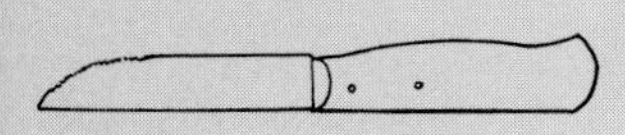

Used to cut and replace blemishes in veneer without tearing. Has serrated blade.

KNURLS (knurling tool)

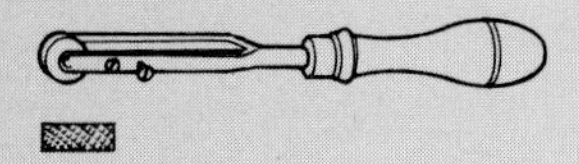

Traditional tool with small wheel filed so as to leave impression when pressed into wood. Used for decorative work.

LEVEL

Instrument for checking horizontal. Old-fashioned type has triangular frame shaped like A with plumb bob suspended from apex.

Bull's eye level

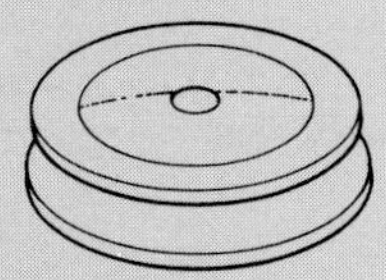

Small, round pocket level used for quick checks.

Line level

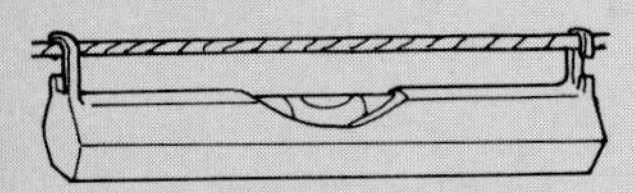

Small level that can be hung on line stretched taut between two points.

Pocket level

Small level, 2 to 4in long, made in a variety of shapes.

Standard spirit level

Made of aluminum, hardwood or plastic. Tube filled with alcohol, oil or chloroform.

Torpedo level (canoe level)

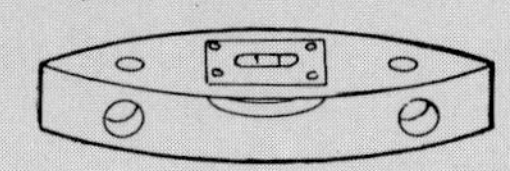

Short level, slightly larger than pocket level made of wood or metal.

Water level

Absolutely accurate leveling device used in ancient times and still in use today. Long flexible tube is filled with water for checking levels at long distances or around corners.

LEVEL SIGHTS

Attached to level for accurate means of leveling from one given point to another a distance away.

MALLET

Wooden hammer used for striking wood or chisel handles.

Carver's mallet

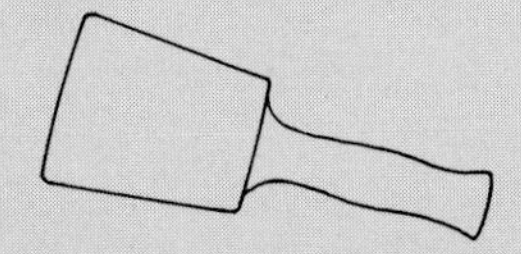

Round head to strike with any part without changing grip. Made of very hard wood like lignum vitae or dogwood.

Dead blow mallet

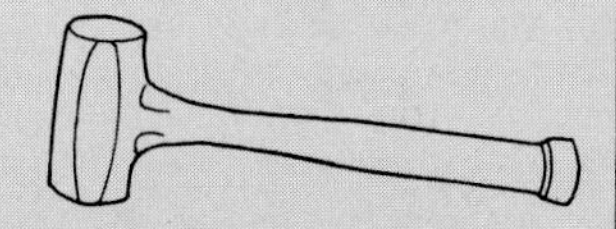

One piece cast plastic mallet with shot-filled head. Has less bounce than rubber mallets.

Rawhide mallet

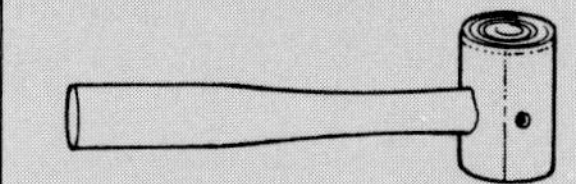

Lightweight traditional mallet made of hide coiled to make head.

Soft-faced mallet

Rubber head on hickory handle. Used in joinery work for knocking joints together.

Square mallet (carpenter's mallet)

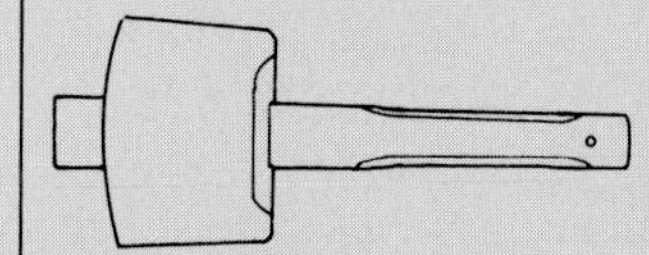

Most common type of mallet made of beech or maple. Used where metal hammer would damage work. Tape cardboard over face to use for joint assembly.

MAUL (beetle)

Large, heavy mallet. Head is elm or oak with iron rings.

Froe club

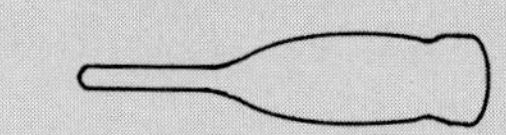

Type of maul used to strike froe.

MITER BOX

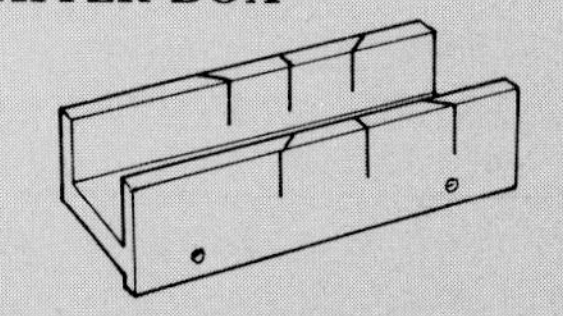

Device for guiding saw to cut miters.

Metal miter box

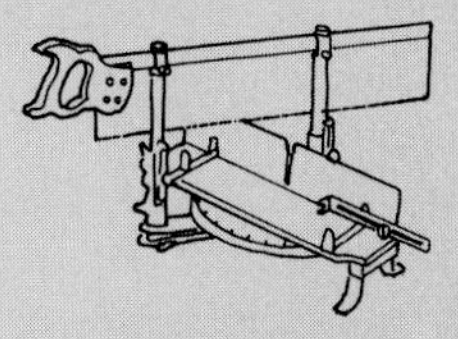

Sophisticated and expensive miter box. Clamps to hold work in position and catches to hold saw.

MITER BLOCK

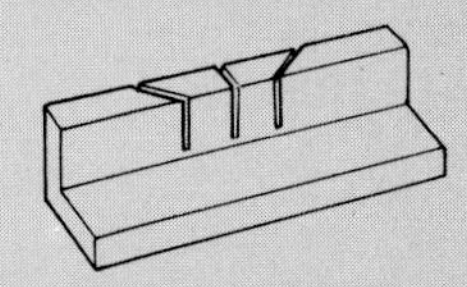

Serves same purpose as miter box but simpler and cheaper. Block of wood with three saw cuts to guide saw.

MITER TEMPLATE

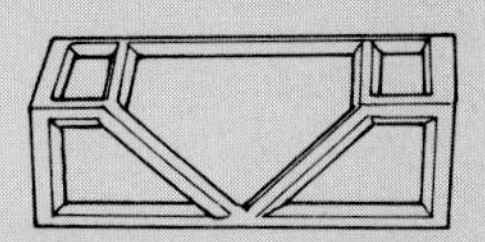

Brass or wooden device for marking moldings at 45 or 90°.

MOLDING TOOL

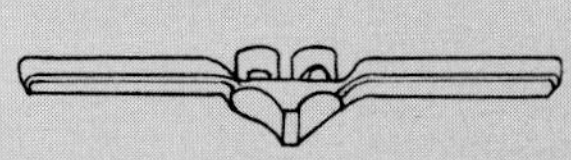

Combination of molding plane and spokeshave. Used for cutting moldings in curved surfaces.

NAIL CLAW (cat's paw)

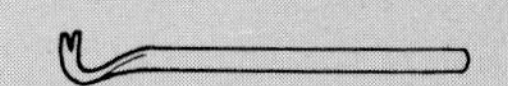

Used for digging underneath heads of nails to pry them out.

NAIL PULLER

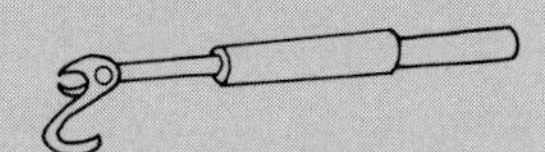

Mechanism to simplify pulling nails.

NAIL SET (nail punch)

Steel punch with concave tip used to drive head of nail below surface. Square or round head in four sizes from $\frac{1}{32}$ to $\frac{3}{16}$in.

Self-centering nail set

Has hollow tip which fits over nail head. Plunger in back is then struck with hammer to drive nail.

OILSTONE

see SHARPENING STONES

PEAVEY (cant dog)

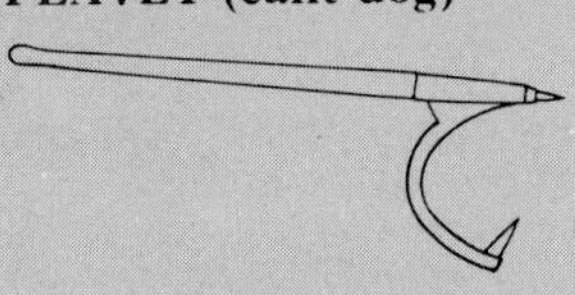

Traditional device for turning heavy logs when trimming undersides.

PINCERS

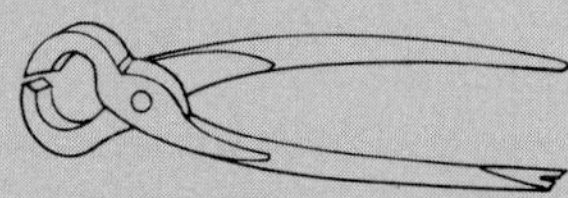

Used to pull nails and cut wire.

PITCH ADJUSTER

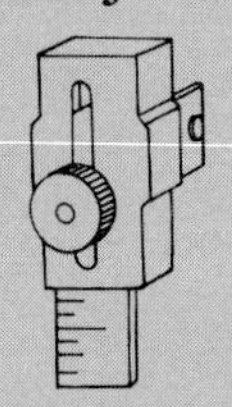

Device attached to a level to determine the pitch of a particular slope.

PLANES

General purpose planes

Fore plane

Compromise between jack and jointer plane, 18in long.

Jack plane

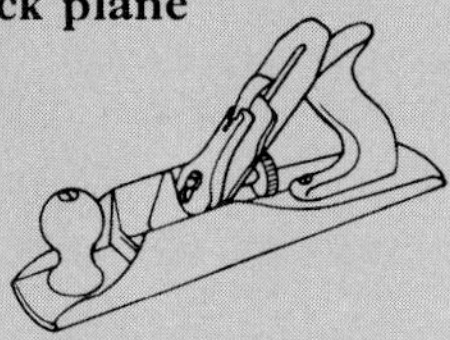

Used for general work. "Jack of all trades" plane, 14 to 15in long.

Jointer plane

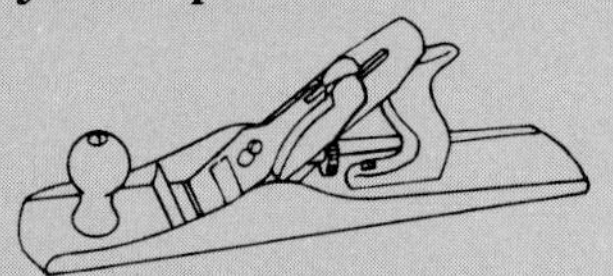

Longest of all planes, 22–24in long. Used to plane up long boards for joining.

Smooth plane (smoothing plane)

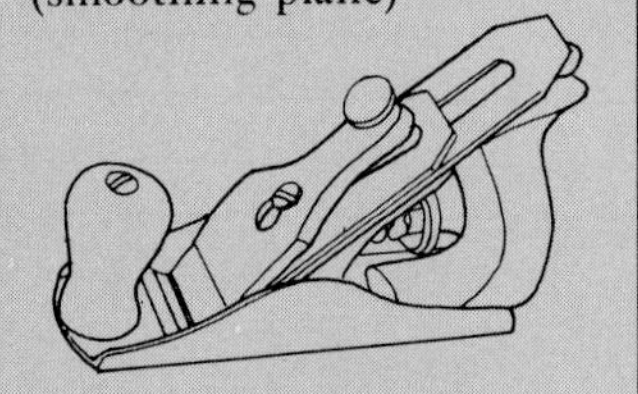

Smallest bench plane, 5½ to 10in long. Last plane used in smoothing process to clean off finished work.

HORN SMOOTH PLANE

Smooth plane with horn in front for guiding plane.

Trying plane (trueing plane)

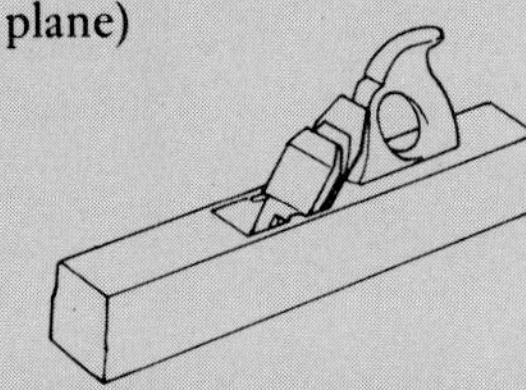

Early type of fore plane with wooden body, 22in long.

Block planes

Block plane

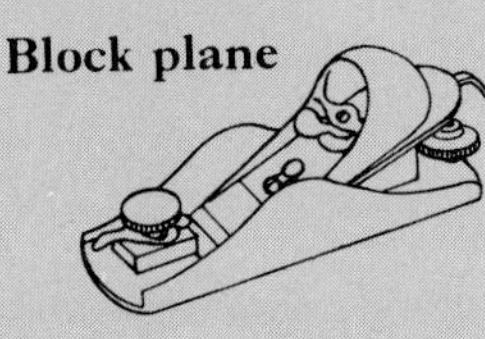

Small, one-handed plane with blade set at a low angle used for planing small pieces and end grain of wood.

Block and rabbet plane

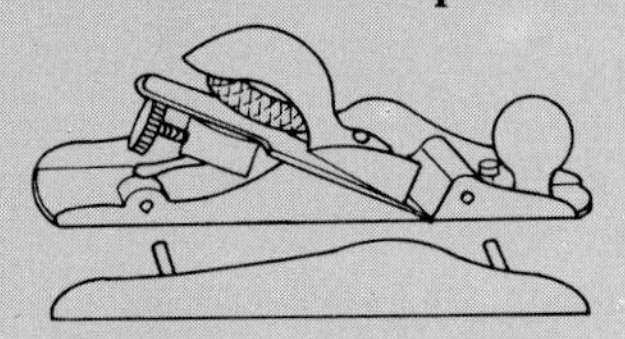

Block plane with detachable side which allows the iron to cut into corner.

Special purpose planes

Beading plane

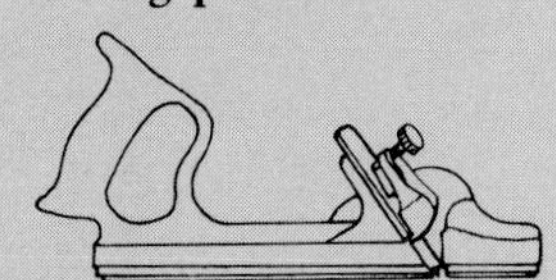

Successor to wooden bead molding plane. Used to cut beads.

Carriage maker's panel router

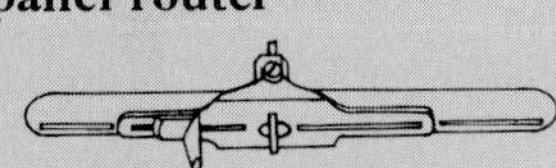

Like beading tool, molding plane and spokeshave in one. Cuts grooves in curved surface.

Carving planes

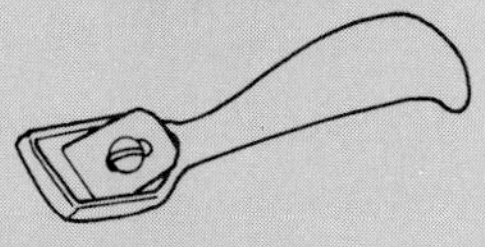

Available with flat bottoms. Lengthwise curve. Look like rifflers.

Chamfer plane

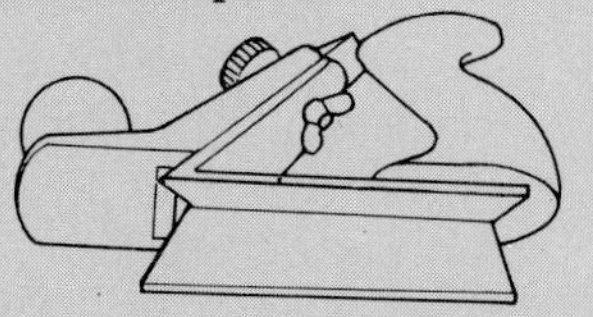

Sole of plane is V-shaped so it fits over edge of wood. Can be adjusted to cut chamfer of varying width.

Combination plane

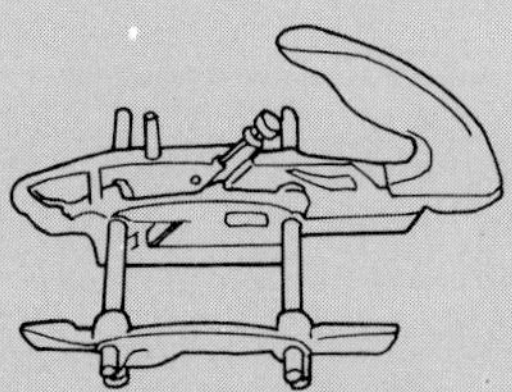

Cuts grooves or rabbets along or across grain. Also beads, tongues and grooves. Simplified version of multi-plane. Spurs score wood to allow cutting across grain.

Compass plane (circular plane)

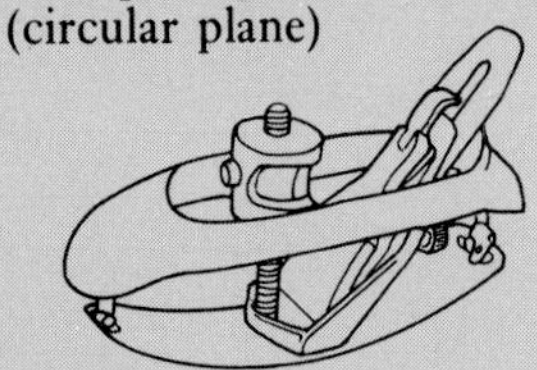

Used to cut convex and concave sections. Curved sole is flexible and adjustable.

Cooper's long jointer

Old-fashioned large plane, turned upside down and mounted on legs. Wood being planed is slid over it.

Dado plane

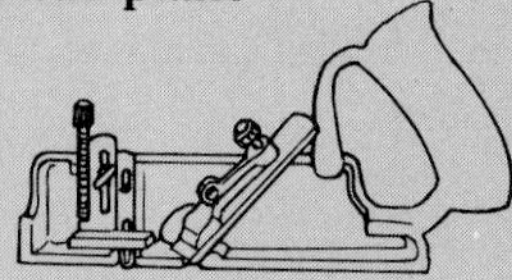

Cuts grooves across grain. Iron set skew, minimizing the tendency to tear the wood.

Dovetail plane

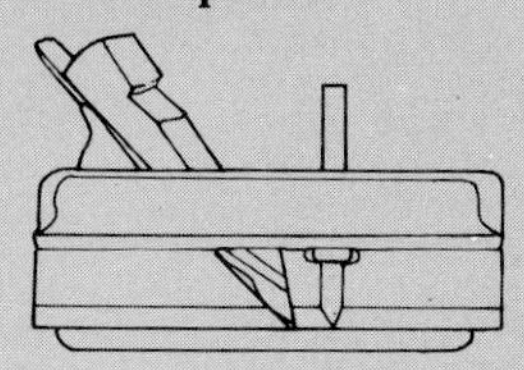

Used after surface is level to make male dovetails across grain.

Fillister plane
(rabbet and fillister plane)

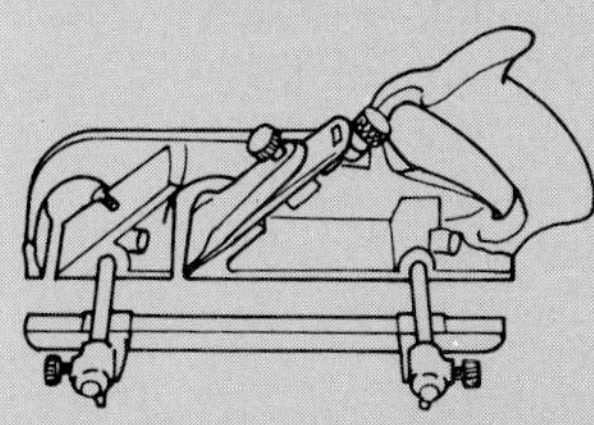

Like dado plane and plow plane. Used to cut rabbet. Fitted with guide fence, depth gauge and spur.

MORING FILLISTER PLANE

Cuts groove in side of wood nearest the user.

SASH FILLISTER PLANE

Cuts grooves in far edge as for windows.

Finger planes

Small planes used primarily in instrument making. Pushing force applied by forefinger in depression in front of plane.

Furring plane

Old-fashioned plane designed for preparing wood as it comes rough sawn from the mill.

Japanese planes

Operate on pull stroke.

BLOCK PLANE

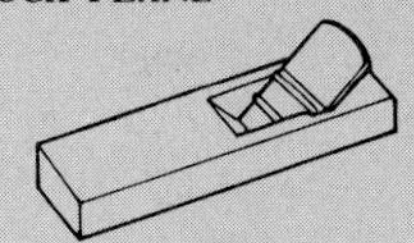

10in long, blade set at 40°.

DRAWSHAVE

Like a spokeshave in use. Cuts flat and concave surfaces.

LONG BLOCK PLANE

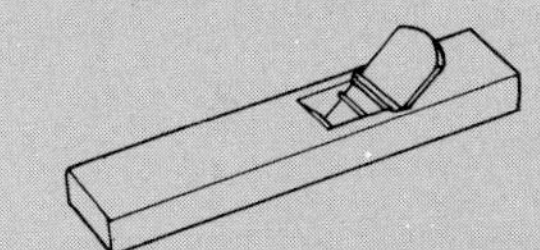

14$\frac{1}{4}$in long, used as a jack plane.

SHOULDER RABBET PLANE

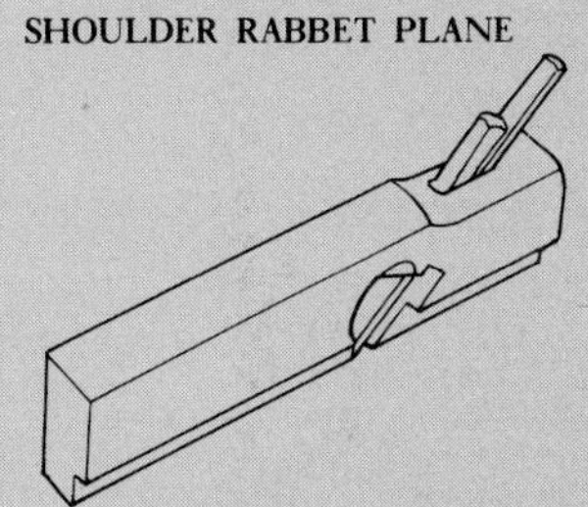

10in long. Used for cutting and trimming wide rabbets.

Gutter plane

Traditional plane but with convex sole, 16in long.

Molding planes

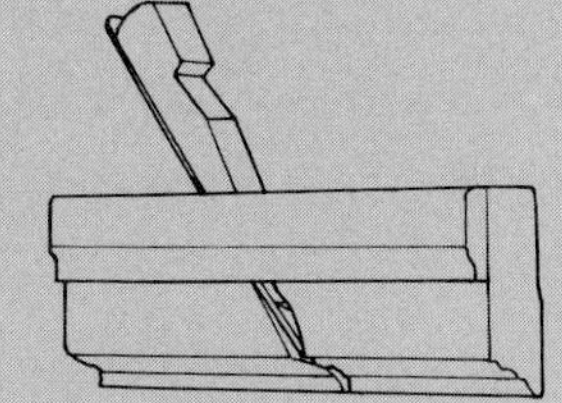

Thin bodied planes with variously shaped soles and blades. Cutting edges for forming molding sections.

Beading molding plane ▼

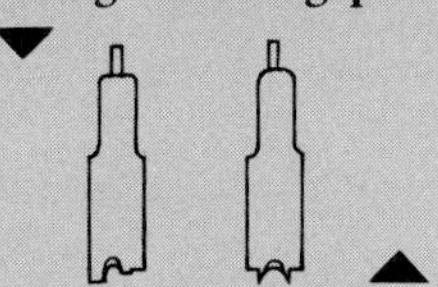

▲ **Center bead molding plane**

Cluster bead molding plane ▼

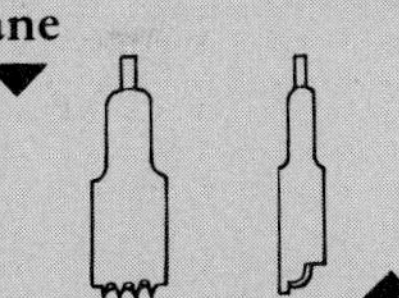

▲ **Cove molding plane**

Crown molding plane

Dado molding plane ▼

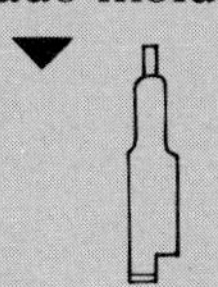

Fillister molding plane

Hollow molding plane ▼

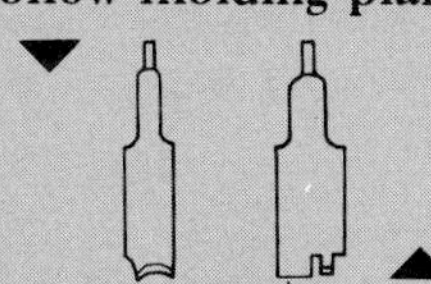

▲ **Match molding plane**
(tongue and groove plane)

Sold in pairs to cut tongue and groove joints. One plane cuts tongue, the other the groove.

Nosing molding plane ▼

▲ **Ogee molding plane**

Quarter round molding plane ▼

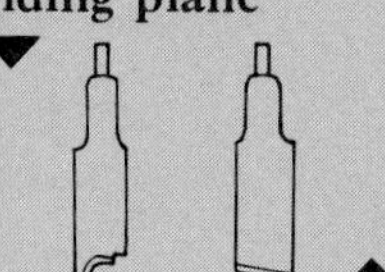

▲ **Rabbet molding plane**

Reversed ogee and square molding plane

Round molding plane
(hollow or rounding plane)
Small version of gutter plane. Sold in pairs to cut matching concave and convex curves. ▼

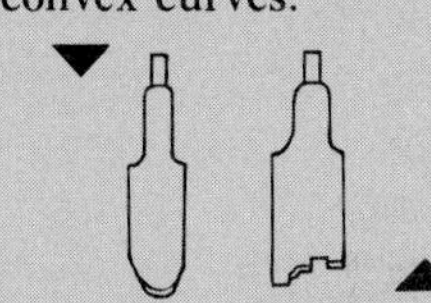

▲ **Sash molding plane**

Side bead molding plane

Snipesbill or quirk molding plane

Table molding plane ▼

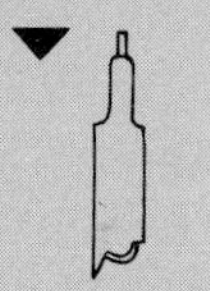

Multi-plane

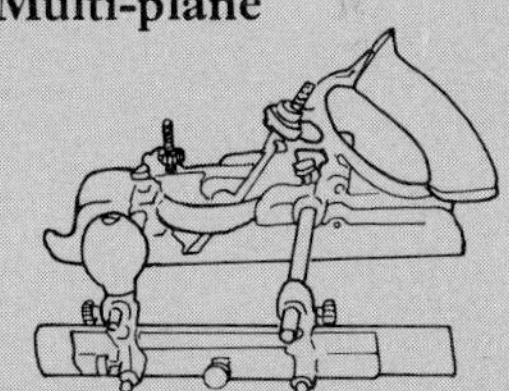

Cuts grooves, rabbets and dados, bead, tongues, ovolos, sash moldings, grouped reeds, hollows and rounds, stair nosings. Combines features of plow and combination plane but with more cutters.

Palm plane

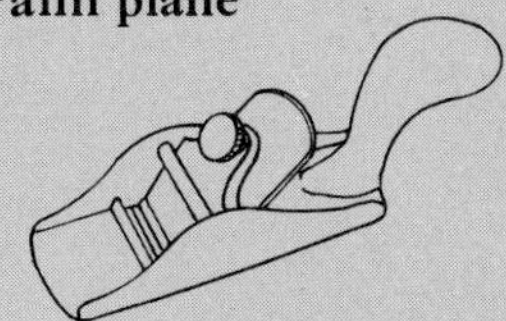

For use in small, difficult to reach places.

Plow plane

Cuts grooves and rabbets along grain up to 5in from edge of workpiece.

Rabbet plane

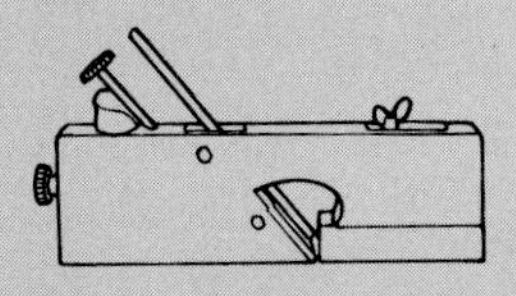

Blade extends across full width of sole to cut wide rabbets. 9 to 13in long.

ADJUSTABLE RABBET PLANE
(cabinetmaker's rabbet plane)

Square sides and bottom can be used right or left-handed. Throat opening adjustable for fine or coarse work.

BULLNOSE RABBET PLANE

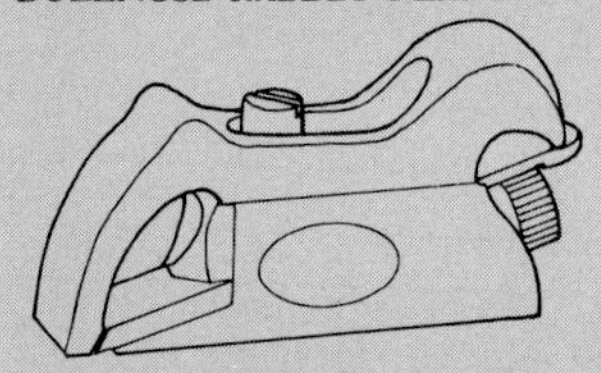

Planes rabbets and, with front removed, planes stopped rabbets. Used as substitute for chisel in hard to reach places.

DUPLEX RABBET PLANE

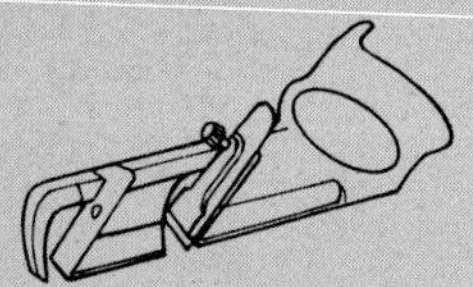

Two seats for the cutter, one for regular rabbeting and the other for bullnose work.

SIDE RABBET PLANE

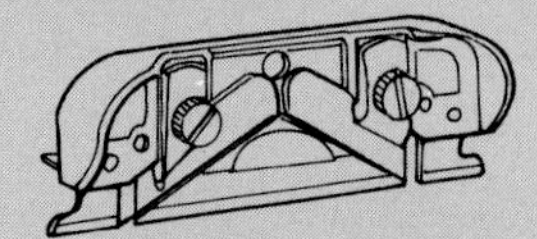

Designed for side rabbeting and trimming dados and grooves. Can be used from both sides.

STOP RABBET PLANE
(chisel plane)

$7\frac{1}{16}$in long wooden bodied plane.

Rasp plane see SURFORM TOOL

Router plane
(hand router)

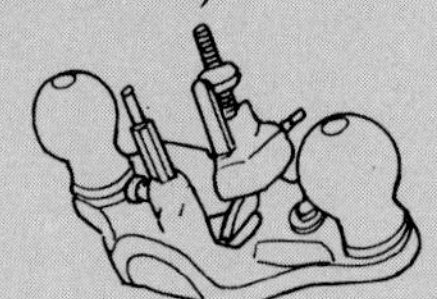

Used to level the bottoms of grooves and dados.

OLD WOMAN'S TOOTH ROUTER PLANE

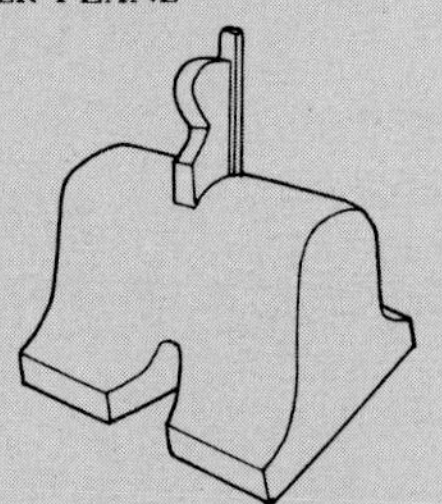

Old-fashioned, wooden bodied, router plane, still in use.

SMALL ROUTER PLANE

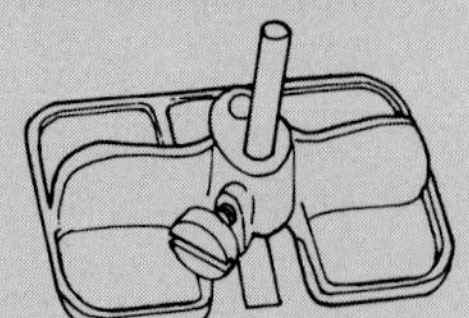

Used for small jobs like inlaying. Cutter can be set for stopped or through dados.

Sanding plane

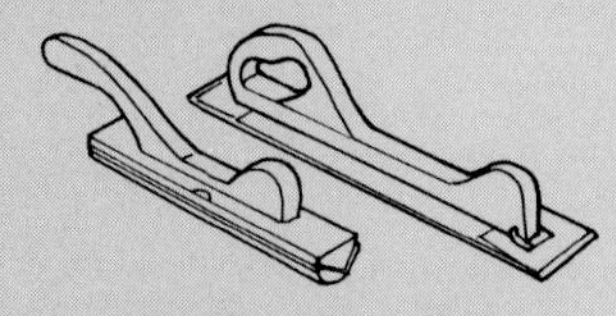

Lightweight plane fitted with sandpaper to sand contours and flat areas. Hardwood body with aluminum base which is either convex, half round or flat.

Scraper plane

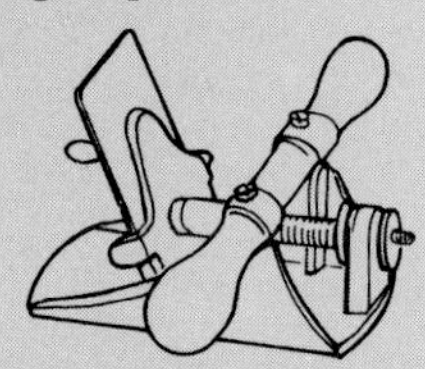

Really a scraper in steel holder. Can be adjusted to desired angle by knurled nuts. For veneer or solid work.

Scrub plane

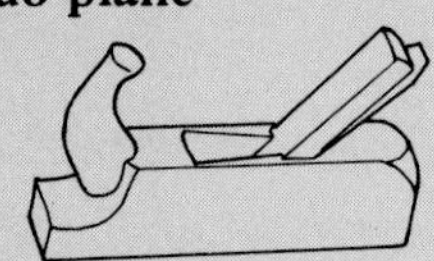

Rounded cutting edge on iron used for removing large amounts of wood before using jack plane.

Shoulder plane

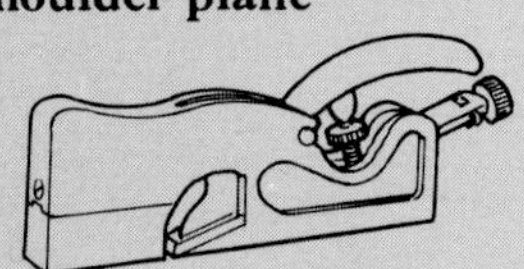

Used to trim shoulders of large joints and to cut rabbets. $4\frac{1}{2}$ to 8in long.

Three-in-one plane

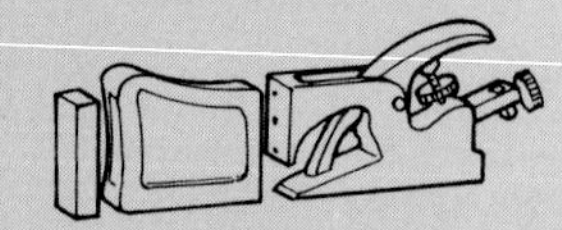

Handles operations of bullnose, shoulder and rabbet plane.

Tonguing and grooving plane

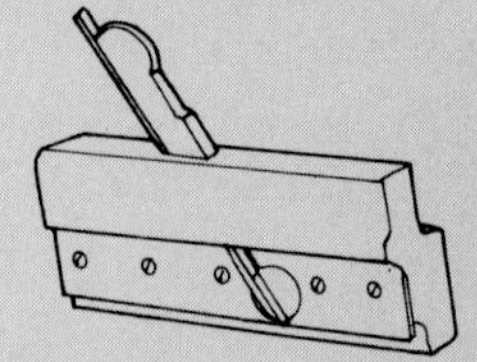

Two blades cut a tongue and then one is covered so that the other cuts a corresponding groove.

Toothing plane

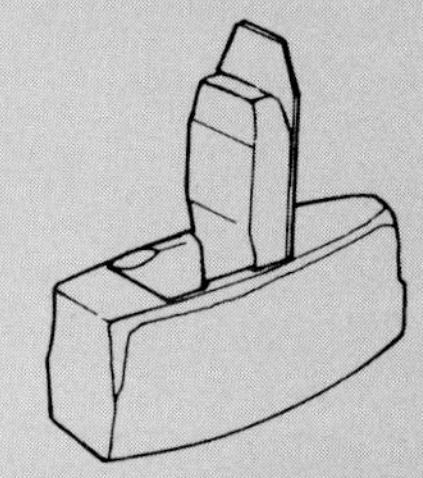

Used to score surface prior to gluing or veneering. Serrated edge pushed along grain scores wood.

Universal plane

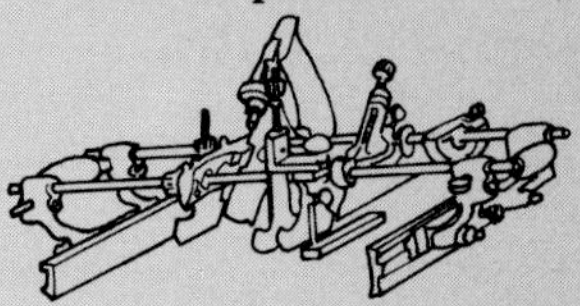

Combination of many plane functions. Comprises plow, dado, round, rabbet, beading, hollow, chamfer and fillister and many molding planes. Removeable arms, depth gauge, fences and different blades.

PLANING BOARD

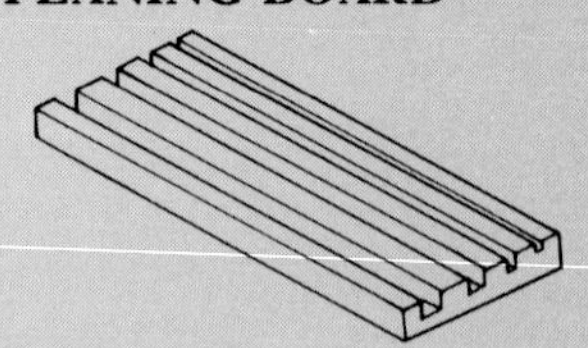

Improvised device with grooves for holding, laths etc. while planing.

PLIERS

Combination pliers
(plier wrench)

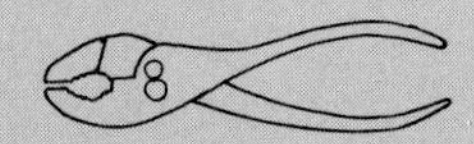

Used as pliers and can be locked onto object like pipe wrench.

Needle nose pliers

Used for gripping small objects.

Slip joint pliers

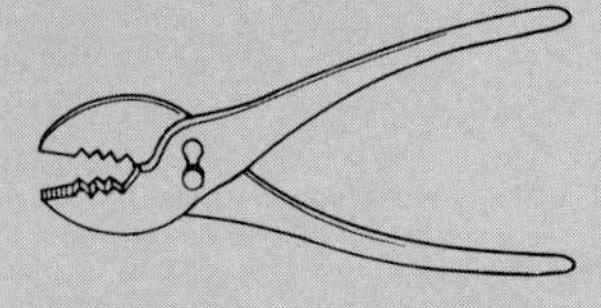

Jaws can be opened in usual way and can be adjusted to positions to grip large objects. Some types cut wire.

PLUG CUTTER see BIT

PLUMB LINE

Weighted string used to mark vertical line.

PLUMB RULE

Long piece of wood held vertically with plumb bob suspended from it. When bob is centered, rule is vertical.

PRESS SCREWS

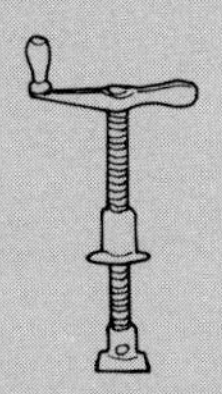

Device for making small veneer presses.

PROTRACTOR

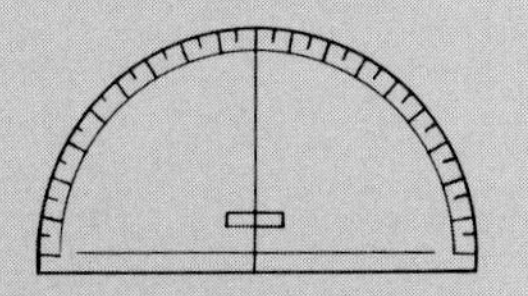

Device for measuring and marking angles.

PRY BAR

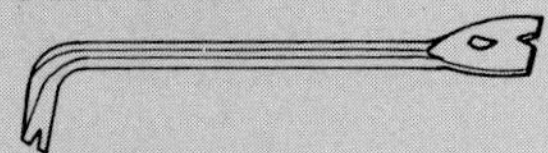

Piece of heavy, straight steel with flat blade at either end. One end chisel-shaped, the other pointed. Used for general prying.

PUNCH see also NAIL SET

Catapunch

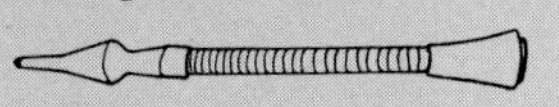

Used to mark centers without hammer.

Center punch

Device used to mark hole centers before drilling.

AUTOMATIC CENTER PUNCH

Center punch with self-centering device.

PURFLING CUTTER

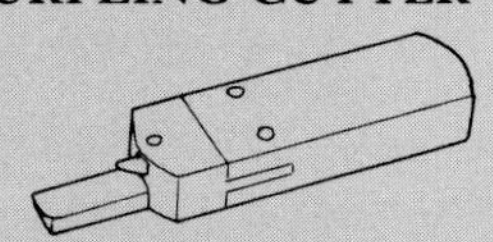

Rabbeting tool for use in instrument making.

PUSH PIN see BRAD DRIVER

RASP

Very coarse file with raised points for cutting. Triangular teeth arranged in rows. Made in different degrees of coarseness and various shapes.

Curved rasp

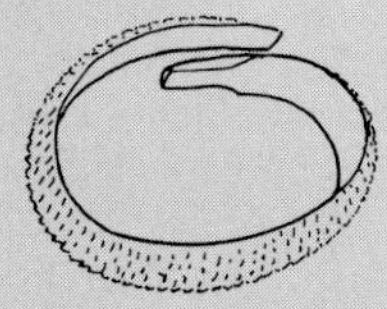

Rasp bent into circle. For making bowls, etc.

Flat wood rasp

Like flat file in section with rasp cut for flat and convex surfaces.

Half round cabinet rasp

Best known wood rasp. Like half round file. One flat and one round edge for rough work.

Half round wood rasp

Used for flat, concave and convex surfaces.

Horse rasp

Largest and coarsest rasp. Often square at both ends. No tang for handle. Used for rough work.

Round wood rasp

To rasp holes or curved surfaces.

Stickleback hand rasp

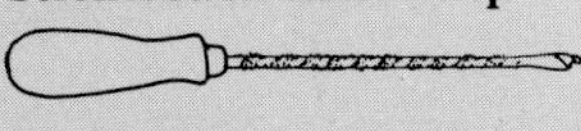

Used for starting holes or as round rasp.

REAMER

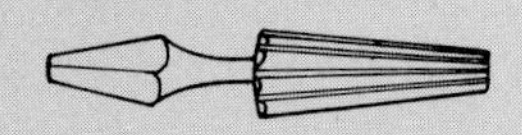

Tool for widening and truing holes. Used in brace.

RIFFLER

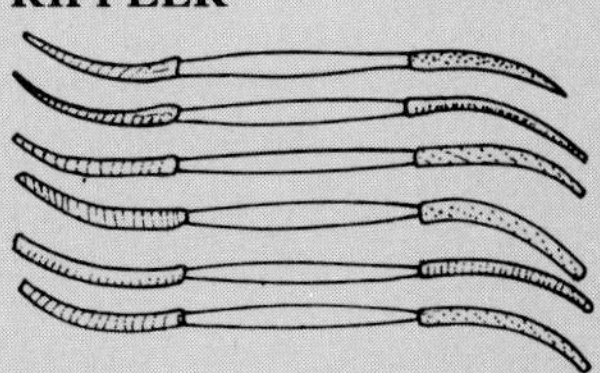

Craftsmen's miniature files. Available in large selection of cut, coarseness and cross section. Used mainly in carving for corners and recesses.

RIPPING BAR
(wrecking or crow bar)

Goose-neck steel bar used for prying and levering. Nail slot at one end.

RIPPING CHISEL

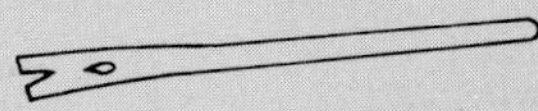

Wider chisel end with nail slot and wedge-shaped hole for pulling nails.

ROUNDING CRADLE

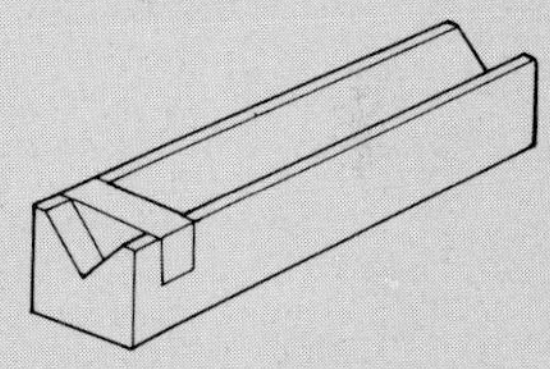

Homemade device to support square stock during rounding or shaping.

RULE

Bench rule

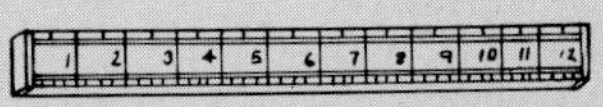

Common type of rule graduated in $\frac{1}{8}$ and $\frac{1}{16}$in. Maple or hickory.

Board rule

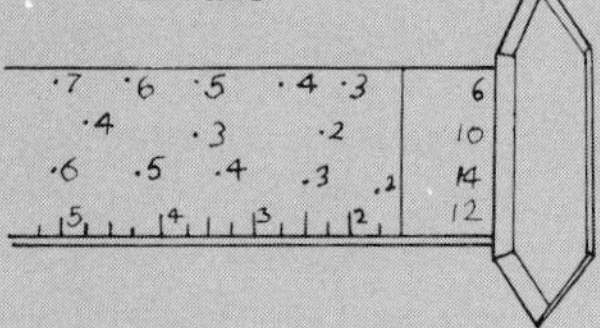

Device for computing how many board feet in a length.

Circumference rule

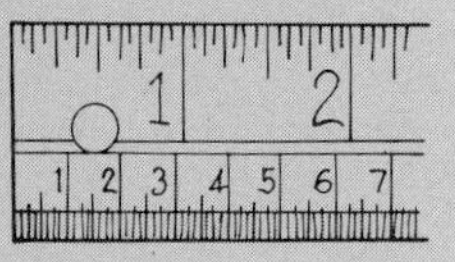

Can be used as steel rule, but also to calculate the circumference of disks and cylinders.

Digital rule

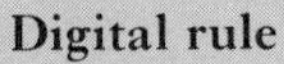

999ft capacity. Rule run over surface and display indicates length. Measures curves as well as flat surface.

Folding rule

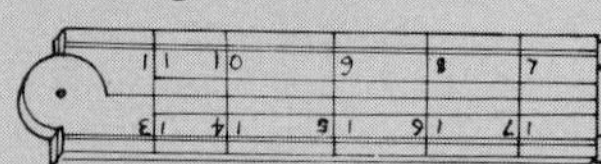

Basic rule is box wood, steel or plastic 2ft long with four folds. US types are graduated left-handed, i.e., numbers begin at right hand end. UK models are the opposite.

Pattern maker's shrinkage rule

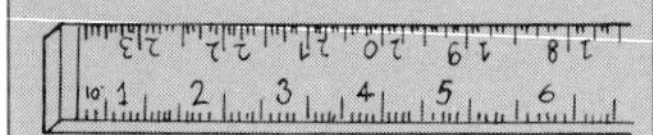

Graduated so each foot is $\frac{1}{10}$, $\frac{1}{8}$, $\frac{3}{16}$ or $\frac{1}{4}$in longer than actual so possible to measure wood that will shrink but still use the nominal finished measurement.

Steel rule

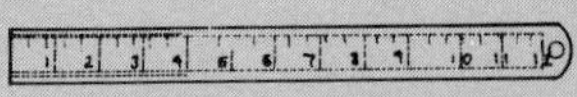

Essential measuring tool. Can be used as a straight-edge.

Tape rule

Flexible steel tape in compact case. Tape retracts automatically when button is pressed in some models.

Zig-zag rule

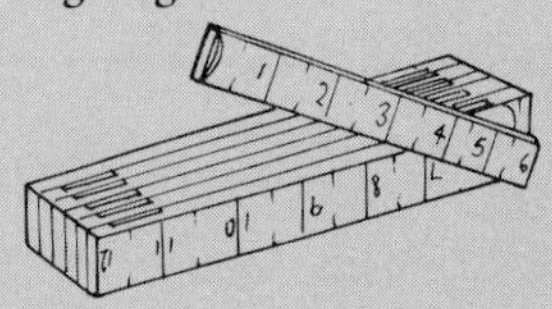

One foot lengths of wood jointed so that they can be extended or contracted.

EXTENSION RULE

Zig-zag rule which includes brass slide for taking internal measurements.

SANDING BLOCK (sandpaper holder)

Device for holding sandpaper on block. Rubber faced type has clamps to hold paper securely.

SAWS

Back saw (tenon saw)

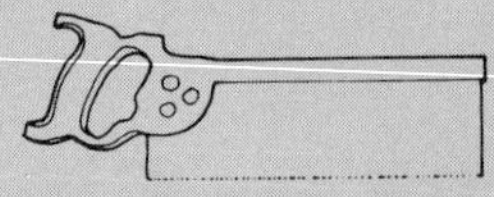

Crosscut saw with blade held rigid by reinforcing piece of steel or brass along top edge.

BEAD SAW (GENT'S SAW)

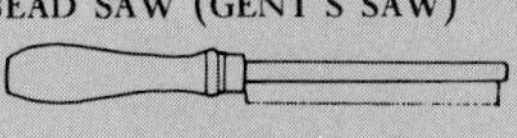

Backsaw with 38 points for very fine work.

DOVETAIL SAW

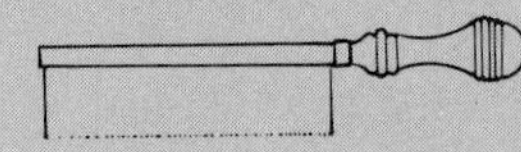

Small back saw with much finer teeth and straight handle. Used to cut joints.

Bow saw (frame, turning, web or cabinet saw)

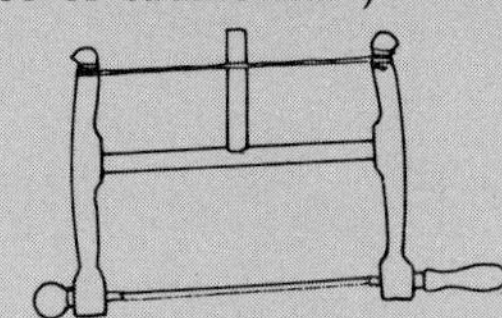

Was once the general use bench saw. Blade can be turned in the frame. Used to rip, crosscut and cut curves.

Buck saw

Similar to turning or bow saw, but blade is fixed.

Compass saw (keyhole or pad saw)

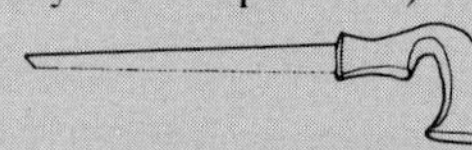

Saw with narrow blade used for cutting curves.

Coping saw (jigsaw)

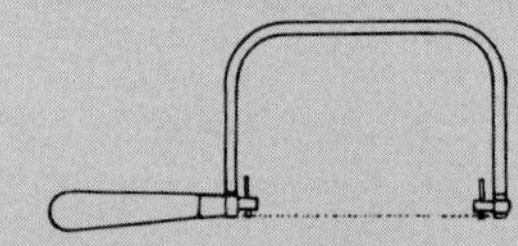

U-shaped steel frame with slender, short blade inserted in slotted pauls. For cutting curves in thin material.

Crosscut saw

Traditional saw operated by one or two men. $2\frac{1}{2}$ to 5ft long. Different tooth pattern such as peg, tooth and gullet, lance and Great American.

Fret saw
(bracket or scroll saw)

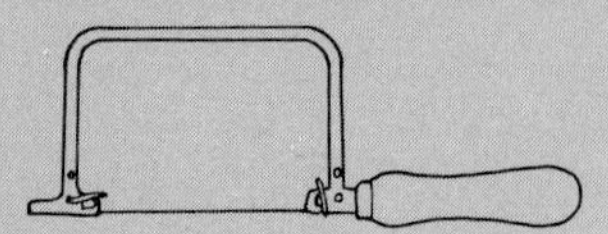

Similar to coping saw, but with deeper frame. Used to cut intricate curves.

Gauge saw

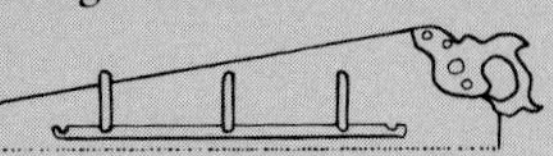

Regular crosscut handsaw with adjustable depth gauge fitted to blade. For tenoning, shouldering, dovetailing, etc.

Hacksaw

Saw designed to cut metal. Used by woodworkers to cut through nails and screws.

Hacksaw

CROSSCUT HAND SAW

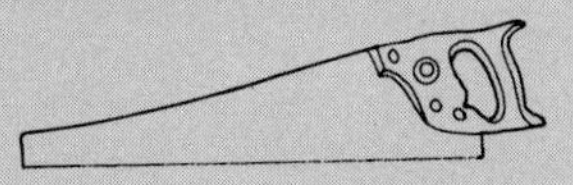

Saw for cutting across grain. 26in long with 8 to 10 points to the inch. Fitted with closed handle.

PANEL SAW

Finer version of crosscut hand saw, 20in long with 10 points to the inch.

RIPSAW

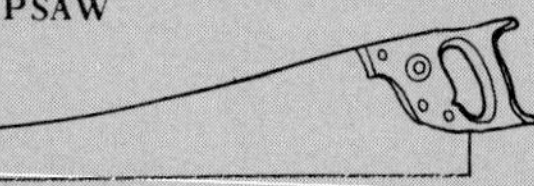

For cutting along length of wood. 5 to 7 points to the inch. Blade is 20 to 26in long.

Inside start saw

Very thin, flexible blade. Front edge has teeth so you can turn saw upside down. Back curved so cut can be started in middle of work.

Japanese saws

Saws designed to cut on the pull stroke. Thinner gauge and setting teeth is not necessary.

AZEBIKI SAW

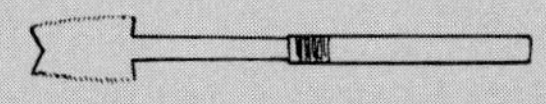

Used to start cuts in center of work as well as to make stopped cuts. One edge with 10 points per inch, the other with 15.

DOZUKI SAW

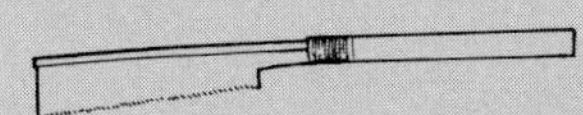

Tension back saw. 23 teeth per inch for dovetails and other fine work.

HANAMARU SAW

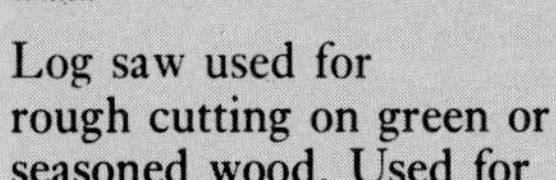

Log saw used for rough cutting on green or seasoned wood. Used for ripping or crosscutting. 6 teeth per inch.

HIKIMAWASHI SAW

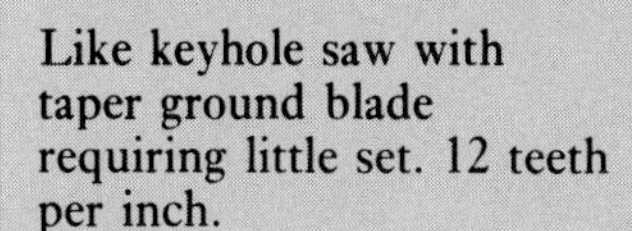

Like keyhole saw with taper ground blade requiring little set. 12 teeth per inch.

RYOBA SAW

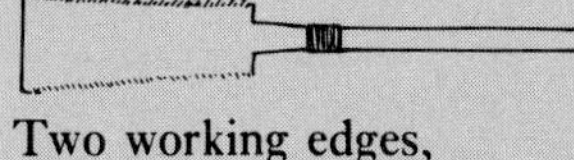

Two working edges, crosscut with 9 points per inch and rip edge with $4\frac{1}{2}$ points per inch. Used like panel saw.

Keyhole saw (turning saw)

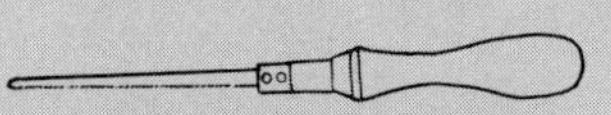

Narrow blade that tapers to a point. Teeth for ripping or crosscutting. Used for cutting curves like keyholes. Often sold as nest of saws with one handle and three blades.

Offset back saw

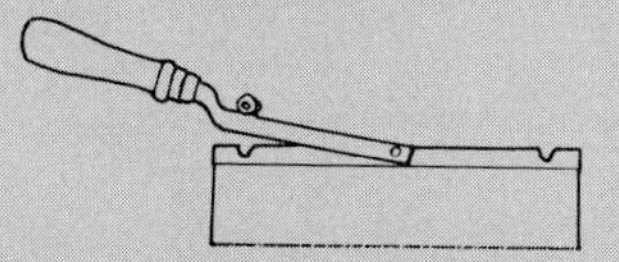

Saw for use in awkward places where straight back saw cannot reach. Blade pivots for right or left-hand use.

Veneer saw

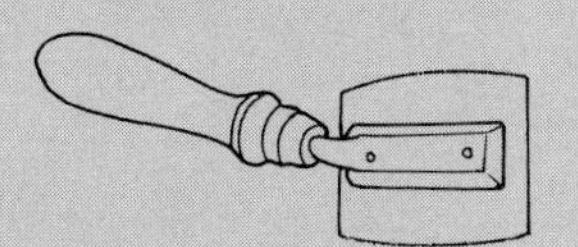

Very small saw with fine teeth for cutting veneer. Two curved serrated edges one for crosscutting and the other for cutting with veneer grain.

SAW CHOPS

Wooden device made of two jaws held by bolt and wing nut, held in vise and used to hold saws during sharpening.

SAW FILE see FILES

SAW JOINTER

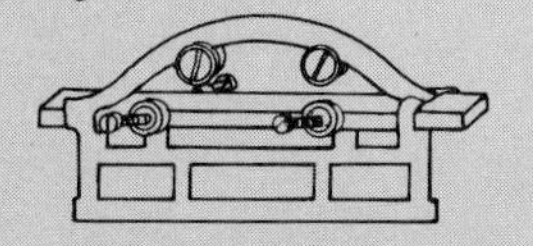

Tool used in first stage of sharpening saw. Accurately "joints" all the teeth to equal length.

SAW SET

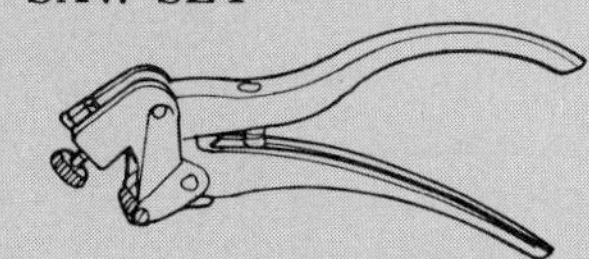

Tool for bending teeth to left and right so saw will not bind in the kerf. Adjustable for different hand saws and fine-toothed circular saws.

SAW SHARPENER, HAND

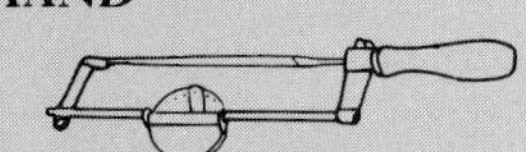

File which is slid over teeth to set to correct angle.

SCORP
(round shave or scoop)

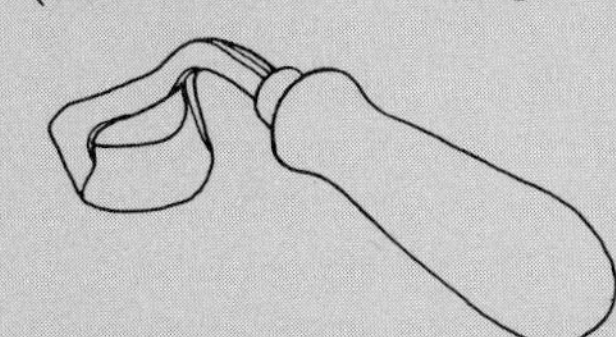

Cross between adz and drawknife. Used for scooping out bowls, etc. Operated one-handed and pulled toward user.

SCRAPER

Final finishing tool before sanding. Flexible steel blade whose edge is turned over to allow burr to cut wood.

Adjustable scraper

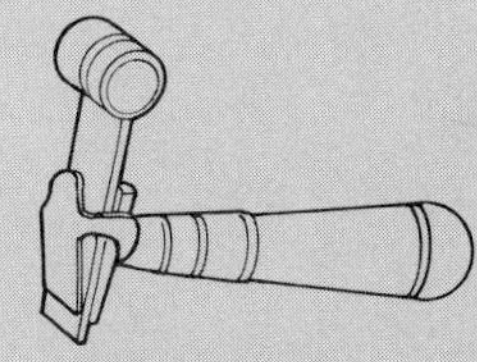

Plain cabinet scraper with an adjustable handle so can be used to last inch.

Box scraper

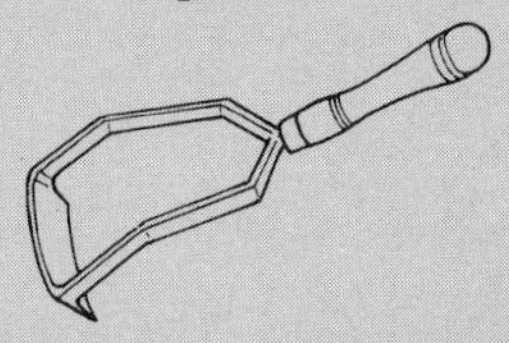

Scraper blade with wooden handle. For rough leveling of wood.

Cabinet scraper

Simple piece of steel with two cutting edges for working flat areas. Leaves cleaner finish than sandpaper which clogs grain with dust. Either straight, rectangular or convex/concave shape.

Double handed scraper (scraper plane)

Used to scrape wood prior to polishing. Curve adjusted by means of thumb screw. Can be converted to toothing plane by changing blade.

Four bladed scraper

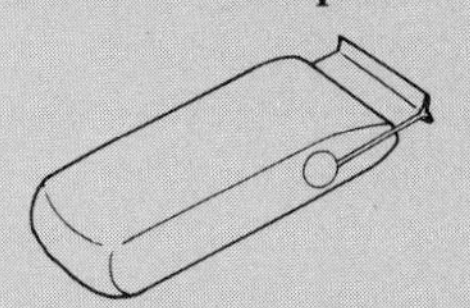

Tempered blade with four scraping edges. Blade is easily reversible.

Molding scraper

Designed for smoothing moldings, therefore made in a variety of shapes.

VERSATILE MOLDING SCRAPER

Eight-edged scraper with variety of points, concave and convex. Used with push, paring or whittling stroke.

Ship scraper

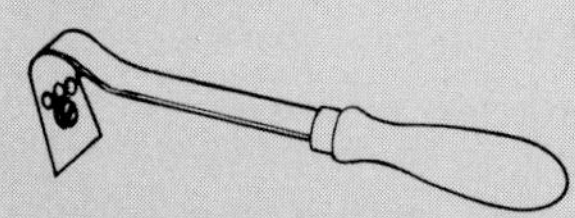

Scraper with wooden or cast iron handle for heavy-duty use.

Swan neck scraper

Scraper for curved work with all-around cutting edge.

Veneer scraper

Capable of fine adjustment. Similar to cabinet maker's scraper but smaller version.

SCRATCH AWL see AWL

SCRATCH STOCK

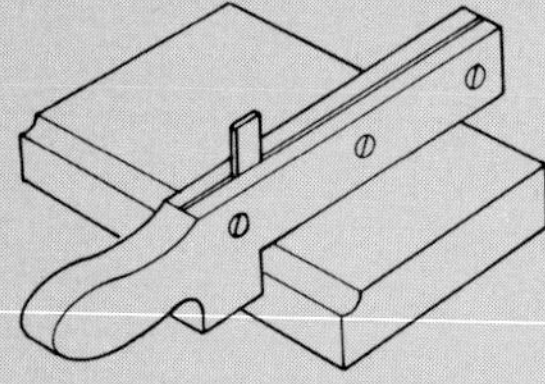

Homemade device for cutting grooves in inlay, bandings, etc.

SCREW BOX AND TAP (woodthreading kit)

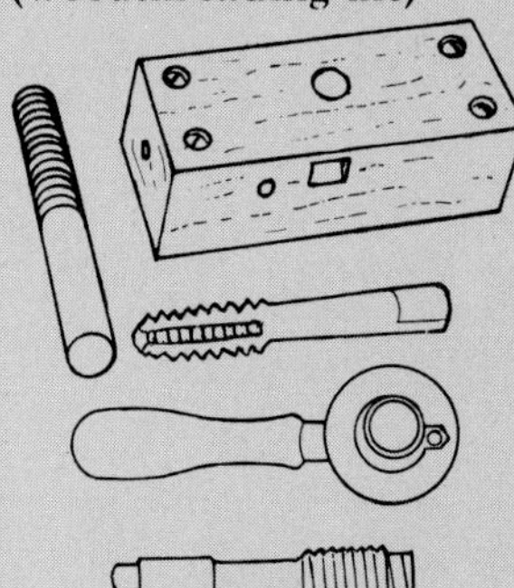

Tool for cutting a thread on a dowel. Tap cuts thread into block of wood to accept screw. Available in range of diameters.

SCREWDRIVER

Most common tip is cabinet tip which is flat and ground square. Blade is flared out and tapered. Parallel tip version does not have flared end. Blade is tapered and ground square for use in restricted places. Phillips head screwdrivers have cross head top, Posidriv have cross head tip with additional hole in center of slots. Other types include reed and prince with cross head tip which comes to a point.

Cabinet screwdriver

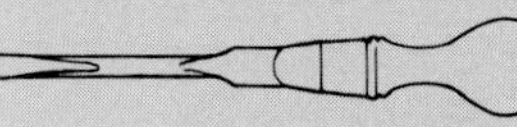

Traditional screwdriver with oval section hardwood handle.

Clock screwdriver

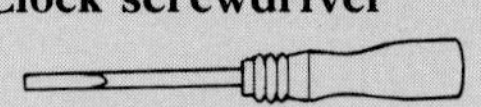

Small cabinet screwdriver. Blade is same width as shank.

London pattern screwdriver

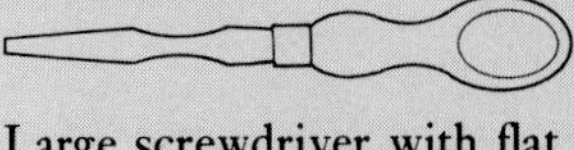

Large screwdriver with flat waisted blade and hardwood handle flattened for good grip.

Offset screwdriver
(cranked screwdriver)

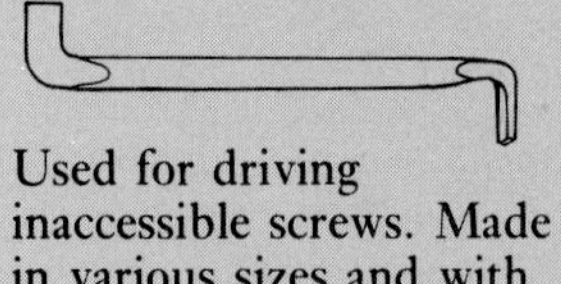

Used for driving inaccessible screws. Made in various sizes and with different tips.

Pocket screwdriver

Small, pocket-sized screwdriver often with retractable blade.

Ratchet screwdriver

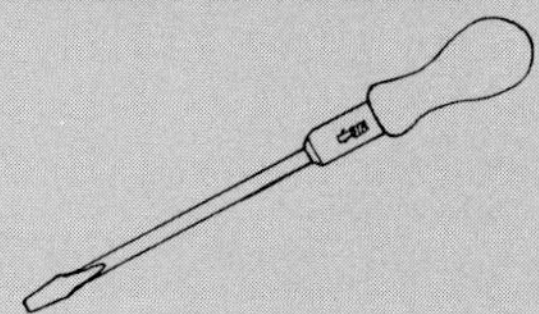

Used to drive screws without altering the grip. Change direction with thumb slide.

Spiral screwdriver
(Yankee or pump screwdriver)

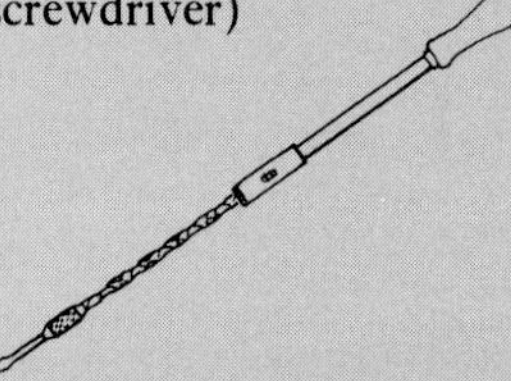

Spiral reciprocating shank. Drives screws by pressure. Can be turned clockwise or counterclockwise by adjusting thumb slide or can be locked to use as standard ratchet.

Stubby screwdriver

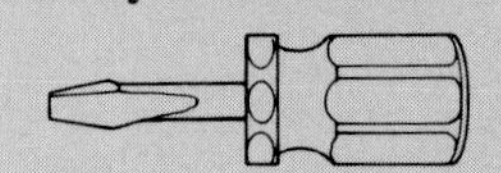

Small screwdriver for driving screws in restricted space. Can be fitted with tommy bar for increased torque.

SCRIBER OR MARKING AWL see AWL

SHARPENING STONES (oilstones or whetstones)

Hard stones used for fine sharpening of edged tools. Lubricated with light machine oil to keep them unclogged. Usually $6 \times 1\frac{1}{2} \times \frac{1}{2}$in or $8 \times 2 \times 1$in. Natural types of stone include hard and soft Arkansas, lily white and rosy red Washita and turkey stone. Artificial stones are aluminum oxide called India stone and silicon carbide called crystolon. Some stones have coarse grit on one side and finer grit on the reverse. The harder the stone, the finer the finish it gives.

OILSTONE BOX

Wooden box with lid to hold sharpening stone. Nail head just protruding in base prevents box from slipping on bench.

Handstone

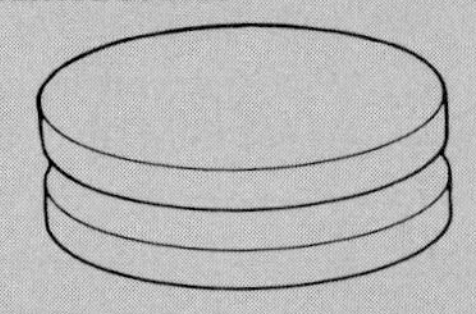

4in diameter, silicone carbide stone used to sharpen axe blades as well as plane irons and jointer blades.

Honing stone

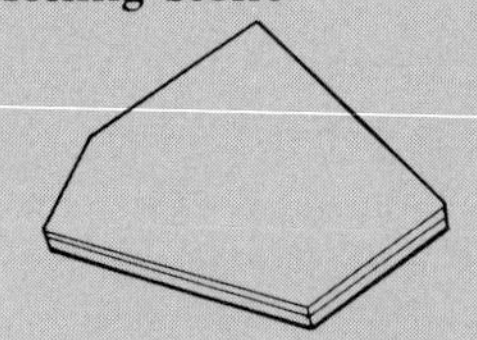

Very fine and soft stone for putting finish edge on carving tools.

Slipstones

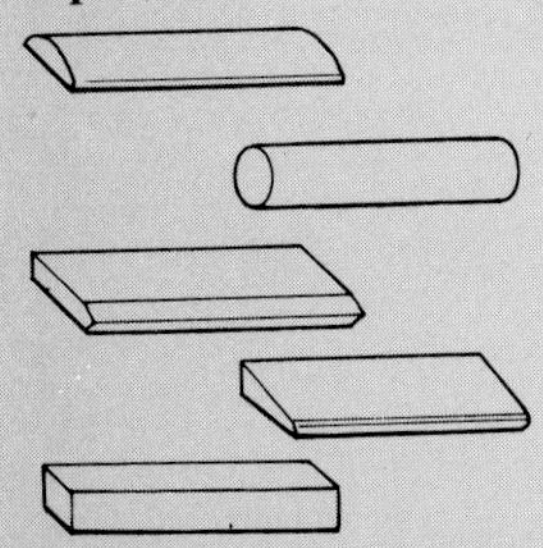

Small, shaped sharpening stones either natural or artificial. Used to sharpen irregularly shaped tools like gouges. Available in rectangular, circular, triangular, tapered and square shapes.

SHAVE HOOK

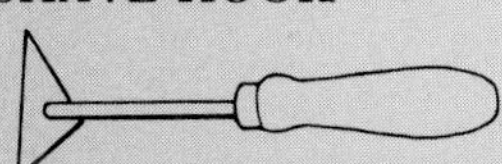

Small scraper with screwdriver-like shank and handle with blades in a variety of shapes: triangular, pear-shaped and combination.

SHOOTING BOARD

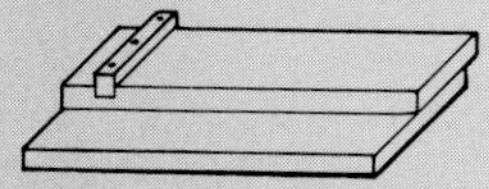

30in long beech jig used to guide plane to cut ends or edges of thin boards. Used clamped to bench. Usually homemade.

Donkey ear shooting board

Traditional design for mitering in thickness.

SLICK
(paring chisel)

Traditional tool with long hardwood handle and large blade used for paring large areas. Used two-handed like a plane. Pushed by hands and shoulders rather than being struck.

Socket slick

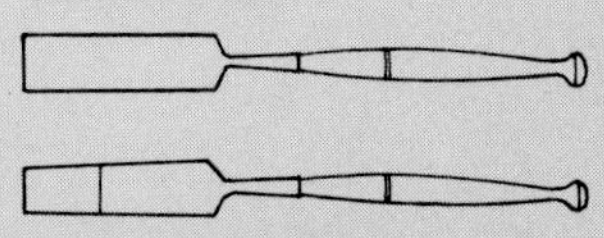

Broad chisel used to level off hand-hewn surfaces and clean out mortises.

SPOKESHAVE

Small plane like a narrow drawknife used to shave and dress wood. Usually operated by being drawn toward user. Traditionally mounted in wood holder, but modern types are often metal.

Adjustable blade spokeshave

Flat spokeshave with two adjusting screws for vertical and lateral blade adjustment.

Chamfer spokeshave

Used to cut accurate chamfer in edges of piece of wood.

Hollow spokeshave
(half round spokeshave)

Used to smooth sections of wood curved in two directions. Concave face and matching blade.

Rabbet spokeshave
(carriage maker's spokeshave)

Sharpened on two sides. Often sold in set of two, one with lengthwise curved face, other with flat face. Each has two blades for use in either direction.

Universal spokeshave

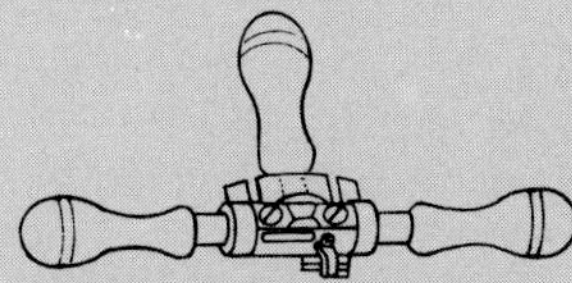

Spokeshave with detachable handles which can be screwed into back or side of stock.

SPUD (bark spud)

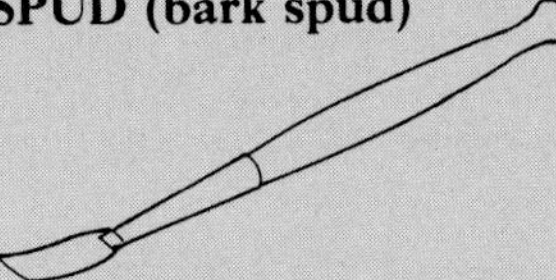

Traditional tool for removing bark for tanning hides. Made in many shapes and sizes.

SQUARE

L-shaped device for checking squareness of wood and measuring and marking right angles.

Adjustable miter square

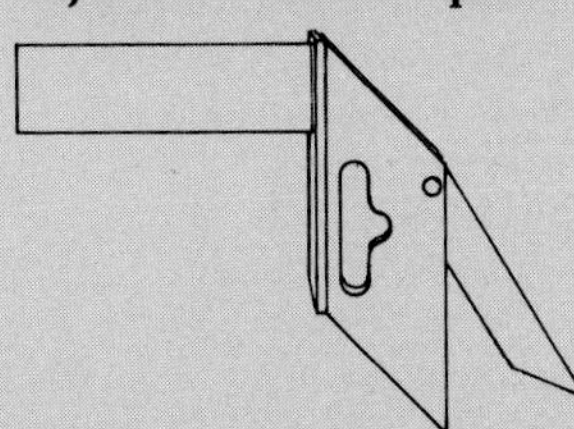

Stock has extra blade which may be used as a bevel. Head and foot of stock are at 45° for miter joints.

Center square
(radial square)

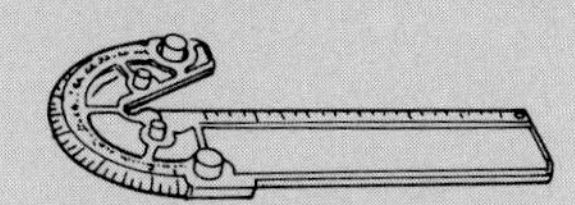

Small, flat metal strip shaped at one end. Used for marking right angles, locating center of circular piece and taking small measurements. May be used as a protractor.

Combination set square

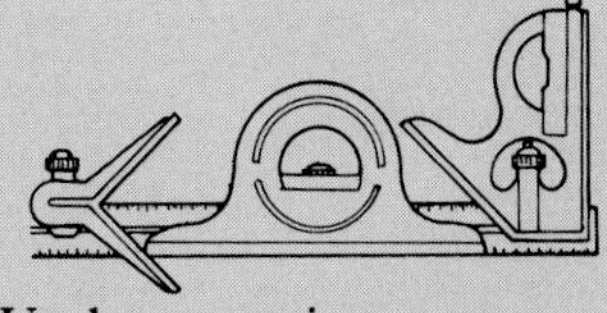

Used as try, miter square, protractor, center square, level and rule.

Miter square

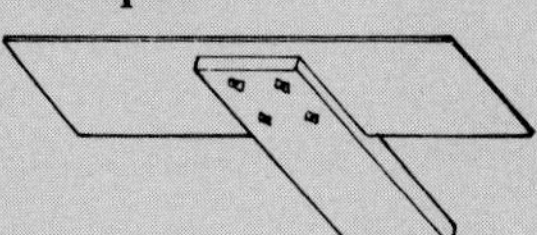

Really a fixed bevel set at 45°. Used to mark out both halves of miter joint.

Try square

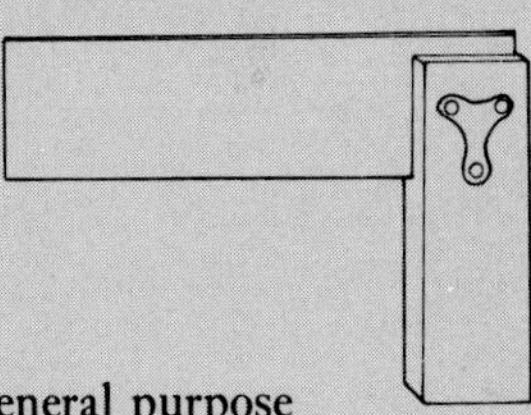

General purpose woodworking square. Wooden or plastic stock and steel blade. Used to mark lines at right angles to an edge and to check that frames are square.

Try and miter square

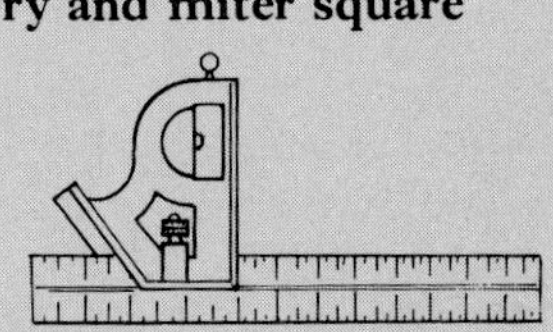

Designed to measure right angles and 45° angles. Includes small spirit level, scratch awl and sometimes a depth gauge.

Engineer's square

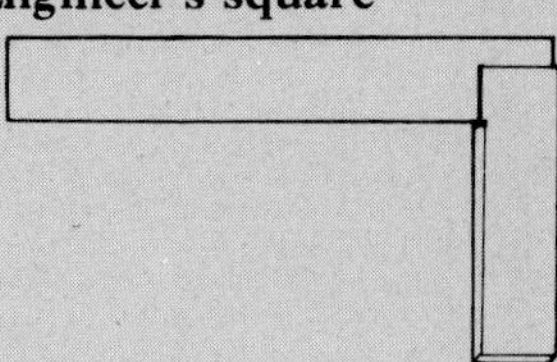

Small, accurate, all metal square used by machinists and engineers but also handy in woodworking to set machines.

SQUARING ROD

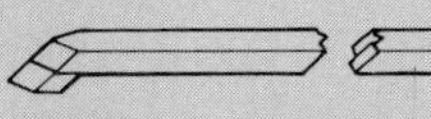

Length of wood 1 × $\frac{5}{16}$in, blocked and pointed at one end used for checking square of carcass openings.

STAIRCASE FIXTURES

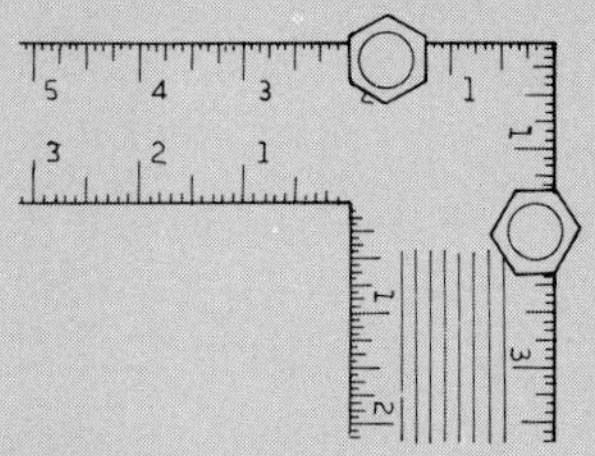

Clamped to steel carpenter's square to make gauge for laying out stairs. Useful for repetitive gauging.

STRAIGHTEDGE

Straight length of steel with parallel sides, beveled on one edge. Used to cut or scribe against or to test flatness. Good ones are expensive and invaluable.

STROP

Base for final honing of fine cutting edges. Dressed with oil and used with fine carborundum powder.

SURFORM TOOL (rasp plane)

Device like a hollow rasp. Has replaceable steel blade pierced with many sharp edges to rapidly cut away wood. Waste does not clog teeth.

Surform flat file

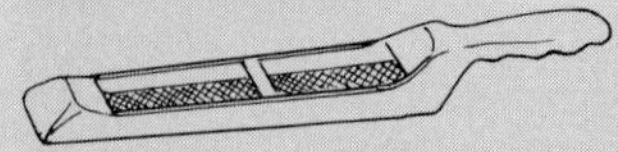

$16\frac{1}{4}$in long for general work, available with flat or rounded blade.

Surform round file

Tube shaped with removable front holding piece, 14in long.

TACK CLAW

Like nail claw but smaller. Used to pull tacks.

TAPE MEASURE

Cloth tapes from 25 to 100ft long in case. For measuring large dimensions.

Retractable steel tape

16ft long steel or fiber glass tape in case.

TEMPLATE FORMER

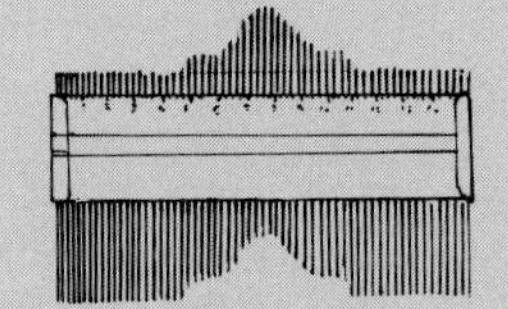

Steel needles pushed through rule to form shape as template for moldings, etc.

TIMBER JACK

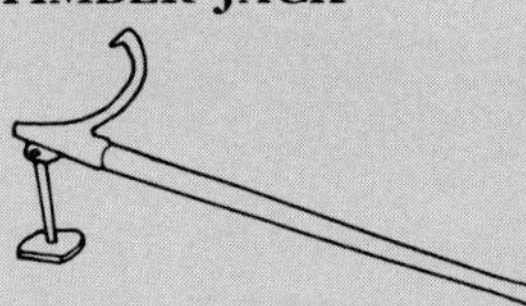

Peavey claw with stand used to lift log while cutting.

TIMBER SCRIBE (raze knife)

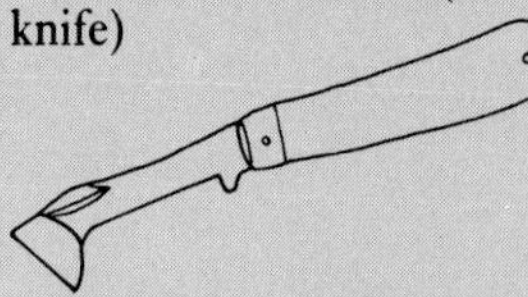

Knife with hooked blade used to score footage on ends of planks or gouge wood to determine type and color.

TRAMMEL POINTS (trammel heads)

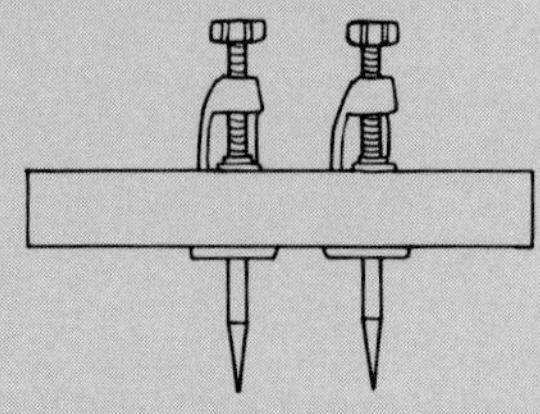

Sharp steel pins which may be attached to rules, steel squares or pieces of wood to describe large arcs and circles. One point can be replaced by pencil.

TWIBILL

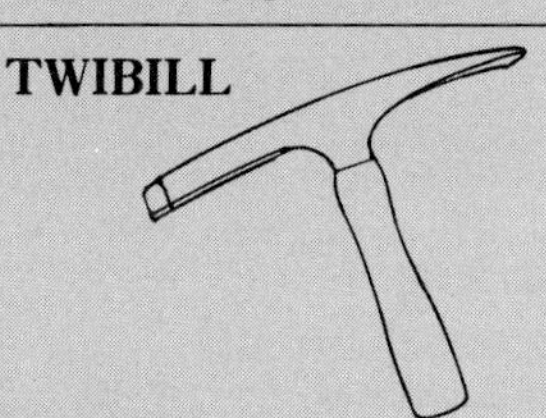

Old-fashioned two-headed axe used for cutting mortises.

UTILITY KNIFE see KNIFE

Veneer edge trimmer

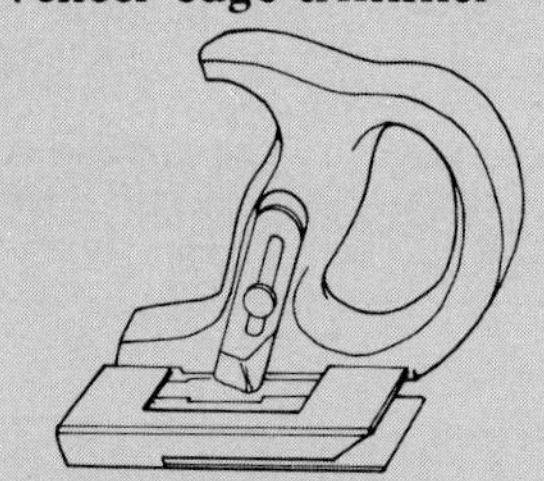

Cuts veneer glued to panel without damage. Cuts forward and reverse, with or across grain.

VENEER INLAY CUTTER

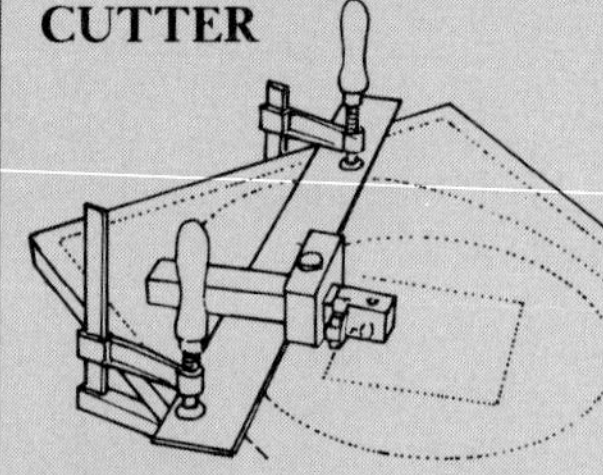

Special purpose tool for quick and accurate cutting of straight, shaped and circular veneer veins of required depth.

VENEER AND LAMINATE CUTTER

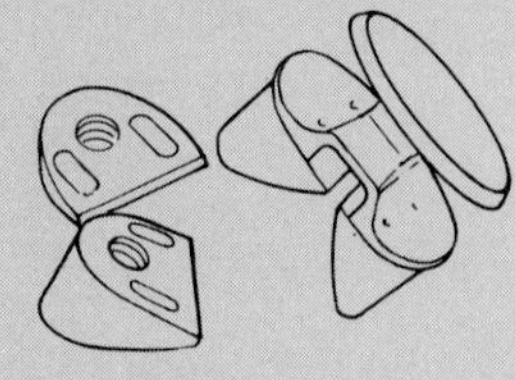

Device used to trim wood, veneer and plastic laminates using system of fences.

VENEER PUNCH

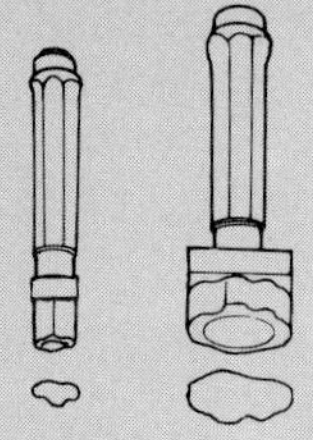

Device for cutting out irregular pieces of veneer for repairs. Available in sizes from $\frac{1}{2}$ to $2\frac{1}{4}$in across.

VENEER ROLLER

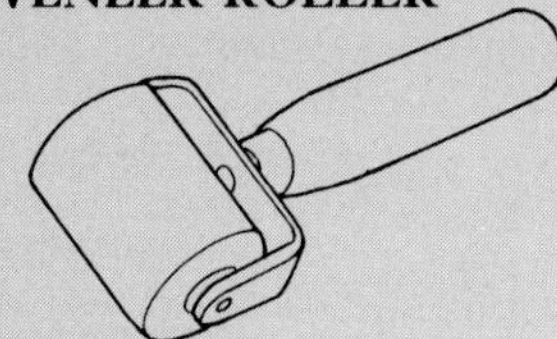

Device used to put even pressure on glue joint. Made of maple mounted on brass bearings with spring steel frame.

VENEER STRIP AND JOINT CUTTER

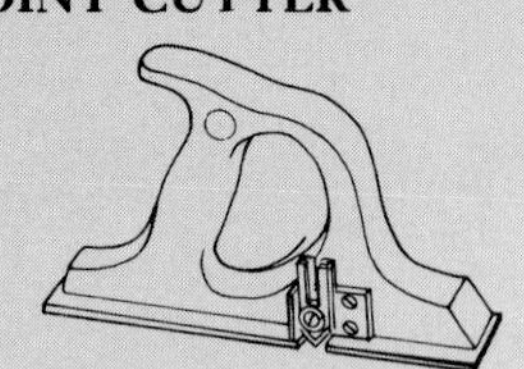

Used with straightedge. Left side for cutting joints with and across grain other side for inlay strip cutting.

VENEER TRIMMING CHISEL

Used to trim surplus veneer off edge of panels without following the grain of the veneer.

VISE

Two jaws, opened and closed by lever, cam or screws. Used for holding work securely.
Woodworking vices have wood lined jaws to prevent marring work.

Clamp-on vise
(portable vise)

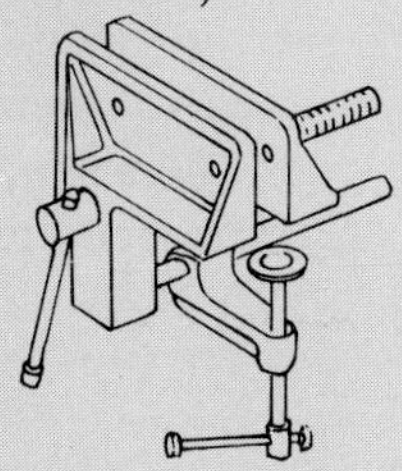

Used to hold lightweight work.

Coachmaker's vise
Woodfaced jaws. Jaws are higher than surface of bench on which vise is mounted.

Edge vise

Fits most benches or table tops up to $2\frac{1}{2}$in thick. Used for holding panels, dowels or turned pieces.

Hand vise

Small, hand-held vise for holding bolts, etc.

Panavise
Versatile vise which holds work and turns and tilts to any position. Can be tilted 90° in any direction and rotates full circle.

Pin vise

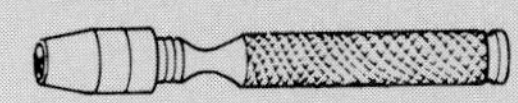

Smallest vise for holding wire, pins, etc.

Saw vise

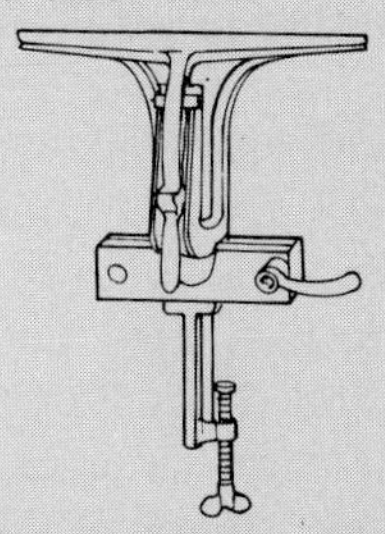

Clamps to bench and holds saw in long jaws while it is sharpened.

Universal woodworking vise

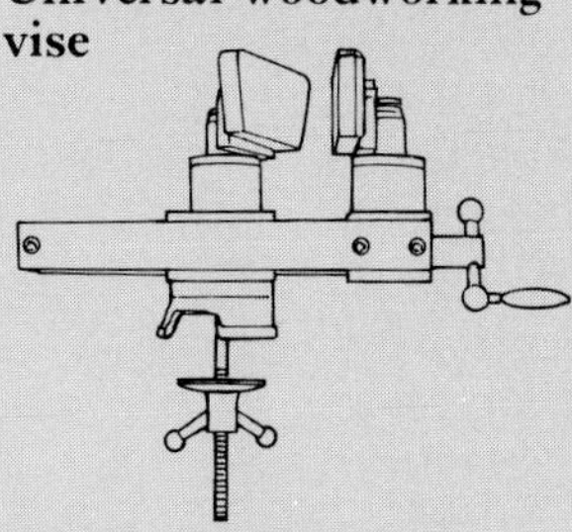

Each jaw swivels 360° independently.

Utility vise

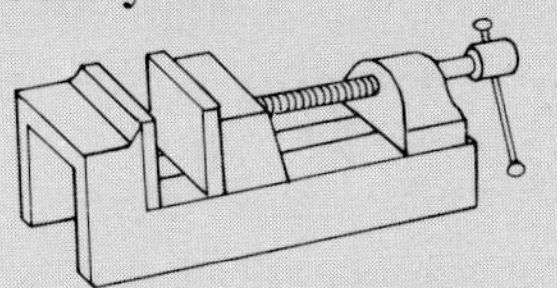

Portable unit with flat base used to hold work at drill press, grinder, etc.

Woodcarver's vise

Has wooden jaws and screws. Installed under front edge of bench at one end, jaws flush with bench.

Woodworker's vise
(standard vise)

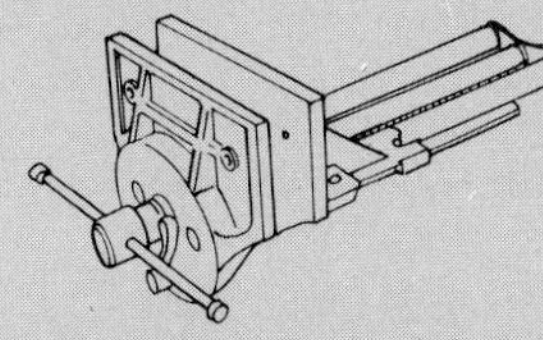

Some models are quick release, others have built-in bench stop.

WEDGE

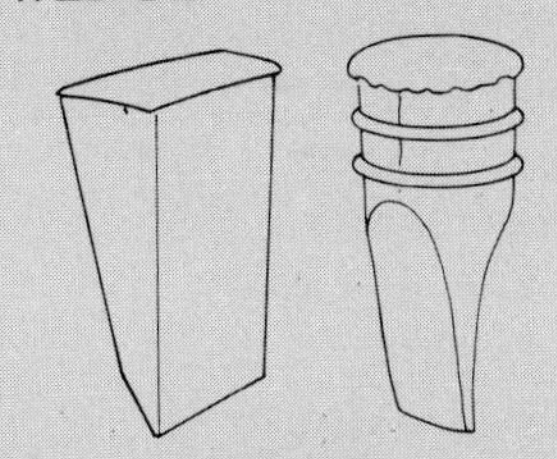

Used for splitting logs. Iron with steel tip. Old-fashioned type called gluts, were oak with iron rings.

Bucking wedge

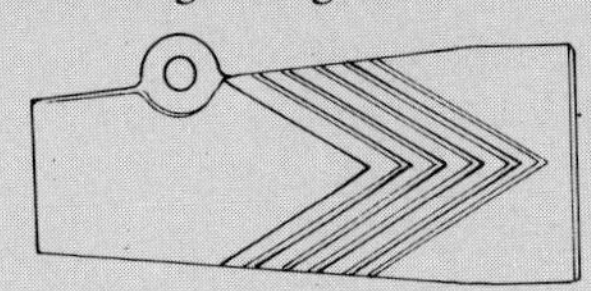

Designed to keep cut open while crosscutting with chain saw, etc.

WHEEL DRESSER

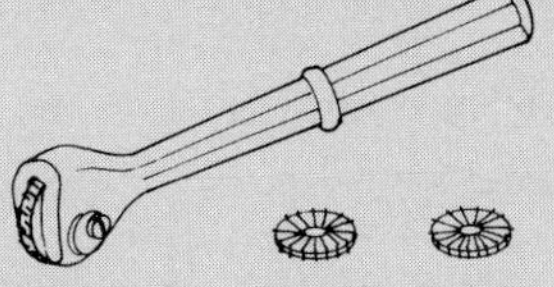

Device for removing glazed surface from grinding wheels.

WINDING STICKS
(winders)

Two straightedge sticks of equal width, laid on opposite ends of board and sighted across to test flatness.

WOODTHREADING KIT
see SCREW BOX AND TAP

WOODTURNER'S SIZING TOOL
(woodturner's gauge)

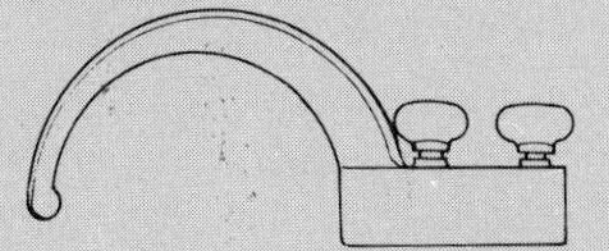

Device designed as outside calipers and to measure tenons for repetitions on the lathe.

WRENCH (spanner)

Used to tighten and loosen nuts.

WRECKING BAR
(crow bar)

Used to remove nails and lever structures apart. One end bent into tight curve with claw for nail pulling. Other flattened into blade for levering.

The Workshop

There seems to be a tendency among woodworkers to buy more tools than are really required. Part of the reason, I suppose, is the beauty of the tools, but another part is the mistaken belief that more tools will make us better woodworkers. This may or may not be true, but it depends very much on the individual.

The point is that despite the impression given by this and most other books, you need very few tools to do good quality work. I have always found that the fewer tools I have, the more I tend to improvise, making the tools on hand do various jobs by inventing jigs and holders to make them more versatile. You can do beautiful work with hand tools and a few portable power tools. Machines simply make it easier to do accurate work more quickly.

Setting up a workshop requires a lot of ingenuity and quite a bit of compromise. We have to learn to make do with what we have and by improvisation, extend the use of the tools. A portable router for example can be mounted upside down in a homemade table and used as a shaper, and a portable circular saw can similarly be turned into a table saw. Throughout this book I've tried to show how to make up jigs and other devices to make the basic tools more versatile and how to get professional results without the use of expensive machinery.

Setting up a workshop, however humble, is very exciting. Many people drift into it, gradually turning a hobby into a profession, but even then, the decision to leave a job for the uncertainties of self employment is not easily taken. I think the one thing required to succeed at it is commitment; to love wood, to love furniture, to enjoy making things. It may sound corny, but a pit of passion is needed to get through the rough early days.

Although circumstances may vary widely there are several rules which I think are universal and worth noting:

1 Charge enough for your work. Almost everyone starting out makes the classic mistake of working for next to nothing in eagerness to get work.

2 (related to 1) Work usually takes twice as long as you estimate.

3 Design flair is, besides the good craftsmanship which is taken for granted, the most important part of the work. Imaginative ideas attract customers and make your work stand out.

4 Keep overheads low. Until you are well established, buy stock or tools as they are required.

5 You'll never have enough clamps.

Premises The size of workspace required depends entirely on the projected scope of the shop. Not everyone is going to set up as a business venture, so extra space may have to be found at home to keep the cost down. For this kind of small, one man shop, a garage or large basement is adequate. Basements do not usually have enough headroom to allow handling large sheets of plywood, and access is also restricted making it difficult to get supplies in and finished work out. Garages are better for they are usually drier, better lit and have more headroom to maneuver sheets. In addition, the doors can be opened in the summertime for extra workspace when required.

But the scope of a home workshop is limited primarily to handwork with perhaps one or two small machines such as a band saw to assist. To turn woodworking into a full-time occupation, larger premises will inevitably have to be found to allow for all the equipment and all the activity that a small business generates. And finding a suitable workshop space, never mind financing it, seems to be the biggest obstacle in setting up a shop, particularly in cities where space is at a premium.

When looking for premises, keep in mind that the space should be high ceilinged and large enough to allow not only for the machines, benches, storage and so on that you already have, but also some extra space for growth and if possible, a small side room kept dust free to be used for finishing. A separate storage room or area for a wood rack is also a must.

Access for delivery of materials is also important, and the shop should have doors large enough to allow moving the finished work out assembled. For these reasons avoid workspace which is not on the ground floor, unless there is a large industrial elevator. And make sure to check that the floors of the shop are capable of taking the weight and vibration of the machinery you intend to install.

All these specifications point to an industrial building which is likely to have the large, uninterrupted space, good lighting and services required for heavy-duty woodworking.

The location of the shop is more important than some people realize. There are advantages of being easily accessible so that you can sell direct from the workshop. You can keep prices lower than they would be through a store because overheads are lower and customers will enjoy seeing how and where the work is done. Of course this is not a hard and fast rule. There are many people working in the countryside enjoying advantages such as low rent, peaceful surroundings, proximity to mills and cheap wood, but I'm sure they'll be the first to admit that it's more of a struggle selling work at first until they are well-known. In short, there are more potential customers in the city and also more opportunity for much needed free publicity (see page 43).

It's difficult to give advice on how to find a space. It takes persistence. Estate agents or realtors will send lists of properties most of which are likely to be too large and too expensive. It's best to phone and let them know who you are and what you want.

Often they'll have premises which need renovation, and if you're prepared to spend the time and money to fix it up, they're usually happy to help out. But be realistic about the time and costs involved. It's probably better to rent a more expensive shop, and use the available capital to better advantage buying tools and stock.

In Britain, local councils are often willing to help craftsmen by letting out low rent space. It is worth badgering them for lists of empty buildings which they may rent for a limited period, long enough to get a good start.

For additional sources, check advertisements in crafts magazines and newspapers, and bulletin boards in craft shops for workshops to rent or for space in co-operatives, where several craftsmen have individual workshops but share in the cost of communal services such as telephones, showrooms, receptionists, etc.

Regulations Before signing a lease or agreement it is important to check that the local regulations allow for light or heavy industrial use of the premises. Taking on premises which have previously been used as a workshop usually presents no difficulties and has the advantage that facilities such as adequate light and heat and power will already be installed.

There are also local bylaws concerning various aspects of woodworking shops. Depending on local laws and the number of employees, the factory inspector may have to be notified and after inspection, be satisfied that all regulations concerning guarding and placement of machines, fire regulations and so on are complied with. The insurance company will also inspect the premises to assess the rate for insurance against fire, theft, employer's liability and so on. Seek the advice of the insurance company early on, as there are many ways that you will be able to cut down on the insurance premium.

Workshop equipment The number of tools, machines and other accessories required obviously depends on the scope of the shop. A small production shop needs machines such as a saw, jointer, thicknesser, band saw and mortiser, in order to make up runs of furniture economically and

Second year students at John Makepeace's school at Parnham House, Dorset.

accurately. Whereas the one man shop may only require a few hand and portable power tools to do individually designed pieces.

Most shops grow gradually adding a tool here or a machine there as the work comes in to pay for them. But it's important to start off with at least a basic selection of tools, hardware, stock etc., to avoid time-consuming buying trips when the work comes in.

Hand tools The type of hand tools required in a shop will vary greatly depending on the type of work being done, for there are many specialists' tools such as carving, veneering or turning tools. The dictionary of hand tools (pages 6–25) includes just about every modern tool available in addition to a large number of old-fashioned tools and devices which may be of interest to those specializing in reproduction work.

In addition to these basic hand tools, various homemade devices such as a bench hook, dovetail template, shooting board and miter box are essential for handwork as are various shop aids such as clamps, benches, sawhorses, etc., described in further detail on page 32.

Portable power tools With a good selection of portable power tools you have the basis for an entire workshop. Good quality tools can be used freehand almost anywhere on site or in the shop and they can also be mounted in stands and homemade devices to turn them into very efficient machines.

To turn the portable saw into a table saw, for example, follow the instructions on page 52. This is an invaluable trick for work on site where the table can be screwed or clamped to a pair of sawhorses. But there is no reason, provided the table is well-made, that it cannot serve perfectly well instead of an expensive table saw in the workshop.

The portable router (page 64) is equally versatile, and like the saw can be used freehand or fixed in a homemade table to serve as a shaper. In the hands of an experienced woodworker, the router can become almost a one tool workshop.

Machinery The output, and in many cases, the quality, of any workshop will be governed largely by its machinery. Those who prefer hand methods may argue with this, but I have yet to find a craftsman capable of rivaling a table saw or planer in accuracy or speed. Good quality machines are indispensable in a large workshop, and perfecting and extending the use of the machines to suit designs is as much a test of the ingenuity of the craftsman as any other aspect of his trade.

Designing for production, even in small numbers, really means adapting the details to the limitations of the machinery available. The beauty of the small workshop is that the designer is also the craftsman and can therefore intuitively work the production and the design aspects together logically.

Of course it is easy to say how convenient machines are and to present mouthwatering lists of equipment. Most people who are setting up will be fortunate to be able to afford one or two machines and will be concerned to use their limited resources to buy the right ones. New machines are obviously best, but costs have risen tremendously in the past few years; an alternative and less expensive source is secondhand machinery (page 41).

The top priority, in my view, is the table saw and planer (preferably an over/under planer, page 106), and a bench grinder (page 142) to sharpen tools. Those concentrating on handwork may prefer the band saw instead of the table saw. A universal machine or a radial arm saw combining the basic functions of several machines has many advantages, especially for the small shop. The choice should be governed by the availability of portable power tools which can, in the beginning, serve as useful substitutes for some machinery.

The complete list of machines in order of priority are the table saw, jointer or surface planer, band saw, thicknesser, shaper or spindle molder, drill press, sander, mortiser and lathe.

Stock It's difficult to invest in wood before it's actually required for a job, but it's a good idea to have a few boards of different species on hand from the beginning. This allows enough time to dry it out thoroughly in the shop before it's required (page 268) and also enables you to show examples of the color, grain, finish etc., to prospective clients. There are many sources of wood besides lumber yards. The portable chain saw mill (page 259), for example, enables you to cut boards from tree butts which are often available free from local farmers and highways and parks departments. Secondhand furniture carefully taken apart, often yields good quality and well-dried hardwood. One particularly good source is old church pews from wrecking companies (try the telephone book). Church pews are usually oak or pitch pine and in long straight lengths, perfect for furniture.

Stocks of hardware such as screws, nails, bolts, knobs and catches are not required in quantity. Start out with nails, pins and a box each of the common screw sizes in steel and brass, and buy any special hardware as it is required. A catalog from a large supplier is handy to give customers an idea of the range available.

Heating and fire precautions When dealing with solid wood, particularly with expensive hardwoods, it is important to control the moisture content carefully so that the finished work will be relatively stable (page 264). Many small shops experience difficulties during the winter months when finished work is moved from an intermittently heated shop into a centrally heated house leading to problems with work splitting or warping. I think it makes more economic sense to spend money on keeping the shop well and consistently heated than to spend time repairing work which has reacted badly to being moved into a warm house.

Some form of central heating system in the shop is obviously the best, but also the most expensive solution. Many modern factories in England run their entire central heating from simple boilers which are fed on shavings and sawdust. A less elaborate model which burns shavings is perfect for the small workshop. It resembles an oil barrel and is set off the floor on a fireproof platform. It burns at a slow enough rate so that it can be regulated to burn throughout the night. The drum radiates heat in all directions and if the flue pipe is channeled indoors for a good distance, it will draw most the heat which would otherwise go up the chimney. Check with the building inspector and with your insurance company to be sure that this system complies with regulations before you install it.

Alternatively, a coal burning stove can be used in exactly the same way. Our workshop is heated very economically using coal burning stoves with exposed flue pipes which act as radiators.

Small localized electric fan heaters are fine but should be directed carefully so that the dry heat does not overdry the work in the shop.

Whatever form of heat is used, a woodworking shop constitutes a high fire risk and it is important that the building, stock, tools, machines and work in progress are insured against fire. The insurance company will advise about the regulations which vary from area to area but usually stipulate that the floors be swept at the end of each day and that smoking is not permitted in the workshop.

It is also important to safeguard against the possibility of explosions or flash fires which can occur when fine wood dust mixed with air is ignited. The use of spray equipment and storage of inflammable materials such as cellulose lacquers is also covered by local laws.

Fire extinguishers of the correct type should be hung on wall brackets within easy reach. The old-fashioned soda/acid type is fine for wood fires but is unsuitable for fires involving chemicals where a carbon dioxide or a foam type extinguisher is required. Fires involving electrical equipment require carbon dioxide or dry chemical type extinguishers. Specialist companies provide fire extinguishers and service them once a year for a small fee.

Miscellaneous items One quickly finds out when setting up even a small workshop that an endless list of small items is eventually required. Besides the tools, machines, benches and the obvious cabinets and drawers needed for storage, you will need different types of racks for storing wood and clamps (page 39), blankets for protecting work on bench tops and in transit, dust sheets or large plastic sheets, baskets or boxes for scrap, a pair of inexpensive scales for measuring resin glues, a broom and shovel and a brush for cleaning off bench and machine surfaces (or perhaps a blower as shown on page 34), and containers to put waste in. You will also need lots of clean rags and soft cloths, containers of all sizes including used cans and plastic tubs for mixing wax and glue, dust masks, ear plugs and a good first aid kit (page 31). Be sure to have the telephone number and address of the nearest emergency hospital near the phone.

The list of items needed in the shop goes on and on and these will accumulate gradually after two or three year's work as the need arises.

A typical small production shop.

Planning a workshop

SAFETY Remember that all machinery is dangerous and there are therefore several safety points to bear in mind when planning a workshop. Allow adequate space between the various machines and benches so that the operations don't interfere with one another even when large pieces such as plywood sheets are involved. Good lighting is very important for safety as well as for doing good quality work. Natural light is the best if it is not too bright, but a general level of fluorescent lighting augmented by drop lights is the best arrangement.

Distribute power points around the shop to use for hand power tools, heaters and so on. It is an excellent idea to install spring mounted drop cords over benches to keep the surface clear of wires. Fire regulations in most areas usually stipulate that all electrical wires be enclosed in metal conduits and that all appliances be grounded.

In planning the layout of the shop, make sure not to build in hazards like overhead storage near machines or benches located behind the table saw. Keep all storage areas well away from machine and assembly areas to avoid the danger of tripping over the inevitable scraps. A messy and cluttered shop is dangerous and usually the result of too little or badly organized space as much as bad habits. Include a rack or box for scraps in the original plan for the shop.

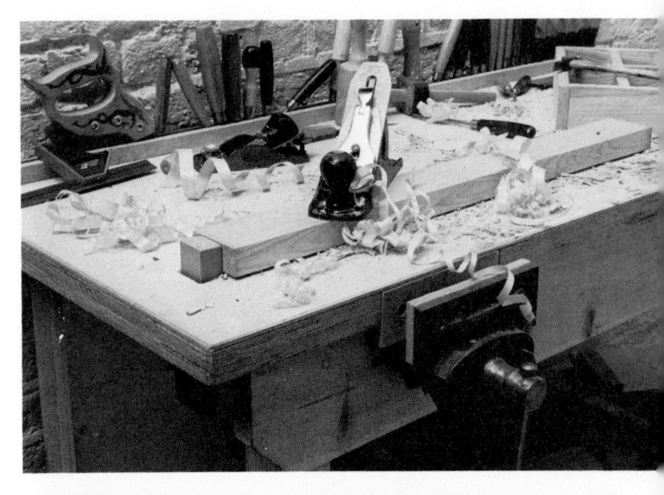

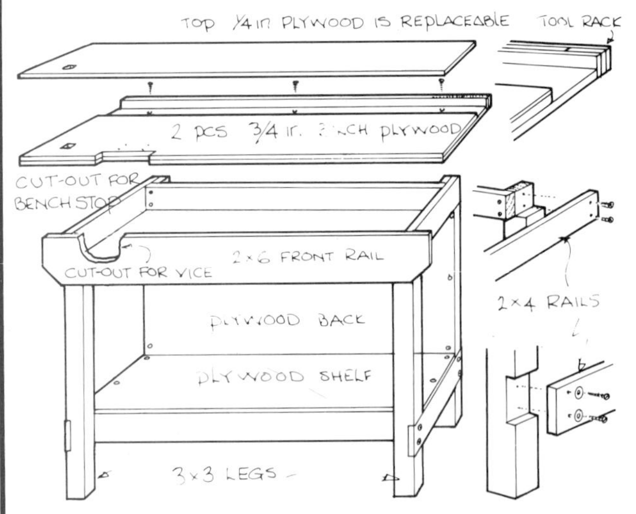

Details of a simple workbench made of softwood with a laminated plywood top.

Individual requirements vary widely, so it's of little use to show plans with specific shop arrangements, the following points give a few useful hints for laying out a shop.

1 Placing the saw diagonally gives much more room, especially when ripping long pieces which can extend through the door if necessary.

2 Make sure to leave a large area clear to be used for cutting up large sheets and also for assembling large pieces.

3 A simple, sturdy 2 × 5ft bench can be made quite easily out of plywood and softwood. Make the top out of double or triple layers of ¾in plywood, then add a ¼in plywood top which can be replaced when necessary. The shelf underneath and the tool holder at the back are both useful for storage. Attach a bench stop and a woodworker's vise to complete the bench.

4 Sawhorses are indispensable. Ideally you should make two pairs, one pair 20in high for general work and another pair with wider tops about 10in high for assembling carcass work.

5 Temporary benches like the familiar Workmate, are marketed to be used by beginners but are, in fact, very useful in a professional shop. They are very handy for holding large, heavy pieces and two together with a piece of plywood will form a temporary work surface.

6 Keep the main bench out of the line of the saw blade both for protection against thrown back wood and also to stay clear of sawdust.

7 Arrange storage flat against the wall away from the machines so that you leave as much width free as possible in the shop. Don't store boards high up on the walls where they are difficult to reach.

8 Keep as few scrap pieces as possible to avoid clutter. A cardboard box or a special rack (page 37) keeps the pieces organized.

9 Special tools not used every day are best kept in cabinets or drawers. Everyday tools should be kept within easy reach, hung on wall racks or placed in drawers.

10 One long fixed bench is useful to have if there is room. It can be used for small machines like the bench grinder, drill press and also homemade jigs such as the table for the router (page 74). Plan the space underneath for storing long pieces of wood if there is no room for a wall rack in the shop (see photograph below).

11 Clamps are expensive and should be kept organized on wall racks within easy reach.

12 A separate closet is the best place for storing paint and other finishing materials. Build a simple wall rack to take all the incidental items which would otherwise clutter up the shop and store them in the same place.

13 Place a temporary bench or a board mounted on a sawhorse next to the saw to support the ends of long boards or sheets when crosscutting.

14 Place a special roller support or a lift-up flap behind the saw bench to support the ends of boards when ripping on the saw bench.

15 A separate place for sharpening stones, oil, slipstones, etc., is very convenient. If possible provide a lift-off cover to keep stones dust free.

16 A first aid box should include at least:

good antiseptic	*scissors and tweezers*
sterile finger dressings	*sterile eye pads*
sterile absorbant cotton	*triangular bandages*
medium and large	*address of nearest hospital*
sterile dressings	*telephone number of*
safety pins	*ambulance*

17 Portable power tools are best stored away from dusty areas. Make a small cabinet with bins or compartments the right size for each tool.

Shop aids

BENCHES Good sturdy benches are essential; hardwood for handwork and softwood ones for general assembly work. Several manufacturers sell good quality hardwood benches, usually in beech, with front vise, and end vise with bench stops for holding boards for planing, etc. (see advertisements in specialist magazines). As a less expensive alternative it should be quite straightforward to make a hardwood bench, out of beech or a heavy maple butcherblock worktop augmented with vises, etc., bought from specialist suppliers. A generous size is 18 × 72in. Opinions vary about the usefulness of a trough or well at the back of the bench. Personally, I think they're a nuisance, inviting inevitable clutter. Working height is generally 35½in but most of us aren't built to general specification, so it's advisable to make the height to suit your own comfort.

Softwood benches for assembly work can be of softwood, say two 3 × 9in sections on either side of a plywood well on top of a sturdy softwood base.

All benches should be well-maintained and on no account should nails or screws be driven into them for they can put deep and difficult to remove scratches in finished work. Assembly work should preferably not be done on the hardwood cabinetmaker's bench. All glue drips which are very hard and sharp when dry should always be cleaned off the bench as soon as possible. An occasional going over with a cabinet scraper will restore the bench top to near new condition.

In addition to benches, sawhorses are absolutely essential. The most useful height is about 24in for all sorts of jobs including stock to be cut to length, plywood sheets to be cut with the portable saw, and a temporary work surface for oversized jobs.

Sawhorses are also essential for gluing up tops. I have found that cutting matching slots about ½in deep in the top of each horse to hold the bar clamps makes it much easier to hold the clamps in position. They otherwise tend to slip and fall over as you tighten up.

If possible, make another pair of sawhorses about 10 to 12in high for supporting tall cabinet carcasses for assembling and finishing.

The portable Workmate is also useful for professional woodworkers and not just the weekend do-it-yourselfer.

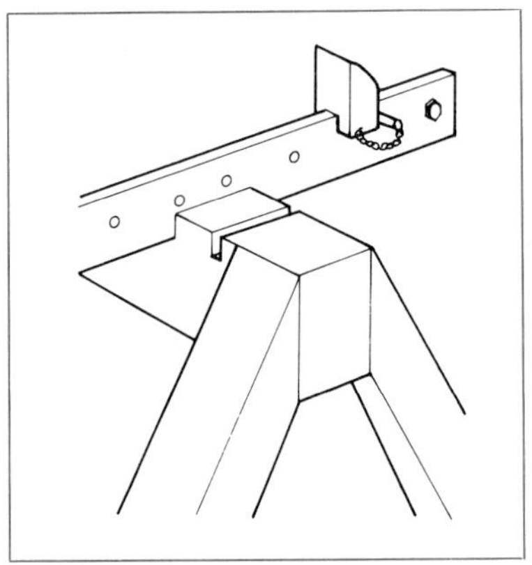

Using sawhorses to glue up tops, conveniently frees bench top space for other work. To keep the bar clamps from falling over, make matching saw cuts in the two sawhorses which will hold the clamps upright.

A good quality hardwood workbench is essential, especially for handwork. The massive vise on the right clamps long pieces between the bench dogs for planing. The woodfaced woodworker's vise is useful for holding pieces for drilling, marking, sawing, etc. Storage underneath is optional.

CLAMPS As mentioned earlier, you'll never have enough clamps, especially bar or sash clamps which are always required in large numbers when gluing up complicated cabinets. The basic kit should include several sizes of C-clamp and hand screws or fast action clamps. The latter are the most convenient especially for holding down work to the bench or holding jigs and patterns to the workpiece when using the portable electric router for example. Old-fashioned wooden hand screws or their modern equivalents with steel screws are useful for clamping stops and jigs to machines close to the saw blade where there is a danger of contact with cutting blades.

The kit should also include as many bar clamps as possible. The ordinary bar clamps are the best, but clamp heads and pipe clamps are also useful. Record clamp heads fit onto 1in wide boards, pipe clamps fit onto plumbing pipes and both make into extra long clamps as the need arises.

JIGS AND TEMPLATES Though not strictly tools, homemade jigs and templates are essential for accurate repetitive work, especially with the portable router. A template made for cutting say, a hinged recess with a router, should be marked and hung up for future use. With time, a wide range of reusable templates will build up.

Left: Two versions of roller supports for extra support behind the table saw, planer, etc. Below: Hand screw used on table saw.

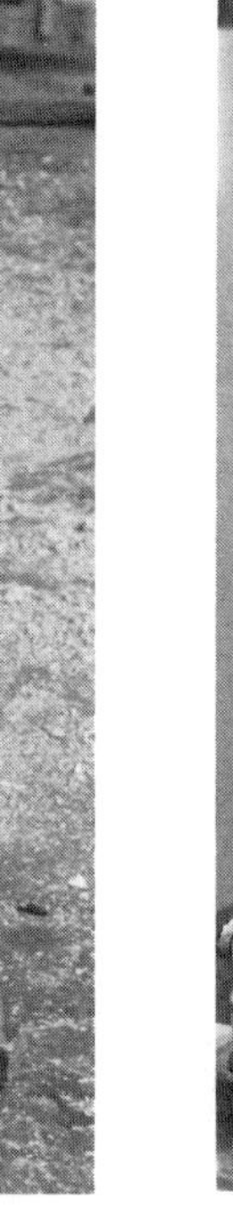

DUST EXTRACTION AND CLEANING Brooms and brushes are fine for cleaning off machines and benches, but a small portable electric blower is even better. It is used extensively in factories for cleaning out machines, and especially where moving parts are involved. It can be changed into a powerful vacuum cleaner if necessary.

Many of the so-called vacuum cleaners are quite ineffective, especially for picking up the large quantity of shavings generated by machines.

A dust extraction system is obviously the best but also the most expensive solution. Central extraction systems, used in factories, are very expensive, but smaller portable units suitable for fitting to one or two machines and for general vacuum cleaning are affordable. But make sure that the system is powerful enough to match the size of machines.

BLOWER/VACUUM CLEANER The Wolf blower which can be converted into a vacuum cleaner is a useful device for large workshops. It is used to clean out the sawdust from machines, a job otherwise done with great difficulty using a brush. Most production shops have a regular maintenance schedule which includes cleaning out, tightening nuts and applying the odd dab of oil or grease to keep everything running smoothly.

One of the most useful applications of the blower in our shop is to clean off the dust from a sanded surface prior to finishing. It does a far superior job to wiping with a cloth.

In the absence of a blower, use the outfeed of a vacuum cleaner which though not as powerful, will do an adequate job.

The blower can be fitted with various attachments for use as a vacuum cleaner, but it is more useful as a simple blower as shown.

Below: Dust extraction equipment varies from expensive permanent installations to portable units which can be moved from one machine to the next.

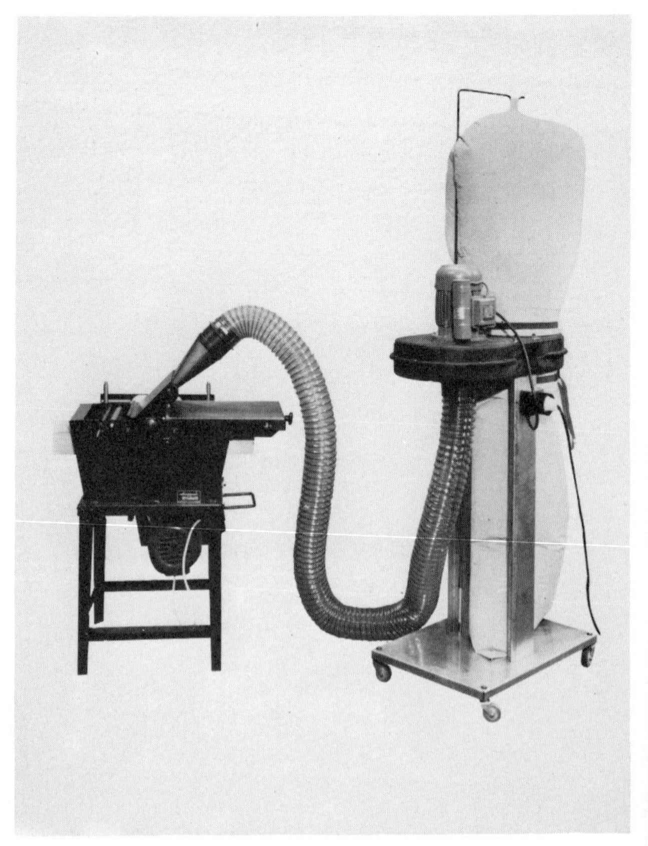

GLUING ACCESSORIES A glue spreader is very convenient where several edge-to-edge joints require a thin, even layer of glue. And instead of stirring resin glues by hand, make a mixer from a $\frac{1}{4}$in diameter rod and fit it in an electric drill to mix the glue thoroughly and quickly (see page 213).

A small press is convenient for gluing up small jobs and for delicate veneer work. Alternatively use heavy weights on top of wood blocks to provide clamping pressure for flat work.

Above: Small book binding press, handy for small gluing jobs. Left: Weights can be special purpose or made by pouring concrete into large cans. Below left: Glue spreaders are available in several sizes. Below right: Bent rod fitted to the electric drill used to mix resin glues. Right: Old-fashioned veneer press solidly made from structural steel sections to give even pressure for gluing veneered panels.

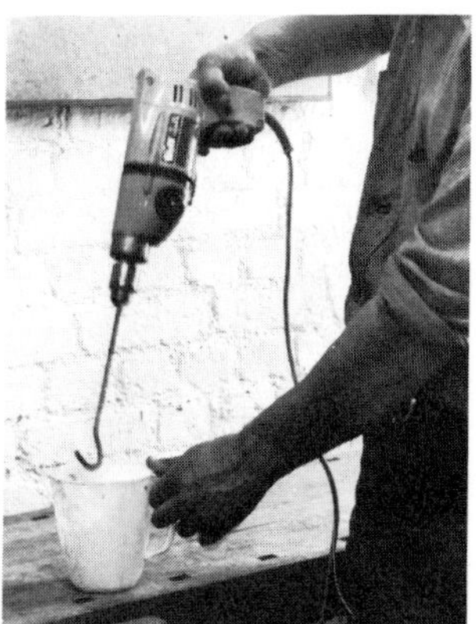

DRAWING EQUIPMENT A drawing board with T-square, triangles and instruments is essential not only for presentation work to clients but also for making accurate scale or full-size drawings of whole or parts of furniture, and of specific details and joints to enable you to study scale, proportion, etc. before setting out. I also find it useful to draw templates directly on pieces of $\frac{1}{4}$in plywood taped to the drawing board.

STRAIGHTEDGE A quality engineer's straightedge is an expensive luxury, but is well worth the money. It is more difficult than most people realize to judge whether a surface is flat or legs are in line or planer tables are parallel, and an engineer's straightedge is reliable because it will, unlike cheaper rulers, remain straight.

MOISTURE METER Anyone specializing in high quality solid wood furniture must control moisture content to match the eventual surroundings of each piece. The moisture meter, described on page 267, is the only simple way of determining the moisture content of wood.

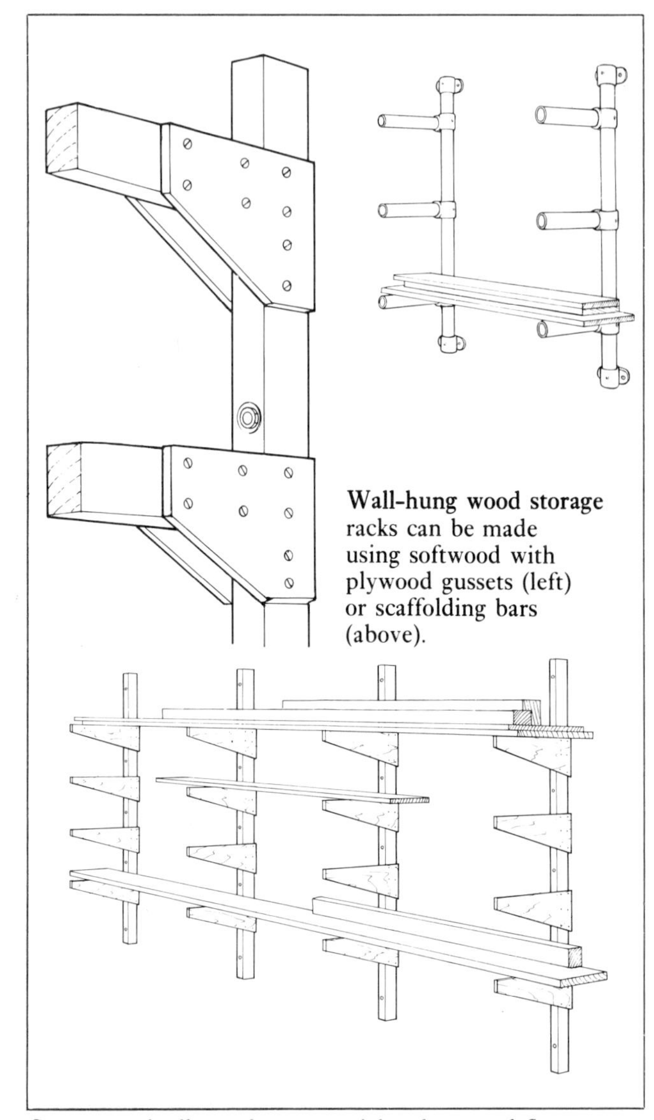

Wall-hung wood storage racks can be made using softwood with plywood gussets (left) or scaffolding bars (above).

Storage rack allows sheet material to be stored flat.

Storage ideas Woodworking, even on the most modest scale, involves a seemingly endless amount of equipment which must be organized effectively in order to maintain some level of efficiency in the workshop.

I have always intended to work out just what proportion of my working time is spent just putting tools away and searching for screws or going out to buy a bracket because someone forgot to order it in the chaos of the last job. The efficiency of a workshop is partly a matter of personality, some people seem to thrive in an environment of controlled disorder and know exactly under which tool to find another. Others work methodically replacing each tool in turn as work is completed.

Whatever the circumstances, efficiency and organization increase with the amount of storage space available, which makes it very important to plan very carefully when starting a workshop.

STORING WOOD Boards should be stored horizontally with adequate support to prevent sagging.

Wood being air dried is stacked with a cover for rain protection on a firm level foundation, with 1in square stickers of a neutral wood such as pine or Douglas fir, at about 18in centers placed directly above one another. Air dried boards with a moisture content of about 16 to 18% can be dried to usable levels of around 6 to

12% by stacking indoors in a moderately warm room or in the shop itself. They are again stickered and piled carefully to allow further evaporation. For further information on air drying see page 268.

General purpose wood racks are best placed against walls, away from machines, and can be made in several ways; by mortise and tenoning, by joining with plywood gussets, or by scaffolding. Supports should be placed at maximum 2ft centers and should support the ends. There is no need to sticker dry everyday wood. Many shops store wood overhead in the roof structure which is relatively safe, if inconvenient. But storage racks built overhead are ill-advised and dangerous since wood can fall onto the working area.

Sheet materials are best stored on edge against a wall. Blockboard should be stored flat or standing on the short edge to prevent buckling.

SCRAPS A rack which takes 4 × 8ft sheets as well as various lengths of plywood and wood scraps can easily be built out of chipboard or construction grade plywood nailed together and placed in an out of the way corner.

It's also a good idea to place two small boxes near the saw – one to take scraps to be thrown away, the other to take usable scraps.

HEATED STORAGE We have found it very useful, particularly in the cold winter months, to store valuable wood and work in progress, and any other stock to be sold, in a small insulated room gently heated (or better yet, de-humidified) to keep it dry. A small storeroom will do, or a part of the shop or storage area can be partitioned off either by making an ordinary stud wall or, better yet by using special panels made up in a hardboard/insulation board sandwich construction.

PAINTS AND CHEMICALS These should be stored in a separate closet. If highly flammable materials such as lacquers and thinners are involved, the insurance company may insist on a special lined closet separated from the shop.

VENEERS Veneers should be stored in a cool room so that they don't dry up and become brittle. Anyone specializing in veneer work needs to put aside quite a large area not only to store but also to allow selection of veneers for close matching. They should be kept out of sunlight to avoid discoloring, so a blanket or other covering cloth is essential.

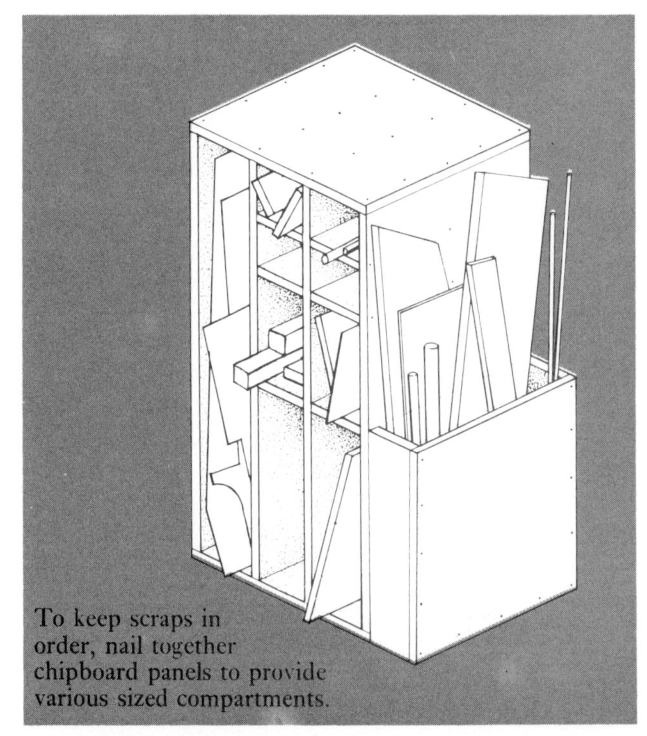

To keep scraps in order, nail together chipboard panels to provide various sized compartments.

It is far more convenient to store scraps end on to make each piece accessible.

HAND TOOLS Everyone has his or her particular preference for storing hand tools. Some prefer putting them away in drawers, others on racks or on walls within clear view and easy reach. Although open storage present a dust problem, particularly if machines are used in the vicinity, I think it's the easiest way to keep to track of tools. The racks can be the familiar pegboard, with special hangers or simply a plywood sheet with nails or dowels to take the tools. In either case drawing the outline of the tool on the wall behind makes it easier to find the right place. Planes are best stored vertically in special cabinets or on shelves with battens under to prevent damage to the blade. Many woodworkers prefer to make special cabinets for the more valued tools, keeping measuring tools in one, planes or chisels in another, etc.

SHARPENING TOOLS The oilstone, oil, slipstones, etc., for sharpening tools, should always be handy. If possible, provide a separate well-lit table or bench area out of the way near the grinder for sharpening and reconditioning tools. A hinged cover over the stones will keep them dust free.

Several tool racks. Left: Chisels and gouges stored conveniently inside the bench cabinet door. Right: A special box for valuable measuring tools with plywood cut-outs for each. Below: Open storage with tools and clamps within easy reach.

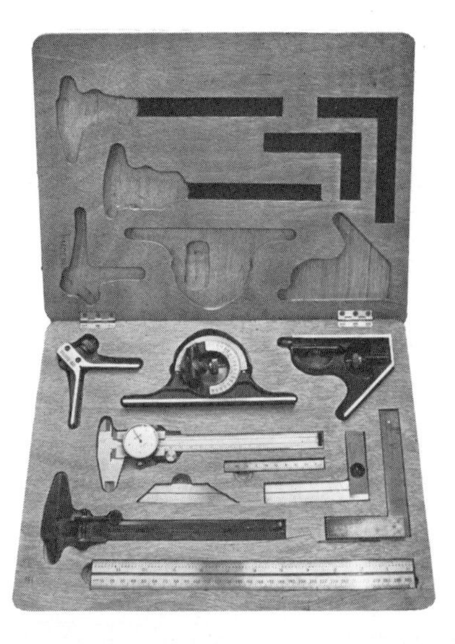

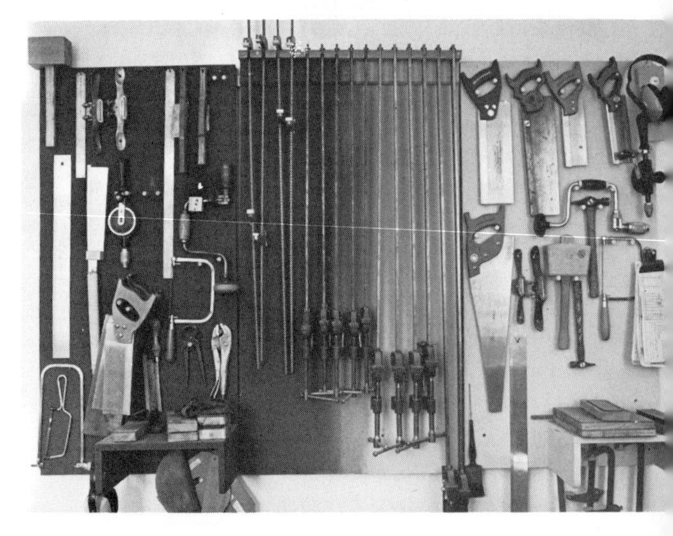

CLAMPS All clamps must be kept in good condition. Dried glue should be scraped off and a light coat of oil applied as part of a regular maintenance schedule. There are few things more annoying than having a bar clamp jam just when you are urgently trying to position it on a cabinet as the glue is setting. It's best to make a special rack by cutting slots in a well supported shelf to take the bar but if necessary, bar clamps can be hung simply on a 2 × 2in batten screwed firmly to the wall. C-clamps can be hung on to dowels extending from a similar but vertical batten. Or one or more free standing units can be made to take several types of clamps and placed at convenient locations.

We found our horizontal burglar bars perfect for hanging C-clamps, fast action clamps, screw clamps, etc. A similar arrangement could easily be made using a few heavy dowels.

HARDWARE Screws, nails and other hardware have to be well organized; stored in drawers or cabinets well labeled and away from the dust. Keeping boxes and bags in a disorganized pile on a shelf leads to a tremendous amount of waste of material and waste of time in searching. There are several plastic or metal cabinets with suitable sized drawers on the market. Since these are quite expensive we found it convenient to make our own out of $\frac{1}{4}$in plywood simply glued and nailed together.

I designed the boxes to be modular and kept a cutting list beside the saw so that any plywood scraps could be conveniently trimmed to size for assembly into a box in a spare moment or two.

Another idea, standard in home workshops, is to screw jar lids to the underside of shelves and store screws, nails, etc., in the jars.

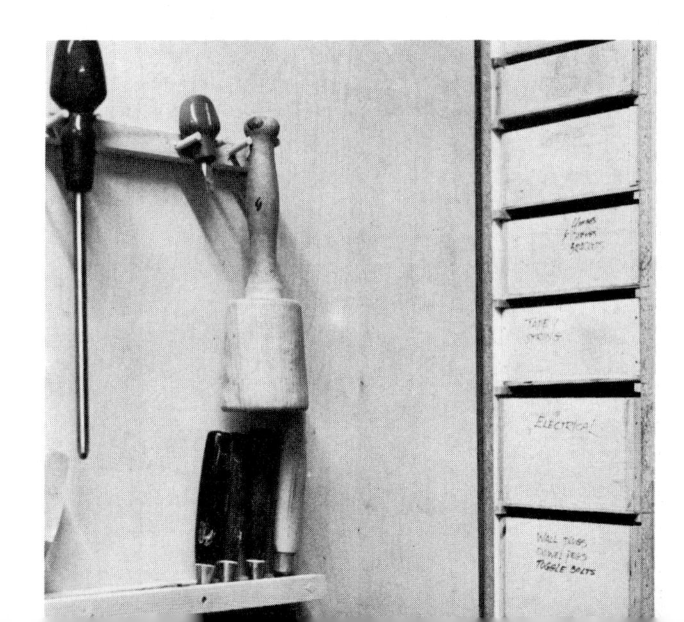

SPECIAL PURPOSE CABINETS The accessories for machines and other tools are best stored in small cupboards or wall-hung cabinets. Drill bits for example can have their own small cabinet, simply made from plywood, and hung near the drill press. Auger bits for the brace can be stored separately from the flat and machine bits for the drill press.

Similarly router cutters should be carefully stored to prevent damage and to keep them organized. A wood block with appropriate holes drilled to hold the shafts is the easiest way to store a few cutters, but for an overhead router, make a cabinet as shown.

It is important to avoid storing machine accessories on shelves above the machine to avoid the danger of having anything fall onto an operating machine.

French head cutters (page 124), like jigs and templates, are made for a specific use, but if stored properly for quick reference, they can be re-used again as the need arises.

Below left: Drawers for organizing French head cutters. Left: Cabinet for holding router heads heads and cutters. Below: Small wood block for router bits. Bottom: Wrenches for spindle molder.

Bottom: Cabinet for storing files and burnishers for making French head cutters.

Buying secondhand machines Old machines are usually much heavier and hence sturdier than new machines. And they are also much simpler which makes fixing them up easier. There is no better way to learn about machines than to have to strip them down and put them together again, rebuilding parts which can't be bought from dealers.

SOURCES The best bargains are from businesses going out of business or moving. This includes builders and shop fitters, as well as woodworking shops and factories. Usually the auctioneers or creditors get there first, but even then prices are usually quite good at auctions. You have to know what you are looking for and also inspect the machines (see below) because there are no guarantees or returns after the sale. Contact auction houses and ask them to keep you informed. Also check newspapers for ads and trade journals for firms going out of business.

Dealers are the main competition at auctions. They buy, fix up and sell the machines. There aren't many dealers in one area, but the good ones soon become known through the trade and may even offer some form of guarantee.

Dealers sell anything from near antiques (which are often the best buys) to nearly new machines sold for just under retail prices.

You must be patient and persistent to find the right machine and you must know what you are doing. If you are not certain, be sure to buy from a reliable dealer who will give you a guarantee on the machine.

WHAT TO LOOK FOR When buying secondhand machines it is a big advantage if you have metalworking skills or the help of someone who does, as you will undoubtedly have to repair or replace some parts that are no longer made.

The first thing to check is the age and overall condition of the machine. But it is like buying a car. It is quite easy for the seller to dust it off and paint it up to make it look good.

It's easy to tell if a machine has been cared for. Bolt heads are not worn round, sliding parts are tight and adjustments are still fairly precise. Check all the moving parts and notice whether there is any play in them or any rust. Turn the machine on and listen for noise and vibration which can be a sign of bad bearings. On a saw, check for wobble of the blade and wear of the grooves for the fences. Check for hairline cracks or warping of cast iron parts, particularly table tops. If not serious, these can be "stopped" by drilling a hole at the end. Try out each machine and find out whether it has been in recent use. Since it is difficult to get spare parts for most old machines, make a thorough check of missing parts such as knobs, nuts and so on. You can probably live with some of these faults, but they are good bargaining points when settling on the price of the machine. Keep in mind that used machines normally have a lot of accessories with them. It's a good idea to look around for boxes of blades, cutters, etc., and to make sure these are included in the price.

Check the cutters. A good sawyer will not use

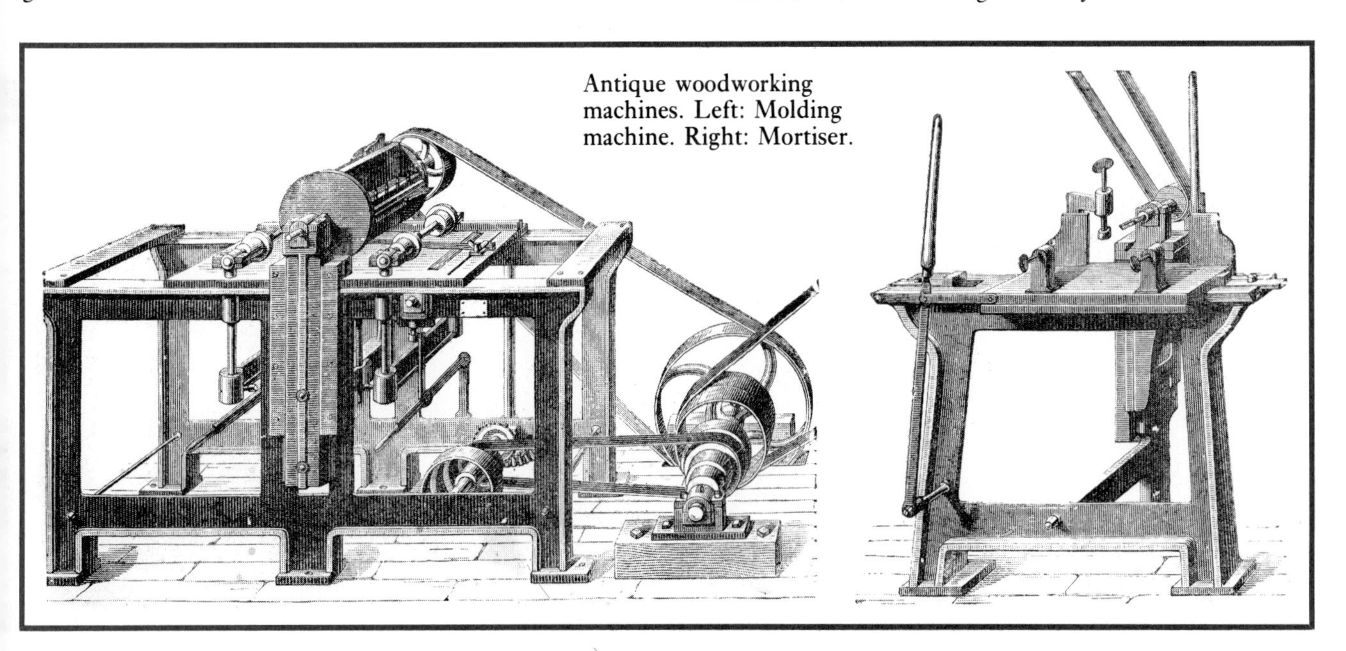
Antique woodworking machines. Left: Molding machine. Right: Mortiser.

a blade which is very dull or burned or has teeth missing, and similarly planer blades with nicks or cracks in them indicate abuse.

Old leather belts can still function perfectly well or be replaced with new endless belts. A small device for joining leather belts is usually available with the machine, and if not, ask for it.

Phosphorous bronze bearings, in use before ball bearings, last a long time, but they should be re-greased and replaced in exactly the same location. If necessary they can be replaced with ball bearings of suitable configuration.

On planers, check that the tables are in good condition and the fence is stable and adjustable. A common fault with old planers is that the tables won't stay parallel as they are raised and lowered.

The bed on old thicknessers can be out of adjustment or difficult to raise. This is due either to worn threads on the main screws or worn or loose gears.

Old band saws are usually out of adjustment and have worn out guides and rubber tires, but as long as the motor and bearings are all right the rest is usually fixable. Wooden guides, often used on old machines, are still preferred by many machinists.

The biggest problem with old machines is usually that they no longer comply with new regulations. They are often driven by belts powered from freestanding motors and these either have to be converted to an integral motor or must have guards built around them.

The guards for machines too have become more elaborate and much safer. Even if no regulations apply, it is advisable to adapt an old machine to modern safety standards. On saws for example, it is quite easy to install a modern riving knife and a two-sided guard, by adapting the fittings. Very old band saws were not guarded and should have metal lined plywood enclosures built on all sides.

Keep in mind that you must add moving costs, which can be considerable, and fixing-up costs, including an allowance for spare parts and for your time.

Below left: Old 30in band saw with metal-lined plywood guard. Notice belts which should be covered. Right: Oil bath switch for thicknesser. Below: Belts for same thicknesser with 10hp motor. Notice joint in flat belt which drives feed rollers. Bottom: Kit for joining flat belts. Each belt end is placed in the device (right) which is closed to drive in the metal prongs. The plastic strip holds the ends together.

Workshop economics The most important consideration in setting up and running a shop is an economic one; where to find the capital to buy tools, machines and other equipment and how to sell your goods to keep the venture profitable. In schools and colleges too little attention is paid to the realism of workshop economics, as if making a profit out of what you enjoy somehow cheapens the experience.

It is vital from the beginning to have a good business attitude, the tougher the better in my view, for I think every craftsman will agree that we give an excellent individual service and do quality and unique work and never charge enough for it.

Most people grow into a business, finding their market gradually, first through friends and relatives. And it becomes easy to plan what further tools will be needed and where more work can be found. But for anyone setting up from scratch it is important to have a fairly clear idea of the scope of the shop and what type of work will be done, whether it will be handwork on individual commissions or small runs of one design sold through stores or perhaps through the workshop.

These considerations govern the size of the workspace and the amount of tools and machinery required. They enable you to plan, within rough limits, setting up and running costs. These give a guide to the turnover or income required to make it pay for itself. We normally plan out each year's projected turnover in January as a rough guide to the amount of advertising, stock, labor, etc. that will be required to reach that figure. It requires considerable restraint, based on last year's figures. Otherwise it's easy to get carried away with hopeful projections. When just starting out, its important to work on minimum figures and revise as you go along.

There are many ways of getting free publicity, ways of advertising and of attracting business. And there are many sources of income which can supplement woodworking, particularly in the early days. The following sections give details on these and other aspects of workshop economics.

SETTING-UP The sources of financing are usually obvious but the costs are not. The capital required for setting up can come from personal savings, from loans from family or friends or from banks.

Another source in Britain is grants from crafts organizations but these are very limited and usually only available after you are already set up, which seems slightly contradictory.

To get a loan from a bank, it's important to make a carefully thought out presentation detailing expected costs, and a realistic estimate of projected income, giving a good reason why they should lend you money. Even then, some form of collateral is usually required. Banks have never been known to be particularly benevolent to unproven craftsmen, but there are exceptions and the climate is changing.

I set up my shop from the proceeds of a book on woodworking and there is undoubtedly a need for good books on the subject, so it may be worth a try if you enjoy writing.

Setting up costs should be realistic and should include not only the cost of the tools, machinery and equipment but also a generous allowance for renovation, decoration and generally outfitting premises with electricity, lighting, extra floor support for machinery, security bars and locks, etc. This is usually more expensive than you think it will be, and it may be worth asking a builder or another woodworker for realistic advice. In addition, an allowance for a certain amount of stock and hardware should be included as should material required to make racks or benches and storage cabinets. It's also a good idea to include an amount to be set aside to cover the first few months' basic wages and costs, as insurance in case it takes a bit longer to get off the ground than planned. It's useful to check with an accountant from the start on what can be deducted against income tax and what can't. If a lot of renovation is required you should negotiate a reduction in rent to allow for it.

Of course the more people who share in the cost, the smaller the financial burden. Co-operatives and partnerships are a good way to run workshops as long as there is a clear and perhaps legal understanding about the proportions contributed and the general direction of the workshop.

PUBLICITY Word of mouth is the normal mode of advertising for many craftsmen, but it takes a while to take effect and there are other more effective ways of letting the world know about your work. Even if you have enough commissions, it gives a nice feeling of confidence when more people are interested and this confidence often allows you to raise prices to realistic levels.

The most effective way to get publicity is through newspapers and magazines who are quite eager to write up a craftsman and print pictures of his work. This channel is not to be minimized unless you don't want publicity. Press like this is much more valuable than advertisements.

Having ascertained who writes the column, phone the newspaper and briefly interest him (more often a her) in yourself and your work. It's a welcome change for them to hear from someone who is not a public relations company. Invite them to visit the workshop and follow up by sending a letter and black and white photos of your work. Good photographs are important, so get a professional photographer to take them unless you can do it yourself.

Then be prepared for all the mail by having something to send out. Photographs are too expensive, so it's better to prepare a sheet giving details of the type of work you do, shows where it has appeared, perhaps a drawing or two, and if necessary a map giving the location of the shop. New copying machines do a decent job of copying photographs so include one or two clear ones and have enough copies run off as they are required.

Many woodworkers and small firms go a step further and prepare a small brochure, which can be a simple folded sheet, a four page version or a more elaborate (and expensive) brochure with individual photographs for each design and a separate price list and information on caring for the furniture. The new quick copy businesses will do all but the highest quality printing for brochures, mailers and headed paper.

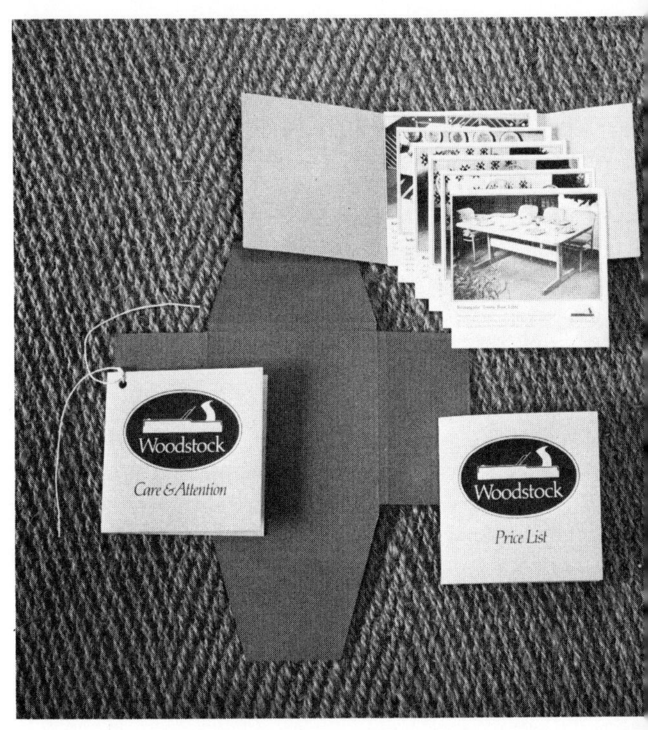

Our brochure with a separate card for each design.

BREAD AND BUTTER WORK It is a big help if you have at least one design (of which you don't mind making more) which you can advertise and sell in moderate numbers. This means you are selling a product rather than a service and customers want something specific to buy – even if they end up buying other designs as well. And making more of one piece enables you to do small runs, saving in time and material.

This doesn't appeal to some who want to make only individual designs, but I think it is the perfect way to run a shop. Allow one or two designs to be the bread and butter work to pay for other less productive work such as prototypes for new work. I find the days set aside for these limited runs are very enjoyable in themselves.

ADVERTISING Selling directly from the workshop (see below) may require some form of advertising other than word of mouth and free publicity. Small ads in national papers are quite expensive; local newspapers are a better place to start. Generally photographs don't reproduce well on newsprint. It's better to prepare a simple line drawing showing one good design, then briefly describe the type of work done. This is perhaps done by implication and the tone of the ad as much as by the words. Putting in a price is optional but may select out the customers, eliminating time wasters. Name and address should of course be included together with instructions to send for brochure or to visit workshop, etc. Newspapers require an accurate layout with all the components laid down or drawn in, and it's important to check the column size with the advertising department before starting. They are usually more than happy to help out someone who is just getting started.

WHERE AND HOW TO SELL Woodworkers normally have a fairly clear idea of whether they want to do individual designs, limited runs, large runs or any combination. I've always liked the idea of "limited editions" making up, say twenty, of one design, and advertising it as such, like a silk screen run of an artist. The output will largely determine the market place. The main outlets for high quality woodwork is **1** direct sales from workshop/showroom, **2** craft shops, **3** quality furniture stores.

SELLING FROM WORKSHOP Depending on how commercially minded you are (or have to be) direct selling is the best method because it eliminates the high markup of furniture stores, enabling you to charge more than you would get wholesale but less than the store price with its 80 to 125% markup. Having customers visit the workshop is a mixed blessing. Most love the atmosphere and seeing how furniture is made, but you can waste half your day just talking to customers. If possible, set aside an area, away from dust and noise, where the work can be displayed. It's useful to have photographs and drawings to present as well as samples of wood, which can be displayed in say 2ft lengths, half finished, half unfinished, on the wall.

I think it's important to find out how serious people are by coming to the point early, giving a very rough price guide for the type of work suggested. Every one of us has spent much too much time chatting about design and sketching ideas without getting an order.

Selling a ready-made design is easy, customers see exactly what they will get (with perhaps variations in wood species) eliminating the need for sketches and drawing, which are usually required for presentation. Even if a job doesn't work out, save the sketches in a book to show other clients.

There are various ways of presenting ideas. Sketches and detailed drawings are usually the best, including sizes and materials, finishes, and incorporating any specific requests from the customer. Some woodworkers actually make up samples of interesting details which can't be well presented in drawing form. But it's best to do a preliminary sketch with a rough estimate to get broad agreement before going into further work. Final sketches should be accompanied by a price, an agreed delivery date and a request for one third to half the total amount as a deposit. The price is usually a fixed one, but occassionally some form of flexibility must be built in allowing you to charge extra if the work takes much longer than estimated. There are always legal ramifications as to who owns the design, what happens if either party cancels or the customer doesn't pay, and so on, but rather than get in legal documents, it's much better to cautiously trust people, requiring a deposit and then final payment before shipment. In my experience almost every customer dealing with a craftsman will be straightforward and agreeable.

I think worrying about rights and payments often invites trouble. But it's important to be firm. If a client wants to change his mind about a detail halfway through, he should either be talked out of it or be made to pay extra for the additional work.

CRAFT SHOPS AND GALLERIES Craft shops will either buy pieces wholesale or will take goods on a sale or return basis. An outright sale is almost always preferable unless you want to be able to recall the work for an exhibition. Sale or return means that the shop pays for it if and when they sell it, which is really quite unfair to the craftsman who has paid for material and put in the time which in effect finances the craft shop. But if it isn't possible sell a piece outright it is, at least, a way for an untried craftsman to show his or her work. The mechanics are simple; phone or write to the various shops, who may be listed with crafts organizations, or advertise in crafts magazines. Include a leaflet or brochure and good photos with a letter, giving name, address and brief details of your aims, design ideas and experience. Also ask for an appointment to show actual samples and if you haven't heard back in two or three weeks, follow up with another call. If possible get them to come to the workshop which allows you to present more work and show wood samples. Prices should be based on accurate estimates (see page 46) but someone running a craft shop will usually have a good idea of what the sales price should be, so the final price may be arrived at by give and take, as long as you can still make enough profit.

FURNITURE STORES Furniture stores or department stores specializing in high quality goods are often interested in selling craftwork. The approach should be the same as for craft shops with an initial telephone call to find out which buyer to talk to and whether they are interested. Stores usually buy outright at an agreed wholesale price. The usual problem is quantity. If they're eager to sell one at a time and mark the prices accordingly it's easy to deal with. If, on the other hand, they want ten or more at a time, you face the choice between stepping up production or refusing the work. Depending on your outlook you can hire people to help you, or perhaps farm some of the work out, having factories make components while you do the assembly and finishing, enabling you to keep control over the quality.

A sale should be followed up with a letter (headed paper) stating prices, delivery date and any other details such as wood species, finish etc. The delivery should be accompanied by a delivery note which should be signed by the store, stating that the goods were received in good condition and followed immediately by an invoice including sales tax and the terms for payment. Normal payment period is net 30 days, which means that having laid out the money for materials and labor, you then have to wait one month for payment.

OTHER OUTLETS There are other ways of selling work such as at crafts or country fairs and through co-operative craft stores and other local outlets. The best way to find out about them is to ask other woodworkers or crafts organizations for advice and to check crafts magazines for notices of crafts work wanted.

Architects and interior designers often require special designs which may range from a one-off desk to outfitting a whole restaurant or lobby. Architects will usually contact you, especially when there has been publicity about work or work on show. Failing that, it is quite difficult to contact them. It is, of course, possible to arrange for appointments to show your work, but this is slightly awkward since it is difficult to find the right person to speak to and they will be unlikely to need anything at that moment. In short, you are unlikely to get much business from architects and designers in the beginning, but with time and word of mouth, they can become a good source of commissions.

OTHER SOURCES OF INCOME It may be useful, particularly early on, to supplement profit from sale of goods with other sources of income, and there are quite a few that spring to mind. Part-time teaching is a well known source, but too often it takes over in importance leaving little energy left for the designing and making your own work. Writing books or magazine articles on design or on woodworking can be profitable and enjoyable. But it can also be very badly paid. Certain publishing houses pay a very small advance on royalties which may take years to amount to decent pay for the time and effort put in. The same is true of specialists' woodworking journals whose pay per thousand words is usually low. Writing is only for those who enjoy it, to others writing deadlines are a terrible burden.

A more enjoyable source of income is from designing pieces of furniture for magazines, such as do-it-yourself, women's and other specialist magazines. The designs are a challenge because they usually have to be simple enough for less experienced people to make, yet interesting and attractive at the same time. This is the type of quick design work most craftsmen don't get a chance to do, so it may be particularly challenging and enjoyable.

The materials for the designs should be easily available, such as plywood or pine and easy to join, avoiding complicated joints such as dovetails or tenons. The best way to get work is to draw or make up a few designs, perhaps for your own or a friend's house, then phone up the appropriate editor and suggest an appointment to show your ideas. Fees are usually standard but often negotiable, and should depend on what is required. This may be just sketches with dimensions and details for them to make the furniture, or it may include making up the piece or even providing photographs of the finished piece. In this case, a separate amount should be allowed for each. Make sure to try to negotiate separately for the design itself though, more often than not they will pay for only the finished piece with the design cost built in. You are not usually required to do any writing, but only to supply sketches and cuttings lists.

It is also important to retain the copyright to the design which allows you to collect a number of these into a proposal for a book. And having sold the rights in one country you can then sell the design to magazines in other countries. Agents for foreign magazines may contact you after first publication anyway if the design is good.

Other possibilities include doing designs and making up furniture for exhibitions in conjunction with magazines and/or manufacturers. And the same manufacturers may hire you to design pieces using their product, such as plywood, veneered panels, paint or shelving systems.

These sources take a while to develop but there is nothing to lose by phoning up and presenting ideas.

Of course a successful design can also be sold directly to a furniture manufacturer, but this can be tricky and it's wise to get the advice of a lawyer so that you can protect your copyright. I know of several designers who have received good royalties for years from a successful design which they couldn't have mass produced themselves.

ESTIMATING Accurate estimating is difficult even for experienced managers, but the general tendency, particularly in the small workshop, is to under-estimate or even ignore the time it takes to design and set out work, to select wood, to do delicate detail work, to glue up and to do finishing work. I am still amazed at my own inability to assess the time a job takes, often under-estimating by as much as 50%.

Everyone develops his own methods for arriving at a price. This may differ from established methods used in production shops and may involve many intuitive fudge factors such as multiplying the material cost by a fixed number.

At first, most craftsmen suffer from undervaluing their time in eagerness to get work, but with experience it should be possible to arrive at consistent prices based on rational methods. The price should include allowances for labor, materials and overheads, plus a sub-contractor's costs and delivery charges, site visits and any other special costs incurred.

MATERIAL COSTS The cost of wood, plywood, hardware, and so on, is either the actual cost or the replacement cost, whichever is higher. A handling cost of say, 15% is usually added to this which does not include delivery, damage or waste factors. Delivery is a direct cost from the supplier, damage and weathering costs are estimated from the losses incurred and usually added as a percentage addition for the particular item. Wastage factors for wood vary greatly from one species to another and those for plywood and other sheet materials from one type of project to the next. You must take account of the use the materials; table tops and the like require more careful selection and therefore more waste, whereas understructures and insides are less critical. Some species like oak, have notoriously high wastage factors, over 100%, when splits, shakes and waney edges are taken into account, but only hard experience will give the right answers. Machining waste in sawing, planing and off-cutting, are normally not less than 50% for fine work in hardwood, but for softwoods which come already planed, it will be less. It's advisable to keep up to date catalogs and price lists of wood, plywood and hardware with your own invoices and records of wastage factors, for various materials so you will be able to estimate more accurately in future.

Miscellaneous materials such as glue, nails, screws, sandpaper, oil and wax should always be taken into account, for their costs add up over a year's work. They can either be added to general overheads as a percentage addition or itemized separately for each job which is probably a better solution.

The final materials estimate is taken from the working drawings which should include precise cutting lists for each component, plus a list of any special hardware required such as hinges, stops or catches.

LABOR COSTS The sum of the labor costs for the job is worked out by breaking down the job into sections and estimating the time for each. Hourly rates are arrived at by adding to the basic wage the yearly costs of overtime pay, health insurance and pension contributions, vacation pay and sickness pay, all divided by the hours worked per year. For the individual self-employed craftsman, it is crucial to allow an adequate basic wage plus these allowances. Too many woodworkers end up working for starvation wages as a result of under-estimating their own time. For individual jobs, the extra time spent selecting, laying out and re-doing work should be included. A typical estimate of time which should be multiplied by the appropriate hourly wage may be as follows:

Designing, consultation, preparation of drawing	5½	hrs
Selecting and crosscutting	3	"
Machining, surfacing, ripping and thicknessing	5½	"
Matching and selecting	2	"
Making and cutting joints	14	"
Gluing up components	5	"
Cleaning off, sanding, scraping	4	"
Assembly/clean up	6	"
Door/drawer making	6	"
Door/drawer fitting	5	"
Final sanding	3	"
Finishing	3	"
Polishing	2	"
Fitting hardware	2	"

A few hours should be added to allow for moving the pieces around, getting help in carrying and holding from other workers. This is a hypothetical case only to demonstrate the principles involved. Each job will vary, depending on the proportion of handwork, the delicacy involved and the general standard of work and finish.

OVERHEADS Overheads are general costs which are divided into two main parts and included in the estimate as a percentage addition to the total cost. The first part includes overheads such as cleaning up time, electricity, heating, and machine running costs, plus the cost of miscellaneous materials.

The second part is general business costs such as rent, taxes, insurance, interest on capital, office expenses, transportation costs and professional fees.

At the outset, most of these are low or non-existent but should be analyzed nonetheless. One year's careful record keeping will give a good projection of the overheads of the following year. These costs are added to the total estimated labor and material costs for the year to arrive at a figure for the total turnover. A small percentage for profit must be added with which to partly finance new machinery and stock.

The overhead, expressed as a percentage of total turnover, can then be simply added to the cost of each job. The percentage may be anywhere from 15 to 40%, depending on the nature and size of the business.

MISCELLANEOUS COSTS An estimate should also include any fixed sums from sub-contractors with a 5 to 10% addition for handling and consultation, plus any direct delivery costs, unless already included.

Portable power tools

PORTABLE CIRCULAR SAW

The portable circular saw is considered by most people as a site tool, to be used on construction sites for rough construction work. But I have seen many clever adaptations of the portable saw used in the workshop. It can, for example, be turned into a table saw just as the portable router can be turned into a spindle molder – by attaching it upside down to a table. This is ideal not only for those who don't have access to a table saw, but also for craftsmen working on site installing, say kitchen cabinets. It will do quite accurate work which can easily be cleaned up with a few strokes of a sharp hand plane.

The portable circular saw gets constant use even in a well-equipped shop. We use one to rough cut boards outside in the yard, prior to milling – the same job done with a radial crosscut saw in factories. And it is indispensable in cutting up large sheets of plywood and chipboard which are difficult to maneuver on the table saw. As with most cumbersome workpieces, it is better to bring the tool to the work than the other way around.

The essential features of a good, versatile saw are a powerful motor, large diameter blade, rise and fall and also tilt mechanism and generally robust construction. Blade diameters vary from about 6 to 10in. The 7¼in model is quite popular and will cut slightly over 2in thick wood and thick plywood quite easily. For heavy use, the 9 to 10in models are more applicable, particularly for rough site work, though these machines are quite heavy and not recommended for continuous use. The smaller models and drill attachments are not suitable for general woodwork, even for the occasional do-it-yourself job. It is far better to invest in a good saw and have it pay for itself in accurate work. Most saws come with a detachable rip fence, which should be reinforced with a wood strip for stability.

One well-designed German saw is part of a system of interchangeable components allowing it to be used as a table saw, miter saw or snip-off saw as well as a freehand saw. The saw can also be fitted with a dust extractor which may be useful if you are working in a spare room and want to keep the work area dust free.

Guide to good practice

SAFETY
The portable saw is very safe to use provided a few basic safety points are observed:

- Disconnect the power before changing blade.
- Make sure switch is turned off before plugging in to power supply.
- Concentrate: keep fingers away from blade and don't look up if disturbed until saw is switched off and blade has stopped.
- Never wear loose clothing.
- Never use saw if bladeguard is not working properly. Have it fixed immediately.
- Use the saw with cautious confidence and keep a firm grip with both hands, if possible.
- Finish cuts in smooth motion, simultaneously shutting off switch. Make sure guard drops in place before putting saw on bench.
- Don't overload motor which can cause permanent damage. If blade speed is drastically reduced and motor gets hot, remove saw and let it run for two or three minutes without load to let it cool off.

Basic operations: freehand cutting In most construction work and rough cutting, the portable saw is used freehand, guided by eye along marked lines to make rough cuts. With practice, fairly accurate work is possible, though it is better to use a straightedge or fence as shown on page 50 to cut really straight lines. Most models include a small notch at the front of the base plate which is used to guide the saw by eye along the marked line. With experience, it is quite easy to cut consistently along the waste side to leave enough waste for cleaning up with a plane.

It is important to support the work well. When ripping, clamp down the workpiece if possible to leave both hands free for the saw. Remember to

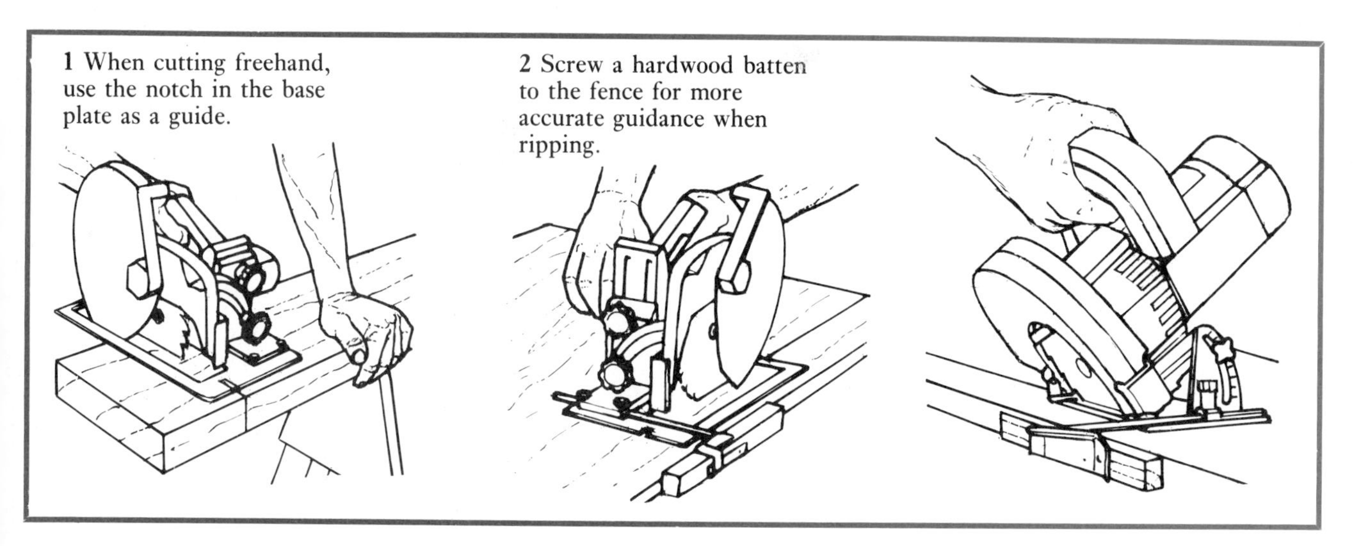

1 When cutting freehand, use the notch in the base plate as a guide.

2 Screw a hardwood batten to the fence for more accurate guidance when ripping.

check underneath that the saw blade will not cut into the workbench, but sometimes it may be convenient to allow the saw to just cut into the sawhorses underneath. The blade should be pre-set before cutting to extend about $\frac{1}{4}$in below the workpiece.

Rough crosscuts are usually made with one hand holding down the board. For continuous work, it's a good idea to nail a batten to the bench or sawhorses as a stop to help support the boards.

For more accurate ripping, use the rip guide, particularly for boards too narrow to be ripped with a straightedge guide. Notice the hardwood strip fixed to the rip guide to give better stability.

Grooving and rabbeting are standard operations with the rip guide. Set the blade depth and the fence distance, then make a test cut before going ahead. Move the fence over to make successive cuts until the groove or rabbet is cut.

To make angled cross cuts, rip cuts or compound angle cuts, simply set the blade angle and make the cut, either freehand or with a guide.

Using straightedge guides Using guides to keep the portable saw in line produces much better results than freehand sawing. Straightedges are particularly convenient when working with large sheets which are sometimes impossible to cut on the table saw. This is particularly relevant today when so much work is done using sheet materials as in making kitchen cabinets out of melamine chipboard, for example.

It must be remembered that the saw blade cuts in an upward motion leaving a rough, splintered finish on the top surface which should be avoided on sheet materials. To prevent this, first of all use a sharp, tungsten tipped or a crosscut blade if possible and always cut with the good face down. Where both faces will show on veneered boards, it will be necessary either to score along the side of the cut or alternatively to carefully place masking tape along the line of cut.

The best material for a rip guide is $\frac{3}{4}$in plywood for it will stay stable. I always prefer to use the manufacturer's edge and it is worthwhile sorting out and finding a sheet with a good clean edge before buying it. Cut off a strip about 6 to 8in wide, 8ft long and to make it even more stable you can cover it top and bottom with Formica neatly trimmed and sanded off. A hole drilled at either end makes it easy to hang on the wall.

The 8ft length is a bit unwieldy, particularly for cutting across the width of a sheet so it is worthwhile to make a 4ft straightedge as well.

To use the straightedge, first determine the offset distance required from the straightedge, that is the left edge of the saw plate, to the inside (left) of the

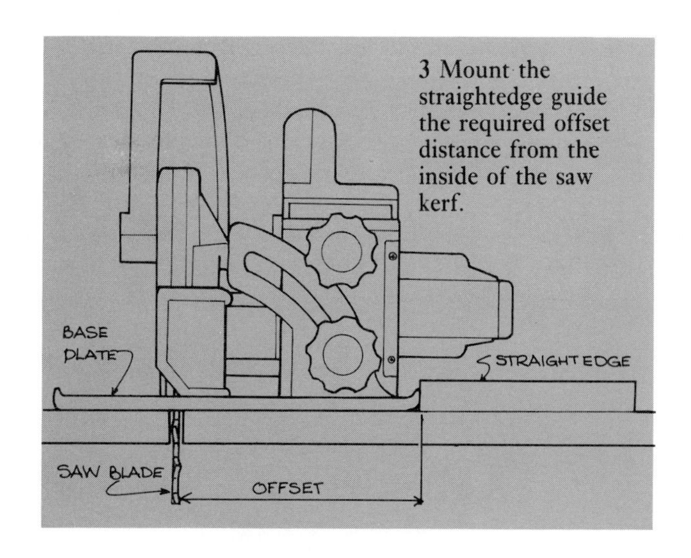

3 Mount the straightedge guide the required offset distance from the inside of the saw kerf.

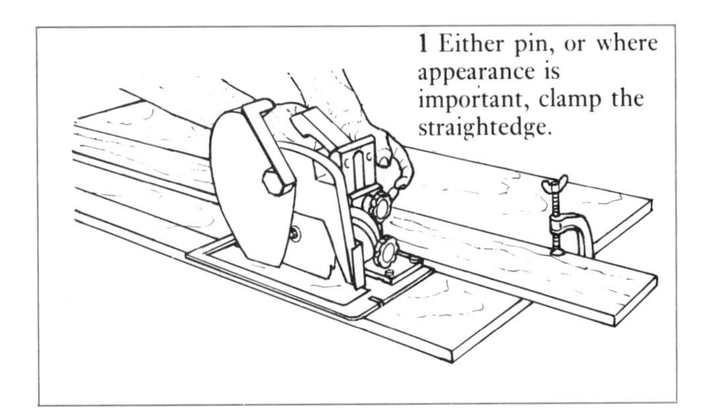
1 Either pin, or where appearance is important, clamp the straightedge.

saw blade. Rather than measuring on the saw, it is easier to clamp the straightedge to a piece of scrap and make a short cut and measure the exact distance to the inside of the kerf. I find it useful to write this dimension on the straightedge for quick reference.

Before attaching the straightedge, mark the exact location of the cut on the workpiece, keeping in mind which side of the line the cut will be. Then mark off the distance to the straightedge and clamp it along those marks. (1) Make sure to keep the base against the straightedge, particularly near the end of the cut where the back end always tends to shift away. Lengthwise rips near the center of 4 x 8ft sheets are awkward because it is difficult to reach. It is best to place two boards underneath on either side of the cut which will allow you to climb up on the board. The two halves will also conveniently tilt away from the cut at the finish preventing any tendency of the saw to jam.

To avoid the need to measure the offset distance each time, screw a thin $\frac{1}{8}$ or $\frac{1}{4}$in board underneath the straightedge which should be $\frac{1}{2}$in thick. (2)

Allow the side of the thin bottom to extend about 10in sideways. Then running the saw against the $\frac{1}{2}$in edge, cut off this excess leaving just the right amount extending to allow immediate location against any cut marks measured out on a workpiece.

To go one step better, use $\frac{1}{8}$in thick plywood covered with Formica which reduces friction against the saw plate. You can also wax the bottom with candle or carnauba wax to give less resistance.

Another idea which keeps the saw in an absolutely straight line involves bolting a metal strip to the bottom of the saw plate as shown. (3)

A $\frac{1}{4}$ x 1in piece of mild steel about 12in long is about right. Drill countersunk holes in it and attach it very carefully with rubber-based adhesive or by clamping to the bottom of the saw plate, making sure it is exactly parallel with the blade by using a block of wood as a spacer. Then drill through the holes to make matching holes in the saw plate. Bolt the strip to the base with countersunk machine screws fixed with wing nuts.

The steel strip rides in a groove in a straightedge made to match. Either rout a 1in wide groove in a length of $\frac{1}{2}$in plywood or create a groove by gluing and screwing two $\frac{1}{4}$in widths on top of a $\frac{1}{4}$in plywood board.

Make the total width about 12in with the groove just left of center, then run the saw along the board, strip in groove, to cut off the right side so that it can be used to locate the cut against marks on the workpiece. Again glue down a piece of Formica to the top and in the bottom of the groove if possible to reduce friction. The maximum depth of cut is reduced by about $\frac{1}{2}$in with this jig, but for a $7\frac{1}{4}$in saw with a $2\frac{1}{2}$in capacity, this still leaves 2in which is more than adequate, particularly for sheet materials such as $\frac{3}{4}$in plywood.

2 The $\frac{1}{8}$in board screwed to the underside of the straightedge lines up the cut quickly.

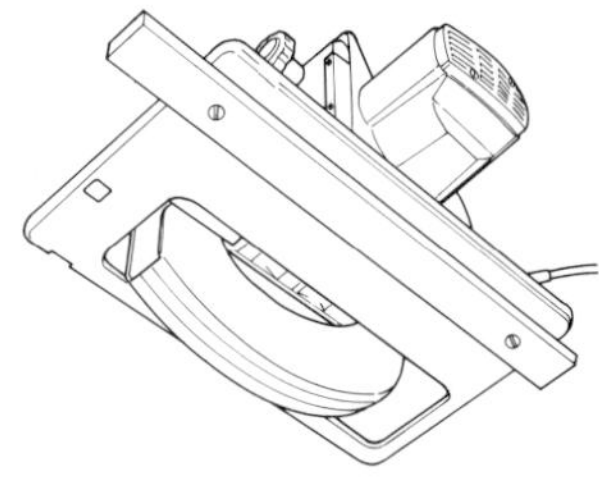

3 This is by far the best system. Clamp a straight board against the saw blade to line up the steel strip so that it is perfectly parallel to the blade.

Making a T-Square

To make accurate and square crosscuts, it is better to use a special T-square guide than to rely on freehand sawing.

Attach a piece of $\frac{3}{4}$in ply 8 x 14in, A to a piece of $\frac{1}{8}$in ply 14 x 14in, B. Then screw and glue a straight piece of 1 x 3in underneath. The 1 x 3in must be absolutely square to the good edge of the plywood. I find it best to glue it down first, perhaps adding one pin for reinforcement and then allow the glue to set after I have checked for square.

After adding three or four screws, clamp the T-square to the edge of the bench and run the saw against the plywood edge cutting off the protruding 1 x 3in and the plywood.

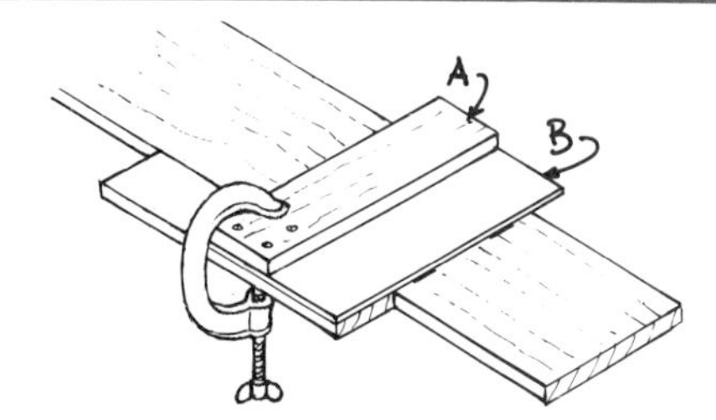

To make more accurate crosscuts, run the saw against a T-square. Notice that the end of the plywood lines up the cut.

The edge of the plywood can then be lined up against marks on the workpiece to make the T-square very quick and easy to use.

It is better to clamp the T-square to the workpiece when sawing, but as a quick alternative, nail two $\frac{3}{8}$ or $\frac{3}{4}$in pins through the plywood and allow the tips to protrude. The tips grab the wood, and when held down with the left hand, the T-square is quite firmly located.

Crosscut jig Cutting many pieces to length is time-consuming with the portable saw since it involves measuring and setting up the T-square repeatedly. To make it easier, make up a simple jig out of $\frac{3}{4}$in plywood and a few scrap pieces. Screw and glue two 2 x 2in battens to a base of $\frac{3}{4}$in plywood, as shown. Make the distance between the battens a touch more than the widest board you are likely to cut. Then make the top part as shown so that it will fit tight over the bottom part as a guide for the saw to make square cuts. Drill the holes for the four locating pegs with the top part clamped in place.

To use the jig, insert each piece to be cut into the jig and run the saw over it along the guides, while you hold the piece firmly against the fence. Simply line up the mark against the saw cut as a quick guide. For repetitive cuts, use a stop block to make all the pieces identical. Remember to set the depth of cut so that it will extend into the plywood base by about $\frac{1}{16}$in.

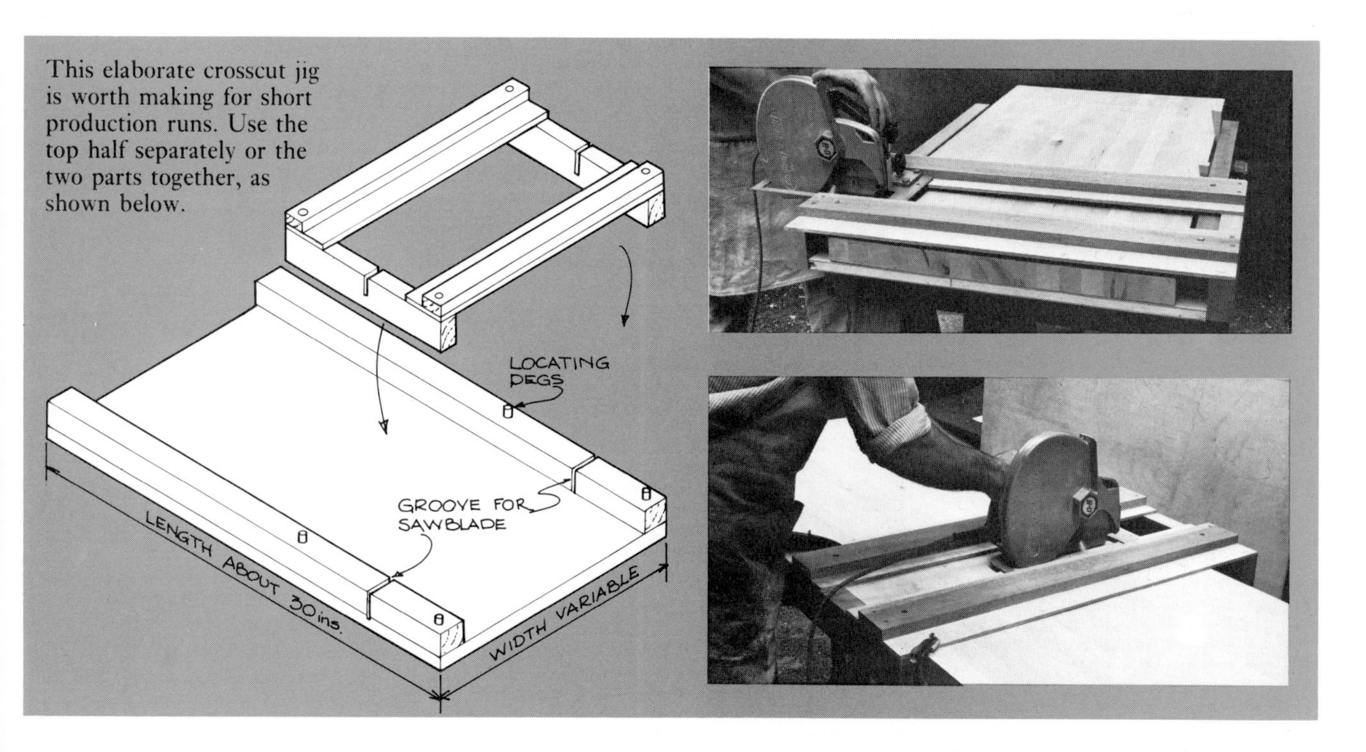

This elaborate crosscut jig is worth making for short production runs. Use the top half separately or the two parts together, as shown below.

The portable saw as a table saw

For the craftsman just starting out and unable to afford a table saw, converting the portable circular saw into one may be a good alternative.

Some manufacturers sell metal saw tables, but most are too small and light to be suitable for all but the lightest work. In my view, the larger, more robust and more powerful the saw, the safer it is, so make sure that the table you buy is sturdy. Generally the tables sold to be used with large 9 to 10in portable saws are intended for construction site use and therefore made with heavier gauge steel.

The model shown, made for a $7\frac{1}{4}$in diameter saw, can be mounted in two ways, either as a snip-off (crosscut) saw or as a straight table saw. It includes several safety features such as a riving knife, a guard (not shown) and an on-off switch separate from that on the saw.

A homemade table, though it doesn't have some of these features, is nonetheless more convenient in some ways and certainly much cheaper. One of the main advantages is that it can easily be made large enough to take sheets of plywood or small enough to be carried to and erected on site. The exact details and sizes will obviously be varied to suit individual requirements and as always the only limit in design and use is the ingenuity of the craftsman.

As a basic model, I suggest making a simple top out of $\frac{3}{4}$in plywood reinforced underneath with 2 x 2in or 2 x 4in depending on the size. This table top can easily be carried (cut out a handle?) to site and nailed, screwed or clamped down to a pair of trestles. With constant use it may be a good idea to cut slots in the trestle tops to take the battens for a quick locating fixing or as shown here, screw battens around the perimeter to fit around the trestles.

Cut an opening near the center of the top, large

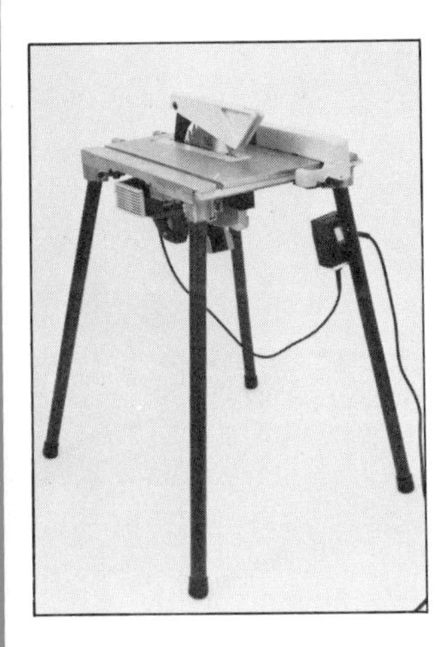

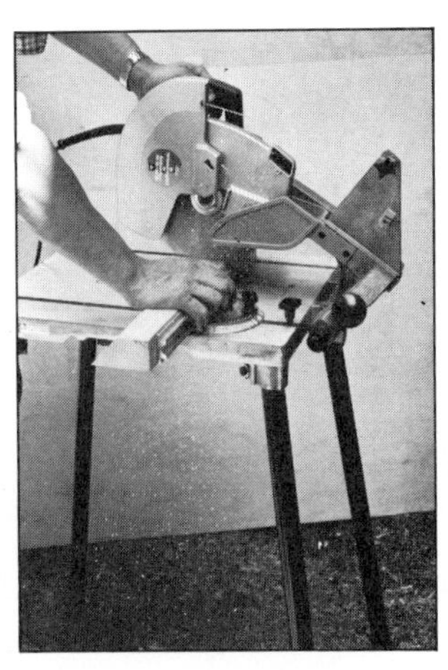

Make sure the table stand accessory is sturdy. This one from Elu also converts to a snip-off saw (right).

1

3

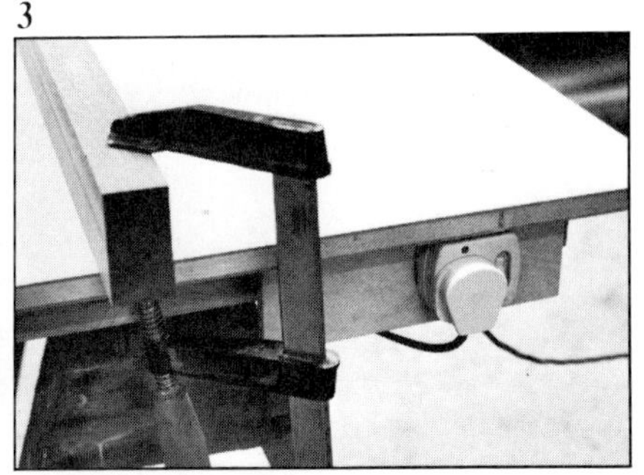

4

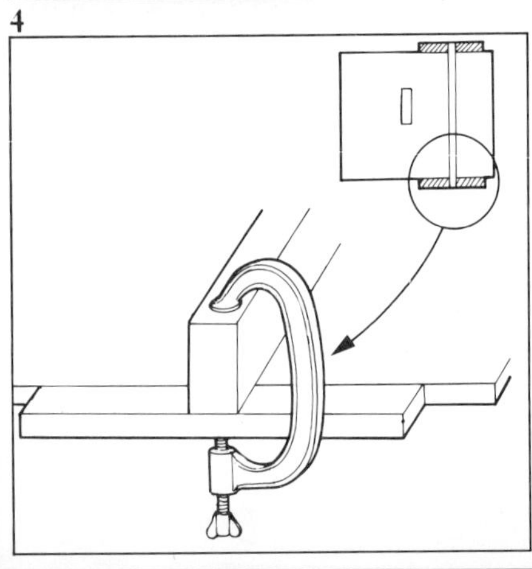

5

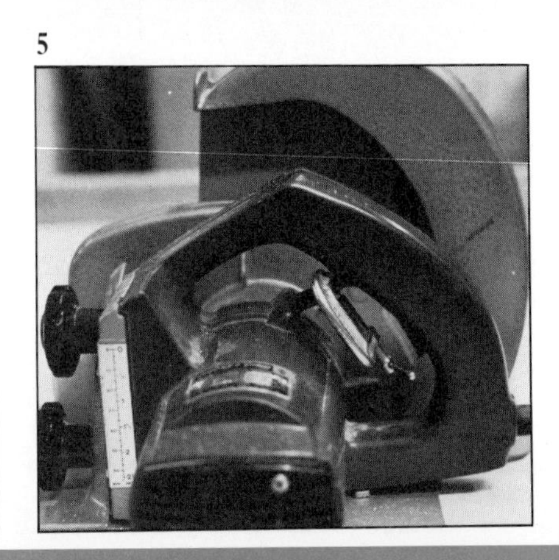

1 The table shown upside down. Notice the batten surround made to fit around the trestles.
2 Clamp a board to the table as a rip fence. Run a T-square against the left edge as a crosscut fence.
3 Run the cord to a switched outlet attached within easy reach.
4 An alternative rip fence with parallel guides top and bottom.
5 Notice the small clamp which holds the switch on so the other switch can be used.

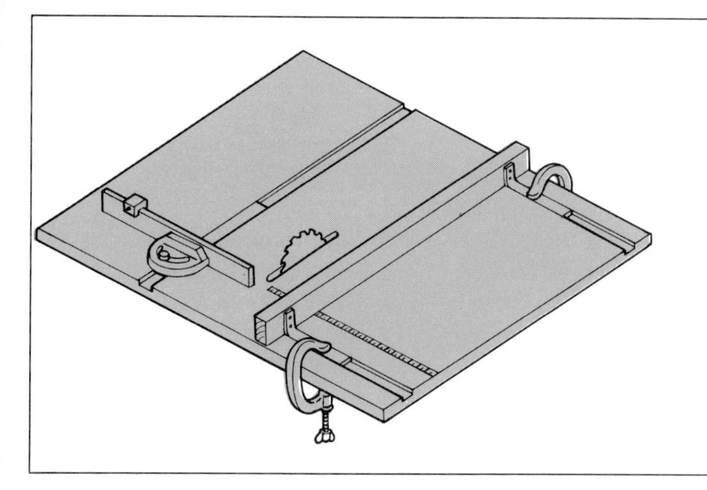

As an alternative arrangement, make shallow grooves in the table as shown, for holding the miter fence and the rip fence. Notice the packing behind the bent steel strip which is used to adjust the fence (right).

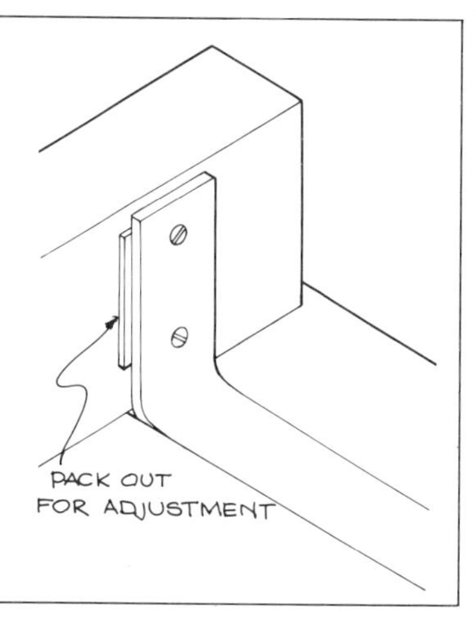

enough for the blade to pass through either vertically or tilted at 45°.

Turn the table upside down and screw the saw to the underside through four holes made in the corners of the base plate. Make sure to locate the saw so that it is parallel to the left plywood edge.

Run the electric cord safely out of the way underneath, tacking it down with cable clips and plug it into a switched plug conveniently located within easy reach. This can be part of a removeable extension cord when used on site.

As a crosscut fence, the simplest solution is to use a T-square guide run against the left edge of the table. Make it robust enough so that it doesn't bend when sawing. Two pieces of $\frac{3}{4}$in plywood about 3 to 4in wide are about right. Reinforce the cross piece with a renewable cover piece nailed or screwed to the plywood edge. This can be made as long as is necessary. It should be installed overlength and cut off by the saw blade afterwards to serve as a backup to the saw cut and as a guide to the location of the saw cut.

In crosscutting, hold the workpiece against the fence with both hands, keeping the T-square firmly against the table edge with the left hand when pushing it through the saw.

The rip fence can be equally simple, consisting of a sturdy and stable board such as a 2 x 4in of mahogany which is clamped to the bench at each end and measured for parallel for each new setting. To make it quicker to locate the rip fence, it may be a good idea to make it into a double T-square with added cross pieces riding against the top and bottom edges of the table. For this to work, the table will have to be square but the fence will still need checking and clamping each time. Remember when setting a fence that the rule of thumb is to set the top end (i.e. rear of the saw) to the left so that it is out of parallel by $\frac{1}{32}$in in 5ft of length.

To construct a deluxe model, mount the saw in the same way but improve the fence operations by routing $\frac{1}{4}$in deep grooves in the table top to take the sliding fences, similar to standard table saws (page 76).

Either make a miter fence out of a strip of $\frac{1}{4}$ x 1in mild steel screwed to a cross piece of hardwood or buy a standard miter fence sold as an accessory to the table saw. Make the groove to match and sand the bottom and sides well to allow smooth sliding. Alternatively, cut the groove slightly deeper and cover the bottom with Formica.

Make the rip fence the same way, making sure the two grooves are perpendicular to the saw blades. Notice that the steel strips are bent up behind the rip fence. This is partly to reinforce it but also to allow adjustment both in parallel to the blade and in square to the table, by packing out between the metal and wood as required. As a further convenience, locate a good quality wood ruler in a groove perpendicular to the blade as a quick setting guide for the rip fence.

Many improvements and safety devices are possible with either model. The top can be covered in Formica to make it harder and smoother. A riving knife and separate guard can be fitted (make the guard out of $\frac{1}{4}$in clear plastic), adapted perhaps from parts of an old saw. And as a more permanent fixture, make a sturdy base and fix the whole table to the floor for added rigidity.

As a final reminder, keep in mind that machines, particularly table saws, are dangerous. Make sure to make it sturdy and safe and if in doubt about any detail, check with someone who has experience with woodworking machines.

PORTABLE DRILL

Portable electric drills have many uses in the workshop, from drilling dowel holes in components to drilling screw clearance holes in cabinets. As with all other portable tools, it is often much easier to bring the tool to the work, particularly when the work is too large to fit into the machine as is often the case with the drill press.

But it isn't possible to drill as accurately with a portable drill as with the drill press. It is nearly impossible to keep a portable drill exactly square to get a truly vertical hole, and steady enough to get a clean finish. A drill stand, which is available as an accessory to most models, makes this possible. In the absence of a drill press, the drill stand is indispensable.

Portable electric drills vary in size and detail from model to model. Most have a chuck capacity from $\frac{1}{4}$ to $\frac{1}{2}$in, to accept a wide range of bits from the ordinary twist bits to dowel bits and flat speed bits. A chuck key to fasten and loosen the drill bit to the chuck comes with the drill and should always be replaced with one of the same size. It's a good idea to buy a small rubber tie to fasten the key to the electric cord within easy reach.

Power ratings vary from about $\frac{1}{4}$hp for small drills to over 1hp for heavy-duty industrial drills. Most models have one, two or four speeds but two are enough; fast for wood and slow for metal and masonry. Hammer action to be switched on for drilling into masonry is convenient if you plan to do general building work as well as woodworking.

Drill bits used with the portable drill

Basically the portable drill accommodates the same drill bits as the drill press with one or two exceptions such as the larger sizes of speed bits and hole saws which are difficult to control with the drill held freehand.

Twist drills – technically intended for metals, these are widely used in woodworking for general purpose work such as drilling screw holes. Twist bits are difficult to center accurately and without spurs do not leave a crisp finish. Available in diameters $\frac{1}{32}$in to $\frac{1}{2}$in in $\frac{1}{64}$in increments.

Dowel bits or spur machine bits – used for dowel holes and elsewhere when a clean, professional finish is required. Sizes vary but generally available in $\frac{1}{8}$in increments from $\frac{1}{4}$ to $\frac{1}{2}$in diameters.

Speed bits – these flat bits, available in diameters from $\frac{1}{4}$in to about $1\frac{1}{2}$in, cut fairly clean holes fast and efficiently. The large sizes have a tendency to "grab" and should be used carefully when drilling freehand.

Masonry bit – carbide tipped to go through hard materials. Sizes vary to match screws, plugs and bolts but go up to over 1in diameters for bolt fixings.

Countersink bit – in several sizes for countersinking screw holes.

Combination bits – for screws, drill clearance and pilot holes in one operation.

Note: For further information about drill bits, see page 134.

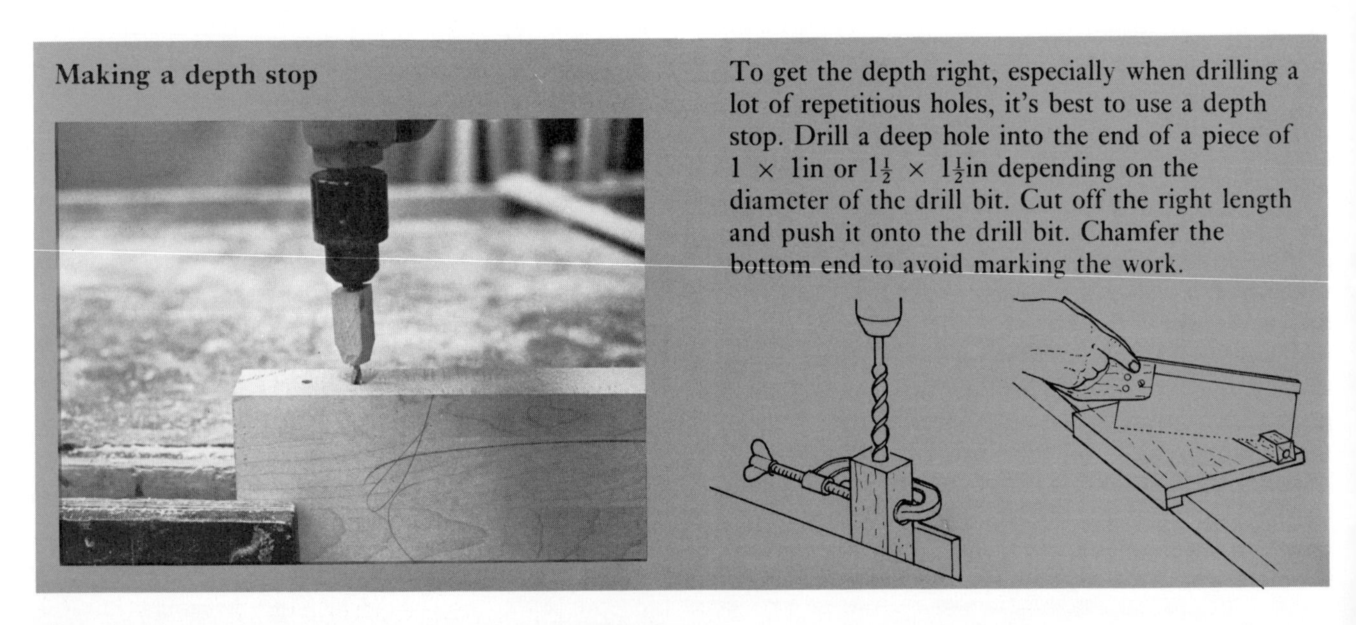

Making a depth stop

To get the depth right, especially when drilling a lot of repetitious holes, it's best to use a depth stop. Drill a deep hole into the end of a piece of 1 × 1in or $1\frac{1}{2} \times 1\frac{1}{2}$in depending on the diameter of the drill bit. Cut off the right length and push it onto the drill bit. Chamfer the bottom end to avoid marking the work.

Using the drill The portable drill is so familiar to almost all woodworkers that it needs no introduction. Drilling screw clearance holes, which seems to be its main use, is straightforward but it requires some care to keep the depth of hole correct. Refer to the table on page 57 for the correct drill sizes for the range of screws in both hardwoods and softwoods.

Keeping the drill perpendicular to the workpiece is very important in many applications such as drilling holes for dowel jointing. It's always best to make a drilling jig to guide the drill bit, particularly for repetitive jobs. See examples below.

But where it isn't practical, the best method is to have someone else sight in one direction while you check alignment in the other direction. Many books advise standing a try square on the bench beside the drill, but I've found this method impossible to use. Remember in dowel jointing edge to edge joints, to make boards into a table top, for example, that it is critical that the holes, and hence the dowels be parallel to the board faces. Therefore stand at the end so you can see along the edge for alignment.

Drilling jigs In industry, drilling is done on a variety of sophisticated machines, such as the multiple horizontal boring machines, where many holes are drilled simultaneously in precise arrangements. For the small shop, a drill press is essential for accurate work, but without one it is best to resort to devising jigs and fixtures for the portable drill to make it as versatile and accurate as possible.

A drilling jig is used for both location and guiding. It should be made so that it clamps easily onto the work. This is best done by screwing on battens which locate positively against the edges of the work. At least two battens are usually required to locate the jig in either direction, but in some cases more will be necessary. The more work done to make the jig accurate and easy to locate, the less work has to be done aligning it on each individual piece. Hardwoods such as maple or beech are best for the jig but one or two layers of plywood makes a good substitute. These are hard enough to take repeated drillings without wearing the hole out of true.

As with all jigs, it's important to take care to make it accurate. Drill the hole with the drill held in a portable drill stand, if possible, to get it nice and square.

Exact details will vary from jig to jig but to make it quick and positive to locate, it's a good idea to attach battens on the back which will locate against the edges or ends of the workpiece.

EXAMPLE 1 We use this drilling jig to drill two holes in the trestle ends of a table base to take the fixing for the cross brace. The holes must be accurate and straight since there is very little tolerance in the bolt and cross dowel fitting. Ideally the job should be done on a drill press but this has proved impractical. The jig, made from 2in thick rock maple, slips over the upright, with plywood sides to hold it in place. Also notice that the jig locates positively in the up and down direction by slotting over the top. This way there is no need for measuring. It simply clamps on for quick and accurate drilling. The difficulty with this and all jigs like it, is that it requires accurate machining to make the workpieces the same size. If there is play in locating the jig, some accuracy is lost.

The matching holes in the ends of the cross brace are drilled with a similar jig which slots over the end of the board. The holes in this jig were actually drilled from the first jig to make them exactly the same spacing.

Right: This trestle end is too awkward to hold in the drill press. Instead it is held in a vise and the jig which locates positively over the top and sides, is clamped to it to enable the pairs of holes to be drilled accurately.

Below: Another jig made from the first to make sure the hole spacing is identical, slips over the rail ends to drill the matching holes in the rails. Yet a third jig, not shown, locates the holes for the cross dowels. See page 200 for more information on this joint.

EXAMPLE 2 This shelving system relies on accurately located holes for its stability and alignment. The holes in the shelf ends are drilled with a portable electric drill fitted with a $\frac{1}{4}$in dowel bit using a small jig which fits over the end. This hole has to be accurate so that the shelf locates firmly against the upright to give the unit stability against rocking.

The holes in the uprights, too, require accuracy. The spacing has to be consistent so that the shelves are horizontal. The drilling jig made from a piece of $\frac{3}{4}$in birch plywood is carefully prepared, the measurements double checked and the hole centers carefully center punched before it is drilled in a drill stand.

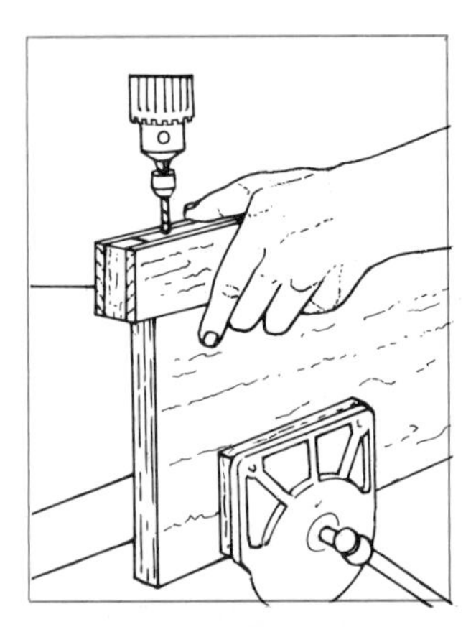

In a production shop, these holes would be drilled on automatic machinery. In a small shop, the best solution is to make accurate jigs which allow one or a limited number to be made. The shelf is held at four corners by a $\frac{1}{4}$in dowel peg inserted through the upright. Accurate alignment is essential, hence the two jigs to line up the holes and guide the drill bit.

EXAMPLE 3 Doweling requires accurate drilling which is best done using a special doweling jig, available with varying details from several manufacturers. It is an excellent and relatively inexpensive jig which makes doweling accurate. It is particularly useful for the handyman and for the small workshop where frameworks are more easily jointed by doweling rather than mortise and tenoning.

There are various doweling jigs for making dowel joints (page 184). The Record Ridgway jig above is useful for solid wood sections, but the model below is better for working on long edges such as plywood panels.

Drill stand The drill stand, which is sold as an accesory for the portable drill, is a useful substitute for the drill press. It is, of course, not as sturdy or as versatile as the stationary machine, but provided it is well-made, it will do the basic job of drilling straight, accurate holes perfectly well. It is ideal for the craftsman with a limited budget.

If possible, buy an industrial stand to go with a heavy-duty drill, for it is much more sturdy than the do-it-yourself models. Most models have a built-in depth stop which is essential.

Some models can be converted into mortisers with the addition of a mortise drill and chisel (see page 138). This is intended for light duty only but will work very well in softwoods.

To turn the drill stand into a horizontal borer to drill dowel holes in the ends of rails, for example, remove the base and hold the shaft firmly in the V-grooves of the Workmate.

The workpiece is clamped against guide fences and stops to keep the hole location consistent.

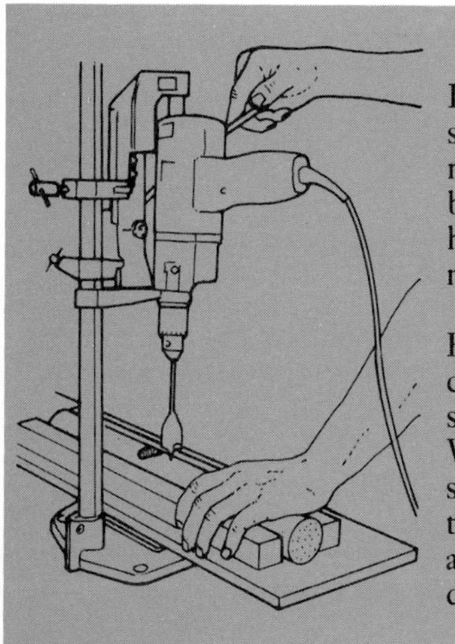

Left: The light duty drill stand shown has the necessary basic features but for continuous use, a heavy-duty, industrial model is better.

Below: An interesting combination of the drill stand and the Workmate. With the base off the stand, hold the column in the vise and use the drill as a horizontal borer for dowel boring, etc.

Table of drill sizes for screws

The table gives the appropriate drill sizes for the pilot and clearance holes. For very hard wood, a slight increase in size may be necessary. I find it useful to pin a copy to the wall near the drill press.

In the absence of the table, the rule of thumb for working out the shank size (clearance hole) is to add 3 to the gauge number which gives the size in $\frac{1}{64}$th of an inch. Thus a number 10 screw requires a clearance hole of $10 + 3 = \frac{13}{64}$ths.

TABLE OF CLEARANCE AND PILOT HOLES

Screw no.	Clearance	Pilot
0	Use bradawl	
1	Use bradawl	
2	Use bradawl	
3	$\frac{3}{32}$	$\frac{1}{16}$
4	$\frac{7}{64}$	$\frac{5}{64}$
5	$\frac{1}{8}$	$\frac{5}{64}$
6	$\frac{9}{64}$	$\frac{5}{64}$
7	$\frac{5}{32}$	$\frac{3}{32}$
8	$\frac{11}{64}$	$\frac{3}{32}$
9	$\frac{3}{16}$	$\frac{1}{8}$
10	$\frac{13}{64}$	$\frac{1}{8}$
12	$\frac{15}{64}$	$\frac{1}{8}$
14	$\frac{1}{4}$	$\frac{5}{32}$

SABER OR JIGSAW

The saber saw makes straight and irregular cuts in all types of material including wood, metal and plastics. It is most frequently used on site for making curved cut-outs, such as for sinks in kitchen worktops. But it does have its place in the workshop too, particularly where a band saw is not available. I find it almost indispensable for a variety of jobs such as cutting out sink holes, rounding corners on table tops and cutting shapes in plywood for special designs.

It is not worth buying a cheap saber saw. Buy the very best saw which will be sturdy enough to cut through 2in of hardwood without having the blade wander and tilt, a common fault with lightweight saws.

The saber saw cuts on the up stroke, oscillating at about 3000 strokes per minute, with stroke lengths from about $\frac{1}{2}$in to 1in. On some models, the base can be tilted to make angled cuts, but I have never found this particularly useful or accurate. The most important consideration is power output, solid construction and, if possible, an oscillating motion feature for the sawblades. The stroke advance can then be adjusted to suit various materials, allowing a large swing for soft materials and a short swing for hard materials, such as steel sheeting. And on the return stroke, the blade is lifted clear of the material to remove the waste and reduce friction resulting in longer blade life.

Guide to good practice

To maintain the saber saw, keep the ventilator openings free from dust and dirt. Check the electrical cables for wear and replace if damaged.

- Oil the appropriate points, including the saw blade guide periodically, according to manufacturer's instructions.
- Inspect the brushes periodically and replace or send in to have them replaced when worn down to less than $\frac{1}{4}$in of contact.
- Send the machine for re-greasing and servicing after approximately every 100 hours or 12 months of use.

Safety

The saber saw is relatively safe and easy to use but keep the following points in mind:

- Worn blades cause overheating and are unsafe. Replace before they are blunt.
- Always check underneath the workpiece to make sure the blade will not cut into the workbench.
- Never hold your hand under the workpiece near the line of cut.
- Unplug when changing blades, and always switch off immediately after finishing each cut.

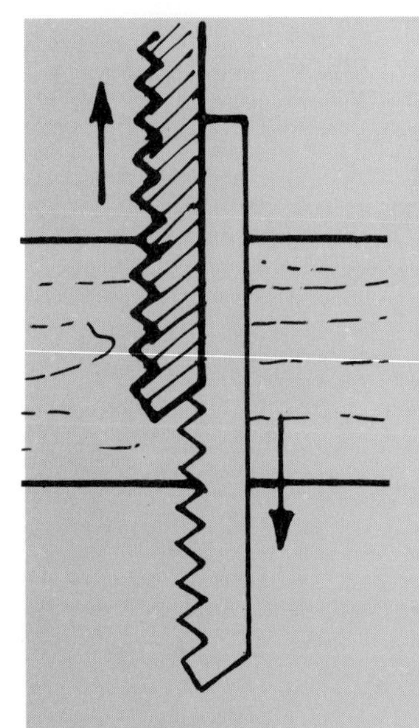

Blades with adjustable stroke advance cut on the upstroke as usual, but the blade moves back slightly on the return stroke to minimize wear. This is intended to make it easier to cut metal and plastic sheet but also makes it easier to cut wood without splintering.

Blades for the saber or jigsaw

FOR WOOD

- 6 teeth per inch for rough work and fast cutting.
- Fine blade for better finish in thinner materials particularly plywoods.
- Hollow ground blade for extra-smooth finish.

FOR METALS

- 10 teeth per inch for thick metals.
- 18 teeth per inch for $\frac{1}{8}$ to $\frac{3}{16}$ in thick ferrous metals.
- 32 teeth per inch for thin ferrous sheet less than $\frac{1}{8}$ in thick.

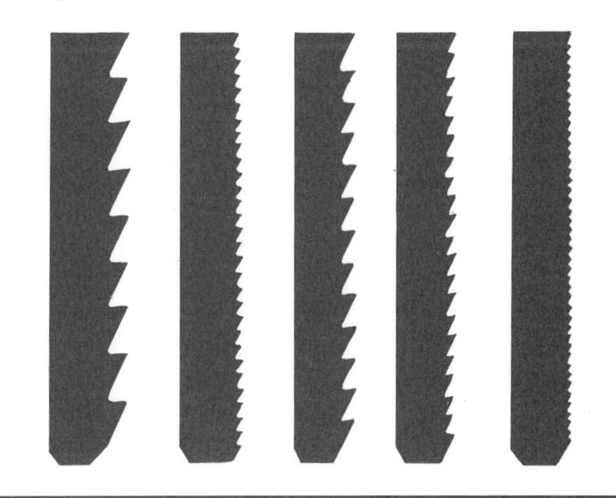

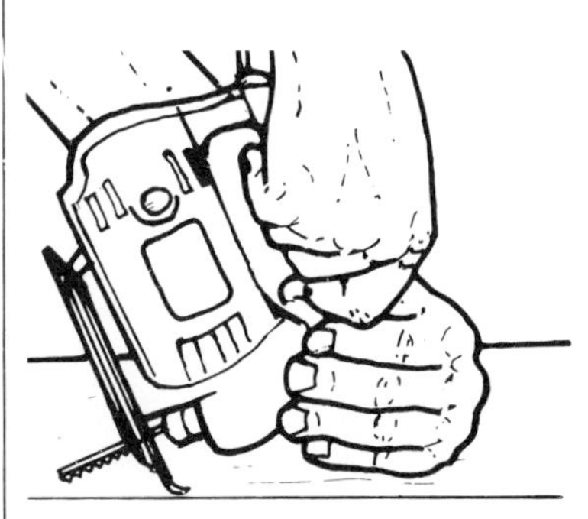

It is safer and easier to start internal cuts by drilling a hole, but a robust saw will make quick plunging cuts in thin wood.

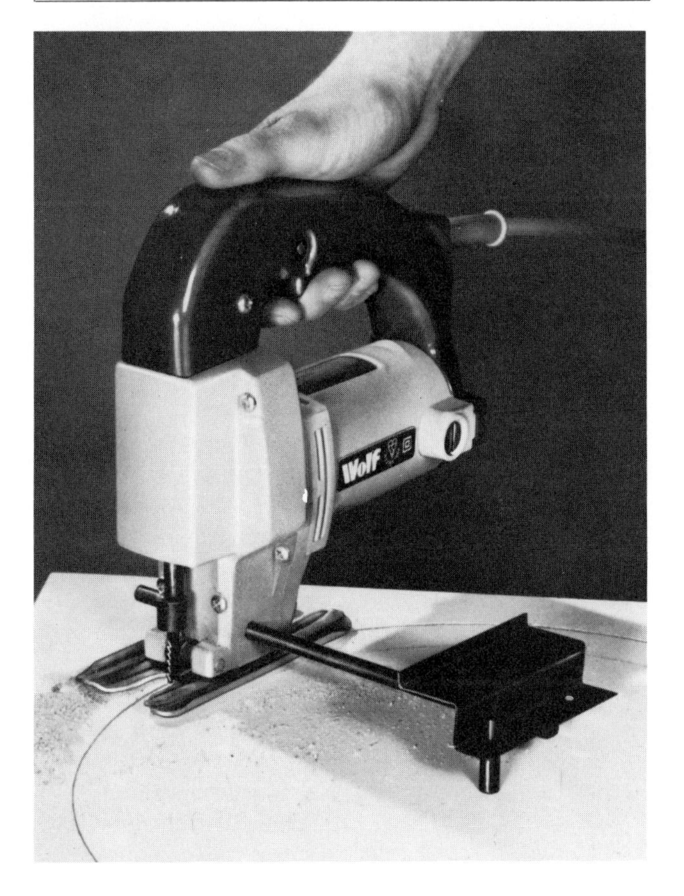

Using the saber saw The blade cuts on the upstroke so it tends to cause splinters and breakout on the top surface. Work with the good face down when finish is important. The workpiece should always be well held down. It is best to clamp it down to the workbench, moving the clamp along as required.

It is standard practice to mark the cutting line, then to cut just slightly on the waste side to allow cleaning up with a plane afterwards.

A rip fence accessory can be fitted to allow ripping along a relatively straight line but this is recommended only if a circular saw is not available.

To start internal cuts, either drill a hole within the waste area or if the saw is robust enough, make a plunging cut by tilting the machine forward allowing the blade to cut through at an angle.

Circles can be cut freehand or by using a circle cutting attachment.

When making the cuts, remember to adjust the cutting speed to suit the thickness of material, in order not to overload the motor.

If the cutting speed is markedly reduced and the motor gets hot, remove the saw and run on no-load for two or three minutes to allow it to cool down. Adjusting the speed by the sound and feel becomes intuitive with experience. One last word – use ear-plugs when cutting metal sheet. The sound is piercing and can cause great discomfort and possibly damage to the eardrums.

The biscuit jointer is ideal for panel construction such as kitchen cabinets where the larger gluing surface holds much better than ordinary dowel pegs. Adjust for offset distance and size of biscuit, then simply hold the fence against the edge and push the handle down to make the plunging cut.

BISCUIT JOINTER AND GROOVER

This small, portable jointing tool made by Elu should be of interest to small to medium capacity firms who make panel construction cabinets such as kitchens. It is basically a substitute for the doweling jig but because of its mode of operation, it makes joints which are quicker, more accurate and considerably stronger than dowel joints.

The machine has a 4in diameter saw blade which is safely plunged into the wood at a pre-set distance from the fence to a depth which fits the "biscuit" dowels. The dowel, made from compressed beech, is glued into its slot and the matching panel is joined into it. The joints may or may not need clamping, depending on the size and configuration. But once the joint is fixed, the glue makes the dowel biscuits swell up for a very strong joint. With ordinary dowel joints, a minimum of two dowels are required to prevent twisting but with the elongated biscuits only one is necessary.

There are three biscuit sizes with corresponding depth settings on the machines. The biscuit jointer is suitable for use in any application where dowels would ordinarily be used. To make kitchen cabinets, for example, each pair of panels is marked and cut in turn using two or three biscuits per joint depending on the length.

Because of their strength, biscuits, used one or more per corner are an excellent substitute for mortise and tenon joints. These joints can be straight rectangular joints or, as shown, for mitered frameworks.

The tool can also be used to cut continuous grooves. For wider grooves, make a series of parallel cuts adjusting the offset distance between each cut.

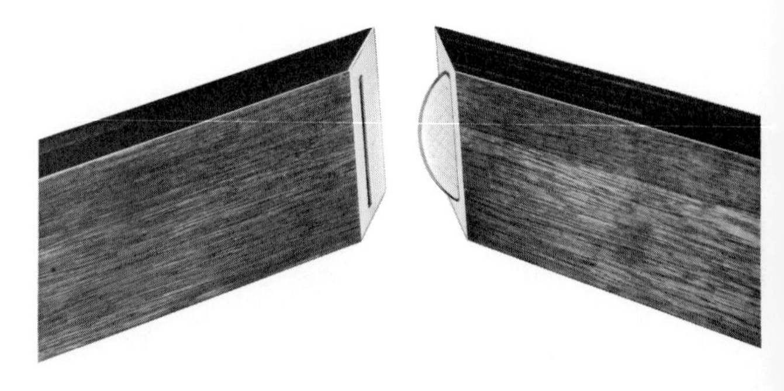

The biscuit jointer is really a superior doweling jig which can be used with one or more biscuits per joint in any mortise and tenon application.

PORTABLE PLANERS

Although they have traditionally had limited use in the furniture workshop, portable planes nonetheless have several useful applications.

Due principally to prejudice, they have been thought vastly inferior to hand planes but they can, with a bit of practice, be used with surprising accuracy for a variety of jobs from trimming laminates to shooting edges.

As with a stationary planer, the accuracy increases with length. The short portable block planes are intended for rough work and for the small jobs such as beveling and laminate trimming. The longer, 15 to 18in models, do more accurate jobs such as trimming a cabinet door.

Cutters, which can be either of high speed steel but preferably carbide tipped, vary in width from 2 to 4in, with 3in the most common. The sets of blades are either spiral or straight and are set as for stationary jointers, level with the rear or outfeed shoe.

To operate the plane, set the depth to zero by checking with a straightedge that both shoes are level. Then use the depth attachment lever to set the depth of cut which should be less than $\frac{1}{16}$in for hardwoods and up to $\frac{1}{8}$in for softwoods, depending on the power of the motor. Feed rate should, of course, be varied to suit the material and becomes intuitive with experience.

Model and framemakers may be interested in a stand accessory which allows the portable plane to be

used upside down like a jointer. Its use is limited to small pieces up to about 3 to 4ft long but since most frameworks are within this size, it can be a useful alternative to an expensive stationary jointer. The high cutting speeds, up to 15000rpm, guarantee the high degree of finish required in this type of work. A special guard accessory should always be fitted when used as a jointer.

For safe operation, the planer should always be firmly held and well balanced on the work. Remember there is no guard underneath, so never put the planer cutters down on the bench.

Because of the high vibrations, make sure to hold the stand firmly to the bench with at least one, but preferably two clamps.

PORTABLE SANDERS

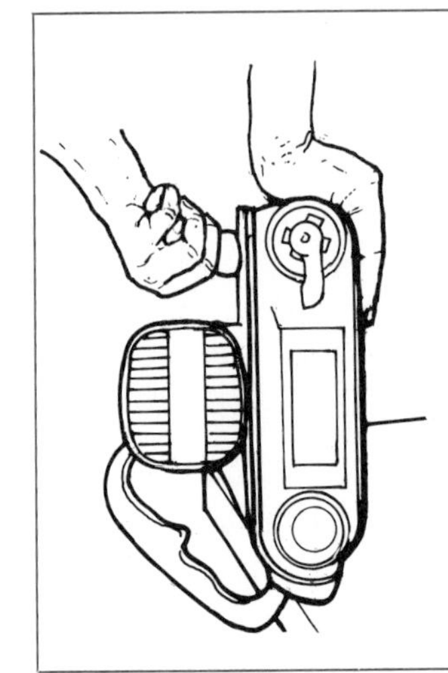

To change belts, decrease the tension by pushing down the front wheel until it clicks in place. Arrows printed inside the belt indicate the direction of rotation so that the seam in the belt will not "pick up" and break during use. Methods of releasing the wheel tension to hold the belt in place varies from model to model.

There are three main types of portable sander. The belt sander and the rotary sander are both used primarily for rough work such as removing paintwork, but the orbital type finishing sander is widely used even in large production shops for final surface preparation.

The belt sander The size, determined by the width and length of the belt, varies from model to model, from the small 2 × 21in size to the heavy industrial 4 × 26in size.

Belts, made with a flexible cloth backing, are available in the usual range of grits, from coarse to fine. But it should be remembered that, generally, coarser grits are used for machine sanding than for equivalent results in hand sanding. The basic kit should include 60 grit belts for rough work, 100 for medium and 120 or 150 for fine finishing.

To change belts, push the wheels together to decrease the belt tension. Then release the spring-loaded catch on the front wheel to hold the new belt in place. Sideways belt adjustment, done with the machine turned on, varies from model to model, but it is usually done by turning a knob until the belt remains near the center.

An integral dust bag is essential, because the belt sander often produces toxic dust such as when removing old paintwork.

As anyone who has ever used it knows, it takes long experience and a delicate touch to do fine finishing with a belt sander, particularly on veneered boards.

I know cabinetmakers who can produce a perfect finish with a belt sander, but I have never achieved good results. It is much too easy to dig in either by tilting it slightly or by dwelling too long in one place.

The sander should be lowered carefully onto the work back heel first, then moved constantly in overlapping strokes at a slight angle to the grain.

It is a very good idea to pin waste strips to all the edges of flat work, flush with the top face, to prevent the sander from taking off too much near the edges.

For rough work such as paint removal, work diagonally across the surface in alternate directions before finishing off along the grain.

Elu has produced a very clever belt sander design which makes it possible to do accurate and delicate work even on veneer work without the usual problems. The belt is parallel to a surrounding frame and can be raised or lowered gradually with a turnscrew to take off small or large amounts. This makes it much like a portable planer. The depth of cut is set with the knob.

This Elu model with a frame and fine adjustment can be used on delicate veneering work.

To get the full use of the "mounted" sander, fit a stop or fence as on the sanders on page 160.

Make sure to brush off the surface regularly to remove any loose sanding grits.

Hold the sander carefully in the vise with softening if possible to avoid cracking the case. This is a convenient arrangement for pointing the ends of dowel pegs.

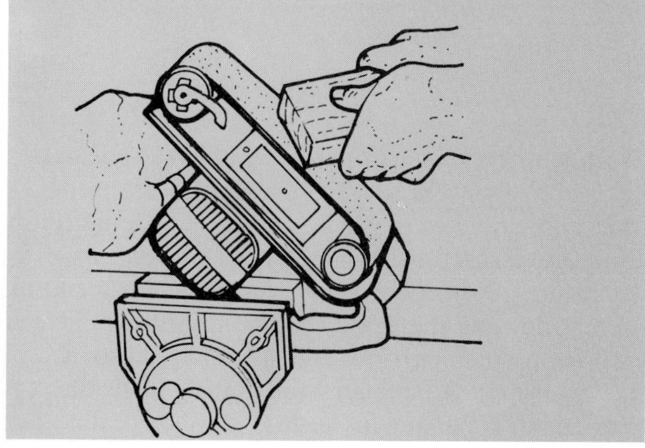

To use the machine as a stationary belt sander, gently clamp it upside down in a vise or where available, hold it in a special stand. The flat part can be used to square off boards, to point dowels and so on, while the rounded back end is useful for curved work.

The finishing sander Finishing sanders work with either an orbital or an oscillating motion. The orbital type which produces faster results, works in a circular motion at between 4000 and 14000, $\frac{3}{16}$in orbits per minute.

The slower oscillating type works in a reciprocal back and forth motion. Multi-motion sanders can be switched from one type to the other.

Sizes vary from the lightweight, one hand models using a quarter standard sheet of paper ($4\frac{1}{2} \times 5\frac{1}{2}$in), to large industrial models, taking half sized paper ($4\frac{1}{2} \times 11$in) and weighing up to 10lb. The larger models are usually filled with a bag for dust free sanding.

Finishing sanders are used for all types of finishing work. With care, it will produce a high standard finish which won't require additional touching up by hand. It's useful to constantly brush away the dust on the surface with a soft brush. Loose grits which get caught under the sander can produce pronounced circular marks which are difficult to remove.

Only light pressure is required, and as with all power sanders the machine should be kept in constant motion. Although not strictly necessary, it is a good idea to move the machine along the grain. The usual procedure is to progress from 60 to 100 to 150 grit of open coat papers but work which has previously been belt sanded may only need the final grit size on the finishing sander. As in all finishing work, make sure to keep the work well lit. I keep a couple of 150 watt clip-on reflectors handy so they can be mounted at work level a few feet away from the bench to show up any small imperfections in the surface.

The rotary sander Rotary sanders have only limited use in cabinetmaking. They can be used, like the belt sander, for taking down old finishes and paintwork but, more often than not, they dig into the work producing circular marks which are very difficult to remove.

They are, however, useful when placed in a bench mounted stand where they can be used with control against a stop to square off board ends and other small sanding jobs.

The rotary sander can also be used hand-held and fitted with a polishing pad to buff waxed or varnished surface to a high polish.

PORTABLE ROUTER

Next to the table saw, the router is undoubtedly the most versatile tool in the workshop. It is useful not only in the production shop for finishing off edges, doing template work and so on, but particularly in the small shop where its low cost, ease of operation and wide range of application make it indispensable. Even in large factories, it is often much easier to use a portable router than the stationary overhead router especially where the workpiece is large and unwieldly to handle.

The beauty of a portable router is that it does a professional job, producing clean, accurate cuts. Its possibilities are endless, limited only by the experience and ingenuity of the craftsman.

The standard operations of cutting grooves, housing and rabbets are well-known, but the router will also cut joints such as half laps, dovetails, tongue and groove and mortise and tenon joints. And used with jigs and templates, it will cut out repeated shapes such as hinge slots or decorative grillwork and drill perfect holes for dowel jointing.

Description The basic router components can be seen in the photograph. The tapered collet which holds the cutter shafts is held by a locking nut. Exact configurations vary from model to model but changing cutters is usually done with two open wrenches, one to hold the spindle, the other to loosen or tighten the locking nut. Some routers come with two interchangeable collets one for $\frac{1}{4}$in diameter cutter shafts the other for $\frac{3}{8}$in diameter shafts. On some models, the top is flat allowing you to stand the router upside down on the bench as you change cutters. This is very convenient since both hands are needed to use the wrenches.

The depth is set with the cutter projecting below the base by the required amount, and for most models the motor must be turned on before the whole router is gradually lowered to make the cut. The more expensive plunging routers have a spring-loaded mechanism which allows you to place the machine on the work and then lower the rotating cutter safely into the wood to a perfect depth. This plunging action is very useful especially for making internal cuts, since the cutter enters the wood without the usual tendency to grab the wood. It also makes it easy to drill perfect, flat-bottomed holes by simply plunging the cutter into the wood at the start of the cut, then releasing the spring to bring the cutter back up.

Router cutters The interchangeable cutters make the router very versatile. Normal cutters are made from high speed steel which is heat treated and finish ground to a sharp edge. These need constant re-sharpening to maintain a high degree of finish and to prevent burning the work. Carbide tipped cutters are used for production work and special applications such as trimming laminates and cutting materials with a high resin content such as chipboard. The tungsten carbide tips are brazed to the alloy steel body. These give a superior finish, smoother cut and last much longer than high speed steel cutters. But unlike the latter, which can be sharpened with a special attachment, carbide tipped cutters must be sent to a specialist for sharpening with a special diamond grinding wheel.

Special carbide cutters which are made of solid carbide are also used in production work. They have the same strength and durability as tipped cutters but eliminate the problem of having the brazed tip break off.

Most manufacturers can supply a wide range of standard high speed router cutters and publish specification sheets of their range. Shank diameters are either $\frac{1}{4}$, $\frac{3}{8}$ or $\frac{1}{2}$in, with the $\frac{1}{4}$in diameter shanks the most common, particularly for general purpose routers. It's useful to buy a machine with will take both $\frac{1}{4}$in and $\frac{3}{8}$in shanks with a change of collets,

Table of standard router cutters

STRAIGHT CUTTERS

Single flute 	For rough cutting of grooves and internal cut-outs, requiring high feed rates. $\frac{1}{8}$ to $\frac{1}{2}$in dia.
Two flutes 	Two flutes give smooth cut. For cutting grooves, rabbets, housings, cutting recesses with templates. $\frac{1}{8}$ to 1in dia.

EDGE FORMING CUTTERS

Rounding over 	One setting produces rounded edges from $\frac{3}{16}$ to $\frac{1}{2}$in radius. Lower setting produces rounded edge with decorative bead.
Beading 	For decorative edging such as table edges and drawer fronts. Radius $\frac{1}{16}$ to $\frac{3}{8}$in.
Chamfering 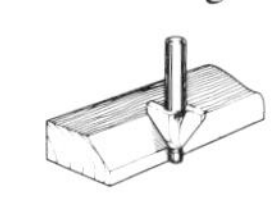	45° cut for decorative beveled edges. Cutter is lowered for larger bevel.
Cove 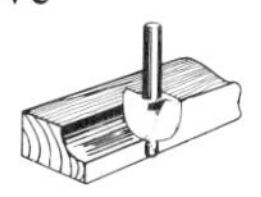	Radius $\frac{3}{16}$ to $\frac{1}{2}$in for decorative edges and for making drop leaf table joint.
Roman ogee 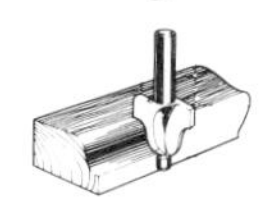	For decorative, period furniture.
Rabbeting 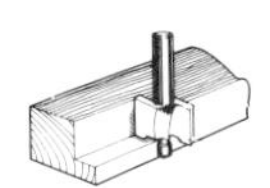	For rabbeting without a guide. Rabbet widths $\frac{1}{4}$ and $\frac{3}{8}$in, depth $\frac{3}{8}$ to $\frac{1}{2}$in.

GROOVE FORMING CUTTERS

V-grooving bit 	For decorative and freehand work such as lettering and for simulating plank construction.
Veining	For making decorative lines $\frac{1}{8}$ to $\frac{7}{32}$in dia.
Core box 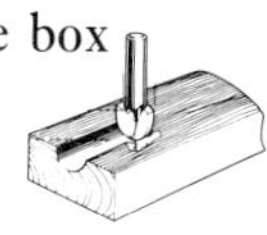	For round-bottomed grooves in decorative work on window frames and for textured internal surface cuts
Dovetail 	Depths from $\frac{3}{8}$ to $\frac{17}{32}$in for making dovetails and cutting dovetail housings.
Stair routing 	Similar to but more robust than dovetail bit for cutting housing for a secure joint.

TRIMMING CUTTERS (carbide tipped)

Flush trimming 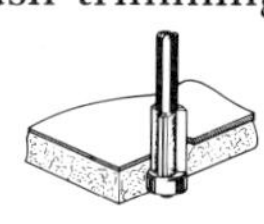	Ball bearing guide pin for trimming laminate edges.
Double flush trimming 	Spacers allow adjustment to trim top and bottom laminates in one pass.
Bevel trimming 	Trims laminate at angle from $7\frac{1}{2}$ to 15°.

OTHER CUTTERS

Trimming saws	Steel or carbide tipped blades for trimming edging.
Slotting cutters 	Tipped blades for cutting edge slots for tongued joints, and for fitting plastic edging and handles.

since for the occasional heavy job such as cutting through thick, tough hardwoods the $\frac{3}{8}$in shank is much less likely to break off.

Edge forming cutters such as rounding over or ogee bits usually have a built-in pin which follows straight or curved edges, obviating the need for a guide. These pins tend to burn the edge, requiring tedious cleaning up afterwards. For production purposes, special cutters fitted with ball bearing pins are used to prevent burning. These can be bought in a variety of configurations, for trimming laminates, shaping edges, cutting edge grooves and a variety of other purposes.

In addition, some suppliers make up "specials" – tipped cutters made (at considerable expense) to a required specification. This may be worthwhile, particularly for production work where they will quickly pay for themselves.

Guiding the router There are several ways of controlling the direction of cut. Making effective guides and jigs for the router is the most important aspect of using the tool.

1 STRAIGHTEDGE This is the most common method used to achieve straight, accurate cuts for grooves, housings, etc. The simplest method involves clamping or nailing a straight board to the work and running the router base against it. To get the correct offset distance, i.e. to the straightedge, use this formula:

$$\text{Offset distance } X = \frac{(D - d)}{2}$$

Subtract the cutter diameter from the diameter (or width if it has straight sides) of the router base and divide by two. Thus for a $\frac{1}{2}$in diameter straight cutter, using a router with a 6in diameter base (which is a common size), the offset distance should be $\frac{6-\frac{1}{2}}{2}$ or $2\frac{3}{4}$in. It's a good idea to memorize or write on the wall the offsets for the $\frac{1}{2}$in and $\frac{3}{4}$in diameter cutters which get the most use. Other straightedge guides such as T-squares and box guides are described on pages 68–69.

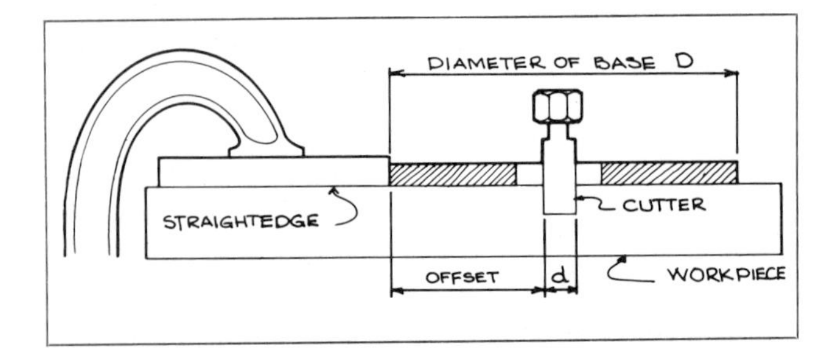

2 STRAIGHT OR CIRCULAR GUIDE This is an accessory which bolts or screws onto the router base. The guide runs against the edge of the workpiece to direct the cutter parallel to the edge. The straight guide which is used for straight edges can, on most models, be converted to a circular guide which makes cuts along or parallel to any regular curved edge. These guides have limited use, since cuts can only be made a short distance in from the edge.

3 Using cutters with built-in PILOT GUIDES is very common, particularly when applying a decorative edge to the workpiece. The pilot runs along the edge under the cut to guide the router. Since the pin is rotating at high speeds, it tends to burn the wood as it heats up. To minimize burning, don't push the cutter and pin too hard against the edge, but gently guide it along. Also keep the cutting speed just right by trial and error, and clean the tip with lacquer thinners and steel wool every time you sharpen the cutter. In production work, use a straight guide with the router set so that the pin just misses the edge. Or alternatively invest in a special cutter with a ball bearing guide pin which doesn't burn the edge.

4 A TEMPLATE OR PATTERN JIG is used in conjunction with a special template guide which is attached to the base of the router. Precise arrangements vary from one model to the next but the basic principle remains the same. The straight cutter protrudes through the template guide which runs against the edge of the pattern or jig usually made from $\frac{1}{4}$in hardboard or plywood. Any shape can be copied by first carefully making a jig slightly oversized to allow for the distance between the cutter and the guide pin. This method is commonly used when cutting repetitive shapes such as hinge recesses or cut-outs for flush hardware. It is also used in conjunction with a special accessory to make dovetail joints.

Cutting around external templates such as this requires care to keep the router from drifting out of line.

5 The TRAMMEL BAR accessory is used to cut circular shapes or grooves, either internal cuts such as cutting out for a sink in a kitchen worktop or external cuts such as cutting out a round table top.

6 Freehand routing requires an experienced and steady hand. The router is guided by eye to cut out decorative shapes or lettering. Using a fine bit in thin materials the router can also be used as a jigsaw.

Holding down the work In general work, it is easiest to clamp the workpiece to the bench. Quick action clamps are the best, but any other clamps will do.

Clamps sometimes restrict the movement of the router, and there are many alternative fixing methods which are more convenient.

The workpiece can be pinned down for example, particularly if the pin holes can be located in a waste area.

As a variation on the pinning method, hammer the workpiece onto protruding pins fixed in a board underneath.

Cams are simple but effective devices and are easily made. They have a variety of applications in the workshop. They are especially useful when routing many pieces of the same size.

Double sided tape, available from art supply stores, provides a very effective fixing particularly for holding jigs and templates. A strip or two in each direction is usually enough.

Vacuum clamps are devices I haven't actually tried, but which look perfect for production runs using templates. In principle it is a very simple notion – the workpiece is held to the template by a vacuum formed between them by sealing the perimeter with neoprene and pumping out the air through a tube fitted to a vacuum pump. One router manufacturer gives advice and sells equipment for setting up vacuum clamps.

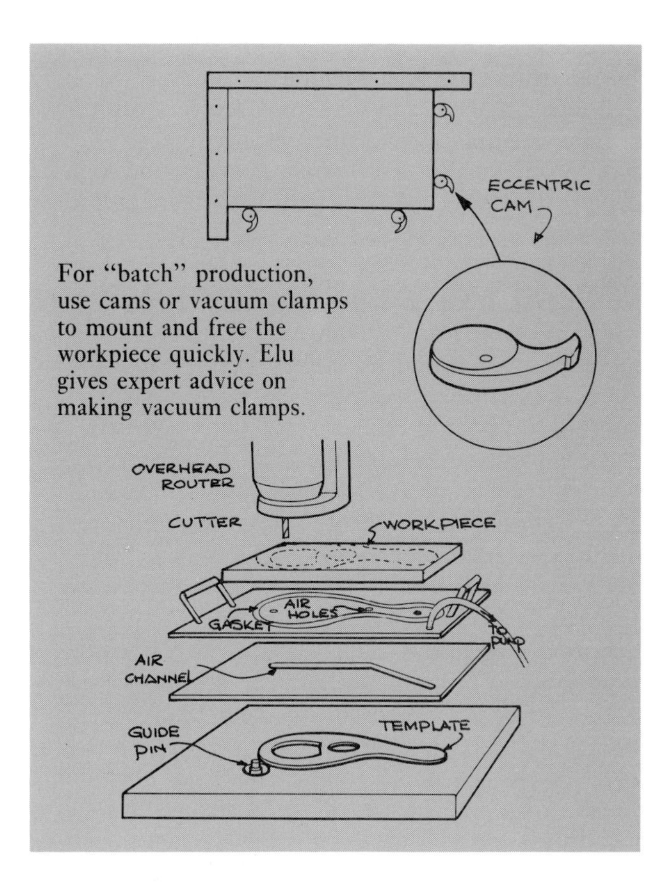

For "batch" production, use cams or vacuum clamps to mount and free the workpiece quickly. Elu gives expert advice on making vacuum clamps.

Guide to good practice The router is probably the easiest and safest portable power tool to use. To get the best results, first of all keep the machine and particularly the ventilating holes free from dust and shavings. It's best to store it in a cabinet or on a high shelf when not in use. Regularly examine the collets for grit or burrs and clean them periodically with thinners. Shut off the router when not in use, particularly when adjusting the depth of cut. Always pull out the plug when changing cutters which should be seated all the way in and tightened firmly but not overly tight.

When using the router, always hold it firmly by both handles and feed the cutter into the work as shown. Feeding the work too fast or making too deep a cut overloads the motor and reduces the speed which is necessary for a good finish. As a rough guide a $\frac{1}{8}$in diameter cutter requires a cutting speed of about 22000rpm whereas a $\frac{3}{4}$in diameter cutter needs only about 18000rpm to maintain the same periphery speed. The feed rate should be based on the sound of the motor, and with experience it becomes intuitive to listen and adjust the feed rate accordingly.

Dull cutter bits also overload the motor and produce bad results. High speed steel bits should be sharpened regularly, every day with heavy use. Some manufacturers sell a sharpening accessory.

Besides cleaning and tightening screws from time to time, the only maintenance required is to occasionally inspect the carbon brushes and replace them when worn out, according to manufacturer's instructions.

Cutting grooves or dados Most of the work done with a portable router involves cutting grooves used in various applications in cabinetmaking and joinery.

Grooves serve many functions. Wide grooves, housing or dados in cabinet sides hold shelves to make bookcases, desks or chests of drawers. Narrow grooves are used to hold cabinet backs or drawer bottoms or in tongue and groove joints to reinforce edge to edge joints. And in joinery, the router is ideal to use in making grooves in stairway stringers to hold the treads and risers.

For cutting grooves along narrow boards, a side fence is fitted to the router and run against the edge of the board. It is important to check the fence setting and also the depth of cut on a trial piece first and to make any minor adjustments before beginning work. In making the cut, the work should be well held down and the fence run firmly against the board's edge. Since the cutter will copy any irregularities of the edge, it's always a good idea to make a quick check by sighting down the edge. To stop the groove before the end of the board, clamp or pin a small stop to the work and finish off the rounded end of the groove with a chisel if necessary.

On wider boards, such as cabinet sides where the groove is too far in from the edge to use a side fence, it is standard practice to run the router base against a straightedge in cutting grooves. For accurate work, the board must be perfectly straight, and most shops have at least two straightedges of different lengths to use not only with the router but for many other applications. The easiest way to make a straightedge which will remain straight is to cut off an 8in width of ½ or ¾in plywood, choosing a good factory edge for the straightedge. Plywood is fairly stable but to keep it even more so, glue a piece of Formica to both faces and trim off the edges. Then sand the good edge carefully to make it smooth and finally drill a hole near either end for hanging within easy reach.

The straightedge which can be pinned, clamped or held to the work with double sided tape, must be located the correct distance away from the groove. The offset distance is worked out by the simple formula given on page 66.

To make the work easy, most woodworkers try to work out the details so that the groove width corresponds to the diameter of one of the standard straight cutters such as ¼, ⅜, ½ or ¾in. Where the width of the groove must unavoidably be wider than one of these, the groove has to be cut in two or more runs of the router with the guide moved over the exact amount to give the correct width.

Housings or dados, which are grooves cut across the board, are easily made using either straight or dovetail cutters with the router using a straightedge held to the work. This operation is carried out so often that it is advisable to make a T-square guide which can be quickly moved along from cut to cut.

On narrow boards, use the fence accessory to cut grooves parallel with the edge. On wider boards where the fence doesn't reach, run the router against a straightedge clamped to the work.

Making a T-square guide The T-square guide, which is indispensable in router operations, is a simple and effective jig used to guide the router when cutting housings or dados. Make it from ½ to ¾in plywood with the cross piece B wide enough (about 6in) for the clamp not to get in the way. To get it absolutely square, glue the two pieces together first and allow the glue to set without clamping together, checking the pieces against a square and adjusting if necessary. Reinforce with four or five countersunk screws after the glue has set. Before using, run the router across the guide making a short groove in the cross piece which is used to line up all future cuts. I think it's best to make several T-squares, one for each cutter so that the "guide groove" stays constant for each one to make it easier to line up.

Housing joints used in bookcases and similar

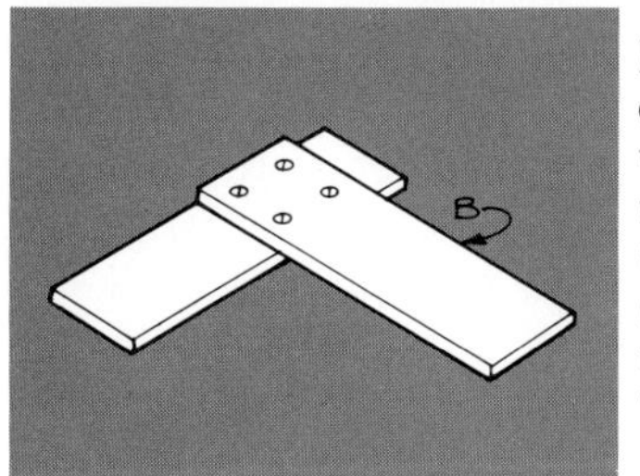

Keep one T-square for each size of cutter. That way the groove made in the cross piece B can be used to quickly line up the cut against a mark measured out on the workpiece.

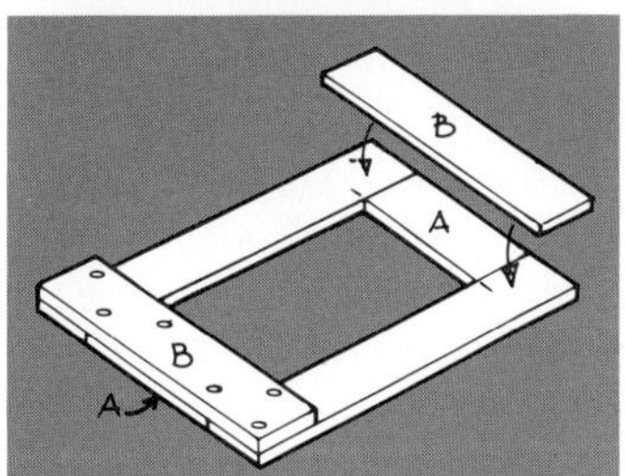

A box guide which controls the cut from all sides is easier to use than the T-square. The only difficulty is lowering the cutter into the work. In this case, a plunging router is very convenient.

work, rely on tightness of fit for their strength so they must be accurately cut. It is inadvisable to make the grooves wider than the diameter of the cutters since the fence must be moved over a small amount and it is almost impossible to get this small amount exactly the same for each groove.

Where wide grooves are required, it is better to use a box guide which controls the movement of the router in all directions. With this, unlike the T-square guide, the router cannot move away from the fence to ruin the work.

For stopped housings, pin a stop to the T-square or better yet, use a box guide, made to measure. The end of the groove can either be left rounded and allowed for when cutting the notch in the shelf, or it can be squared off by hand with a chisel.

For through housings, it is best to clamp a waste strip to both edges of the board to prevent breakout as the cutter enters and leaves the board.

Box guide The box guide which controls the movement of the router in all four directions has many applications, such as cutting stopped housings, making internal cuts for slots for through tenons, or recesses for flush fitting hardware.

For repeated use, make it out of $\frac{1}{2}$in plywood strips about 4in wide, but for one time use, softwood scraps can be substituted.

The box guide must be dead accurate, which is achieved by cutting the pieces straight and locating cross pieces A accurately. Hold the rectangle together by overlapping two end pieces B, glued and screwed. Screw a removeable guide underneath at one end for use as a T-square. I always find it easier to make a quick sketch noting down the size of the cut-out and the offset distance to the router base, depending on the size of the cutter. It is then simply a matter of adding up the dimensions in each direction to get the internal dimensions of the rectangle for the guide, two offset distances plus the width of the cut-out. As with all jigs, it is essential to try it out for accuracy on a scrap before cutting the actual workpiece. If it is inaccurate, it is better to throw it away and start from scratch again, rather than trying to shave a bit off here and there.

Cutting rabbets Cutting rabbets is a standard operation of the portable router, using either the special rabbet bit or a standard straight bit.

The rabbet bit is guided by the pilot which runs against the edge of the board so that no fence is required. But the rabbet width is pre-determined by the bit – usually $\frac{3}{8}$in – and limited in depth to about $\frac{1}{2}$in depending on the cutter. The guide will also tend to burn the side of the workpiece.

Cutting rabbets with a straight bit is exactly like making grooves as shown. The fence is set to the exact width required and the bit, which should be larger than the rabbet, is lowered to the rabbet depth required. In hardwoods, it may be necessary to make two runs. Again the cut should be tested on pieces of scrap before beginning work.

Rabbets can be cut by various tools such as the table saw and by hand, but with a portable router the rabbets can be cut after the framework has been assembled. This is done either with a rabbet bit or a straight bit with a fence. But since the pilot of the rabbet bit tends to burn the edge, particularly at the difficult to reach corners, the latter method is usually used. The router is fitted with a special right angled fence attached to the router base as shown. The cutter leaves a rounded corner which is easily cleaned up by hand, or left as a decorative touch.

Template routing This is another elegant use of the router which makes it easy to cut repeated shapes from a master template or jig. Template routing has many applications from cutting out recesses for fitting hardware to cutting decorative shapes and grillwork. We use templates constantly for setting in shrinkage brackets in table rails, for cutting hinge recesses, and for fitting a variety of hardware such as flush brass handles on drawer fronts.

Each new template is marked and hung on the wall for the next time. Cutting dovetails with a router, as described in detail on page 197, is another timesaving use of template routing.

The technique for template routing is as follows:

1 Choose the straight cutter and a template guide to match. The smaller the bit, the tighter the corner radii and the less work in squaring off the corners by hand afterwards. The internal diameter of template guide must obviously be larger than the cutter so that the cutter can rotate without rubbing against the guide. Sizes of template guides vary from model to model but it is usual to supply guides for $\frac{1}{4}$in and $\frac{1}{2}$in diameter cutters. Of course a smaller bit can be

Cutting circles To me, this is the most elegant operation of the router and one which cannot be equaled by any other tool.

The router is attached to a trammel bar accessory or to a homemade jig, which allows it to be rotated around a center to make circles of various diameters and to cut decorative patterns in concentric circles. Most manufacturers sell a trammel bar which cuts small circles. A cheaper alternative is to bolt the router base to a $\frac{1}{4}$in thick piece of plywood on which small center holes have been drilled to correspond with various radii. The trammel bar has a built-in center pin, but for the plywood jig, use a small nail at the center. To cut out table tops, work on the underside to avoid leaving a pin hole in the top, but where it is necessary to work on the top side, fix a piece of plywood at the center to take the pin. This can be attached with a few strips of double sided tape, or if necessary, by spot gluing.

Fitted with a straight cutter, the router can cut out circles in material as thick as the depth of the cutters, but in all but the thinnest materials it is best to make a number of cuts, successively lowering the cutter until it cuts through.

On circular table tops or on articles such as dinnerware or cutting boards, the edge can subsequently be shaped by fitting the router with a molding bit and rotating it around the same center.

For small circles, use a trammel bar accessory.

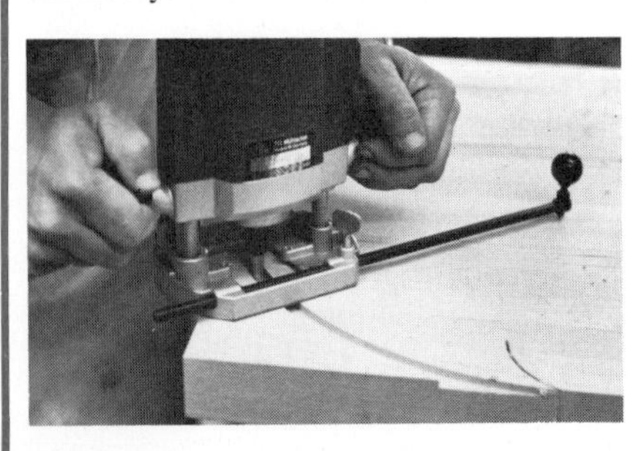

A simple jig for cutting a large, 6ft diameter table top.

Bolt the $\frac{1}{4}$in plywood strip to the base with countersunk machine screws.

used in a large guide so that a $\frac{1}{8}$in diameter bit can be used in a $\frac{1}{2}$in guide, for example.

2 Fit the cutter and the guide. The cutter is fitted as normal and the guide is fixed to the base in any number of ways depending on the model.

3 Fix the template or jig to the work, by pinning it lightly, by clamping it if there is room, or by using double sided tape.

4 Set the cutting depth and lower the router onto the work to cut out the area within the template. The guide runs against the side of the template copying the outline (and any imperfections!).

Making jigs and templates Pattern making is very precise work. Jig and pattern makers are specialists and always have high status in any workshop. Most templates are rectangular and relatively easy to make but those involving irregular shapes require an exact eye and delicate touch to get the shape right, without any irregularities which will be copied by the template guide.

The template opening must be larger than the opening to be cut to allow for the guide and the following formula gives the extra amount to be added to get the overall template opening:

$$Z = (X - Y)$$

where X is the outside diameter of the guide, Y is the diameter of the cutter, and $\frac{1}{2}Z$ is the amount of overlap on each side. As an example, in cutting out a

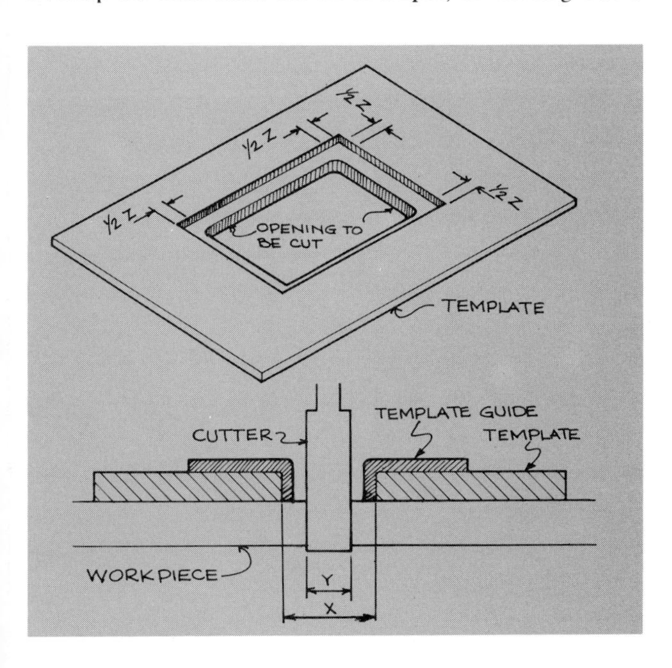

1

2

3

4

1 Making a template to make a template: pin straight, sanded strips to the template and cut out with router. 2 The template clamped on. 3 Finished hexagonal table. 4 Leg detail.

recess for a flush handle 3 x 6in, using a $\frac{3}{8}$in diameter cutter and a guide with an outside diameter of $\frac{1}{2}$in the template would have to be $3\frac{1}{8}$ x $6\frac{1}{8}$in to allow $(\frac{1}{2}-\frac{3}{8}) \div 2$ or $\frac{1}{16}$in overlap each side for the template.

The best material for templates is $\frac{1}{4}$in plywood or hardboard. For rectangular and other straight sided shapes, I find it best to use the router itself in cutting the template. I make a template by pinning pieces of $\frac{1}{4}$in plywood, which have been cut straight on the saw and edge sanded, to the piece which will become the template. Basically it's making a template to make a template, but it works well since it avoids having to make the cuts with hand tools which are not as accurate. When cutting out, don't forget to put a piece of waste plywood underneath to protect the workbench.

More elaborate template shapes can be made using a combination of straightedges and freehand cuts or circular cuts.

Factory made templates for specific applications can also be purchased, and one manufacturer sells a clever jig which is adjustable, for cutting out recesses for various sizes of hinges in production workshops. A plunging router is particularly convenient in template work since it is easier to feed it to the work when small templates are involved. Drilling straight and perfect holes for dowel jointing, for example, is always cumbersome with a drill and doweling jig, but easy with a plunging router and a template.

Shaping edges and making moldings In production shops moldings are shaped on expensive automatic machinery. In smaller shops, spindles or shapers are more common and do the same job although more slowly. But these too are expensive, and for the craftsman an inexpensive alternative is to use the router, either freehand or fixed to a table to serve as a lightweight spindle. In our shop we use the router constantly for putting decorative "professional" touches on our furniture. After a few years, you build up a wide selection of shaped cutters which can be used singly or in combination with one another to shape table edges and drawer fronts, to bevel or round off legs and rails, and so on. Changing bits is easy and the pilot pin on the cutters guides the cutters around the edge.

As with all router cutting, the router is fed so that the bit cuts into the work, which means running it clockwise on internal openings and counterclockwise on outside cuts such as rectangular and round pieces.

The pilot does tend to burn the edge, but this can be eliminated by using the fence or by buying special bits with ball bearing pilots.

When shaping rectangular pieces such as drawer fronts, remember to make the cuts across the two ends first so that the breakout at the end of each cut will be cut off when shaping the two long sides with the grain.

Alternatively, clamp a waste strip on either side to prevent the wood from breaking out when cutting across the grain.

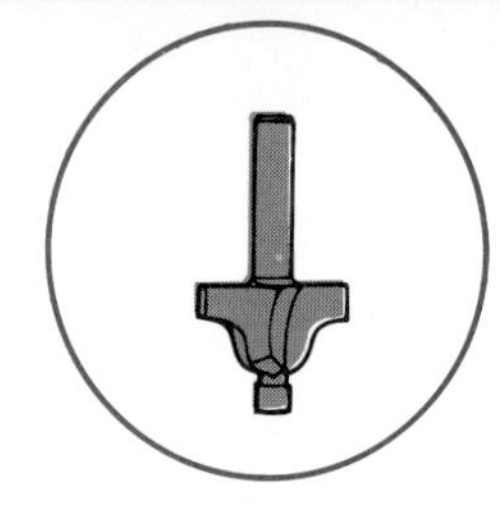

Molding cutters can be used singly or in combination to decorate edges. With the router used freehand, the pin guides the cutter. A better solution is the router stand (right) with a feather board to keep the work against the fence.

Making moldings This is best done with the router fitted upside down in a special table, bought as an accessory or made from $\frac{3}{4}$in plywood (see page 74).

Using the router as a spindle, first cut the wood into strips on the saw, then feed it through with a guide holding it tight against the fence for a consistent cut. One sweep cuts the rabbet; then replace the straight cutter with a shaped cutter and feed all the strips through to cut the shape to make picture frame moldings, for example. For compound shapes using the same cutter or several cutters in combination, the cutters and the fence must be precisely adjusted for each cut. It is advisable to have a few extra lengths available to allow for trial and error in adjusting.

Making moldings with the router hand-held is possible too, although not as accurate. But instead of cutting the wood into strips before the molding is cut, this is left until after it has been shaped. The procedure is to shape the edge, unclamp the board and feed it through the saw to cut off the molding before remounting it to repeat the procedure.

Miscellaneous operations

DECORATIVE WORK Decorative panels can be produced using a straight cutter, a V-groove, veining or core box bit in any number of ways. To make paneling out of plywood sheets for example, the router is fitted with a V-groove bit and run against a straightedge to produce a series of vertical, horizontal or diagonal lines which simulate wood paneling.

Similarly, it is easy to simulate paneled doors by cutting a shaped groove around the perimeter using a fence to keep the edge distance constant.

TRIMMING By using a special tungsten carbide cutter fitted with a ball bearing guide, trimming laminate worktops is almost automatic. Depending on the type of cutter, the trimming can be flush or double flush, where a double cutter trims both top and bottom laminates at once, or the cut can be beveled to varying slopes by using a bevel cutter.

To trim the laminate around an internal hole, such as a sink cut-out, use a combination hole and trimming cutter. With this cutter, you simply drill through the laminate which covers the pre-cut hole and trim around the opening.

DRILLING AND PLUG CUTTING A plunging router makes perfect holes, much crisper than those made with a drill, and they are always perpendicular which is difficult to achieve with a hand-held drill. Also special matching bits are available which remove knots or marks in plywood or wood and seal the drilled hole with a tight fitting plug.

Another useful drilling operation for the plunging router is to drill out special sized holes for fitting modern round hinges frequently used in kitchen cabinets for 180° opening doors.

GROOVING Edge grooves have many applications such as in spline joints and in fitting edge trims or hardware to chipboard panels. Grooves can be cut with the board on edge using a small straight cutter, but the best way is to use a grooving bit fitted with spacers and ball bearing guides. Since the router runs on the face of the board, it is much more stable than when resting on the edge. For grooves wider than the cutter thickness the cutting depth must be adjusted.

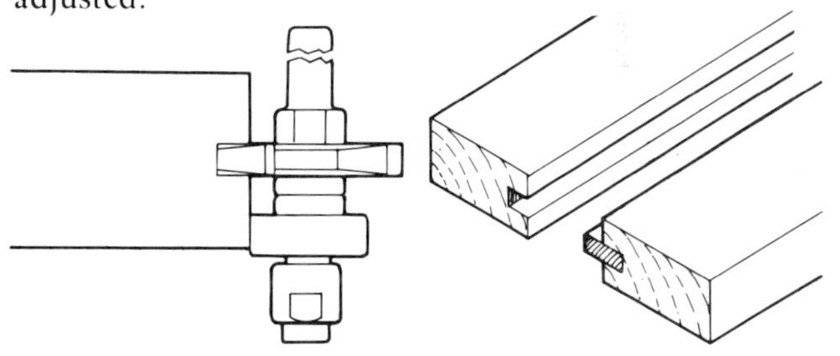

MORTISING AND TENONING To cut mortises, either use a jig with a template guide, which is more suitable for plunging routers, or the fence against a pair of blocks clamped on either side to keep the router stable. For full details on cutting mortise and tenon joints with the router see page 177.

Turning the router into a shaper

With some models you can buy a matching table which holds the router upside down to make it into a lightweight spindle molder or shaper. It is a good idea to check it carefully to make sure it will stand up to continuous use. It should resist racking, that is any tendency to sway sideways when being forced. And the fittings should generally be robust enough

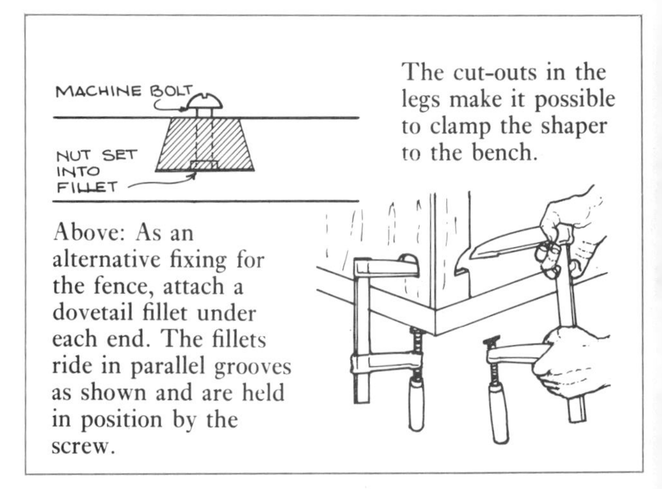

The cut-outs in the legs make it possible to clamp the shaper to the bench.

Above: As an alternative fixing for the fence, attach a dovetail fillet under each end. The fillets ride in parallel grooves as shown and are held in position by the screw.

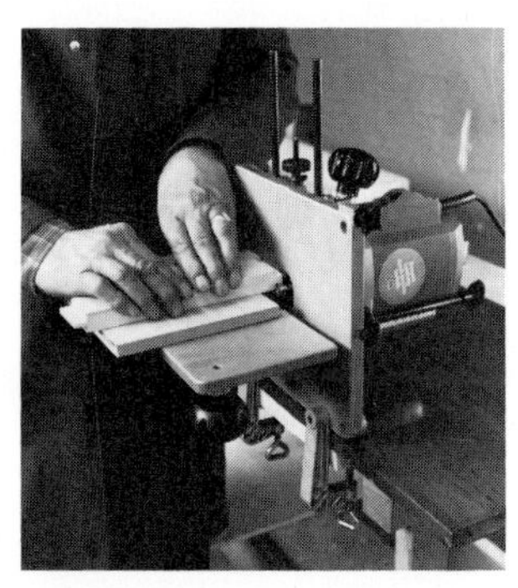

This table can conveniently be mounted vertically or horizontally depending on the application.

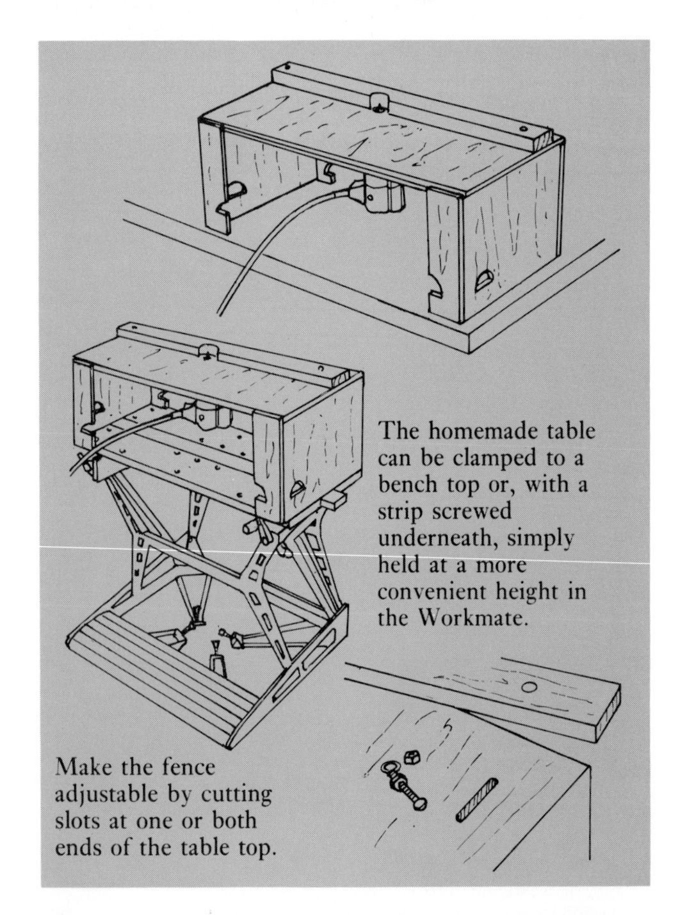

The homemade table can be clamped to a bench top or, with a strip screwed underneath, simply held at a more convenient height in the Workmate.

Make the fence adjustable by cutting slots at one or both ends of the table top.

to withstand the vibrations from the high speed of the router.

I found it better to make my own for a fraction of the cost from $\frac{3}{4}$in plywood.

Screwing the base to the underside of the plywood top does have the disadvantage of minimizing the amount of cutter that can be exposed above, but if this is a problem it is easy to cut a recess about $\frac{3}{8}$in deep into the underside to take the base.

The fence has to be adjustable to vary the width of cut. Make the fence out of a good quality, stable hardwood such as mahogany about $1\frac{1}{2}$ x 3in and cut out a slot behind the cutter, to allow the chips to clear. Otherwise they accumulate between the work and the fence to produce bad results.

There are a number of ways to make the fence adjustable. The simplest is to bolt it down at both ends, with one bolt fixed as a pivot and the other free to rotate in a slotted hole cut in the top. However, if you want to use a miter fence which slides in a groove, it is important that the two fences are perpendicular to one another. In this case, slot the holes at both ends of the fence and set it each time against a T-square before tightening. An alternative fixing which allows adjustment of the fence, yet keeps it fixed in direction, is to screw dovetail fillets to the underside. These run in dovetail grooves one on either side cut into the table top. To fix in place, attach a small machine screw, nut underneath, through the fillet and tighten with a screwdriver.

Notice the cut-outs in the legs which make it easy to clamp the table to the top of the workbench. As an alternative, attach the table to a Workmate which is lower than a workbench and leaves the spindle table at a comfortable working height.

To make stopped cuts, clamp stops at either end of the fence to hold the work at the start and end of the cut. Fit a longer fence, if necessary.

Using the table router For small runs of work such as molding edges, cutting rabbets or grooves, that is cutting along the board, fit a straight or shaped cutter as required and adjust the fence and the height of the cutter to give the desired cut. Always check with a trial run or two. When molding edges, I find it best to adjust the fence so that the pilot guide on the cutter just misses the workpiece to avoid burn marks.

As a safety device, use a push stick at the end of each piece. To hold thin boards firmly against the fence, use a feather board or any other tensioned device such as a diagonal board clamped to the top.

Grooves, bevels and shapes, can be stopped before either end of the workpiece by using guide marks on the fence or more safely by using stop blocks fitted to a long fence. To begin the stopped cut, the workpiece is carefully pushed onto the cutter at the right point and then stopped and gently pulled away at the other end. A common application is the stopped bevel, applied to the inside of door frames in traditional cabinetmaking.

Making cuts with the workpiece held short edge against the fence requires the use of a backing block or a miter fence. A backing block is in my view better, because provided the block is square, the cut will be square. With a miter fence, the two fences must be perpendicular and fine adjustments are often time-consuming. The backing block also backs up the cut, preventing breakout. Notice the screw-on handle which is easily relocated onto another block.

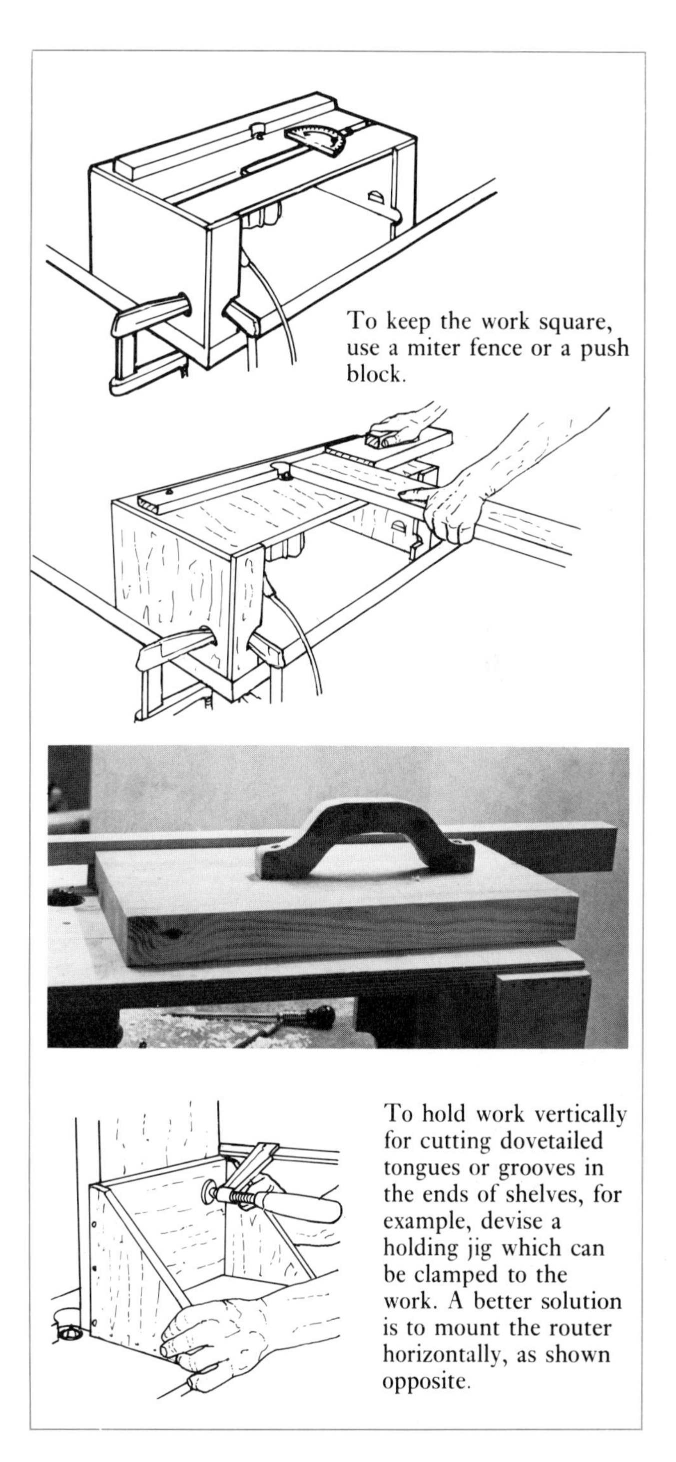

To keep the work square, use a miter fence or a push block.

To hold work vertically for cutting dovetailed tongues or grooves in the ends of shelves, for example, devise a holding jig which can be clamped to the work. A better solution is to mount the router horizontally, as shown opposite.

In this way it is easy to cut tenons for tongue and groove joints, end tongues including dovetailed tongues for housing joints, and to shape the edges of small pieces such as drawer fronts.

To cut grooves or tongues on the ends of long boards to fit into housing joints, for example, it is important to clamp the work securely upright to a holder. Alternatively, mount the router horizontally.

Woodworking machines

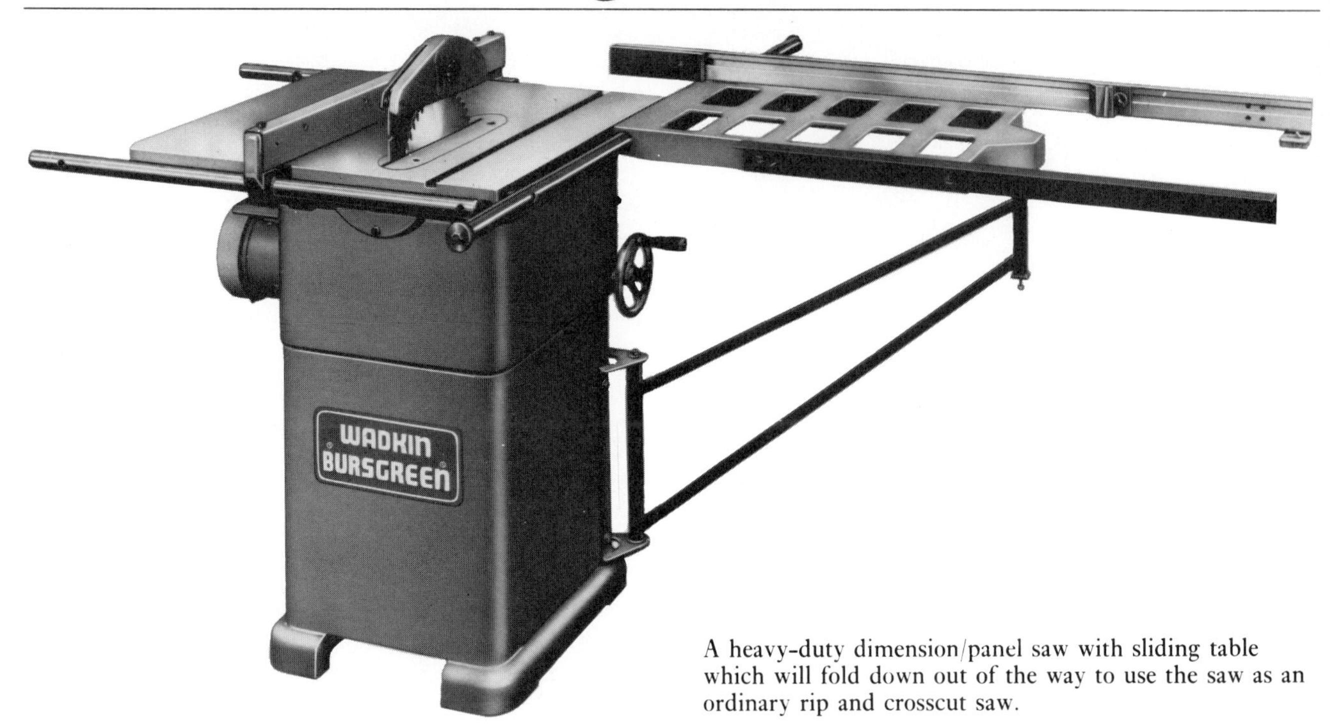

A heavy-duty dimension/panel saw with sliding table which will fold down out of the way to use the saw as an ordinary rip and crosscut saw.

TABLE SAW

The table saw, also known by other names such as the bench saw, saw bench or circular saw is the workhorse of any shop. It is, I suppose, the tool most woodworkers would choose to be stranded with on a desert island, for it not only rips and crosscuts but also performs a number of other functions. Like so many tools, the limitations of its use is in the ingenuity of the user, and necessity motivates craftsmen to invent new jigs and holders to solve specific problems for which no other tool is available. Perhaps some enterprising publisher will someday produce a book of all these jigs, shortcuts and shopkinks to be added to for each new edition by suggestions from woodworkers.

The basic requirements of a table saw is that it be robust and powerful – I feel that the more powerful a saw is, the safer it is. There is less chance of the saw biting into the work and kicking it back. A 10 to 12in, 2hp saw is about right for general work. The rule of thumb is 1hp for each inch of depth normally required to be cut. Estimating depth of cut as one third the diameter, the 12in diameter blade will make 4in cuts. Most large saws will handle 4 × 8ft sheets, with extra tables or roller supports placed around the saw. But for workshops where the predominant work is panel cutting as, for example, in making kitchen cabinets, the best choice is a special panel cutting saw. These have sliding tables capable of taking large sheets. Some panel saws have an extra small blade in front of the cutting blade to score veneered panels and prevent chipping.

Ripsaws capable of re-sawing deep stock are more powerful and diameters vary from about 18 to 24in. These can be fitted with special powerfeed devices which hold down the work and feed it through at pre-set speeds. Taking this one step further, the multiple or gang saws are powerfed saws with several blades, capable of ripping a board into several widths. They are frequently used to square edge waney edge boards since they don't require a straightedge or fence.

Dimension saws are designed for accurate ripping and crosscutting. They frequently incorporate a sliding table on ball bearing runners which makes continuous crosscutting operations much easier. Although it is basically just a table saw, the dimension saw does incorporate several features to make it capable of accurate joint cutting, as shown on page 88.

Above: 24in rip saw. Below: Power fed gang saw.

However, the general purpose 10 to 12in table saw is more than adequate for most shops, particularly if a band saw is available for ripping deep stock. It should be very sturdy with an accurately machined, heavy cast table which will remain absolutely flat.

Some saws are surprisingly badly designed, and it is a good idea to ask around among other woodworkers for recommendations of models which have proved to be safe and easy to use. There are all sorts of little details which can be maddening in use, such as hard to reach adjustment knobs, fences which don't stay adjusted or table inserts with slots which are too wide or which interfere with the tilted blade.

It is absolutely useless to buy a cheap machine (unless it is a good secondhand one – see page 41), for it will not be as safe and it will probably ruin a lot of good wood.

Remember when positioning the saw in the workshop to allow enough room so that the saw can handle 4 × 8ft sheets comfortably and also allow enough run in front and back for ripping long pieces. In small workshops, it is often possible to place the saw diagonally, allowing ripping from corner to corner, or in line with a doorway which can be opened for the occasional extra long board.

Maintenance All machines need regular cleaning, lubricating and inspection. Mount a maintenance schedule on the wall with dates to cross off for maintenance checks. In many shops, the saw is cleaned every day using a small portable blower (see below), or an attachment from the central dust extraction system. But a small brush used at least once a week is adequate. Many parts such as sliding tables with ball bearing guides, are particularly sensitive and need more frequent attention. The rise and fall and the tilt mechanism should always be kept clean to function properly. Parts such as wheels on sliding tables, which are exposed to dust, should be lubricated with grease free graphite lubricants.

Modern motors are usually sealed for life and need no lubrication. But other moving parts such as pulleys may need occasional lubrication according to manufacturer's instructions. On belt-driven machines, the belts should be checked for wear and tension as part of the maintenance schedule.

▶ Check the blade and fences regularly for alignment and square. The fences on some models tend to go out of square and need constant re-setting against a metal engineer's square.

For setting the rip fence in line with the blade, remember the rule of thumb of giving a lead to the fence. That is, setting it out of parallel by $\frac{1}{32}$in in 5ft, with the top or back end set closer to the blade. Also set the back or top end in line with the back of the saw blade to allow ripped work to lie free after the cut.

▶ Finally, keep the top clean and never put bottles or cups on it since they will leave permanent rings. Brush it off regularly and rub candlewax on it for easier working. A good car wax applied in a thin coat and well-buffed should be applied once in a while to make the wood and the fences slide more smoothly. It makes an amazing difference.

Cleaning out the saw with a powerful blower (see page 34 for further details).

Safety

More accidents occur on the table saw than any other machine (about 40%), probably because it is the most frequently used. Accidents occur either because the equipment is unsafe or because the technique is unsafe. Professional sawyers who run a saw all day long, maintain a quiet and fearful respect for the saw and rarely vary from their habits. There are too many stories of losing fingers to serve as reminders of what happens when you take chances or lose your concentration. The following safety rules, taken from specialist safety journals, are very important and should always be adhered to.

Equipment

SAW BLADE Keep the saw blade sharp and in good condition. A dull saw or one which has cracks is unsafe.

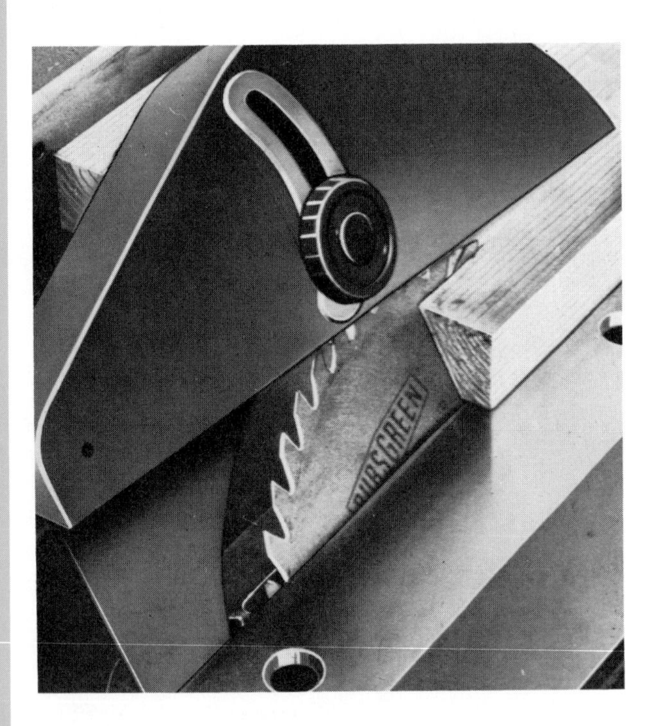

GUARD The adjustable guard should be of sturdy construction and should cover the top and both sides of the blade. Exact methods of attachment vary, but it should be positioned for each cut so that it is no more than about ½in above the workpiece. Some guards have an adjustable front piece for quick setting. Most sawyers object to this since it obscures their vision, hence the see-through plastic guards. Most good saws also have a guard under the table to safeguard against accidents when clearing sawdust.

RIVING KNIFE When ripped, wood often releases in-built stresses by springing as it is cut. Riving knives or splitters prevent the wood from closing on the back of the saw which would cause the wood to be thrown upward and forward with great force. The riving knife must be rigidly attached in such a way that it is raised and lowered with the blade. It should be in line with the blade, and of slightly more (10%) thickness than the saw plate with a rounded front edge. It should be replaced for a new saw blade with a different kerf and should not be more than ½in in back of the blade at table level. The highest point of the riving knife should not be less than 1in below the top of the blade. Generally, a knife with a straight up and down edge is not suitable, for it should follow the contours of the blade. But this rule is rarely followed in the less expensive machines.

TABLE INSERT The insert or mouthpiece fitted into the table around the blade should have a narrow slot so that small, narrow scraps will not get caught between it and the blade. On larger saws, replaceable wooden fillets are fitted tight to the blade.

Techniques

- Don't operate the saw without the guard in position. On the few occasions when it must be removed, take extra precautions. Clamp down the workpiece and concentrate.
- Keep your fingers away from the saw *at all times.* Use the push sticks to rip narrow pieces and to clear away scraps left near the blade.
- Support the workpiece. Place supports behind the saw or beside the saw when ripping or crosscutting long pieces.
- A helper behind the saw must never pull the piece through the saw.
- Keep the area around the saw clutter free. It's easy to trip on scraps, especially dowels. And boards leaning against the wall can fall onto the saw.

▶ Don't stand behind the blade in case the workpiece is thrown back.

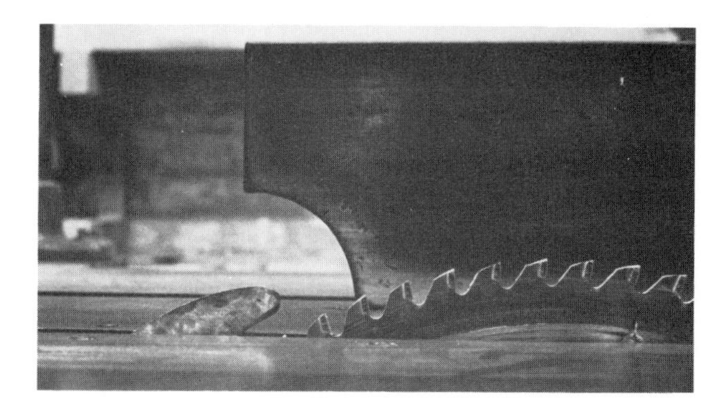

▶ Align the back end of the rip fence so that the workpiece clears it after the rip cut. Scraps which get caught between the blade and fence can be thrown back with great force.

▶ Larger saws have brakes to stop the blade. If no brake is fitted, wait until the blade stops to make adjustments.

▶ Don't wear loose clothing which can get caught in machines. Long hair should be tied back.

▶ Wear safety glasses and if no extraction system is fitted, a proper dust mask for fine-grained woods such as mahogany.

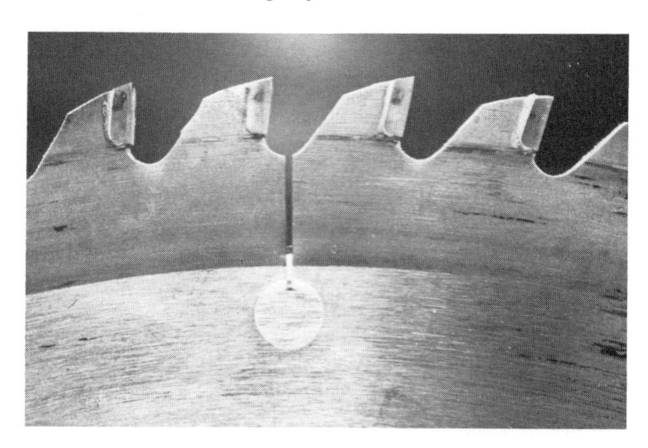

▶ Large tungsten tipped sawblades without set should have anti-whistling notches to high-pitched sound. If not, wear ear protection.

▶ Most of all, *concentrate.* If you are overtired or drowsy, it is better not to use the saw. And if someone interrupts, don't look up until you have finished the cut and shut off the machine.

Adjusting the saw bench

Good quality saw benches have accurately and well-stiffened table tops with grooves which are simultaneously milled to be parallel to each other. But with use, components will creep out of adjustment, so it's a good idea to check them constantly and re-align when necessary. I find it best to make all adjustments against the blade rather than against grooves or table edges.

THE SAW BLADE

Check the blade with a metal engineer's square against the body, and not the teeth, of the blade. Combination squares are notoriously inaccurate and should not be used. If the blade is out of square, clean out the tilt mechanism and check the 90° setting again. If it is still out, reset the 90° mark following the manufacturer's instructions. If the blade goes repeatedly out of square, the saw arbor may be out of line and possibly loose and should be checked out and fixed immediately.

MITER FENCE

First use a good plastic draftsman's triangle held against the fence and the blade to make sure that 90° on the miter scale is exactly that. If not, reset the gauge. Then with the miter fence in position and set at 90°, hold a piece of small diameter dowel against a saw tooth clearly marked with a crayon or chalk.

Slide the miter fence along, making sure to keep the dowel absolutely firm so that it doesn't shift position (clamp if necessary), then check it against the same marked tooth at the back of the saw. The table top may have to be adjusted according to manufacturer's instructions.

RIP FENCE

With the table top properly aligned with the blade (see above), the rip fence can be set against the grooves in the top. But I prefer to set it against the blade itself, in the same way as for the miter fence, using an engineer's square or small block of wood against the same tooth front and back. Follow the maker's instructions to undo the fixings. Afterwards check the scale which indicates ripping width, by actually ripping a piece and measuring it to check against the reading. Then check that the fence is square with the table surface, which is essential for square cuts, especially when re-sawing.

Sawblades The shape and precise configuration of teeth and gullets determine the use of a blade.

Fine points of saw blade design are of interest primarily to production shops and to students for examination questions, but a basic understanding of the principles is important for every woodworker.

The most important factor of saw blade design is the shape of the teeth and gullets and these and other details and nomenclature are shown in the diagram. Other factors to take into account are the peripheral or cutting speed, the set of the teeth, the shape of the blade plate and the tensioning of the blade.

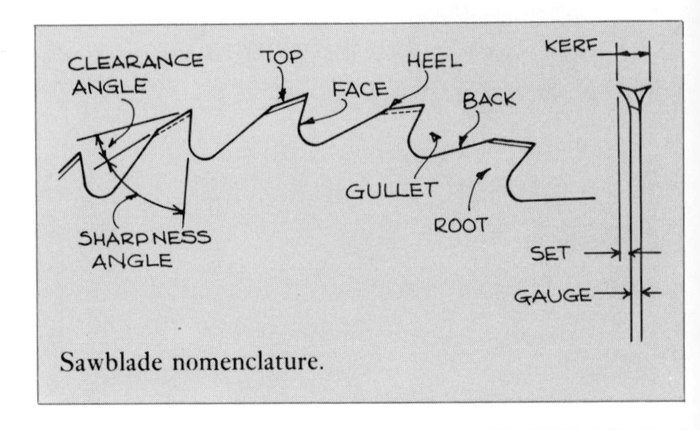

Sawblade nomenclature.

Saw teeth and cutting action Saw blades are designed to rip or crosscut. The shape and precise configuration of saw teeth are different to make them suitable for one of these operations. Ripsaw teeth have chisel-like edges to cut the wood cleanly with the grain, and deep gullets to clear the large particles easily. The teeth lean forward with a positive hook.

Crosscut teeth, on the other hand, have needle point teeth, also alternatively set, which lean away from the wood. But the gullets are smaller since the fine dust from crosscutting clears easily.

Combination and dimension blades are a compromise between these with several crosscut teeth for each rip tooth.

Tungsten carbide tips, which are brazed on give a kerf of about $\frac{1}{8}$in. The tipped blades are more expensive, but last longer and cut smoother than other types.

RIP
CROSSCUT
COMBINATION
FINE CROSSCUT
TIPPED

Rip saw has large, alternatively set teeth. Crosscut saw has fine teeth with small gullets, again set alternatively. Combination or dimension blade has both types of teeth. Fine toothed blade is used for cutting veneered boards. Carbide tipped blade has more or less teeth depending on use.

Blade types In summary, the basic blades are the ripsaw blade with large, alternatively set teeth and deep gullets, the crosscut blade with fine alternatively set teeth and small gullets, the combination blade with alternatively set teeth of both types and tungsten carbide tipped blades with brazed on tips for use as a combination rip and crosscut blade. Tipped blades are excellent for plywood and chipboard with high glue or resin content which tends to blunt ordinary blades more quickly.

The hollow ground planer saw blade produces a very fine finish and is therefore used primarily for dimension sawing where a smooth, accurate cut is required.

Other blades include plywood blades, non-ferrous metal blades and aluminum oxide or silicon carbide wheels for cutting plastics and other materials. In addition, the table saw can be fitted with molding cutters and with dado blades for cutting grooves, rabbets and housings in one stroke.

Special purpose blades used on the table saw: dado blades (above), molding cutters (below).

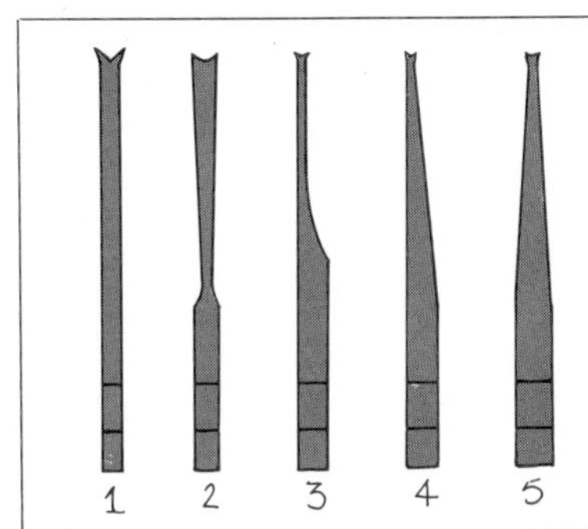

Saw plates:
1. Standard plate saw
2. Hollow ground saw
3. Ground off saw
4. Swage saw
5. Taper saw

Saw plates The shape and thickness (gauge) of saw blades also vary. The standard design is the plate saw (1) where the two faces are parallel. This is used for most rip, crosscut and for tipped saw blades. The hollow ground blade (2) has a neck-like profile, thickening out at the center and the perimeter. Other blades are designed to reduce kerf waste, hence they are ground thin toward the outside rim. These are the ground-off saw (3), the swage saw (4), and the taper saw (5) all of which have specialist applications.

Blades of tungsten tipped blades are generally thicker than the rest which is a disadvantage when ripping quantities of stock because of the large kerf waste involved.

Cutting speed In ideal conditions, a blade is designed for peripheral speed and to take into account the type of material it is to cut. The ideal peripheral speed for wood is about 8,000ft/min for ordinary plate blades and 10,000ft/min for carbide tipped blades. The correct speed is especially important in operations where a good finish is required. Anyone experiencing difficulty getting good results on the saw should, after going through the normal checking procedure for sharpness, etc., calculate the peripheral speed from the formula

$$P = \pi \times D \times R$$

Where $\pi = 3.14$, D = diameter of saw blade in feet and R = speed in rpm of saw blade. R is equal to the motor speed on direct drive saws, but has to be worked out as shown on page 107 for saws with belts and pulleys. If the speed is out by more than say 10%, another size blade giving roughly the right speed should be fitted.

Care of saw blades Although the technical points of saw blade design has little relevance to the average woodworker, the care of saw blades does. The teeth must be sharpened, set and ground to an exact circle and the blade must be "tensioned" to function properly. Many shops do their own saw doctoring using a flat second cut mill file with round edges to file the teeth, a "gate" saw set to set them, a setting gauge to check the set which will vary from 0.3 to 0.4mm depending on the type of wood, and a grinding wheel to "stone" down the saw to an accurate circle. But in the absence of specialist knowledge and experience, it is better to send the saw blade to a specialist saw doctor who has the right equipment and knowledge to do the job.

Tensioning the blades involves making the blade perfectly flat and pre-stressing it to neutralize the tensile stresses built up from heat and centrifugal forces during use. It is still a somewhat mysterious art done by hand with a dog hammer and crowned anvil. Checking proper tension is done with one of a set of special straightedges held on the blade and also by sound and feel, intuitive to an experienced hand. Proper tensioning is important to give the blade stiffness and prevent rattle or vibration.

Tipped blades are usually sent out to be sharpened by specialist equipment. Broken tips are also replaced and usually costed per tooth. Tungsten blades have more or fewer teeth, depending on use, with fewer for ripping. Thus an 18in blade will have as few as 20 teeth for rough ripping and as many as 100 for fine crosscutting with about 60 to 70 teeth as a good compromise for general work.

Many shops do their own saw sharpening and setting but even with specialist knowledge, hand sharpening is not as accurate as machine sharpening. The automatic saw sharpener shown guarantees that the blade stays circular.

Basic operations

The flow of work in all shops revolves around the table saw. Wood is rough cut to length, then planed or edged and faced, then moved back to the saw for ripping to size before being thicknessed and crosscut to exact or slightly overlength. After any joints are cut (many on the table saw), the pieces are cut to final length before sanding and assembly. This pattern obviously varies according to the equipment and the type of work, but the table saw is usually the center and should be positioned accordingly in any shop.

Crosscutting Crosscutting can be to rough size or to exact dimension called dimension sawing as detailed on page 88. Where a radial arm crosscut saw is not available, a portable circular saw can be used. To cut long lengths on the table saw, it is important to use a roller support (see page 87), a temporary table, or a sawhorse to support the other end. Fix a wood batten onto the miter gauge as additional support and get a helper to move the other end for heavy pieces.

There are many devices for supporting long ends. This one can be made in a few minutes.

Cutting ordinary stock square to exact length is the most important function of the table saw. Always check the miter fence by making a trial cut and checking with an accurate square. A wood fence screwed to the gauge will better support the workpiece and back up the cut, to prevent the rough, splintered edge at the saw blade's exit. If the batten is worn at the end, use a temporary back up piece, which can be shifted along with use.

Small, difficult to hold pieces should be held very carefully with the hand gripping the miter gauge for support. It is better to clamp the work to the fence or, if in doubt, discard end pieces and cut from a longer board.

It is usual practice to square off one end, then measure each piece with a tape measure and cut to the mark. But where exact accuracy is not so important, it is convenient to glue or set in an old ruler or part of a steel tape measure to the wood fence as a measuring guide. Periodically test the rule by making a few cuts and adjust it as necessary using slotted holes in the ruler for adjustment. Dimension saws have in-built measures which can be used with accuracy.

Above: Cutting piece too small to be handheld.
Below: Tape measure attached to miter fence.

Block clamped to the rip fence as a stop.

Block clamped to table as a stop.

L-shaped block for cutting halvings.

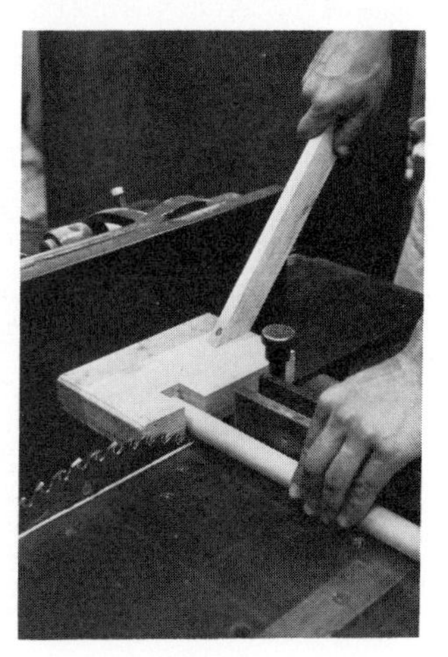

Above: Block clamped to fence as stop.
Right: Handled push block for holding small pieces.

Repetitive cuts It is very important to be able to cut many pieces to the same size, accurately and quickly. A typical cutting list may contain two rails Xin long, four legs Yin long and two panels Zin long and the accuracy and ease of assembly of every job depends on these pieces being accurate and square.

The rip fence can be used as a stop to cut several short pieces to exact length. To avoid having the scraps catch between the blade and fence, a small block is clamped to the rip fence well in front of the line of the saw blade.

Set the distance by measuring from the blade to the block, then test the accuracy by making a trial cut and make final adjustments as necessary.

A variation on this method involves clamping a wood block to the table itself as a stop. Again the block has to be placed well in front of the blade so that the alignment against the block is done before the piece is pushed through the saw, held tight against the miter fence. These blocks can be located either on the right (rip fence) side or on the left side of the blade, or both at the same time. If, for example, the cutting list calls for two different lengths to be cut, one block can be set for each length, so that both can be cut without setting up again.

Another use for two blocks is in cutting halving or cross lap joints. After the pieces have been cut to exact length, the two blocks are clamped, one on either side to allow the end cuts of the halving joint to be made. It is then quite simple to remove the waste in between with several runs of the saw. If many halvings of the same size are to be cut as is common with, say 1in thick or 2in thick battens, it is best to make an L-shaped block which when clamped to the top is used to make both end cuts, the left hand cut using one part of the block and the right hand cut using the other part of the block.

The most frequently used device for repetitive cuts is a small stop block clamped or bolted to the batten which is attached to the miter fence. Since the wood fence can be extended for two or three feet if necessary, this method allows moderate lengths to be cut against the stop. For even longer lengths, it is possible to clamp a block to a temporary table set beside the saw, but I've never found this too successful and prefer to measure each piece separately and cut to the pencil mark.

Cutting off small pieces such as dowel ends for knobs or for dowel joints is always a problem because the scraps tend to be thrown back or get caught in the slot between the blade and the table inset. One solution is to use a notched push stick with a handle screwed on to push them clear.

Cutting miters Cutting perfect miters on the table saw is not as easy as it sounds. Picture framers use a special foot operated guillotine which makes a clean, perfect cut each time, but furniture shops either rely on a fine crosscut saw held in a frame or the table saw for cutting miters.

The problem with the table saw is first that the miter fence has to be accurate. For important cuts, it's best to check the miter setting against the blade using a plastic draftsman's triangle. To check accuracy, make trial cuts in two pieces and hold them together tightly against a try square to see if they form a 90° angle. The second problem is that the workpiece tends to creep as the cut is being made. To prevent creep, nail pins through the back of the wood fence so that the tips just protrude to hold the work.

For frequent miter cutting, it's best to make up a jig out of $\frac{3}{4}$in plywood with two stop battens screwed to the top and two wood or metal battens screwed to the underside to ride in the slots in the table top. To allow the stops to be adjusted, attach them with washers under the round head screws fitted into oversized holes. Refer to page 188 for further information on cutting and jointing miters.

Cutting large panels Crosscutting pieces of plywood is a problem since the width of board allowed between the front of the blade and the miter fence is limited. One solution is to turn the miter fence around so that the workpiece is pushed against the fence. This allows you to cut boards up to say, 30in wide, but for boards wider than that it becomes unsafe because it is difficult to control. For boards too wide to use with the rip fence, it is best to either cut them with a portable circular saw or alternatively to use the edge of the table as a guide. Clamp a straightedge to the board in the correct location, then run the straightedge against the table edge to make the cut. The table edge can be used in the same way for getting a square edge on rough edged boards.

For ripping boards with uneven edges, pin a piece of $\frac{1}{4}$in plywood to the workpiece with the straightedge extending beyond the board to run against the rip fence. Make sure the nails holding the plywood to the work aren't in line with the saw blade.

In workshops where the predominant work is with large sheets, it may be worthwhile to install an extension table beside the saw at exactly the same height as the saw table to make it very easy and safe to cut 4 × 8ft sheets

Jig for cutting miters.

To cut moderately large panels too wide to fit between the blade and the fence, turn the fence around and push the sheet against it.

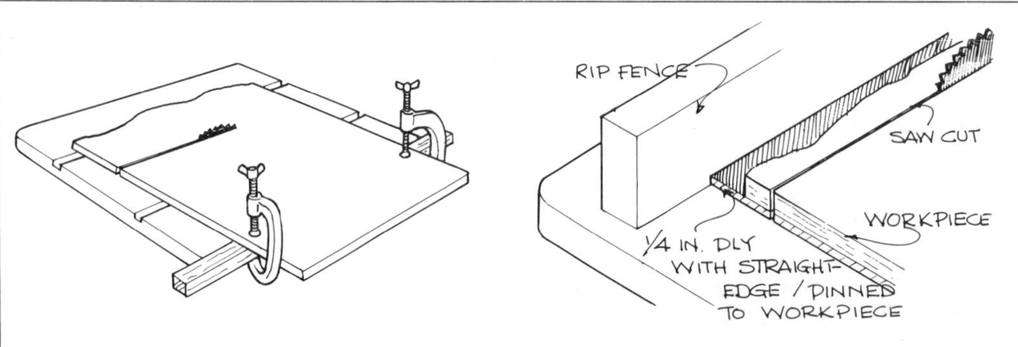

Two ways of trimming sheets: Left: This method is especially useful for cuts too wide to use the rip fence. Right: Pin a piece of plywood near the uneven edge to trim it straight against the fence.

Ripping For large workshops, special ripsaws are used to cut large sections into smaller widths before these are fed through the thicknesser for final processing. In most workshops, one saw does both the crosscutting and the ripping operations and general purpose saws are sold with both a rip fence and a crosscut or miter fence.

Ripping is generally more difficult and more dangerous than crosscutting because the workpiece is more difficult to hold and the operator's hands are brought closer to the blade.

It is important that the rip fence is set up properly as shown on page 79 so that the blade doesn't burn the cut. It is also important to make sure all the safety devices such as the riving knife (splitter), and the guard are in position during ripping. Ideally, the height of the guard should be set so that the gap between it and the workpiece is only about $\frac{1}{2}$in. Also set the height of the blade so that it extends about $\frac{1}{4}$in above the board to be cut.

Before ripping, set the fence to the width wanted, making sure to check the setting by cutting a scrap piece. Some saws have a built-in scale to make the setting easier, but the scale should be checked periodically for absolute accuracy.

When ripping, push the board with the right hand and guide it firmly against the fence with the other. Don't stand behind the saw blade in case the board is thrown back. In ripping, it is important to control the workpiece firmly at all times. Keep it moving and keep it firmly against the fence. It is only when you loosen the grip that the boards tend to bind between the saw and fence and get thrown back.

The rip fence should ideally end near the back of the blade so that the board is free of the fence when the cut is finished. On saws where the fence is longer and extends to the back of the table, it is useful to clamp or bolt an auxiliary wood fence to it with the front end in line with the back of the saw blade. This auxiliary fence is useful for many operations so it may be a good idea to tap holes in the metal fence to make it easy to attach and detach with machine screws, countersunk below the surface.

The basic thing to remember in ripping is to keep the hands away from the blade. When ripping narrow pieces, for example, it is dangerous to use the hands to push the workpiece through. Instead, either one or two push sticks are used to guide the narrow piece. Make two or three good push sticks out of a hardwood such as maple or beech and hang them near the saw within easy reach. I always place it on the table so that when I get to the end of the cut it is within easy reach for finishing the cut. This is better than groping for it while trying to hold the board still in the saw.

For increased safety during ripping, always remember to use push sticks to keep fingers from getting too close to the blade. Push the work clear at the end of the cut. Make sure the back of the fence lines up approximately with the back of the blade. If not, fit an auxiliary fence, as shown above. The guards have been removed in most of these photographs for clarity.

Re-sawing Re-sawing boards into thinner boards is a special operation most often done on the band saw. To re-saw on the table saw, the guard may have to be removed so it is important to set up guides or "hold downs" so that the board is held firmly in place during cutting. There are several ways of arranging hold down devices and more sophisticated saws provide special fittings and accessories for this. The best alternative to this is to make feather boards out of hardwood or good quality $\frac{3}{4}$in plywood.

Use the band saw to make a series of curved cuts about $\frac{1}{4}$in apart to create "feathers" or springs. The feather board is used one or two at a time not only on the saw but also on the jointer and spindle molder. For re-sawing, clamp a large piece of plywood to the rip fence. Then clamp the feather board to it. Push it down slightly onto the workpiece as you clamp it so that it will provide slight pressure as the work goes through. It's usually necessary to adjust it several times to get the pressure just right. To provide side pressure, it is possible to use another feather board clamped to the table top in the same way. Always make sure that the feather boards are clamped down really firmly to avoid the dangerous possibility of having one come loose during use.

If the board is deeper than the maximum depth of cut, make the cut in two runs. Set the saw blade to just over half depth and, after making one cut, turn the piece over end to end so that the same face is against the fence and complete the cut.

The feather boards can be used in other situations too. When cutting rabbets in thin boards, for example, it is important to hold the board down firmly so that the depth of cut stays constant. They are also used when cutting grooves for boards for tongue and groove joints as shown on page 131.

Above: Re-sawing using feather board to hold the work down and push sticks to feed it through.
Below: Ripping at an angle with auxiliary fence clamped to rip fence.

Ripping with the blade at an angle for beveling edges, for cutting triangular battens, for cutting raised panels and for many other applications is easily done by tilting the blade to the appropriate angle and ripping in the normal way. When the blade tilts toward the rip fence as is usual, it is best to fit an auxiliary wood fence to the rip fence and, after setting the width, gradually raise the blade so that it just cuts into the wood fence. Raised panels can be cut this way or as shown on page 131, but again, great care must be taken to support the panel well since the saw guard must be removed.

Making a push stick

Good push sticks are essential. Make them out of any tough hardwood, one with about 3in of wood in front of the notch so that the work is held down as well as pushed forward with the stick.

The other with a simple notch in the end is held in the left hand to guide narrow boards against the fence while pushing them through with the first stick held in the right hand. Make both about $\frac{1}{2}$in thick at the end, tapering up and forming a comfortably shaped handle. I find that it's a good idea to make three or four at a time by clamping or pinning the pieces together before cutting out on the band saw or saber saw.

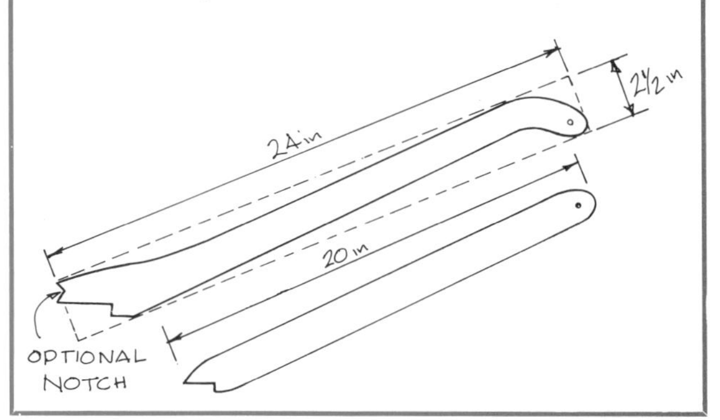

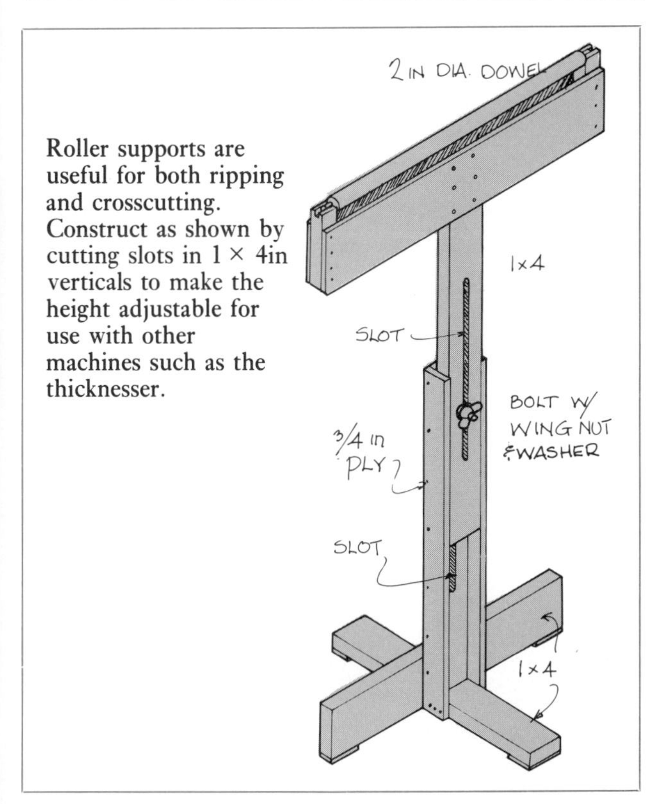

Roller supports are useful for both ripping and crosscutting. Construct as shown by cutting slots in 1 × 4in verticals to make the height adjustable for use with other machines such as the thicknesser.

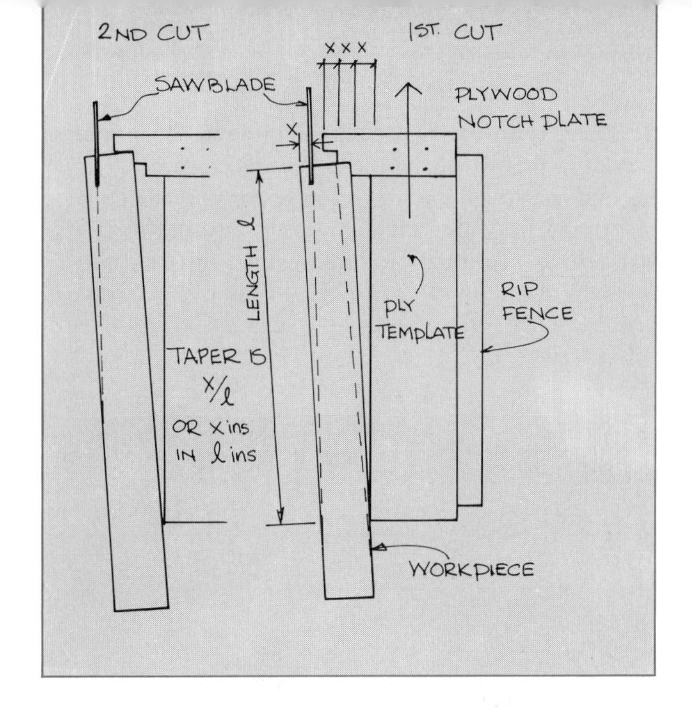

Cutting tapers Taper cutting is occasionally required in furniture making for making tapered legs for example. Although more elegantly cut on the jointer, as shown on page 115, it can be accomplished on the table saw with a simple jig which holds the workpiece at the appropriate angle as it is pushed through the saw against the rip fence. Taper cutting should be carried out very carefully with the guards in place. Work out the taper, keeping in mind that both sides may be tapered. The taper width, X, shown on the diagram is the taper for each side. Make the plywood template about 3ft long and nail or screw to it a removeable piece of plywood which is notched once for a single taper or twice for a double taper.

Cutting wedges Wedges are frequently required both in furniture and in joinery for mortise and tenon joints, stairway work and many other applications. The easiest way to cut wedges is to use a snip-off saw or a radial arm saw but on a table saw use the jig shown.

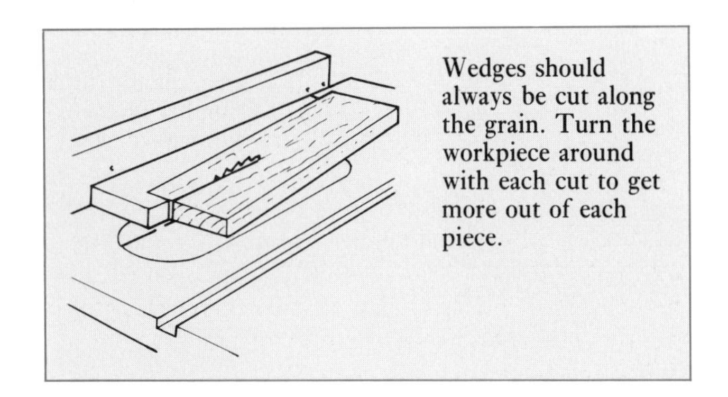

Wedges should always be cut along the grain. Turn the workpiece around with each cut to get more out of each piece.

Cutting tenons Some manufacturers sell special tenoning jigs which hold the workpiece firmly vertical as the cheek cut of the tenon is made. Since the guard must be removed, this is a somewhat dangerous operation and in several countries safety laws do not allow it. A better and safer way to cut tenons using the rip fence as a stop is shown below and on page 176.

Pattern cutting This does not have many applications but it is a very elegant solution which would probably be used more often if it were more well-known. The operation is similar in principle to pattern cutting on the spindle. For cutting repeated shapes, a pattern is made out of say, $\frac{1}{2}$in stock and is pinned to each workpiece which has been roughly pre-cut to size on the band saw. An auxiliary fence is fitted to the rip fence leaving a gap underneath, slightly larger than the thickness of the workpiece to be cut. Set the blade exactly in line with the edge of the fence and raise it so that it just cuts into it. Feed the work through the saw so that the pattern on top runs against the fence and the blade cuts off the workpiece flush with the pattern.

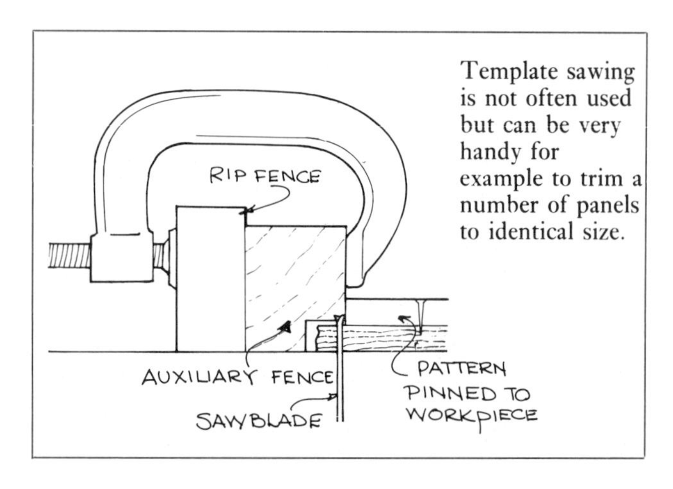

Template sawing is not often used but can be very handy for example to trim a number of panels to identical size.

Dimension sawing

The dimension saw is an accurate multi-purpose saw which can be used not only for precise cutting to length to any angle, but also for convenient joint cutting in addition to its use as an ordinary table saw. The dimension saw is an expensive, heavy-duty machine used by factories for production work and is therefore occasionally available second-hand as factories renew or go out of business.

Blade diameters are usually 15 to 18in. The blades are usually carbide tipped, but can be a combination blade. Motor power varies from 3 to 5hp and is usually direct drive, with the blade mounting directly onto the extended shaft of the motor which tilts with the blade for bevel cuts. The sliding table which rides on ball bearing rollers is what makes cutting easy, for the crosscut or miter fence mounts directly onto it and all crosscuts are made fairly effortlessly by sliding the table forward. The combination of power, accuracy and ease of use makes the dimension saw a pleasure to use.

The operation of crosscutting a number of components to identical length is basic to all projects. The crosscut fence is fitted with sliding stops which can be tilted out of the way. After the stop is accurately set by trial and error or, on modern machines by the scale on the fence, it is flipped out of the way to enable the overlength piece to be cut off at one end. This is then turned round with the cut end against the stop to cut the second end. The ability of the stop to fold out of the way is

Accurate stop fitted to extended miter fence is essential for quick, repetitive cuts. Notice that the stop flips out of the way to trim first end, then drops in place as the piece is turned around to trim the other end.

all important, otherwise all the ends would have to be squared first before fitting the stop.

If long stock is to be cut into a number of smaller pieces, both the crosscut stop and the rip fence stop are put into use. On new machines, a detachable stop is bolted to the rip fence and set up by reading the scale on the table to give the correct length. The procedure then is to square the end and run off all the pieces until the last piece. This is turned around to be cut off against the crosscut stop set to exactly the same distance. This is quick and with the sliding table, almost effortless.

Where two lengths are required, two stops can be fitted to the crosscut fence, the one not in use being flipped up out of the way.

Using the same technique of the rip fence stop and the crosscut stop in combination, it is easy to cut halvings and tenons.

To make the cut-out for halving or cross lap joints, each stop is set for the last cut on the left and right of the cut respectively and the waste in between is quickly cleared with a few runs of the saw. Alternatively, one stop and the rip fence are used.

To cut tenons, the tenon length is set on the rip fence and the shoulder width by the depth of cut. The first cut is made with the workpiece against the fence to make the shoulder cut and the waste is then removed with successive runs over the saw. As a final cleaning up, the piece can be carefully pushed sideways over the top of the blade. Use the same setting to cut all four shoulders.

Below left: Compound fence which gives dimension saw its versatility. It can, for example, be used to make the three-way joint as in the table shown here. Two stops are set up, one for each last cut of the joint. The waste in between is cleared by successive cuts.

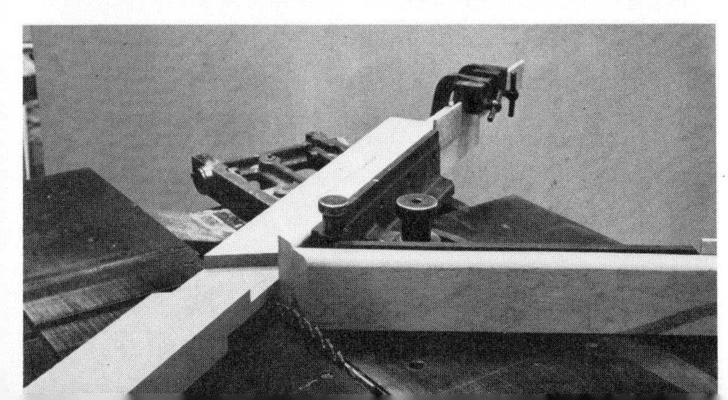

BAND SAW

Many woodworkers prefer the band saw to the table saw and vice versa so it is worthwhile to point out the relative merits. Each has its advantages and there is undoubtedly a need for both in a busy workshop. The main advantages of the band saw are its safe and easy operation. A continuous band of cutting edges is a logical arrangement and safer than the circular saw because the thrust is downward against the table. This arrangement allows several other advantages; first the straight run of band saw blade allows a more gentle and far deeper cut than the table saw and with less power. With an ordinary 1hp, 20in band saw, it is possible to rip a board, say 8 to 10in deep whereas even an 18in diameter table saw is only capable of about a 6in depth of cut.

Another advantage well-known to woodworkers and mill operators, is the narrow kerf of the band saw. Ripping with the table saw produces $\frac{1}{16}$ to $\frac{3}{16}$in of waste per saw cut, which when you are re-sawing valuable wood is a considerable percentage. With a well set up band saw, the kerf is less than $\frac{1}{16}$in, which is one of the reasons logs are milled with band saws rather than circular saws.

The last advantage, which is also important, is the price. Band saws are much cheaper than table saws and considering that, probably give better value for money.

The main advantage of the table saw is the straight and accurate cut it produces which requires very little cleaning up afterwards. With the table saw, it's quite easy to cut joints such as halvings and tenons. The tremendous power and blade size of a large table saw makes ripping large stock quite easy especially in production work where speed and accuracy is essential.

Production ripping wouldn't be possible with the band saw for although it may cause less waste, the cut is rippled and therefore requires much more work in final thicknessing. And unless the blade is perfectly sharp, evenly set and properly guided, the band saw blade tends to wander which reduces the advantage of the narrow kerf.

Undoubtedly the main disadvantage of the table saw is its violent and dangerous mode of operation. More accidents occur on it than any other machine, probably because it is most used but also because there are more chances for accidents. More force is required, often leaving the operator momentarily off-balance, and the workpiece can be violently thrown back with tremendous force. Against this, the band saw is a gentle thing, still dangerous, for a blade can break or a hand can slip but on the whole, much safer and easier to work with.

Band saw features Band saws come in many sizes from the gigantic wide blade models used in mills to cut logs into planks, to the small 10 in wheel models, used for hobby work. The essential features are the same; the two wheels, top and bottom, are rubber or cork tired to hold the blade. The top wheel can be adjusted up and down for setting blade tension and tilted to keep the blade running in the proper location on the wheels.

The blade is kept in position during sawing by several adjustable guides. The ball bearing support attached to the guide post above the table keeps the blade from being thrust backwards during sawing. New machines have an additional ball bearing guide below the table for further support.

Sideways movement of the blade or "drifting" is limited by guide pins set with fine clearance on either side of the blade, both above and below the table. The accurate alignment of these guards plays an important part in the proper operation of the band saw. The old-fashioned wooden block guides still preferred by many have been replaced by modern ball bearing guides.

The table, which on most models can be tilted up to 45° for bevel cuts, is slotted at the front for blade replacement and has a removeable insert or a throat plate around the blade entry.

When buying a band saw it is important to match the machine to its requirements. Most furniture and joinery shops require heavy-duty 30in band saws with the capacity for re-sawing deep boards and cutting the occasional log into boards. For small, one man shops, a 14in model will probably be large enough. However, it should be kept in mind that the larger, more robust and powerful a band saw, the more versatile it is. A 30in band saw will cut small curves as well as a 14in model and the extra throat width and depth of cut makes it much more versatile. Before buying, check the guides and the tensioning devices which should be well-engineered for easy and precise adjustment and solid support. Check the table for solidity and check that the wheels are easily adjusted. A brush fitted to the lower wheel to keep dust from building up is essential. A rip fence and a miter gauge should be available, and on some larger models a useful foot brake should be included. Power is important and for general use the motor should be over 1hp and preferably about 2hp.

A 30in heavy-duty band saw. Notice the large wood block which is used to stop the blade.

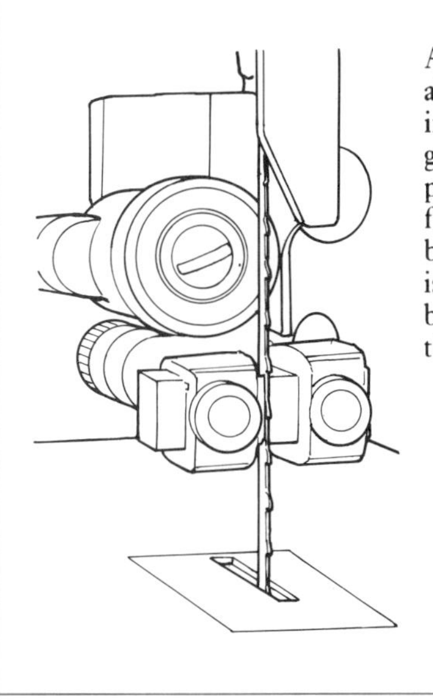

Adjustable guides above the saw table include the two side guides which are set a paper wide gap away from the blade. The back ball bearing guide is set just behind the blade and rotates with the thrust of the blade.

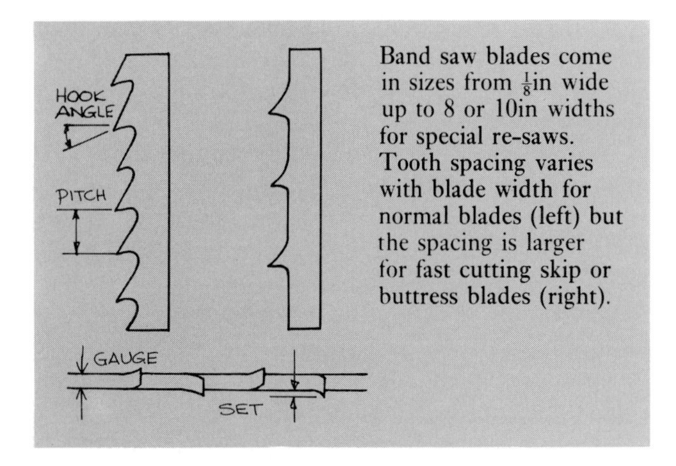

Band saw blades come in sizes from $\frac{1}{8}$in wide up to 8 or 10in widths for special re-saws. Tooth spacing varies with blade width for normal blades (left) but the spacing is larger for fast cutting skip or buttress blades (right).

Band saw blades Band saw blades are made from bands of high strength steel and are available in various widths, thicknesses and in sizes and set of teeth. Widths vary from $\frac{1}{8}$in to about $1\frac{1}{2}$in for general purpose band saws, though the practical sizes for most saws are $\frac{1}{4}$in for fine and small diameter cuts in wood and plywood, $\frac{3}{8}$ or $\frac{1}{2}$in for general purpose use for rough cuts as well as curved work, and $\frac{3}{4}$in for heavy-duty work such as re-sawing. Wider blades up to $1\frac{1}{2}$in are used on heavy-duty machines.

As with hand saws, the spacing and set of teeth determines fineness of cut but the spacing is generally related to blade width, so the narrower the blade, the smaller and hence the more closely spaced the teeth. The thinner, narrower blades such as $\frac{1}{8}$, $\frac{3}{16}$ and $\frac{1}{4}$in widths have up to 12 teeth per inch and the thicker, wider $\frac{3}{4}$in blades have as few as 2 teeth per inch. The narrower the blade, the smaller the radius it is capable of cutting.

The norm for standard pattern tooth blades is between 2 and 6 teeth per inch but skip or buttress tooth blades with wide tooth spacing are popular for fast rough cutting.

Changing blades and making adjustments To operate successfully, the band saw must be carefully set up and maintained and the blades must be sharp and properly set.

To remove the blade, first disconnect the power, then decrease the tension until the blade is free. Remove the table insert, and undo the screw or bolt which covers the table slot at the front. Folding a band saw blade into a small, three loop bundle seems to be a kind of rite of passage in the woodworkers' world, but it is also essential for storing the blades safely. Before hanging it away, I always wipe it with a light machine oil to keep it from rusting but this has to be wiped off again before re-fitting.

While the blade is off, it's a good idea to give the insides a good cleaning and to check that all nuts and bolts are tight. The moving parts, such as the upper wheel tensioning, adjusting screws, tilt mechanism and other points described in the owner's manual require lubrication. Without the guidance of an owner's manual, use SNE 20 oil on these moving parts. On most models the wheel bearings are sealed for life, but on older band saws, it may be necessary to periodically re-grease the bearings or replace the oil in the gear case. Also check the wheel brush for wear and alignment and occasionally inspect the rubber tires on the wheels for scoring.

The tires should be good for life, but if slightly scored, they should be dressed down with coarse sandpaper while the wheel is hand rotated. Badly worn tires are dangerous and should be replaced by a specialist. Although dull or broken blades can be sharpened, set and re-soldered in the workshop they are best sent out to be renewed by a specialist.

Before replacing the blade, loosen the ball bearing support behind the blade, then insert the blade, teeth downward and increase the tension to take up the slack. The important part is to get the tension, wheel alignment and guide adjustments just right. Hand rotate the wheels to check alignment. If the blade stays in a constant position on the wheels, the alignment is right. If it tends to drift, tilt the wheel until the blade stays in position. Some operators prefer to have the blade near the front edge, but it is safer to keep the blade centered.

After tracking the wheel, adjust blade tension. The rule of thumb is that the blade should deflect about $\frac{1}{2}$in for a 12in unsupported length when applying moderate thumb pressure.

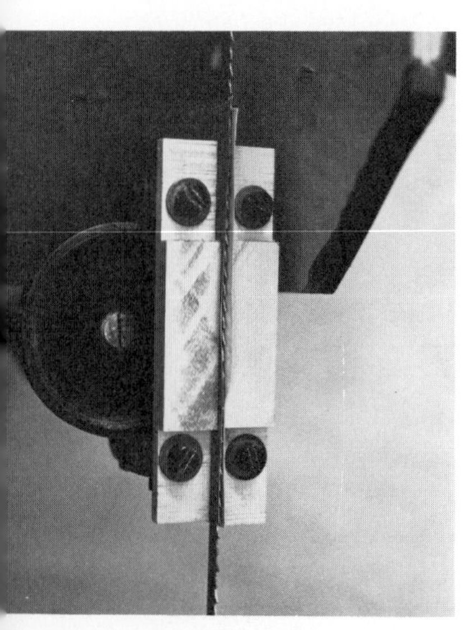

Old band saws usually had wood guides which are still popular with many sawyers today. Any hard, close-grained wood can be used and birch plywood is excellent. The blocks are longer in length than modern guides to give added support to the blade. Notice the pieces of fine sandpaper used to set the blocks.

Setting the guide pins is best done with strips of coarse paper. Loosen the guides to allow the blade to run free. Then bring them together, holding the paper strips between the blade and guides while tightening the fittings. The teeth should be in front of the guide pins so as not to cut into them while running. It is important not to push the blades sideways away from its free path when adjusting the guide pins.

As the final adjustment, make sure the space between the back of the blade and the ball bearing support is about $\frac{1}{64}$in. The wheel should not touch when running free but as the load is applied, it should hold the blade in position by rotating with the blade. Too much contact will tend to case harden the back of the blade, causing the steel to become harder and more brittle, which results in the formation of hairline cracks and eventually in the blade breaking dangerously. An experienced operator will know when the blade is suspect and will throw out a blade which is too worn to be repaired.

Finally, test by running the wheels by hand, to make sure all the components are adjusted correctly. Then replace the guards, table insert and slot cover, and try the saw on a scrap.

Using the band saw

FREEHAND CUTTING Freehand crosscutting is common with the band saw for cutting shapes, rough cutting to size and for odd jobs like cutting dowel lengths, jobs more easily done on the band saw than on the table saw. Generally a miter gauge is used to guarantee a square cut but with practice it is possible to make passable straight cuts which require only a quick cleaning up with a plane or shooting board for joining. Here again, modern factory practice differs considerably from traditional methods. Formerly shapes, tapers, etc. were all done on the band saw and required a considerable degree of skill acquired with years of practice. But today, the same work is usually done on the overhead router with a jig to make the cutting of shapes almost automatic, requiring only a semi-skilled worker to do the work. It is, I suppose, yet another case of technical progress superseding human skill, for no matter how skillful, the band sawyer can't hope to do the work as perfectly as the router; nor with the same high degree of finish. In the small shop, however, band sawing is still the most convenient way of cutting shapes and the skill required in freehand cutting is soon acquired and enjoyed.

The upper guide should always be adjusted before

each cut so that it is about $\frac{1}{4}$in above the workpiece. Generally cutting is done along or beside a marked line so it's a good idea to fit a light to the saw to show up the work clearly. Band sawing is quite safe and you can, when required, bring your hands carefully to within an inch or so of the blade.

Occasionally the blade will bind, in which case, it is necessary to back off the cut slightly and then proceed again a bit more slowly. Feed rates are intuitively adjusted to the work by the sound and feel of the work.

If the blade gets stuck, the work should never be backed out of the cut a long distance without first shutting the machine off. There is no guide preventing the blade from moving forward off the wheels. Instead, shut off the motor, then work the blade loose and by wedging the cut open, bring the blade back out of the cut to begin again.

There are several hints worth mentioning. First, be aware of the limitation of the throat width. Some cuts are not possible because the left machine guard gets in the way. Generally, the throat width is about $\frac{1}{2}$in less than the wheel diameter so that for a 14in saw, for example, crosscuts for boards over 13$\frac{1}{2}$in in length would always have to be made from the right side.

Again it becomes second nature to take this into account after getting used to a machine. It means in some cases marking the cut on the underside to be able to turn the piece around to cut from the right side. This applies too to large sheets which unless you have a 30in band saw are best cut with a hand saber saw.

Freehand work should always be clearly marked and for regular patterns, it is best to make a template out of cardboard, drawn with French curves, compasses, etc. Symmetric patterns such as the rockers on the cradle shown are drawn half size to the center line, then folded over so that both sides are cut simultaneously. The shape can be traced onto the workpiece or more simply the template can be pinned, glued or double sided taped in place.

Similarly, to duplicate shapes, it is better to cut out the first, then attach it over the second so that the two will be identical.

When making long curved or compound cuts, it's best to make a series of crosscuts to the lines first, so that the cut will be broken up into several smaller cuts. Intricate curves need a narrow blade and even then, sometimes several passes are required for tangential cuts. Tight internal curves may be difficult to do smoothly, and it is wise either to drill small turning holes first, or to make a series of straight rough cuts near to the line, then clean up the cut later with hand tools.

More complicated cuts should be cut out to rough shape on the first run, then with easier access, the intricate pattern is more easily followed.

Cutting and gluing is a useful trick often used in making arched window and door frames. A curved piece, for example, would ordinarily be cut from a very wide board. To save material, particularly where many components are to be made, one curve is cut out of a narrow board. The piece just cut is glued to the other side leaving an arch shape which requires a second cut on a parallel curve jig to get the two curves concentric.

Freehand ripping is also quite easy on the band saw. But the guides must be correctly located and the blade sharp and evenly set to prevent the cut from drifting to one side. Drifting is caused either by incorrect guide setting which is easily corrected or by the teeth of the blade being improperly set.

Drifting is temporarily corrected by pointing the cut slightly to one side but this causes undue wear on the blade and guides and should not be continued for too long.

Provided the saw is powerful and large enough, uneven stock or logs can be sawn into boards by first "chalking" a line on the surface, then freehand ripping along this line. Long stock requires two people, one to guide and one to support the end.

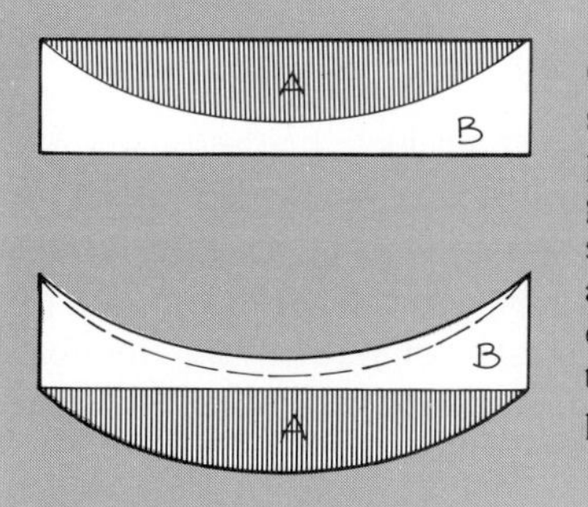

Cutting and arching saves material. The pieces are cut and glued together as shown to create an arch. Make another cut at the dotted line to make the sides parallel.

To cut shapes such as the cradle runners which should be identical, pin or clamp the pieces together.

Chalk line

A chalk line, used in construction work, has many applications in the workshop. It is basically a long string held in a chalk-filled container which when unwound, stretched taught and snapped, leaves a chalk line on the workpiece underneath. No straightedge is required to make a perfectly straight line, for example for marking a line on a log for band sawing into boards or for marking plywood sheets.

One person can operate the chalk line by hooking the end onto the edge at the first mark, then stretching it to the other mark. To mark the line, first stretch it very taught, hold down the end firmly on the mark and with the other hand lift up the taught string and let it go with a snap.

Crosscutting with a miter gauge Most band saws come with a miter gauge which fits in a slot in the table similar to the table saw. With practice, joints such as tenons, halvings and even miters can be cut this way. By using, say a $\frac{1}{4}$in blade with fine teeth (about 6 to 10 tpi), the cuts are fine enough to require little or no cleaning up, depending on the quality of the job.

As with a table saw, the crosscut miter gauge can be used with stops for cutting repeating lengths. Clamp the stop to the table or, as here, to the rip fence fitted on the other side of the table.

On older band saws, which have neither a miter gauge or a rip fence, accurate crosscutting is best done by first clamping on a board as a rip fence then using a square block against it as a backing block for the workpiece. A few strokes of wax on the wooden edges make the fences slide more smoothly to aid in ripping.

Use a pivot block firmly clamped to the table to make it easier to adjust for drift.

Ripping Accurate ripping depends almost entirely on a well set up machine and a sharp, properly set blade. In the small workshop, the band saw is invaluable as a ripping tool for it does it well without undue kerf waste. Make the cut about $\frac{1}{16}$in on the waste side and plane off the wavy or washboarded finish to the line afterwards.

The rip fence can be used for ripping straight boards. Mark the cutting line and feed the work through slowly. Remember that the upper guide should be set about $\frac{1}{4}$in above the workpiece. It is important to keep the work firmly against the fence as you guide it along. If several boards are to be ripped, it is worthwhile to clamp a feather board (see page 86) or any other holding device such as a spring clamp or a pair of battens to the table to hold the work against the fence.

The rip fence doesn't allow room for changing direction when the saw drifts. A better guide for ripping, particularly for re-sawing, is the pivot block, made from a piece of hardwood. It is rounded at the end which allows the direction of the workpiece to be shifted slightly while still maintaining the ripping width.

Right: Crosscutting with miter fence held in groove in auxiliary table. Far right: A push block used to keep work square against wood fence clamped to the table.

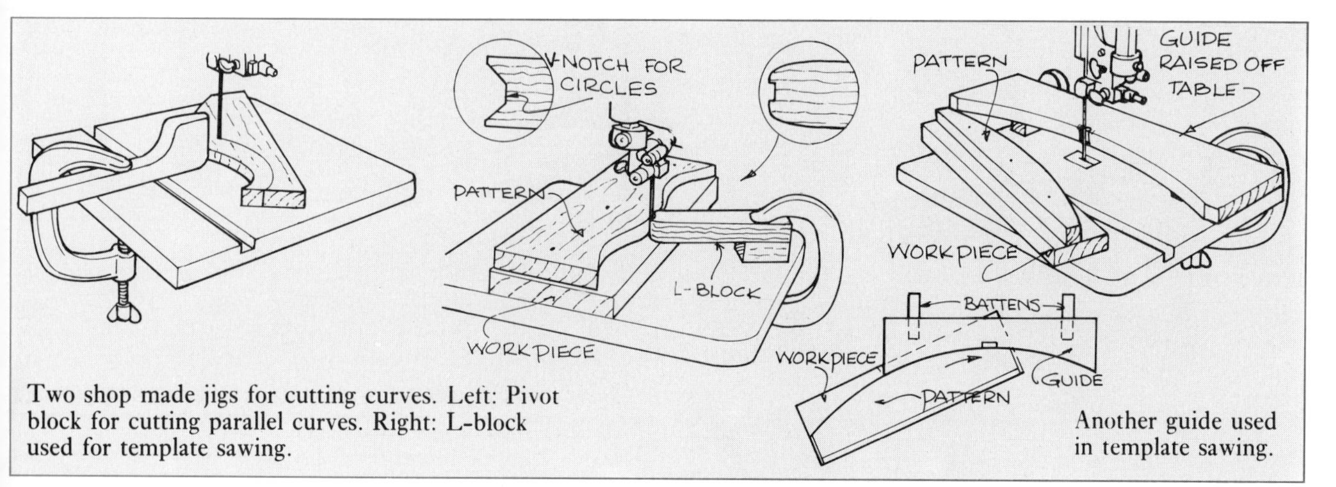

Two shop made jigs for cutting curves. Left: Pivot block for cutting parallel curves. Right: L-block used for template sawing.

Another guide used in template sawing.

Jigs and patterns As with all machines, the usefulness of the band saw can be extended by devising jigs and templates.

Cutting parallel curves is facilitated by use of a pivot block which can be the same one as that used for re-sawing.

Pattern sawing is convenient where many identical pieces are required. It is similar in concept to pattern cutting on routers or table saws. There are several ways of using patterns.

The simplest uses an L-shaped pivot block which runs against the pattern attached to the top of the workpiece.

A small recess is cut into the tip of the block for the blade, so that the pivot can run against the pattern without pinching the blade. The pattern can be nailed, clamped or more conveniently double sided taped to the workpiece for easy removal.

A similar jig can be devised to cut convex patterns such as circles. The tip of the guide is cut into a V-shape and set so that the two edges run against the pattern to cut the required curve.

The standard pattern guide method employs a longer guide for more speed and greater accuracy. The guide is raised off the table by battens to allow the oversized workpiece to pass under. Again the blade is recessed slightly into the guide. A straight guide is used to copy straight patterns, but regular curves can also be cut using a guide cut to a curve to match that of the pattern.

Cutting duplicates is often done by nailing or clamping several boards together before cutting out the shape which is either drawn onto the top boards or is provided by a template pinned to the top. The reverse of this method involves first cutting out the shape out of a block, then ripping that into several thin slices.

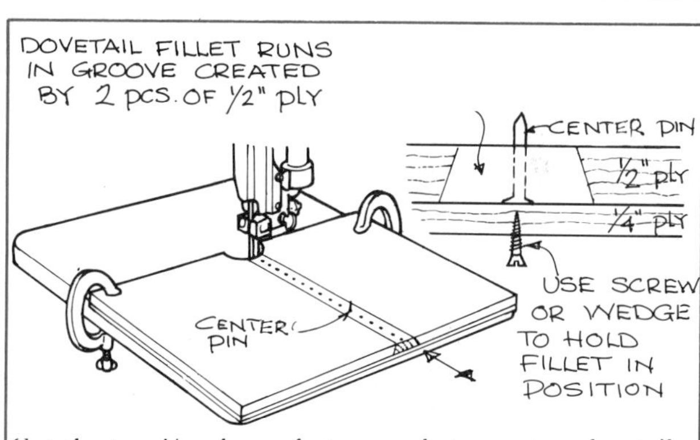

Cut the top $\frac{1}{2}$in plywood at an angle to create a dovetail groove which allows the batten to be adjusted for various diameters.

Cutting circles Cutting circles for table tops, stool seats, etc. can be done by router but for production work it is better to rough cut the circle on the band saw, and finish off with a router.

Several manufacturers sell circle guides which are bolted to the top guide of the saw and are used, pivot above the work, to cut small circles. The easiest method, however, is to clamp a plywood jig to the band saw table. The jig should be large enough to take the largest circle you are likely to cut even if a side support is required for the other end. Make sure that the pivot arm is set perpendicular to the blade and exactly in line with the teeth.

To use the jig, clamp it to the table. Then set the pivot distance by selecting the appropriate hole and by moving the hardwood slide. Lock the slide in position with a small wedge at the end of the housing or by screwing from underneath.

Place the workpiece, good face up, so that the center hole drilled in the bottom fits over the guide pin. Then, carefully feed it into the blade, taking care at the beginning to allow the blade to adjust.

RADIAL ARM SAW

Above: Crosscut saw with bench on either side.
Below: Portable router fitted in special jig.

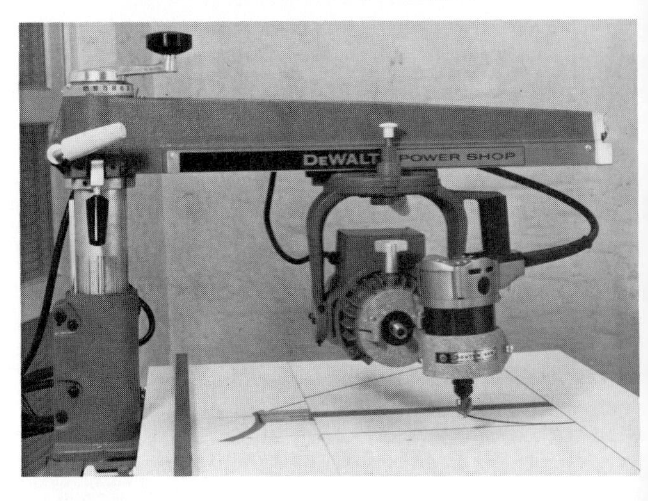

In production and joinery shops, the radial arm saw is strictly a crosscut saw, but for smaller operations with limited space it serves as a sort of universal woodworking machine, capable of a variety of operations.

The basis for a universal machine is sound, for the main difference between the various machines is the axis of rotation, and it is ingenious to have one machine, using only one motor with an axis that can be changed to suit the use.

However, the radial arm saw suffers the disadvantage of all multi-purpose tools in that it takes time to change operations. For busy production shops the various machines are used simultaneously or in quick succession and the radial arm saw is used for crosscutting operations only. But it's for the individual craftsman that the full value of the radial arm saw can be exploited, and the saving in initial cost and in space undoubtedly outweighs the disadvantages in these situations. It should be emphasized, however, that the radial arm saw is considered quite dangerous in some operations such as ripping, molding and horizontal sawing, where the mode of operation and the lack of proper guarding leaves the blade dangerously exposed.

The radial saw consists of a direct drive motor with a spindle to hold the various cutting attachments, such as saw and dado blades, molding head, sanding drums and drill bits. The motor assembly is held by a yoke attached to the radial arm, which in turn is attached to the column at the back of the machine. This configuration gives the saw its versatility. The rear column acts as a pivot for the radial arm so that it can be rotated from 90° right angle crosscuts to any other angled cut. In crosscutting operations, the motor is pulled forward along the arm by the operator so that the saw blade cuts from above with the thrust acting safely downward and towards the back. Most models have various indexed positions where the arm clicks in place at, say 90° and 45° for quick setting square and miter cuts respectively.

The axis of rotation can be changed completely by rotating the motor itself. The yoke allows it to be swiveled and locked 90° in either direction for ripping operations. Normal ripping is done with the blade facing the rear in the "in-ripping" position. The "out-ripping" position, with the blade facing outwards is for ripping wider pieces.

In addition the motor can be "over-turned" so that the blade is horizontal for various operations such as horizontal sawing, molding, drilling and sanding.

Finally, the whole assembly can be raised or lowered by a crank located at the column top on some models, and at the front of the table on others. The crank moves the column up or down for varying the depth of cut, for example.

The machine is never turned on until all the components have been securely locked in position by the various clamps or knobs located near the pivots of each.

Table On most models the column base is bolted to a metal table, but some are sold to be fitted to bench tops or even to sturdy trestles for site work. The replaceable front table, which consists of a piece of plywood or chipboard, should be well held down to the fixings provided in the metal base. The rear board, spacer board and fence are held in place by two rear clamps which wedge them against the front table. These components are moveable to suit the operation and also the size of the workpiece, and should be replaced when badly scored or worn out.

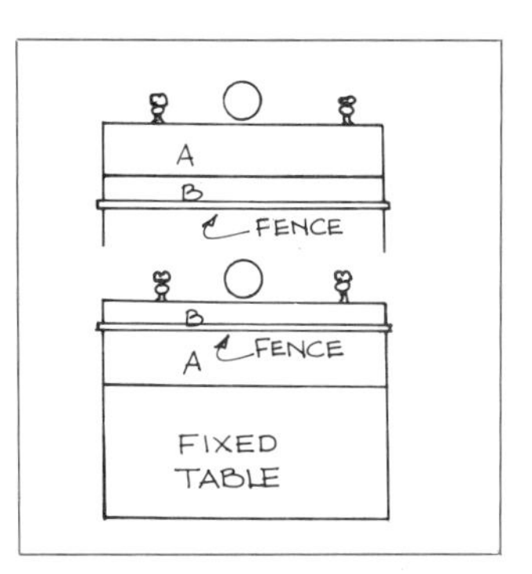

The front part of the table is fixed rigidly to the base. The two rear pieces and the fence are moveable. When ripping wide pieces, for example, move the fence back by putting the large piece A in front.

Size and power Industrial radial arm saws vary in size from 10 to 18in diameter blade, with 10 to 12in suitable for general use and 18in for industrial heavy-duty.

The effective cutting depth is about $\frac{1}{3}$ the diameter, so a 12in, $1\frac{1}{2}$hp is a good general purpose saw powerful enough to handle 4in thick stock. Power output for larger machines is up to 7hp but this is for heavy and continuous crosscutting operations only.

Safety features

As mentioned earlier, the radial arm saw is safe for crosscutting operations where the blade is well covered and the hands are away from the blade, but relatively unsafe for other operations, especially ripping. The blade guard and the retractable guard should be properly fitted at all times. For ripping, where the thrust is towards the operator there is always the danger of the wood closing up behind the blade causing it to be thrown back dangerously. With the anti-kickback fingers fitted $\frac{1}{8}$in below the surface of the workpiece, and with the splitter in place this danger is minimized. During ripping operations, it is important to tilt the guard forward so that the front edge just clears the work. This covers as much blade as possible and also directs the sawdust away from the operator.

The radial arm saw is quite dangerous for some operations and should always be guarded very carefully. In some of these photographs, the guards have been removed for clarity.

Adjustments and maintenance A new radial arm saw usually arrives in boxed components which must be assembled. In addition, all the mechanisms must be accurately adjusted. And because of the adjustable nature of the radial arm saw, you will periodically have to check the accuracy of the settings, and make adjustments as required. Most manufacturers supply complete instructions for setting up, adjusting and maintaining the machine.

Adjustment schedule

1 Table level and flat

2 Blade square with table at 0°

3 Fence parallel with blade at 90°

4 Motor travel perpendicular to fence

5 Saw blade perpendicular to fence

6 Adjusting in-rip and out-rip scales

7 Removing play in sliding mechanism

The saw should also be kept in clean and dust free condition. Each manufacturer specifies the points which should be lubricated. Direct drive motors have sealed for life bearings which need no greasing, but the various moveable parts should be lubricated with powdered graphite. As with all machines it's best to stick to a periodic maintenance schedule, where all the nuts and bolts are checked and any adjustments and lubrication carried out.

Basic operations

Crosscutting There are four basic crosscutting operations, and the radial arm saw does them all well: **1** crosscut or groove, **2** bevel crosscut, **3** miter, **4** bevel miter.

In crosscutting, the arm is set at the appropriate angle and the motor and blade are pulled across the workpiece which is held tight against the fence with the other hand. The thrust of the blade is down and towards the fence so there is little danger. On through cutting the height of the blade is set so that it just cuts into the table and for this reason an auxiliary $\frac{1}{4}$in or $\frac{3}{8}$in thick plywood top is spot glued down, then replaced when worn out.

To help line up the work quickly, it's best to score the fence and the auxiliary table with the three lines representing the two 45° miter cuts left and right, in addition to the most commonly used 90° line for making square cuts.

It's also useful to build long benches on either side of the saw at exactly the same height to help support long boards. Alternatively, use props such as trestles or those shown on page 82 to support long boards in both crosscutting and ripping operations.

The radial saw is especially useful for gang cutting where several boards are cut at once. Working against a stop, several pieces can be offcut to identical length, or halving joints can be cut using perhaps a dado blade to cut both halves at the same time.

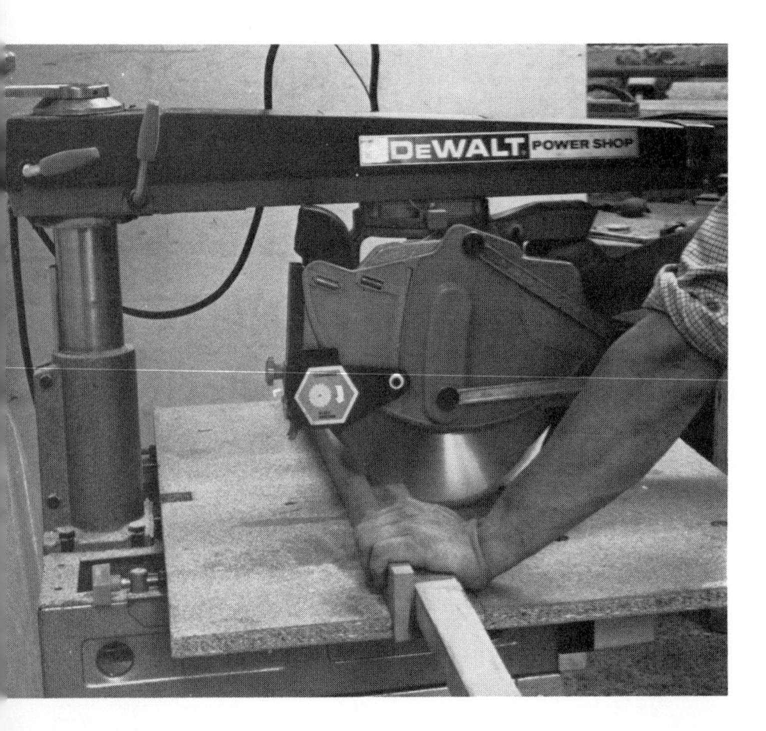

Crosscutting operations. Left: In making standard crosscut, hold the work against the fence and pull the saw out. The thrust is safely toward the fence. Above: Other operations such as cutting miters are equally easy.

The radial saw as a dimension saw

Fitted with an overlength fence and one or two stops as shown below, the radial arm saw will serve admirably as a dimension saw, crosscutting components to square and accurate lengths. With a cutting list mounted on the wall behind, you could, for example, set each stop to one particular length allowing you to cut either one as the length of the board allows. Squaring off the first end the board is to the right of the blade. This end is then moved to the stops and held with the left hand as usual. This is an advantage over the dimension saw for which the board usually has to be turned around after the first cut.

Similarly, halving joints or tenons can easily be cut using one or both of the stops as guides for the last cuts on either side of the cut-out. Remove the wood in between with repeated passes of the saw blade. It's best to set the stops and the depth of cut using a couple of extra boards of the same stock and to try them for accuracy before going ahead.

Miter cuts are made in the same way. Set the angle on the miter scale at the top of the column and, as always, test the cut for accuracy on a piece of scrap which is checked against a miter square or a bevel gauge for accuracy.

To prevent the workpiece from moving slightly during miter cuts, use an auxiliary fence fitted with screws from the back whose protruding tips act as anchors to hold the work.

Forty-five degree miters for frames must be exact, and opposite sides of the frame must be identical for the four joints to close properly. Accuracy is ensured by using stops so that each pair of sides is identical. Rectangular sections can be cut with the saw remaining at the same setting, flopping them over end for end for the second cut made against the stop. Shaped picture frame sections which must be held with the square "rabbeted" side against the fence are moved along to the stop after the first cut, then with the arm angle shifted to 45° left, the second cut is made.

Bevel cuts are made by tilting the angle of the saw, then locking it in place with the bevel clamp. Making tongued miter joints which is a cumbersome operation on the table saw is quite straightforward on the radial arm saw.

After the first end of all the pieces is bevel cut, the pieces are lined up against a stop for cutting the second end. Always cut an extra piece which then allows you to test the grooving cut which is made on all the pieces at once by simply raising the saw and moving the stop.

Similarly, compound miters can be cut by setting the arm and the saw rotation to the appropriate angles.

If possible, fit extension tables on either side, but failing that, extend the fence only and use stops to make repetitive cuts automatic.

Cutting splined miters is straightforward. Cut both miters and raise the blade for the grooves.

Special crosscutting operations Cutting wide grooves for housings etc., is most easily done with dado blades, set up to cut the exact width of the grooves. To cut a series of housings, as for a bookcase for example, mark the location of the grooves on one piece, then line up the boards to cut both or all the pieces in one pass. After making the first cut, place a scrap in the groove to keep the pieces in line for successive cuts.

Stopped housings are cut the same way, with a clamp on the radial arm to stop the travel of the saw in the right position. The curved ends of the groove have to be squared off by hand with a chisel.

Dado blades are also suitable for cutting tenons, halving joints, etc., where a few passes of the blade quickly clears the wood. As usual, make the left and right cuts first before clearing the waste in between.

By turning the saw horizontally and spacing out the cutters with spacing collars, tenons can be cut in one pass, similar to the method used with the Whitehill blocks on the spindle molder. Place the blades on either side of the collar with the chippers above and below these for clearing the rest of the wood.

Ripping For normal width boards, rotate the saw 90° to the in-rip position. It's a good idea to check periodically that the blade is parallel to the fence and adjust if necessary to avoid binding or burning the work. And if either the table top or the fence are so scored that they may impede the motion of the workpiece, replace them immediately.

With all adjustments checked, set the width of cut by moving the saw along the arm, and set the height of the blade which should protrude into the auxiliary table by about $\frac{1}{16}$in for normal through cuts. Remember that the workpiece is fed against the rotation of the blade. The guard should be tilted forward so that it clears the workpiece by about $\frac{1}{8}$in. Then the anti-kickback device is adjusted so that the fingers are about $\frac{1}{8}$in below the top of the workpiece to enable them to dig in and hold the board, should it be thrown backward.

When ripping, hold the board firmly against the

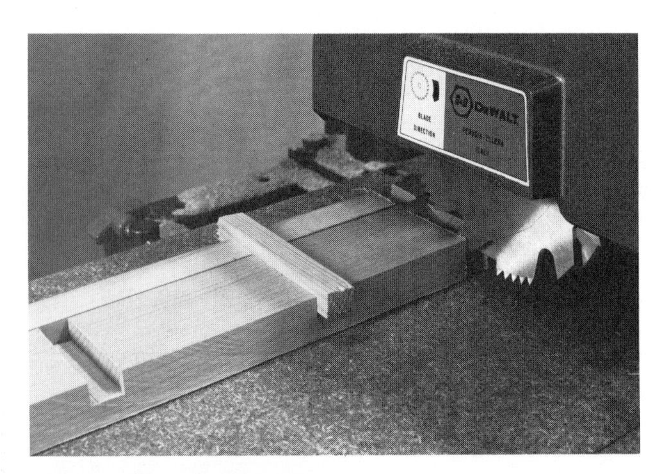

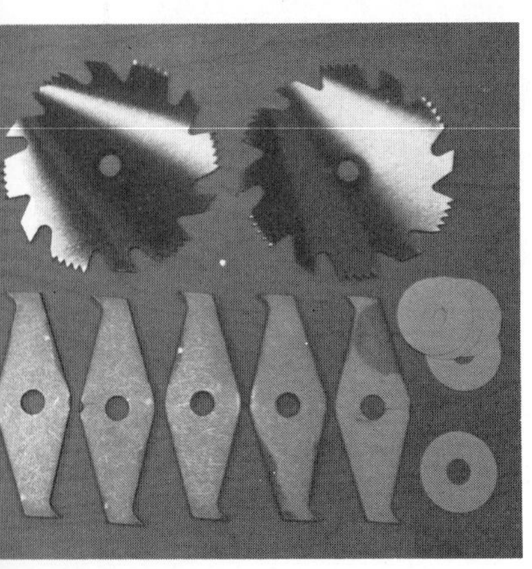

Above: Using the dado blades to cut wide grooves. Notice the strip placed in the groove in the fence to line up successive cuts.

Left: The components of the dado blade. The two cutters at the top and the various thickness chippers and paper washers below.

Use cutters on either side of plywood washer to cut tenons. Use a special fence with blades protruding and push block to increase safety.

fence and feed it slowly keeping the pressure towards the fence with one hand as you push with the other.

To avoid getting the hands near the blade, always use a push stick near the end of the cut. Place it nearby before the cut begins within easy reach.

All but the shortest boards need rear support from a bench or prop to prevent having them tilt up into the sawblade at the end of the cut.

For wide boards, swing the saw to the out-ripping position to get more distance between the blade and the fence. Then because the blade is rotating in the other direction, feed from the opposite side to that used for in-ripping.

To gain even more ripping width, move the fence further back by placing the rear part of the table in front of the fence.

Another way of ripping very wide pieces is to use the front edge of the table as a ripping guide. First check that the edge is parallel to the blade, then clamp a straight batten to the underside of the workpiece and run it along the table edge to guide the work. This is also a useful device for trimming off uneven edges on plywood sheets when the fence can't be used.

Above: In-ripping with two push sticks.
Below: Out-ripping for wider pieces.

Special ripping operations Bevel ripping for making shaped legs and various joints is simply done by tilting the blade to its required angle and ripping as usual. Narrow pieces which may be difficult to push through with the push stick should be treated carefully, fitting a low auxiliary fence behind the work if necessary, to get more access behind the tilted blade.

To cut rabbets, make one cut first, then re-set the ripping width and the depth of cut, if necessary for the second cut.

Grooves along the length for holding door panels, cabinet backs etc., should be made with repeated passes of the saw. Start with the first and last cuts setting the ripping width carefully to marks for each cut, then remove the waste in between with successive runs through the saw. As an alternative method, use the dado blades set to the exact width of the groove to cut it in one pass.

Re-sawing A table saw or a band saw is much better for ripping deep stock into narrow boards. To do re-sawing on the radial saw, first fit a higher fence to adequately support the side of the workpiece. Use a feather board if necessary to support the other side, then rip as normal. Boards too deep to rip in one cut can be cut in two passes. Cut just over halfway, then turn the board over end to end keeping the same face against the fence to finish the cut. But never attempt to rip very narrow boards (1in or less) on the radial saw.

Cutting tapers is done with a taper jig made as shown on page 87. It is used exactly as for the table saw setting each notch to the taper of each side.

A similar jig is used to cut wedges, but it should be emphasized again that these special ripping operations are more safely done on the table saw where more control is possible.

Using a feather board for sawing deep stock.

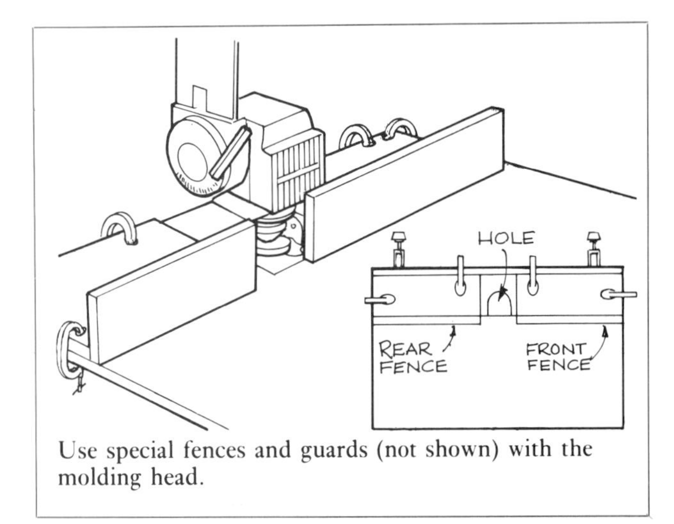

Use special fences and guards (not shown) with the molding head.

The fences are arranged in line for all work except where entire edge is being planed.

Top: Use clamping devices such as the two sprung strips to keep work tight against the fence. Above: For long, thin pieces pull the work at the end of the cut. Right: The molding head and cutters.

Using the molding head To cut molded shapes, rabbets, tongue and groove joints etc., on the radial saw, fit a molding cutter head and special guard to the motor and use it as a shaper. However, if you find that you are using it frequently for shaping operations it is better to buy a separate shaper (see page 120) which is designed for the job with special safety features difficult to incorporate into the radial arm saw.

To set up the table for shaping operations, first move the fence to the rear and cut a hole in the rear table, as shown, before it is clamped in place. The hole allows the head to be lowered to make cuts flush with the table top. Then clamp a pair of identical fences on either side of the cutter head, in the same way as for the shaper (page 129). Position these fences exactly in line for most molding operations except those where the entire edge is cut. For these the outfeed fence should be moved out in line with the cut edge as for the overhead planer.

The molding heads are fitted with two or three knives depending on the type, before being mounted on the saw arbor with the locking nut. A wide variety of knife shapes are available, from straight cutters for edge planing, rabbeting and grooving, to tongue and groove cutters and various molded shapes for decorative work.

To shape the edges of boards, feed the board slowly by the cutter, keeping it tight against the fence with one hand. Always use a push stick or block at the end of the cut, particularly for short pieces. As with the shaper, try to use hold-downs, such as feather boards or spring clamps to keep the workpiece in position. All cuts should be made on the bottom edge if possible so that the workpiece is on top for added safety.

Drilling Fitted with a drill chuck, the radial saw can be turned into a vertical or horizontal drill which has certain advantages over a drill press, since there is no limitation on the length of the workpiece for horizontal boring.

To use the drill make an auxiliary fence to raise the workpiece off the table, since the motor doesn't allow the drill to be lowered full depth.

For face boring, simply push the motor towards the work. A clamp attached to the radial arm acts as a depth stop. For repeated pieces, as in doweling, clamp one stop to the fence for lining up each hole.

Drill ends of boards by turning the drill 90° to face sideways. The workpiece is pushed into the hole while being held against the fence and table. A stop or mark on the table can be used to indicate depth.

For dowel joining, where accuracy is essential set up the drill and the table carefully and make test cuts on a scrap before beginning work. For two or more holes, the drill is moved out after each hole has been drilled in all pieces.

Top: Face boring. Center and below: For dowel boring, arrange table to hold work for straight drilling or with a 45° push block for drilling miters.

Drum and disk sanding

DRUM SANDING Sanding drums come in various sizes, the largest about 3 × 3in being the most useful. Check the manufacturer's accessory list for the availability of drums and abrasive sleeves.

The drum can be used horizontally and diagonally, but is most often used vertically with a hole cut into the table to allow it to be lowered for edge sanding against the table.

Also use the drum for all types of freehand sanding including sanding the insides of circular cut-outs.

For horizontal sanding, to sand the face of narrow boards for example, turn the motor horizontally and raise it to the right level so that you basically thickness each board as it is passed under the drum.

DISK SANDING Mount the disk on the saw arbor, and fit the appropriate abrasive disk to the metal plate. To use the sanding disk, make an auxiliary table, as shown, which allows sanding at the right level. Place the table absolutely square with the disk, so that it can be used to square off board ends for joinery. A groove made in the table by spacing out the two pieces which make up the top, allows you to use a miter gauge for square ends and also for miters. Notice that all sanding is done on the right side so that the force of the wheel is downwards towards the table.

JIGSAW

Jigsaws, which may be bench mounted or freestanding, are used to cut intricate shapes in wood, metal, plastics and other thin material. The early foot treadle models are still available today and are particularly useful for marquetry work, where a very fine blade is used to cut the veneer shapes. Smaller jigsaws are used by modelmakers, toy makers, sign makers and anyone doing small, intricate work, and these are readily available and competitively priced. More powerful and robust jigsaws are used in production shops for a variety of jobs, from cutting scribed moldings in window frames to making accurate patterns and templates for repetition work.

The size of the jigsaw as with the band saw, is determined by the throat width, that is the horizontal distance between the blade and the arm, which governs the largest size of panel it can take. The maximum depth of cut is also a factor though the stock thickness is rarely above $\frac{1}{2}$ or 1in.

The method of holding the blades varies from model to model, but they are usually clamped in jaws top and bottom and tightened by a thumbscrew or Allen wrench. Blades are installed, teeth pointing downwards, at the bottom end first, then at the top end, slackening off the tensioning device fitted at the top, to get the blade in. The tension in the blade should be enough to avoid having the blade bow or go out of line but not so high as to break the blade too readily.

The blade is supported by a blade guide to prevent the fragile blade from twisting. Exact configurations of guides vary from the simple slot in the table insert, to a universal guide which can be rotated, with a slot which accommodates a particular size blade. Only the back of the blade rides against the support when under load. The teeth should not touch the sides of the guide. On some saws, both the upper and lower holders can be rotated 90° to change the direction of blade to enable the saw to handle long boards, though this is rarely required.

Additional features fitted to most jigsaws are the hold downs and the continuous blower. Hold downs

Old treadle jigsaws are still used for marquetry and other fine work.

are adjusted to the thickness of the workpiece and used, as the name suggests, to keep the work tight against the table against the uplifting force of the blade. The blower, from an air pump operated by the driving motor, is very convenient since it is directed at the cutting area to continuously clear the sawdust for a clear view.

Using the jigsaw is quite straightforward. The blade tension and alignment of the blade should be checked by hand turning the wheel before switching on, and the hold down adjusted accurately to suit the workpiece. Coarse cuts using a wider blade limit the intricacy possible, but very fine blades are capable of near 90° turns to make it capable of cutting the most intricate shapes. The work should be fed smoothly, guiding with both hands and turned only as sharply as the blade will permit. For intricate work use fine blades, slowing the feed rate accordingly particularly at sharp turns.

Bevel cuts can be made with the table (and hold downs) tilted at an angle. But for bevels, each cut can only be made from one direction, otherwise the blade will cut a bevel in both directions on different parts of the workpiece.

Internal cuts are possible by removing the top fixing and threading the blade through the hole in the workpiece before re-fitting it, similar to a coping or hand fret saw.

Straight cuts are generally not very successful, but with a wide blade it is possible to clamp a straight guide to the table to cut in a fairly straight line, as in cutting out letters in signmaking, for example.

Duplicate pieces are best made by piling up thin layers and cutting through all at once, as with the band saw.

Coffee table made by cutting up plywood and staining pieces before re-assembling.
Designer/maker: Ewing Paddock

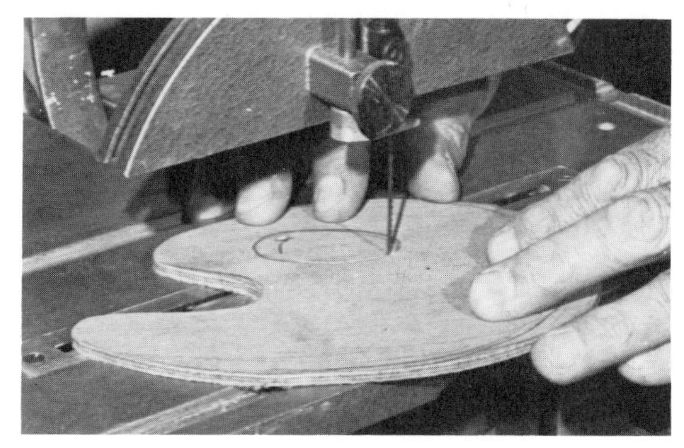

To make internal cuts, undo the blade fixing at the top and thread the blade through the hole.

JIGSAW BLADES Jigsaw blades, like most saw blades, have coarser or finer teeth depending on their use. Lengths are usually 5, $6\frac{1}{4}$ or 7in, but most saws adapt to suit different lengths. When selecting the pitch of tooth, remember that the general rule is that at least two teeth should be in contact with the work at all times. Thus if sheet metal is, say $\frac{1}{8}$in thick, there should be two teeth for every $\frac{1}{8}$in of blade or 16 teeth per inch, as a minimum. Blades should also be selected for thickness and width to suit different materials such as wood, steel, plastic, or cardboard. Coarseness ranges from 7 to about 32 points, thicknesses from .008in for fine blades to .028 for coarse blades. Each manufacturer issues lists of blades with their characteristics and suitable uses.

A selection of jigsaw blades varing from very fine to coarse.

MACHINE PLANING

Planers are essential to all woodworking. They are the second step after sawing the log into boards in the gradual process of making the surface straight and smooth. Although the spindle molder, fitted with a straight cutter is capable of planing, there is really no practical alternative to machine planers in the workshop. Hand planing is a substitute, of course, but only for very carefully handcrafted work which is too slow and inaccurate for most workshops.

There are two types of planers. The jointer or surface planer is used to plane one face and one edge straight and square with one another. The work is usually hand fed, though power feeds are available to be fitted to the jointer for production work. The thicknesser is used to get the second face and edge parallel with the first and at the same time to "thickness" a board down to a specified dimension. This is the most important function of a thicknesser: to enable many boards to be machined to exactly the same dimension which is critical for accuracy in all cabinetmaking and joinery. Thicknessers are always power fed and the feed rate is adjustable on most machines to give a better finish, depending on the type of wood.

Since the basic cutting operation is the same for both types, the jointer and thicknesser are often incorporated into one machine, which has surfacing tables above the cutters and the thicknessing function as usual, below the cutters. This jointer/thicknesser, or over/under planer as it's called in the UK, saves money and space and is suitable for most shops where both funtions are not needed simultaneously.

The two machines differ on several details. The thicknesser is more complicated since it must feed and hold the wood as well as keep it flat at all times while the cutting is taking place.

The size of jointers is determined by the length of cutter (i.e. width) and by the length of the table. A 6in jointer, 36in long is a small machine suitable for light duty work where the board width is unlikely to be more than 6in. Larger shops use a 12in model, with a table at least 5ft long. In principle, the longer the table, the easier it is to do accurate work. The size of thicknessers is given by the width and depth capacity. Thus, a 9 × 6in machine is capable of planing 9in wide and 6in thick boards. In practice, 6in is more than adequate depth capacity, but it is convenient to be able to pass through wide, glued up boards up to say, 18in in width. So the wider the capacity the better.

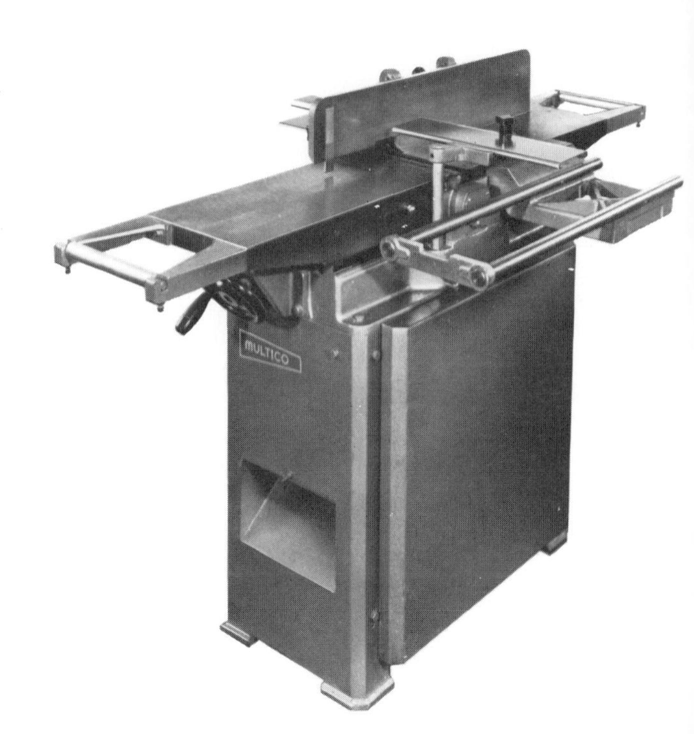

Surface planers or jointers (above) should have tables as long as possible or extensions as shown for accurate work.

Thicknessers vary from small, lightweight models to the large, heavy-duty machine (below) for thicknessing wide panels.

Cutting theory

Planer cutters, which are made from high carbon steel, cut by chopping or chisel action which is the basic cutting action of all woodworking tools. The blade, which is mounted in a rotating block, travels in a circular path around the block. Each blade on the block (and there may be two or more blades) strikes the wood diagonally and like a chisel, it splits and breaks off a length of wood to form a small arc shaped cut.

The spacing or "pitch" (P) of these ripples determines the smoothness of the surface. It is governed mainly by the following factors:

1 The speed of the block (R) in revolutions per minute.

2 The feed speed (F) in feet per minute.

3 The number of cutters on the block (N).

The faster the board is fed through, the further the wood will travel between each cut resulting in a larger pitch, i.e. a coarser finish.

The general grading standards for planing is as follows:

Pitch (marks per inch)	*Grade*
Under 8	Poor work
8 – 12	General joinery
12 – 18	Interior joinery
18 – 25	High class joinery and furniture

The number of cutters N is fixed for each machine, but the block speed and feed rate are variable. The block speed can be varied by altering pulley sizes according to the formula given below and the feed speed is varied either by switching gears in the case of most thicknessers or by slower or faster hand feeding in the case of jointers.

The motor speed and the block speed vary inversely with the diameters of the pulleys:

$$\frac{R}{r} = \frac{d}{D}$$

Transposing this equation, the speed of the machine is given by

$$R = \frac{d}{D} \times r.$$

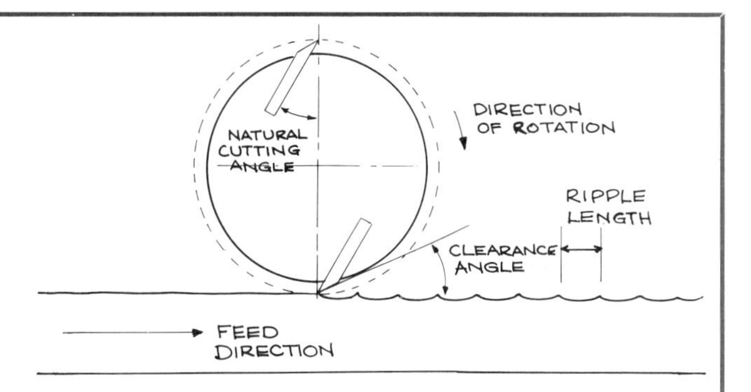

So, for example, if the motor is listed at 3500 revs/per minute, the motor pulley and the block pulleys diameters are 8in and 5in respectively. Then the block speed $R = \frac{8}{5} \times 3500 = 5600$ rev/per min.

r = speed of motor (rpm) Printed on motor label

d = diameter of motor pulley

R = speed of cutter block on machine (rpm)

D = diameter of pulley on machine

The pitch P is calculated by the following:

$$P \text{ (marks per inch)} = \frac{NR}{12F}$$

N = number of cutters on block

R = speed of block, revolutions per minute

F = feed speed, feet per minute

12 converts feet into inches

For example, to calculate the pitch of a planer rotating at 4500rpm, feeding at 50 ft/min, with 2 cutters mounted on the block:

$$P = \frac{2 \times 4500}{12 \times 50} = 15 \text{ marks per inch}$$

which is fine for interior joinery work.

To make the surface smooth enough for furniture P = 25 would require a decrease of feed speeds of $\frac{15}{25}$ or 0.6 × 50ft/m = 30ft/min.

In practice, most small shops don't bother with these calculations. Many small machines don't have a gear change facility for varying feed speeds, and changing pulleys is not really a practical option in everyday work. (To be accurate, a deduction (5 to 10%) should be made for belt slip.

Hardwoods should, however, be fed at a slower speed than softwoods, and for shops where many wood species are used, a variable speed thicknesser is a necessity and these calculations are useful to control the quality of work.

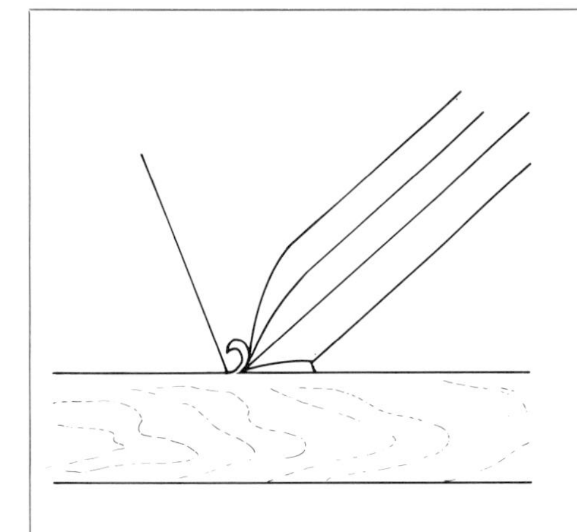

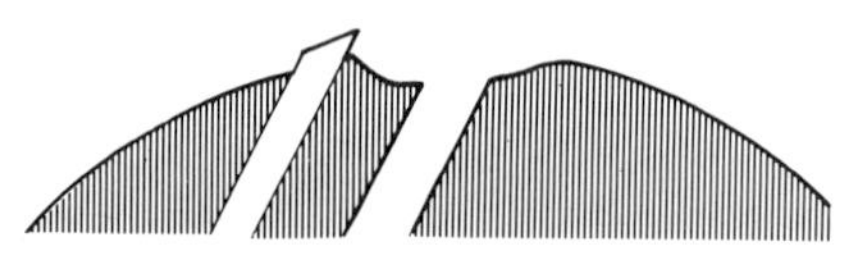

The chips lifted by the chisel-like action of the cutter are prevented from digging in along the grain by the shape of the block itself (above) similar to the action of the cap iron in the hand plane (left).

In addition, the spring-loaded chip breaker (below) helps to prevent splinters.

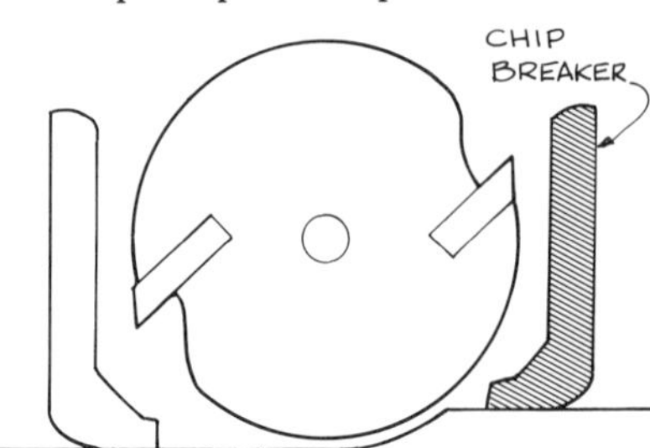

Chip breaking In hand planing, the plane iron digs into the wood wedging it apart. The cap, mounted directly above the cutting edge, breaks up the wood shaving, preventing it from continuing into the wood in front of the plane. Machine planers work on the same principle. The chips are prevented from becoming a long splinter by the shape and configuration of the cutter block which, like the hand plane, will cause the partly severed chip to crack and bend back on itself. On the thicknessers, the chip breaker which is a spring-loaded bar located immediately in front of the block also helps to prevent splintering by holding down the wood in front of the cutting edge.

Cutter blocks The cutter block, which is mounted on a shaft rotated by the motor, holds the cutting knives in position. Blocks can be circular or square for thicknessers but must by safety regulations be circular for overhead planers. The number and manner of fixing of blades varies from one model to another.

The square blocks, which are not in frequent use today, usually hold four cutters. Circular blocks commonly hold two, three or four cutters, but on some modern machines, up to 12 cutters are possible.

The formulas given on page 107 for calculating the coarseness of surface or pitch depends on all the cutters being mounted in such a way that they all cut to the same depth. Although theoretically convenient, this has proved practically impossible, for even with the most sophisticated adjusting apparatus, it can't make the cutting circles identical. As with most power tools, one cutting edge does more cutting than the other.

The method of holding the cutters firmly in place varies. On newer planers, the wedge bar lock is most prevalent. The cutter and a wedge bar drop into a dovetail groove cut in the block and are held in place by screws which when turned, clamp the cutter in position. On older cutting blocks, the slab method may be used. Here the block is made in three parts with the two caps or slabs bolted onto it with recessed bolts to hold the cutter in place. Adjustment is either by special screws on more recent models or by drift slots cut in the head on older models.

The blocks are designed so that the cutters will be set at an exact angle to the block called the cutting angle. This is different from the clearance angle which is the angle that the ground edge makes with the wood on impact.

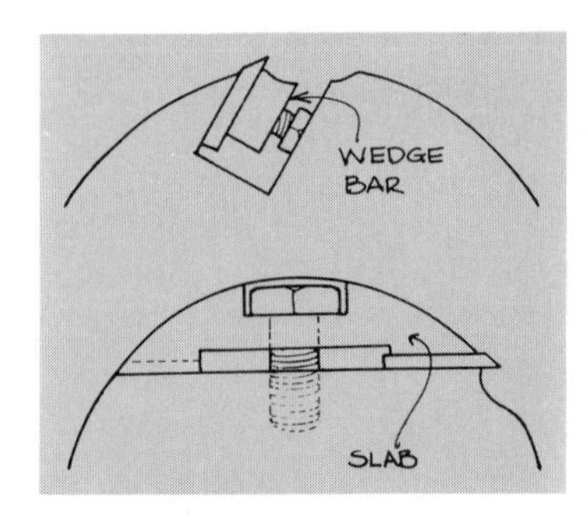

Above: Tightening the nuts on a slab block with a socket wrench. Left: Wedge bar block (above) and slab block (below).

Surface planing

No wood can be joined or finished until it is planed so that it is straight with square edges. Planing is the most basic operation of the whole process of making furniture or joinery. It establishes a datum from which all other measurements and cuts are made. The modern practice of feeding the work through large belt sanders instead of thicknessers is the exception but even here, the wood must first be surface planed to establish a straight face and edge.

Softwood can generally by bought ready planed, but even this may require a pass or two over the planers after it has been allowed to move as it dries out. Most hardwood, however, is bought sawn, that is, in boards which have come off the milling band saw. Boards sawn through and through have waney edges which are best squared off freehand on the band saw. But sawn boards are often bought with square edges which can be brought directly to the surface planer or jointer for planing.

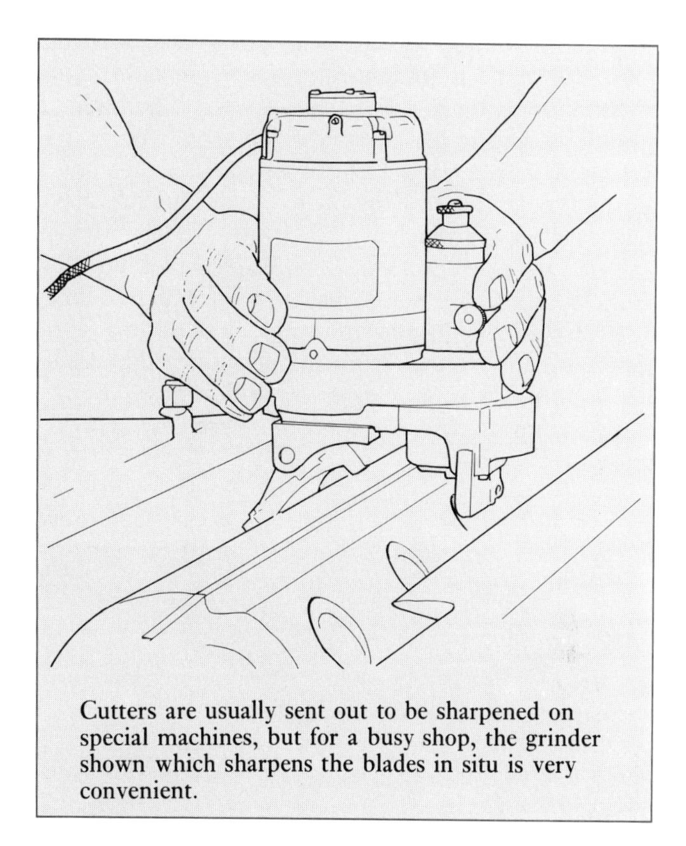

Cutters are usually sent out to be sharpened on special machines, but for a busy shop, the grinder shown which sharpens the blades in situ is very convenient.

Description

The jointer consists of a cutting block, motor and pulleys, two tables which are adjustable diagonally up and down, a fence which is also adjustable and a guard which covers the rotating cutter block during use. The width of the cutter block (usually from 4 to 12in) determines the capacity of the planer, and the length varies from about 36in up to 8ft for more accurate production machines. The infeed table is adjusted up or down for depth of cut from a fine $\frac{1}{32}$in cut to approximately $\frac{1}{8}$in maximum cut, depending on the hardness and also on the brittleness of the wood. The outfeed table, which is also adjustable, supports the board as it comes off the cutters. It is set at exactly the same height as the top of the cutters, that is to coincide with the cutting circle.

Maintenance

Several factors are critical for perfect results. The cutters must be sharp, and they must be set accurately into the block after each sharpening. The adjustment of the outfeed table is critical too, as shown on page 111.

It is important to develop a maintenance schedule based on hours of use or on, say a monthly schedule. Sharp cutters give better results which saves time in sanding, so it's false economy to let them become dull.

Depending on the frequency of sharpening, it may be worth buying a special grinder which allows you to sharpen the blades without removing them from the block. These motorized grinders, which are sold with some modern machines, are run back and forth along the stationary block to sharpen each cutter in turn. Since they use the block itself as the base, all the cutting edges should, theoretically, be parallel and at exactly the same level. It does, however, leave a slightly hollow surface from the round grinding wheel. This method is obviously convenient to productions shops where the "down time" of removing and fitting replacement blades is very costly.

But for the small shop, the most convenient way is to send the blades out for expert sharpening in a wet trough automatic grinder which does a perfect job and removes only up to 0.01in of surface, leaving a highly polished surface. The danger of overheating exists, especially in unsophisticated workshop jigs, and may result in micro-cracks in the cutter edge or in slight bowing, both of which would provide a faulty surface. Cutting angles vary from 30 to 35° with the former more suitable for softer woods. Cutters are usually $\frac{1}{8}$in thick flats of high speed steel, which is carbon steel alloyed with various other elements to give a hard-wearing, long lasting edge. Production factories and large joinery shops frequently fit tungsten tipped cutters which although very expensive, last longer and give much better results.

While the cutters are out being sharpened, it's a good idea to clean out and lubricate the machine and to check belts, fence and guard for wear. The high vibrations tend to loosen nuts which should be tightened as necessary. Lubrication varies depending on the type and age of the planer. Older models may have grease cups which are filled with grease and given a turn every week or so to feed small amounts to the bearings. On newer machines, greasing is done by grease guns onto nibs and some models have sealed for life bearings requiring no lubrication at all. It's important not to overgrease the bearings. They don't use up much and too much grease may actually cause overheating.

Two methods of lubrication. On old machines give the grease cups a turn every week. New machines have nipples for a grease gun.

Setting the cutter blades

There are several clever devices on the market to aid in setting the cutters accurately on the jointer, but the simplest way is the wooden straightedge method. Set the first blade in place, adjusting by eye to make it approximately right. Set the outfeed table slightly below the top level of the cutter edge. Set the straightedge onto the outfeed table so that it extends over the cutter block. Make a mark exactly at the edge of the outfeed table as shown. Then turn the cutter block slowly (make sure the power is switched off) and allow the straightedge to be lifted and carried a short distance by the cutting edge.

Now make another mark at the table edge. The cutter is parallel to the table when both ends of the blade carry the straightedge exactly the same distance. That is, having set one end of the cutter, now place the straightedge on the other end, first mark against the table edge, and rotate the block. Adjust the blade height until the second mark falls exactly at the table edge. Before final tightening, check both ends once more, since adjusting one end often moves the other end slightly. Then lock the cutters very firmly in place and adjust the other cutters in exactly the same way so that they correspond to these two marks.

It's important to tighten the fixings very firmly so that the blades have no chance of working loose. I have heard of this happening and it is a very disturbing thought, since a razor sharp cutter flying through the air at high speed is a highly lethal object. Exact methods of tightening the bolts varies with every shop. The method I was taught is to tighten the bolts in order a little at a time so that no distortion is built up in the blade. Then after *normal* hand pressure can't move the bolt any more, give them one final tightening with a quick pull of the wrench.

Another point of safety is worth mentioning here. It is a firm rule in all shops that once the job of removing or replacing the cutters has started, it should be finished, without interruption. This is to prevent someone from being interrupted and leaving the nuts only half tightened. Another worker, not being able to tell, could start the machine and the cutters and nuts would fly off. If an unavoidable interruption

The planer blade lifts the strip and carries it forward a short distance.

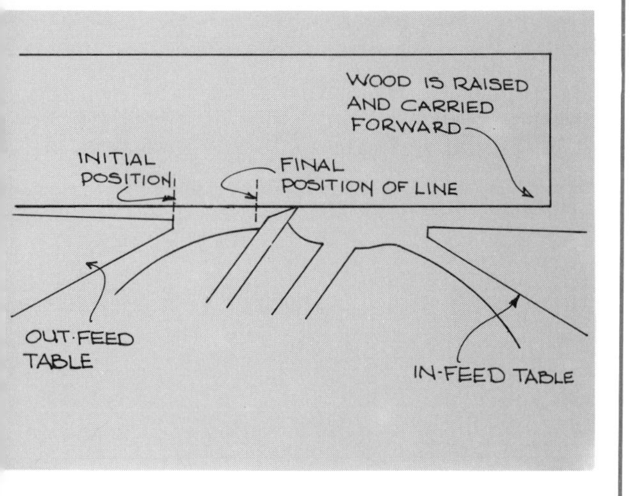

does occur, completely remove the nuts and the cutters so that there is no danger. And remember to keep track of the nuts. Each one should always be put back in the same socket so that the seating, worn in over years, will be right each time.

Adjusting the outfeed table After setting the cutters, adjust the outfeed table so that it is exactly level with the cutting edges. Use the straightedge again and not a metal rule as some books suggest. Place it on the outfeed table and rotate the block. The blade should just touch the straightedge with a slight scraping sound. A bit of chalk rubbed on the straightedge is a good indicator for adjustment.

Incorrect adjustment of the outfeed table is the most frequent cause of bad surface planing. The function of the outfeed table is to support the planed surface as it comes off the block. If the table is too low, there will be a gap between it and the board which will tip forward as it moves forward. When the back edge of the board leaves the infeed table, it drops down leaving a "jump" or "dip" in the board. If the table is much too high, the board will completely stop against its edge. If it is fractionally too high, the work will ride up so that it will be cut for half its length before tipping forward leaving the rear end unplaned. This results in an uneven or a tapered surface.

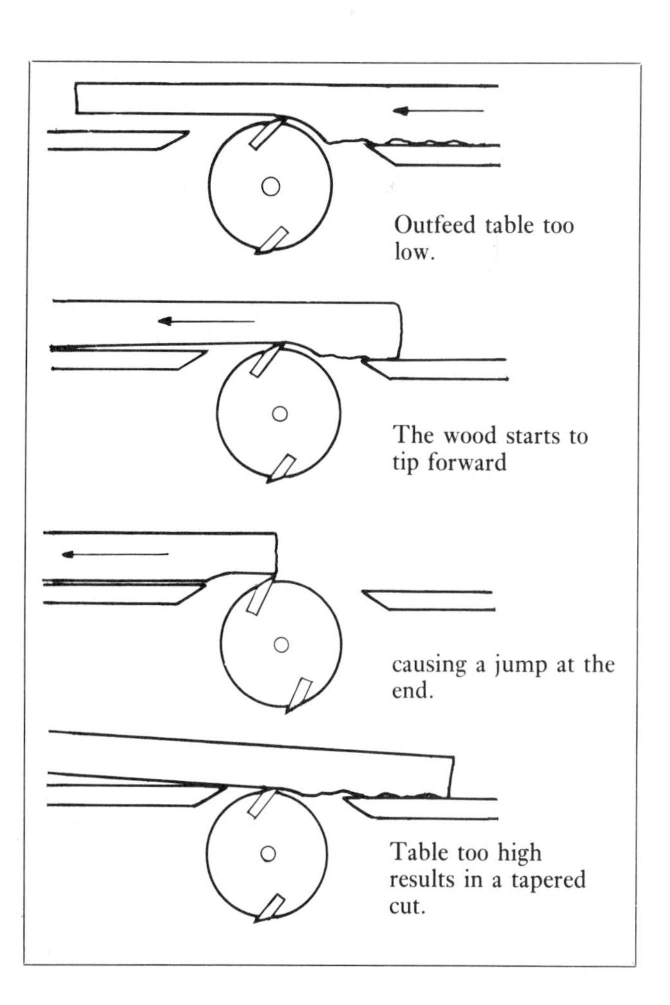

Using the jointer The jointer works by progressively removing the high spots on an uneven board until the surface is flat. Fairly straight boards are easy to plane. Select the depth of cut depending on the wood species and on the finish required. The normal procedure is to start with a heavy cut and then to change to a fine cut for the last stroke or two. The aim is usually to get a straight surface which can be fed through the thicknesser without jamming. For this purpose, it isn't necessary to remove every last bit of sawn surface, only enough to give the board enough straight surface to ride on the bed of the thicknesser or on the table or fence of the table saw. The face is planed first before the edges which are marked, face and edge, for the next process.

Guard configurations vary. American jointers are usually fitted with a spring-loaded plate which pivots out of the way as the wood is fed through and springs back as the end of the wood passes by the cutters. In Europe, where safety standards are usually more stringent, the guard is the fixed bar type which is adjusted and fixed for each cut. For face planing, there should be no more than $\frac{3}{8}$in space between the guard and the board and the guard should extend to the fence. For edge planing, it should be lowered all the way down and extend to within a maximum $\frac{3}{8}$in from the board. Another similar guard should be fitted on the other side of the fence to cover the block when the fence is moved.

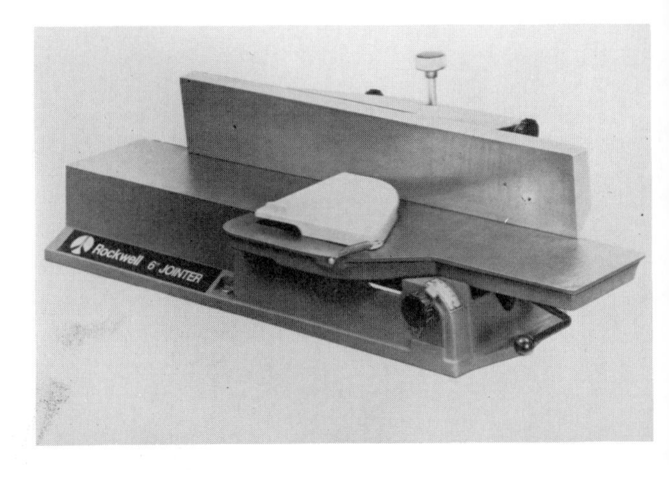

Above: Spring-loaded plate guard which springs back automatically at the end of the cut. Below: Fixed bar guard which must be adjusted for each piece of wood.

Below: Surface planing a heavy board.

Feed the board with the grain to prevent the blades from digging in and leaving a badly marked, splintered surface. One pass will tell you whether the feed direction is right or wrong. The sound of the grain being picked up is immediately apparent. Reversing the direction will remove the marks in the next sweep or two.

Planing bowed or badly warped boards requires a bit more care. For bowed boards, turn the concave face down so that the ends ride on the tables, planing these first. Twisted or warped boards have to be handled carefully. The object is to let them ride on the same three points for the first two or three passes until an easily found surface is established.

Inexperienced operators often press down and try to guide too much. For all but thin boards, it's best to let the weight do most of the work, with a minimum of hand pressure. This should be distributing the weight evenly from one table to the other as the board is fed through.

The most difficult job for beginners seems to be planing up small squares which are often rhombus-shaped in section after drying out. The difficulty lies in keeping the first planed edge firmly flat against the fence while planing the second edge. The left, guiding hand must push from the side as well as keeping the workpiece down with the thumb. After the second run, the workpiece will run square without much trouble.

Planing is done by feel and sound and an hour's practice on a few warped boards is a better teacher than any book.

Rhombus-shaped piece held tight against the fence to plane it square.

Safety The planer is a dangerous machine partly because it doesn't seem dangerous. Safety statistics have determined that approximately 25% of all shop accidents happen on the planing machines. The guards should always be in place and properly adjusted for each cut, where applicable. The most frequent cause of accidents is failure to remove the hand from the back end of the board as it passes over the cutters. On edge planing wide boards, this isn't a problem for the hand is far from the blades. But when surface planing, the left (forward) hand should be pushing the board down and forward onto the outfeed table and the right hand should be moved forward away from the rear end.

Thin boards require firmer pressure up to the very end which is best accomplished with a push block made from a straight board or a piece of plywood about 4 × 12in, with a batten glued into a groove near the back edge leaving about $\frac{1}{4}$in extending. A handle and front knob are essential for a proper hold.

Rabbeting requires the removal of the guard and is therefore considered unsafe by the factory laws in several countries including the UK.

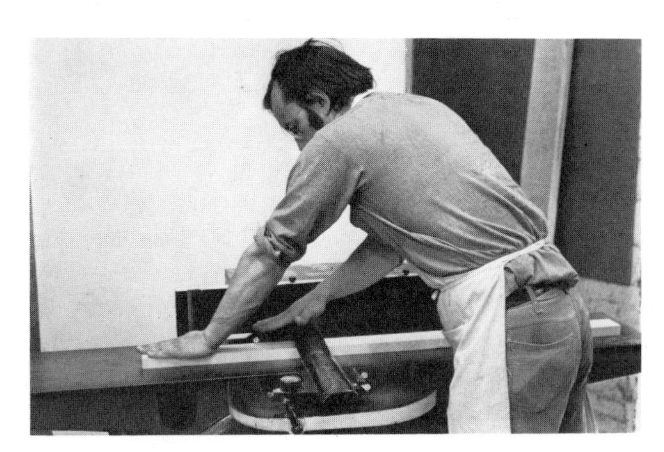

Above: Keep pressure in front of cutter for safety.

Below: Use a push block on thin pieces.

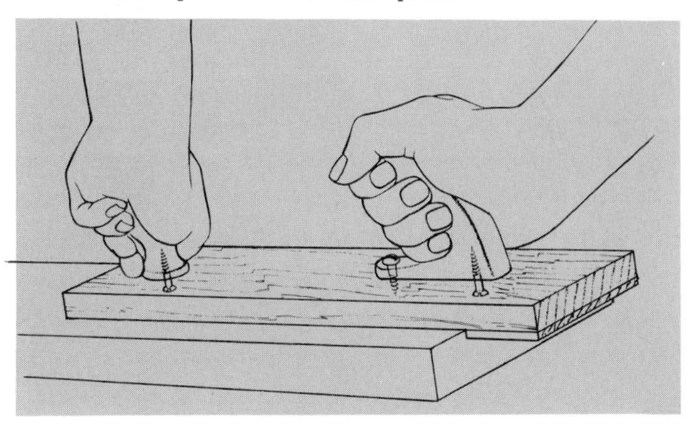

Special applications

THICKNESSING Generally only one face and an edge are surfaced on the jointer but with skill, it is possible to thickness on it as well. After surfacing, mark the thickness with a gauge around the four edges, then plane the other face to this thickness. Boards which are warped are most difficult since they are often wedge-shaped after surfacing. There are no hard and fast rules; it's a matter of trial and error, feeding the board part-way here and there to remove the obvious high points and gradually working to the lines. Sometimes a stroke or two with a jack plane is necessary afterwards.

Left: Thicknessing to line by trial and error. Below left: Planing a bevel. Above: Stop block attached to board for stopped chamfers. Below: Stopped planing to make feet on trestle base.

BEVELING AND CHAMFERING are easily done by tilting the fence to the required angle. Set the fence with a sliding bevel read against a drawing or protractor, and remember to hold the work firmly against the fence.

STOP CHAMFERING, which is often done on traditional furniture, requires the lowering of both the infeed and the outfeed tables to the same level. Adjust the table heights to suit the depth of cuts. Clamp starting and stopping blocks onto the fence or use a piece of plywood with the blocks nailed in the correct locations, as shown. To start the cut, hold the piece against the rear "starting block" with the front end raised. Gradually lower the workpiece onto the cutter, keeping a firm hold on it as the cutters enter. Then feed the piece along until it meets the forward "stop block" and carefully raise the back end off the cutters.

Use the same technique, this time with the fence square to make a recess cut, for example, for the bottom of a table trestle, to leave the two uncut ends as table feet.

CUTTING A TAPER Tapers can be cut on the table saw or band saw, but are more cleanly cut on the overhand planer. Determine the amount of taper, mark it on the end, then lower the infeed table to exactly this amount, working a short trial cut in a piece of scrap to check if necessary. Mark the beginning of the taper on the board, then lower this mark onto the edge of the outfeed table just in front of the cutters. Push the board along, holding it firmly against the infeed table until the taper is cut. The cutters will make a progressively deeper cut as the board is fed by. Large tapers are cut in two passes as shown.

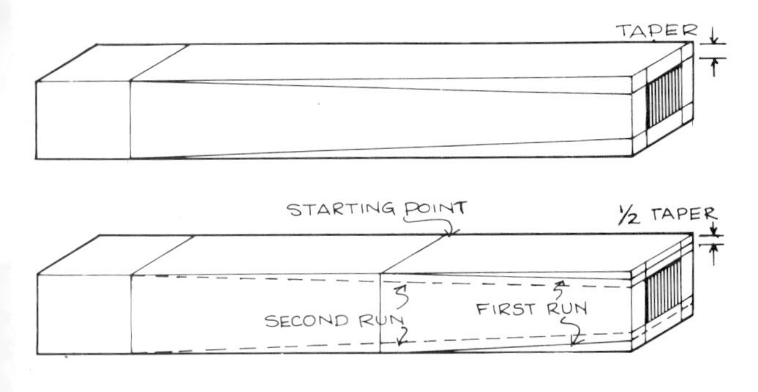

Above: Starting taper cut.
Below: Lowering front table for taper.

Where the board is longer than the infeed table, divide the taper into two parts. Start the taper cut from the center line and set the table to half the depth of the full taper. The first cut will establish the taper line; the second cut will cut the full taper the entire length. For even longer boards, divide the cut into three or more parts in exactly the same way. To make stopped tapers, follow the same procedure as before, but stop the cut against a stop block and raise the workpiece from the table at that point.

PLANING END GRAIN is not recommended on the jointer since the clearance angle between the blade and wood is set for ordinary, with the grain, work. Where a table saw or shooting board is not available, end grain planing should be carefully done. On short, wide boards use a push board jig clamped to the board to keep it square and to back up the end which would otherwise leave a splintered blade exit.

Thicknessing

The configuration of a thicknesser is more complicated than the jointer. The cutter block is the same though often longer to accommodate wider boards and glued up surfaces. The pre-planed surface of the board rides good face down on a flat bed which can be adjusted up and down to accommodate various thicknesses. The board runs under the block so that the top surface is planed. In normal practice, several boards are fed through one or more at a time depending on the design, and the flat bed is raised slightly after each pass to allow the next cut to be made. The depth of cut is controlled by the amount the lower bed is raised and a depth gauge mounted on the side gives the thickness of the board as it comes out of the planer.

Thicknessers are power fed. The configuration of rollers and bars mounted in front and back of the cutter block are critical to proper functioning.

The serrated infeed roller in front of the block is motor driven, usually by the same motor that powers the cutter block. The rate of feed on larger machines can be changed by varying methods to cause the infeed roller to rotate faster or slower. As the wood is fed in, the serrated teeth grab the wood and feed it into the cutters. The backward force from the cutters is tremendous, so the job of the infeed roller is also to keep the board from flying back toward the operator with tremendous force. The infeed roller is set about $\frac{1}{16}$in lower than the cutting circle but the teeth marks left on the wood get planed off by the cutters. As the wood leaves the cutters, the smooth outfeed roller which is also power driven and set slightly higher than the infeed roller, helps the wood through the machine. The chip breaker is located between the infeed roller and the block. It presses down on the wood and prevents the wood from breaking into long splinters as the cutters chisel into it. The pressure bar immediately behind the block is similar to the chip breaker in shape, but its function is to keep the end of the board from tipping up as the wood leaves the planer. The two rollers and two bars are spring-loaded so that their own weight, plus the force from the springs holds the wood down but at the same time allows them to rise slightly to accomodate a slight variation in thickness.

Immediately below the top two rollers and set into the bottom flat bed are the two smooth anti-friction rollers. These are adjusted so that they extend just slightly above bed level to reduce the friction on the board caused by the downward pressure from the rollers and bars above.

These are the essential elements of the thicknesser and must be aligned perfectly for safe and proper operation. New machines have built-in adjustments which allow the top rollers and bars to be set exactly. On older machines no adjustment may be possible. Each element may be sealed in a machined housing which stays fixed, and because of the heavy castings formerly used for the main bodies, these housings are usually accurate. If wear has caused misalignments, it is possible to place thin machine shims under the bearings to adjust the level but it is usually better to replace the worn part, if a replacement can be found.

As an additional variation, the infeed roller (and sometimes the chip breaker) may be of one solid piece or of a sectional construction. With a solid roller, only one piece can be fed at a time for if two pieces were fed, the thicker one would push the roller up leaving the thinner one free to be thrown back by the cutters. A sectional roller is divided into many small segments each of which is sprung internally to allow many pieces to be fed through at a time.

One comical effect of this is the requirement by British safety laws either to fit the expensive sectional roller or to put a sign which costs a few pennies on the machine to read "Feed only one piece at a time".

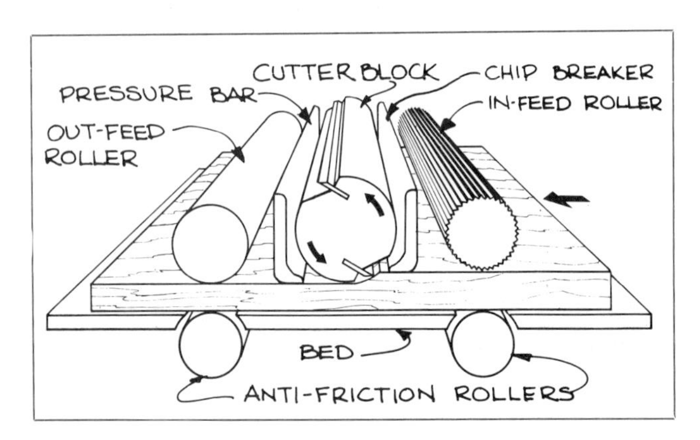

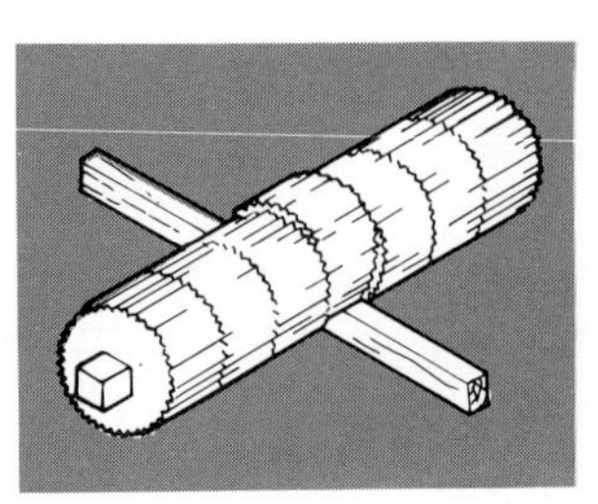

Infeed rollers. A solid roller (left) and a sectional roller (above).

Machine setting On newer machines, it is not usually necessary to adjust the height of the top rollers and pressure bars but only to set the blades and bottom rollers correctly. The forces and vibrations involved in thicknessing are considerable and with continued use, one or more of the elements will eventually go out of alignment. So it's important to understand the mechanics and to be able to deal with any problems that arise. There are various telltale signs which enable the operator to tell from a board going through exactly which part is not working properly. These are listed in the table overleaf.

The first step is to set the blades parallel to the bed and all to the identical cutting circle. On old thicknessers, there may be no vertical adjustments on any of the bearings.

Assuming the rollers and bars can be adjusted, start by setting the blades. There are many sophisticated devices available for blade setting, most of which work off the cutting block. Others use the pressure bar as a datum. The old-fashioned way in using two identical wood strips is, in my view, still the best and certainly the cheapest. The two hardwood strips about 1 × 4 × 24in long, are planed at the same time to guarantee identical dimensions and put in a safe place for future use.

Lower the anti-friction rollers by adjusting the set screws under each end. Then stand the wood strips one on either side under the block. Place the cutters in approximately the right location by eye, then adjust the bed height until the blades just touch the strips when rotated. Now do the fine adjustments until each blade just brushes the wood at each end. A fine dusting of chalk on top of the strips, helps to gauge the setting. If the cutter comes up with fine wood particles, the blade is too low in the block; if it has a bit of chalk on it, it is set just right and if there is nothing, it is too high.

As an alternative use only one stick, and move it from side the side, guaranteeing an identical height.

The next job is to set the height of infeed roller and chip breaker $\frac{1}{16}$in below the cutting circle, that is the tips of the blades. To do this, leave the sticks in place and lower the table $\frac{1}{16}$in. If an exact gauge is not fitted on the machine, use a $\frac{1}{16}$in stip to gauge the space between the stips and the cutters.

Then release the adjusting screws at the bearings either side of the infeed roller and chip breaker and let these just rest on top of the wood strips. Lower the set screws a bit more before slowly raising them until they just make firm contact with the bearings. Finally tighten the lock nuts to lock the set screws at this position. Don't overtighten since the screws may need readjustment.

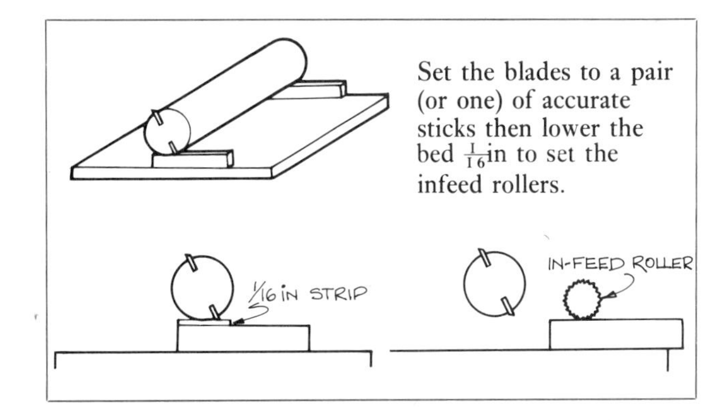

Set the blades to a pair (or one) of accurate sticks then lower the bed $\frac{1}{16}$in to set the infeed rollers.

Tension springs on old machine.

The rollers and bars hold the wood by a combination of their own weight and spring pressure. Adjust the spring tension by first unloosening the screws to free the springs, then tighten. From the point where they first start to compress, tighten the springs about five to six turns depending on the spring. A trial run will tell whether more or less tension is required. Make sure to adjust both ends to the same tension.

The outfeed roller and pressure bar are set just about $\frac{1}{64}$in above this point (i.e. just over $\frac{1}{32}$in – below the cutting circle). So raise the table $\frac{1}{64}$in and repeat the procedure for the back bearings.

Finally set the anti-friction rollers. These, too, must be perfectly aligned. Raise the table to get more access underneath, and adjust the set screws until the rollers are just above table level. Test both ends of each roller either by holding a steel straightedge over them as you adjust or by raising them until the wood strips will just move the rollers as they are pushed along the table with downward pressure. Theoretically they should be less than $\frac{1}{64}$in above the table but the only way to finally check is by a trial run.

Planing faults: cause and effect
The following indicators will allow you to pinpoint faulty adjustments in the thicknesser by observing a trial piece being planed. In practice these faults will also occur during operation, and a bit of running adjustment here and there is usually necessary.

Starting at the infeed end in the order of which the wood meets the various components:

EFFECT	CAUSE
Wood isn't grabbed by infeed roller.	1 Table too low or 2 Infeed roller too high.
Wood stops. Infeed roller doesn't rise to take extra thickness into account.	1 Infeed rollers too low. 2 Pressure too high. 3 On old machines the feed drive is by belts which may be too slack or slipping.
Roller grabs wood but feed stops.	Front anti-friction roller not high enough.
Wood turns diagonally.	1 Usually anti-friction roller is not parallel with table. Failing that: 2 The feed roller is out of parallel.
Wood fed by roller but stops immediately afterwards.	1 Chip breaker too low. 2 Chip breaker pressure too high. 3 Feed roller fractionally too high. 4 Anti-friction roller fractionally too low.
Wood meets cutters but chatters and leaves deep cuts at front of board.	Pressure on chip breaker too low, and anti-friction roller too high.
Wood see-saws slightly.	Anti-friction roller too high.
Wood is cut then immediately stops.	Pressure bar too low.
Wood cuts but sticks soon after.	1 Pressure bar spring tension too high. 2 Outfeed roller too low or pressure too high.
Wood feeds badly, especially at end.	1 Outfeed roller too high. 2 Outfeed roller pressure too low.
Wood chatters at end and leaves deep cut marks.	1 Pressure bar too high. 2 Pressure bar tension too low.
Wood see-saws slightly at end, causing ripples at end of board.	Back anti-friction roller too high.
Wood turns diagonally.	Back anti-friction roller not parallel with bed.
Wood feeds well but shows ridges near front and back ends.	Both anti-friction rollers too high.
Wood feeds and cuts well but board faces are not parallel.	1 Cutters not set parallel to table. 2 Bottom table out of alignment. (This is prevalent on old machines).
Cutters produce bad, chipped finish	1 Cutters dull. 2 Board fed the wrong way round. 3 Cut too heavy.
Small indentations appear on the board, usually on the top surface.	Wood chips are not clearing properly. Those that fall down onto the board get squeezed between the board and the outfeed roller.

As a final test, run a board through the planer. The board should be smoothly cut to a good finish. Then feed the the board through again without adjusting the table. The board should feed through without cutting and without leaving any teeth marks.

Using the thicknesser Once the machine is properly adjusted, planing is very simple. Simply select the cutting thickness, adjust the table height accordingly and feed the board through.

Even though this is straightforward, there are still several points to watch. Long boards should be supported at the beginning until they are well into the thicknesser, and again be supported when they come out. Some machines incorporate supporting rollers at the front and back but if not, a temporary support (or "dead man") can be placed at the outfeed end.

One common problem occurs when an unsupported board leaves the planer and tips forward so that the back end gets caught by the top part of the planer. The board following then gets wedged in between this tilted board and the table and can't be released until the machine has been shut off and table lowered.

Always feed in the wood with the grain for a smooth surface. If you are feeding a lot of pieces at the same time, make a habit of placing each piece nearby so that it points in the right direction. That way, you just have to think about grain direction the first time the board goes through.

Stand to one side of the board when feeding it in to avoid the danger of kickback and never, never look straight in behind the piece to check its progress.

Usually several or many pieces are thicknessed at the same time to bring them to a common dimension, according to the cutting list or drawings. All the pieces are first trued up on the jointer or overhand planer and piled up to go through the thicknesser. The difficult part is to determine the setting for the first cut, because the boards will vary considerably in thickness along each board and from board to board. Basically you have to find the thickest part of the thickest board and use that to determine the setting. This often means that many boards go through first time without cutting or feeding.

Each operator develops his own methods of dealing with this, sorting the boards out by experienced eye as they come off the jointer. To be safe, it's better to feed the pile through once extra rather than having many of the pieces get stuck because they are too thick. After the first run or two, most of the boards should be cutting smoothly.

The depth of cut varies from next to nothing to about $\frac{1}{8}$in depending on the species and on how much has to be taken off. It is good practice to get close to the eventual thickness with coarse cuts then to make the last cut with a fine cut for a better finish.

The thicknesser, set correctly, will usually deliver a better finish than the jointer, so it's a good idea to turn the board over, once the second face has been established, to plane the original face as well. If a lot of wood has to be removed, the wood should be turned to cut alternative faces, so as not to leave the two faces with a different moisture content, which is often not evenly distributed through the board.

As the board is fed in, it may swivel sideways, which is an indication of badly set anti-friction rollers. When the swivel is slight, it is possible to tap it straight until it grabs, but make sure not to get fingers caught under the board and correct the fault as soon as possible.

Remember that only one piece should be fed at a time, unless a sectional infeed roller is fitted.

Planing very thin stock, say $\frac{3}{8}$in or less in thickness, should be done by first placing the board on top of a well-planed backing board which runs on the bed with the thin board on top, both pieces going through together. It's a good idea to rough up the top of the backing board slightly with a saw blade to make sure the grip between the two is firm.

As the first piece goes through without support, it may drop and get wedged against the planer body by the next piece. The machine should be shut off at once to free the pieces.

SPINDLE MOLDER OR SHAPER

Vertical spindle machines are called shapers in America and spindle molders in Britain. They differ in detail but are essentially the same; vertical spindles driven at variable speeds mounted under a table on which the work is fed into cutters. There are many types of cutters and various spindles or "heads" which hold the cutters. In Europe, the "French head" spindle is prevalent and extremely convenient, since it allows the operator to grind cutters to almost any shape quickly and cheaply. The equivalent cutters in America would be the slotted collar pairs which also allow grinding to suit individual requirements, but these are slightly more cumbersome to prepare since two identical shapes must be ground.

Whatever the spindle is called, it is probably the most useful, if under-used, machine in the workshop. I say under-used, because outside the furniture trade where it's use is vastly extended by much ingenuity and jigging, it's versatility isn't generally appreciated. For it will not only cut shapes and rabbets on edges, but it will also plane straight or curved surfaces, cut tongue and groove joints, dovetails, box joints, tenons and even do planing and thicknessing.

Of course for those just setting up a shop, the shaper doesn't have top priority. As a useful substitute, the portable router can be turned upside down and mounted under a table, as a lightweight spindle as shown on page 74.

The basic components of the spindle molder are the motor, belt and pulleys, the vertical shaft or spindle with a cutter holding device on top and the table with various grooves and threaded holes to hold guards and fences.

Motor The power of the motor varies from about 1hp for the lightweight models widely available in America, to about 5hp for the large, heavy-duty single spindles used in production work. Spindle speeds, which must be varied to take into account the diameter and the balance of the cutter, go from about 3000rpm for large square cutter blocks, up to 15000rpm for special, small diameter cutters. The speed is usually changed by moving the driving V-belt to a new pair of pulleys, allowing four forward speeds. The spindles rotate counterclockwise and are fed from right to left, but on some machines the direction can be reversed to allow feeding from the other direction.

Spindle The drive pulleys are mounted directly to the vertical shaft which is carried on heavy ball bearings. Larger machines feature a brake for quick changing of cutters or for safety use and a locking mechanism whereby the spindle is held to allow tightening of the locking nuts.

The precise arrangements for changing spindle heads and for mounting the cutters in these heads vary from model to model. On American models, spindles of different lengths and diameters can be fitted by removing the tie rod nut at the base of the

The underside of a heavy-duty spindle molder shows the various controls. The lever at the top is a brake which can be operated by the knee, if necessary, to stop the machine in an emergency. The main handle in the front locks the spindle for tightening nuts, etc. when mounting the cutters. The right handle locks the up and down travel of the spindle.

Left: The fence mechanism on a heavy-duty spindle molder. The entire fence slides front to back, controlled by a dovetail-shaped nut held in the dovetailed groove of the table. The knob on the left allows fine adjustment after the fence has been locked in position. Right: The locking nut on the same spindle molder.

spindle assembly. On industrial models, the heads are frequently changed to alter the use of the machine. The removeable spindle is locked onto the vertical shaft and held in place by a locking collar on some models, but again precise details vary.

There are two basic spindle types. The plain spindle has a threaded top to take the locking nut. This type takes any type of cutter or cutter holder which has a center hole of matching diameter. This may be ready-made molding cutters, slotted colllars, square blocks, Whitehill blocks and a variety of other cutters, including small saw blades for grooving work. The cutters or blocks are held tightly in place by tightening pressure from the locking nut. Spacer blanks or ball bearing collars, explained below, may also be fitted above, below or between the cutters.

French head spindles of various diameters, on the other hand, hold the cutters in a vertical slot in the center of the shaft. The shaped cutters are ground from bars of high speed steel of varying width and thickness, which must fit into a spindle slot to match the thickness which may be $\frac{3}{16}$, $\frac{1}{4}$ or $\frac{5}{16}$in, or their metric equivalents. French head cutting is not only quick and cheap, but because the cutters cut at a 90° angle to the wood, the true profile can be ground on them saving time in preparation. For a comparison between French head cutters and other cutters such as slotted collars, refer to the diagrams on page 123.

Table The sturdy cast iron table top is similar to the table of the saw. Some have built-in grooves for using a miter gauge and various other sliding holders. All have facilities for mounting straight fences, which are usually in two parts with separate adjustments, and also for holding the range of guards and hold downs necessary to operate the spindle molder safely.

Above: Tightening locking nut. Below: Two types of spindle with slotted French head spindle on the left.

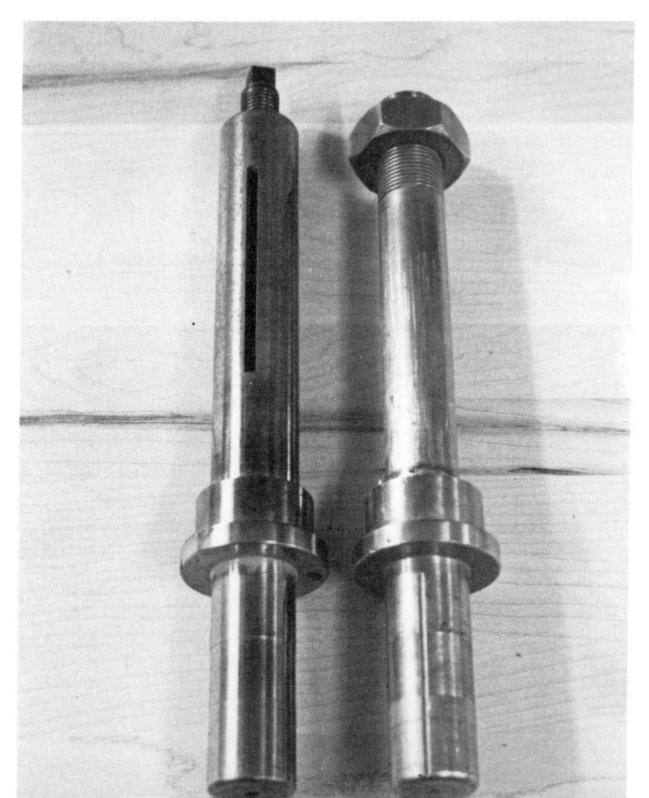

Cutters

1 FACTORY MADE CUTTERS These are machined for perfect symmetry and balance and can be bought in many shapes and sizes from specialist suppliers. The exact types available vary from country to country. *Three lip cutters* are most common and are relatively safe to use. A complete set gives a vast range of possibilities. They are either high speed steel or carbide tipped. The former are easily sharpened by face honing as the profile is ground on the back side of the cutters. Each cutter is set radially to the spindle center so the cutter profile is the true cut.

Two wing solid profile cutters are similar but used primarily by production shops since their high cost must be offset by repetition work.

Grooving blades or saws, used to cut grooves, rabbets and, in sets, to cut box joints are fitted to the solid spindle and held in place against spacers by the locking nut. Exact configurations vary from the two or four wing, tipped grooves to the small diameter fine tooth sawblades.

Factory made cutter blades are sold to fit into various holders or heads. They are used in sets of two, three or four depending on the arrangement. Square head cutters, used less frequently today, are fastened to a large square block with bolts held in dovetailed slots.

Smaller pre-made blades fit the Whitehill disk type holders which are popular for tenoning and grooving. They, also fit slotted collar holders in which matched pairs of cutters can perform a variety of jobs from molding and rabbeting to straight planing.

Molding cutters used on the three knife safety cutter head are also factory made. Although the head used with the shaper is smaller than those used for the table or radial arm saws, the knives are interchangeable.

Throw away tips which fit into special blocks are now widely used by production shops for rabbeting and especially for trimming plastic laminate edges. The small carbide blades have four sharp edges. When one edge is dull, the blade is turned 90° until all four edges are blunt. It is then thrown away.

On first costs it is more expensive, but considering the savings in down time and sharpening, it is quite economical for production work.

Right: Square head block with factory made cutters. Square heads are quite dangerous, but cut quickly and smoothly for production work. Below: Whitehill block can take ready-made or shop ground cutters which are mounted in pairs on opposite sides.

Solid three lip cutter and some of the profiles available.

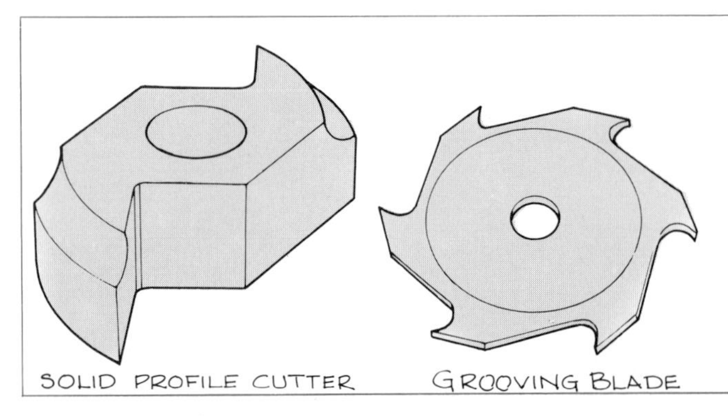

Other factory made cutters. Two wing solid profile cutters (far left). Grooving blades (left) and three knife safety block and cutters (right).

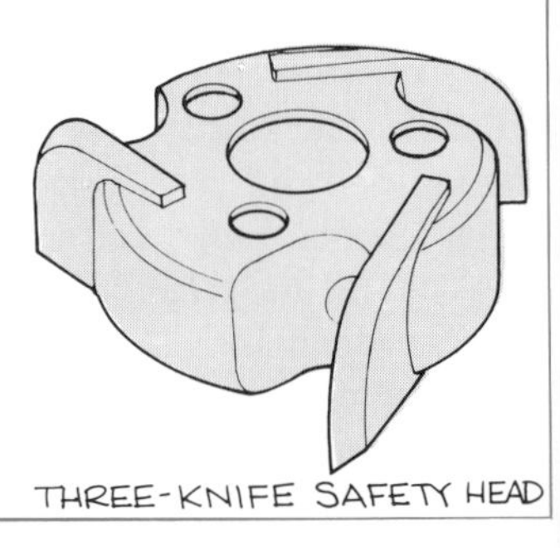

2 SHOP MADE CUTTERS Slotted collar knives and French head cutters (shown above) and also Whitehill cutters can be made in the workshop to any shape and are therefore much more interesting to the individual craftsman. Almost any shape is possible, so that in restoration work, for example, an old molded shape for which there is no standard cutter can be accurately copied. It is important to point out that both types are considered quite dangerous to use and take experience and care in mounting correctly.

The three types work differently. The French head cutters meets the wood at 90° top dead center so it cuts a true shape. That is, the shape of the cutter is reproduced exactly on the work. The slotted collar and the Whitehill knives, however, cut at an angle to the work so the exact shape is not reproduced. There are geometrical methods of working out the cutter shapes but most often the experienced eye can cut the shape adjusting it intuitively to take into account the slight distortion.

The cutting action differs too. The French head cutter cuts like a scraper. This limits its depth of cut somewhat but produces a very fine finish. The edge must be carefully ground and burnished before the cutting edge is turned over as for a scraper and needs constant renewing to stay sharp. The slotted collar and Whitehill knives on the other hand cut like a planer blade, so the sharpening procedure is grinding, and honing as for a chisel.

Steel blades can be bought in various widths and thicknesses for either type of cutter. In America, knife blanks for slotted collar cutters are usually sold in 24in lengths in high carbon steel or in special hardened alloy steel which although more difficult to sharpen keeps its edge much longer.

French head steel has a lower carbon content to be easier to grind. It is sold in the UK in longer lengths and priced by weight. Thicknesses vary, $\frac{3}{16}$, $\frac{1}{4}$ or $\frac{5}{16}$in being the most common.

Two additional types of cutter. The dovetail cutter (left) fits in a special spindle and is used with a jig to cut dovetails or dovetailed housings. The similar stair string trencher also mounts in a spindle and cuts a straight or a beveled groove for housing treads using a special template.

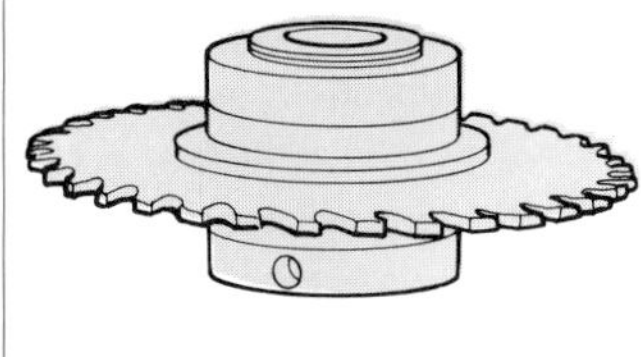

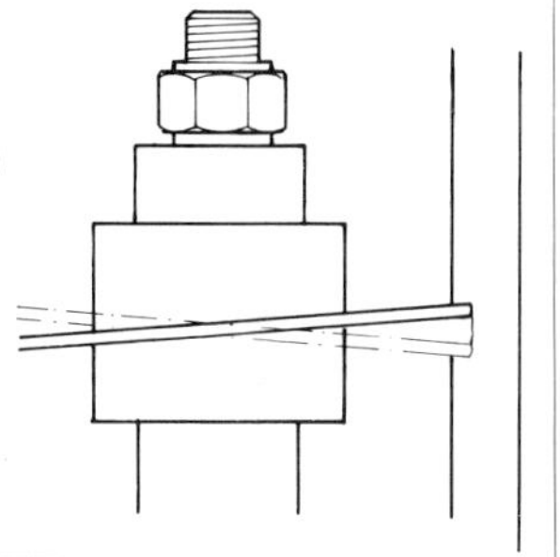

Wobble or "drunken" saws can be adjusted to cut grooves of various widths.

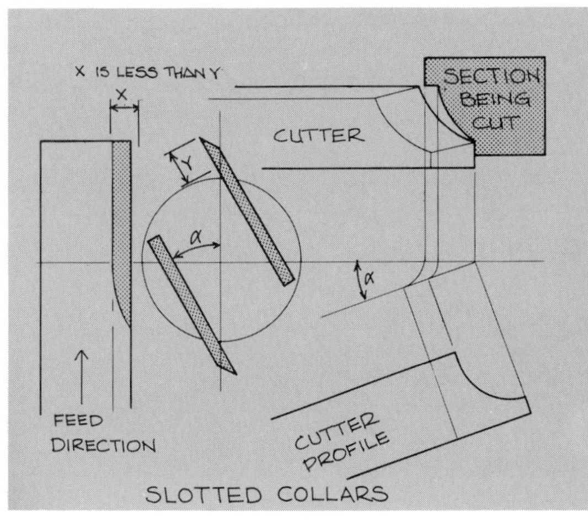

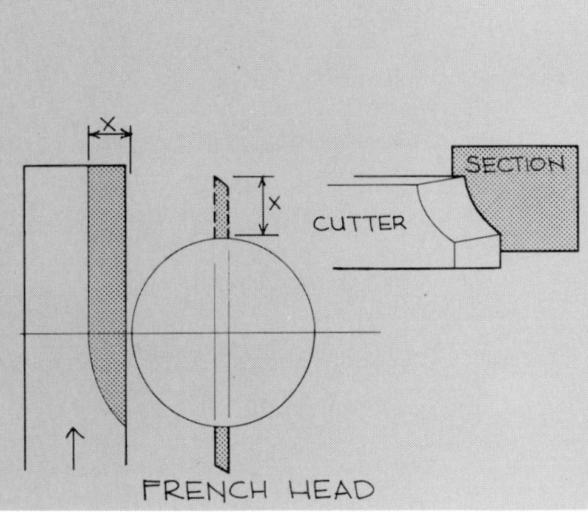

Working out cutter profiles for the slotted cutter or Whitehill cutters can be done by trial and error or geometrically, as shown. French head cutters meet the wood at 90° top dead center and therefore cut the true shape.

Safety

The spindle molder is considered one of the most dangerous machines in the workshop. Records show that many accidents occur due to the cutters flying from the spindle and this is a danger not only for the operator, but for anyone in the shop. Without proper training and experience, it is better to use the solid factory made cutters, such as the three lip cutters which mount directly onto the spindle without any danger of flying off. When using slotted collar or French head cutters which are held by friction, it is important to take precautions such as rejecting deformed or worn out cutters and regularly inspecting the seatings and the threads on the bolts holding the cutters. Make sure they are not deformed from excessive tightening and check tightness periodically during a long run.

Another inherent danger of spindle molders is the difficulty of properly guarding the cutters during certain operations. Where it isn't possible to fit the guard, more care must be exercised by using push blocks to feed the work, working the cutters on the underside of the workpiece and using fences and springs to hold the work in place.

Keep in mind that no single type of guard will work on all types of spindles and cutters, and for proper protection, which is necessary for production work, various guard designs will have to be considered to suit the particular circumstances.

Auxillary fences and even an auxillary plywood base, provide additional protection from the cutters and should be used whenever possible. Small pieces should be held in a jig or holder to keep fingers away from the cutters.

When cutting against collars or using a template, extra care must be taken, first to attach the workpiece to the pattern properly and second to feed it carefully, particularly at the beginning of the cut. The cutters are very difficult to guard for those operations, but use of safe holders minimizes the danger.

Serious accidents have happened as a result of the workpiece working loose from the template. Make sure the pins or spikes project at least $\frac{1}{8}$in into the workpiece and that it is securely held.

In conclusion, as with all woodworking machines, the best safety measure is common sense and alertness. A good, healthy respect for this machine usually guarantees that the user will operate it safely.

Making French head cutters Shaped cutters, mounted in a French head spindle, are cheap and quite quick to make. An experienced man can make a cutter and set it up in the spindle in less than half an hour. French head cutters cut at 90° to the wood, hence their shape is reproduced exactly, unlike the slotted collar knives.

In brief, the procedure is as follows:

1 Draw the shape on a piece of cardboard or plywood.

2 Determine the length of bar required and cut it to length.

3 File the face of the cutting end with a second cut millsaw file.

4 Grind the shape with the correct clearance angles freehand on a grinding wheel constantly comparing to the drawn profile.

5 File in any sections which were too delicate to grind with shaped files.

6 File over the entire bevel with a second cut millsaw file until smooth and perfect fit to drawing.

7 File over the face of the cutter and the bevels with a no3 half round Swiss file.

8 Burnish cutter on face side to harden edge, then turn edge over from other side.

It's best to draw the shape clearly on plywood. The drawing should really show the entire cutter worked out to the correct length, as shown below.

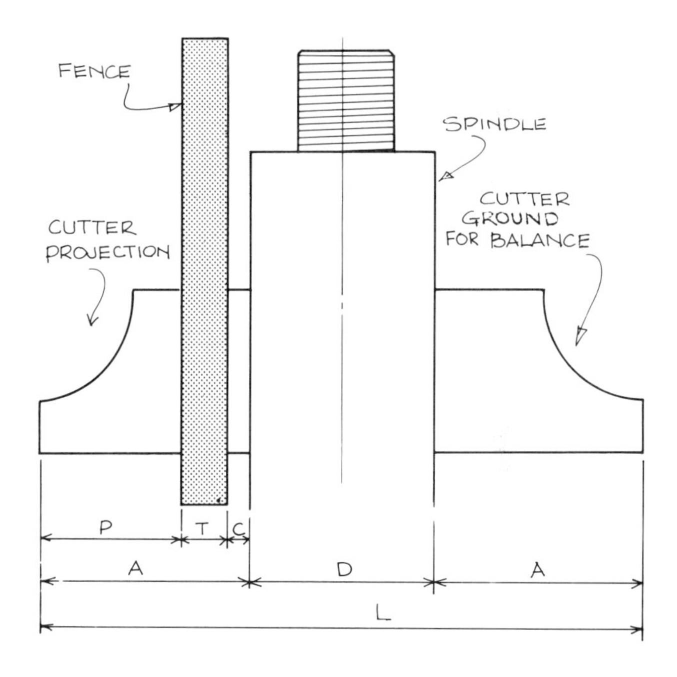

To describe these steps in more detail, the first procedure is to draw the profile on a piece of plywood. This can be a standard section or it can be a completely original section to match a design or perhaps an outline traced from an old molding to be copied.

Then determine the length of cutter. The cutter has to extend through the slot in the spindle and through the wood fence before the profile is exposed to do the cutting. And it must extend out the other side of the slot an identical distance for balance. So the fence thickness T and the cutter projection P is added to the clearance C required between the spindle and the fence. (Usually taken as $\frac{1}{4}$ to $\frac{3}{8}$in). This sum, A, is doubled to allow balance projection and added to the spindle diameter D.

Length $L = 2A + D$, where $A = T + P + C$.

Note: When working directly off the head no clearance C or fence thickness T is added.

Choose the steel of the right thickness to match the spindle slot and of sufficient width for the profile. To cut off a length, score it deeply on the back side with a hacksaw or grinder and snap it off in the vise.

French head steel comes in lengths of various widths and thicknesses. After working out the exact length required, score it heavily on the back and lightly on the front. Place it in a vise and snap it off. A firmly mounted metal vise is more suitable than the woodworker's vise shown.

Notice that the profile is ground with a clearance angle varying from 45° on vertical sections to about 10° on horizontal sections.

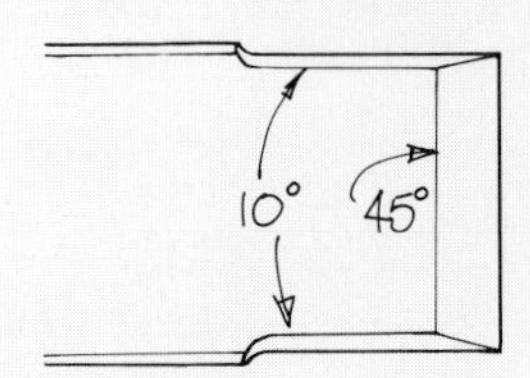

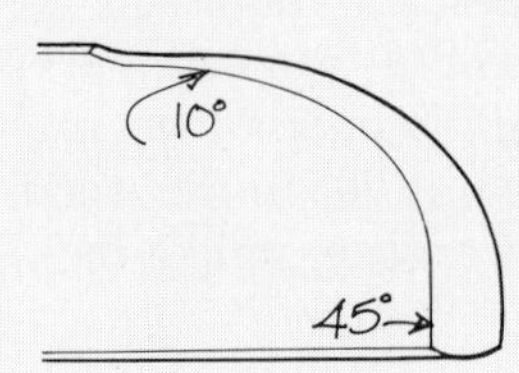

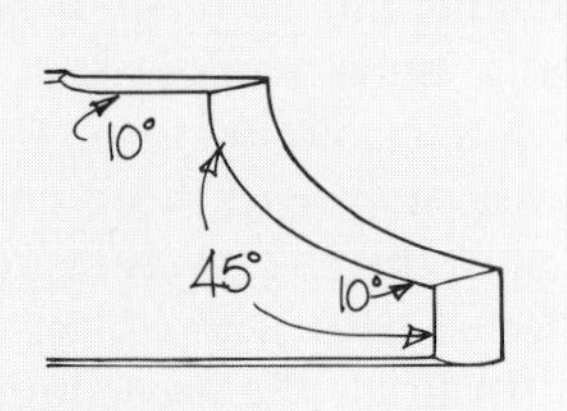

The sequence of making French head cutters: first grinding, then smoothing with millsaw and Swiss files and finally burnishing and bringing over the edge.

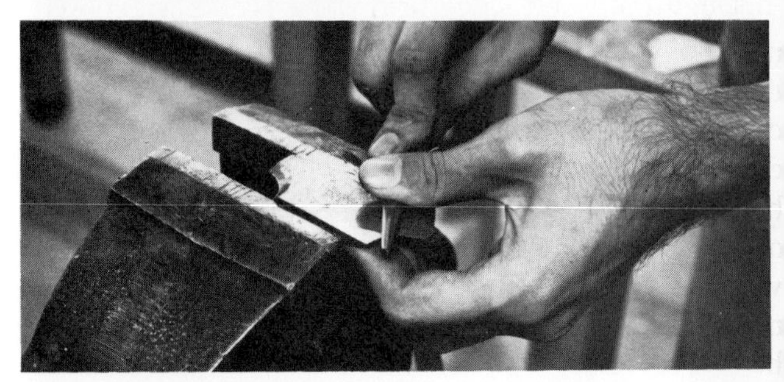

Grinding the profile can be done on almost any grinding wheel but a medium to fine grade (about 60F) is about right. Before grinding, file the face of the cutting end with a second cut millsaw file.

Place a bucket of cold water next to the grinder so that the cutter can be dipped regularly to avoid overheating the tool edge (and to make it possible to hold the cutter).

It helps to draw the profile on the steel as a rough guide, but the final shape should always be matched to the accurate drawing. Most of the waste is taken off quite quickly, but take care when getting near the profile.

The edges are ground at an angle to give clearance behind the cutter. The clearance angle behind the vertical edges should be about 45° and that behind the horizontal edges about 10°. The latter is required partly to aid in cutting through the auxilary fence as the spindle is set up.

Both angles are formed in one smooth motion as the profile is cut. Beginners always approach grinding very cautiously and make small, tentative cuts, but an experienced man does it quickly, like magic as is usual with years of practice. The best way to learn is to practice on a few cutters until you relax. Hold the cutter to the wheel at a 45° angle so that the vertical parts are cut at that angle. Move the edge back and forth smoothly, and transfer the cut to the side of the wheel gradually as the the cut approaches a horizontal part, and for this part tilt the cutter sideways slightly for the 10° clearance angle.

In the beginning it will be difficult to make the transition smoothly from the two edges with their different angles, but with practice, the motion becomes quicker and almost automatic. Don't forget to quench the steel frequently to prevent blunting the edge.

Very fine beads or sharp corners are best filed by hand using flat, half round or rat tail second cut files until the section matches the drawing exactly. Remember to keep to the same 45° and 10° angle when hand grinding.

The other end is cut to roughly the same shape for balance only. This edge won't cut, but at the high speeds required, fine balance is necessary. So the projections either side of the spindle should be just about equal in weight and shape. The balance edge is always cut a bit more, so that the shape will not interfere with the other cutting end. On all but the largest cutters, this balancing shape can be just a rough approximation.

Filing the face and edges smooth is done with the cutter held in a vise, first with a second cut millsaw file, then with a no3 half round Swiss file. Having filed the entire ground edge with the various second cut files, hold the cutter horizontally and file over the face to a fine, smooth finish with a no3 half round Swiss file. Work carefully to get rid of all the marks. Then with the cutter held vertically, and working with the round part of the Swiss file, polish the entire cutting edge. The object is to achieve a smooth edge without any rough file marks so that the burnished edge will be perfectly even.

The final step is the important part. With the face nearly mirror smooth, run the burnisher hard against the flat edge to harden the edge. Finally turn the cutter over and with the burnisher held at half the clearance angle, turn over the cutting edge. It should be regular like a hooked edge without any scratches or jagged edges. The action of the burnisher is to work harden the edge and to smooth down any micro-scratches from filing. Burnishers can be bought but are best made from old Swiss files, whose teeth are removed by grinding before the steel surface is highly polished to produce one thin rounded edge.

The turned over cutting edge will need renewing after a while by burnishing the edges to produce a new cutting edge.

MOUNTING THE CUTTERS The solid section, factory made cutters are the simplest to mount. This includes the three lip cutters, grooving saws and solid profile cutters. These simply slip onto the spindle and are tightened down with the locking nut. The collars which can be placed below, above or in between cutters have other uses besides acting as a spacer. The edge of the wood can ride on the rotating collar as a guide while the cutter above or below does the cutting. Plain collars often burn the work. As a substitute, the more useful ball bearing collars can be fitted and this eliminates the burning.

Most large spindles have locks operated by a lever from the front which engages into a notch in the shaft which keeps the spindle still as the locking nut is being tightened. As an alternative and less suitable arrangement, the top of the spindle is squared off so that it can be held with one wrench while tightening with another.

Slotted collars require more care in fitting. The danger exists that if the two knives are not exactly the same width or are not set or tightened properly, they can loosen and be thrown from the block with great speed. The knives are held in place by friction from the downward pressure of the locking nut.

First, make sure the slots are clean. Place the bottom collar on the spindle then the two knives in their grooves so that they project the correct amount. On American knives, the top edge of the steel bars

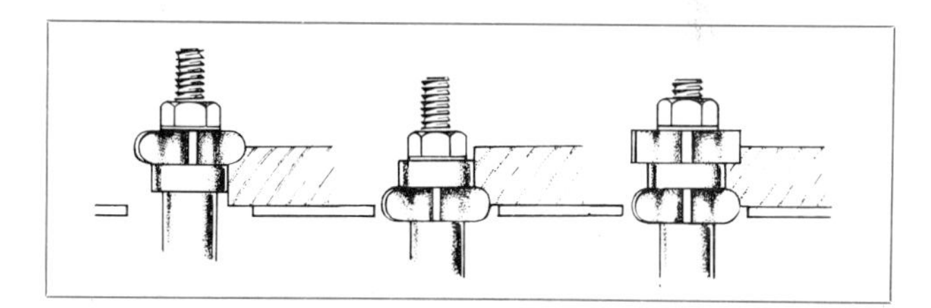

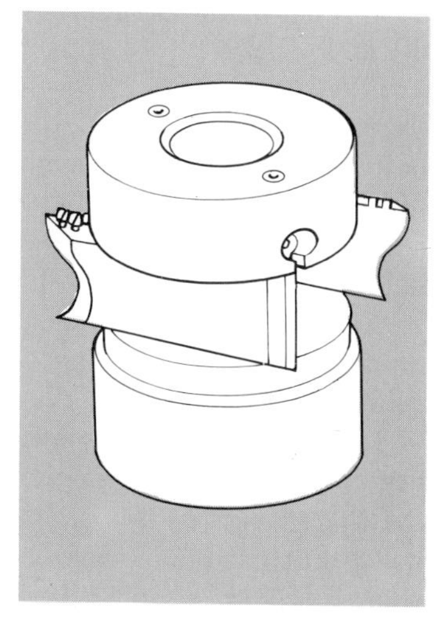

Slotted collars, whether plain or ball bearing, hold the pairs of cutters in place. The collars can also be used as a guide when working "off the head". As shown above, cutters can cut below or above or both at the same time but part of the edge must be in contact with the collar in all cases.

have serrated edges which fit into matching holes in the top collar and thus reduce the danger of the knives being released in use. European cutters have straight edges.

Place the top collar onto the cutters making sure they are properly located in the slots. Adjust the cutters by using an Allen key with the American type and by tapping from behind on the European one. With the cutters properly located, tighten the locking nut first by hand, then by wrench so that it is very tight.

It is important never to use short knives which don't cover at least $\frac{2}{3}$ of the length of the slot. There is always the temptation to break this rule particularly in the hurry of a job that must be done, but I know someone with a missing finger who is sobering proof that it isn't worth it.

French head cutters are mounted in the slot in the spindle so that an equal amount extends both sides, but just slightly more on the side with the cutting edge.

Enough rings are removed around the spindle hole to create a hole large enough so that the cutters can be lowered below the table surface without hitting the table.

The cutter is like the slotted collars held by friction from the downward pressure of the locking bolt. In addition, a spacer made from the same steel section has to be placed between the bolt and the cutter to protect the top of the cutter and to even out the pressure. Score both edges of this spacer with a crisscross pattern using the hacksaw or sharp file and also score the top edge of the cutter to provide a better key.

Tighten the top bolt with the wrench provided, using both hands to provide all the force you can, not by pulling but by bending your arms. When the bolt won't move any further from this, give it one final twist with a short, firm jerk. With this, as with other loose cutters, it's always advisable to start up the machine gradually, stopping and starting, and to place a large wood block in front of the cutters if you are in doubt. The sound of a French head cutter working loose is well known in the trade and the clatter warning causes everyone to dive to the floor in an instant.

Above left: French head cutter mounted in slot with two spacers above. Above center: Removeable rings to accomodate various cutters. Above right: Cutter and spacer with score marks. Right: Tightening locking nut on French head spindle with special wrench. Below: Whitehill block in use.

Other types of cutters such as those for the Whitehill block, are ground and sharpened in the same way as for slotted collar knives. They are designed to cut like planer blades with a clearance angle of about 45° with a final honing to the ground edges with a stone. Whitehill cutters, whether straight or profile ground, are held in the block by locking nuts as shown in the photograph on page 122.

Using the spindle molder or shaper There are several ways of guiding or controlling the work:

1 With a straight fence.

2 With solid or ball bearing collars running against the edge of the workpiece to control the sideways movement, hence the cut.

3 With patterns, attached to the workpiece, running against a collar or ring fence.

4 With a template guide as in dovetailing.

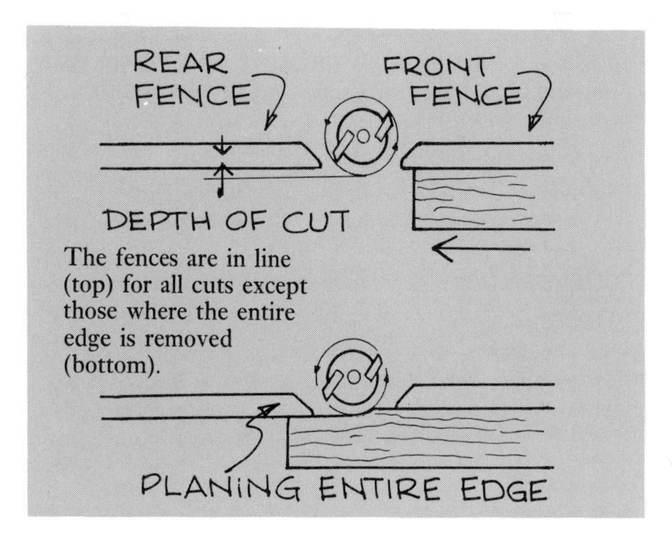

The fences are in line (top) for all cuts except those where the entire edge is removed (bottom).

Below: The thin, auxiliary fence is pinned in place to both fences and the French head cutter protrudes through.

1 SHAPING WITH THE STRAIGHT FENCE Shaping straight edges is the most common way of using the shaper. The fence is adjusted so that it provides a support for the work and also so that it controls the depth of cut. This method is used for cutting moldings, rabbets, grooves, for cutting finger or box joints and for straight planing.

Exact configurations vary from model to model but the fence is usually divided into two halves with separate adjustments, one on each side of the spindle.

For most work, the wood is run against hardwood facing blocks screwed or bolted to the metal fences. The facing strips are cut out of a stable hardwood about $1\frac{1}{2}$ × 6in in section, cut at a 45° angle so that the ends create a narrow opening around the cutter.

For safety, it is important that the work is well supported near the cutter for it has a tendency to grab at the work and pull it into the hole.

Adjust the fences by trial and error to give the right cut, also raising and lowering the spindle as necessary. Again it's always safer to cut the bottom edge so that the cutter is covered by the workpiece. This way gives better results too since any tendency for the workpiece to lift up results in a bump rather than a gouge in the work and a bump is easily removed with another pass.

Make sure that the two fences are exactly in line so that the work is supported when it comes off the spindle. The only exception is when the whole edge is being planed off in which case the outfeed fence has to be set out as on a jointer to support the planed edge.

When cutting with French head cutters, an auxiliary wood fence is pinned to the hardwood facings. It should be about $\frac{3}{8}$in thick and span from the left to the right edge of the table. The normal procedure is to bring the two fences forward of the spindle and attach the fence with short finishing nails or panel pins to make it easy to pull off later. Hammer the heads flush, then lower the spindle to below its normal cutting height. Start the machine and allow it to reach full speed before slowly pushing the fence onto the cutter, allowing it to cut through the fence. At approximately the right projection, lock the fences, then raise the cutter to just above the correct height, then let down slightly to the right cutting level. The 10° clearance angle cut in the horizontal sections of the cutters is there partly to make it possible to cut the fence without burning. Then test the cut with a scrap and adjust as required.

As a safety procedure, an auxiliary plywood base is sometimes clamped to the table top or alternatively nailed into wooden fillets placed in the table slots. The cutter is raised so that it cuts through it and the other fence at the same time.

GUARDS AND HOLD DOWNS When edging stock, the work is held tight against the fence by hand pressure if the cut is light or with various hold downs for heavy cuts.

Spring hold downs (page 124) are common as is the Shaw guard shown in the photograph below.

Alternatively use feather boards to hold the work. See page 86 for instructions on making them.

To meet safety standards, the cutters should also be covered with a cage or bonnet such as the universal guard which allows access to the cutters at the front and for chip extraction at the back. Of course most of theses features are standard only for industrial models but they give a good indication to operators of light duty machines of the safety standards required of this potentially dangerous tool.

The greatest danger is for the rear, feeding hand to come into contact with the cutters. To prevent this, a push stick or backing block is used, particularly for short pieces which can't be pulled through from the outfeed side.

Shaw guards (left) can be arranged in various patterns to press downward and inward. Feather boards (above) are a useful shopmade substitute. In all cases, the head and cutter should be thoroughly guarded (right).

Making a backing block

Make a backing block out of a piece of the same material as is being fed through so that the thickness and width is the same for fitting into the pressure clamps. Screw a pre-made handle to it with no10 screws which should penetrate only about $\frac{3}{8}$in. Place the handle at a slight angle so that it can be pushed into and along the fence. It's a good idea to have two or three handles at hand for making new blocks as they are needed. Make a new backing block for each new set up – removing the handle before throwing the block away. The block is used not only for feeding but also to back up the cut when cutting end grain.

To mold the edges of drawer fronts, for example, the two end grain cuts are made first, with the block backing up the cut to prevent breakout. Then the long sides are cut to cut away any small breakout that may occur. One hand feeds with the block while the other controls the workpiece. Similar guides and holding devices can be made to hold large or tall and other awkward shaped pieces on the spindle as well as other machines.

The tenoning jig from the table saw can be used to hold a piece vertically.

Cutting finger or box joints on the spindle molder is straightforward. The two wing cutters are mounted with matching spaces (left). To use a shop made guide block (below), first mount a thin auxiliary wood fence and adjust it so that the cutters protrude the correct amount. Special sliding table accessories used in production work are safer and quicker (bottom).

For machines fitted with a groove parallel to the fence, a miter gauge fitted with a wood block can also be used as a push block.

On more sophisticated machines, the sliding table tenoning attachment holds the board firmly in place while short tenons are cut by a pair of Whitehill blocks separated by a space equal to the thickness of the tenon.

The spindle molder is the proper machine for cutting finger or box joints. A series of identical two wing, tipped cutters are placed on the spindle with identical thickness spacers between. All the sides are cut in one sweep, then turned around for the second cut. Notice the push block which is used as a substitute for the sliding table accessory.

Small diameter saw blades fitted to the spindle are used to cut edge grooves for tongue and groove joints, rabbets or housings with the wobble collars, or for making raised panels. It is safest to fit an auxiliary plywood base and have the blade cut on the underside of the workpiece. A batten attached to the table raises the panel to the right angle for the cut.

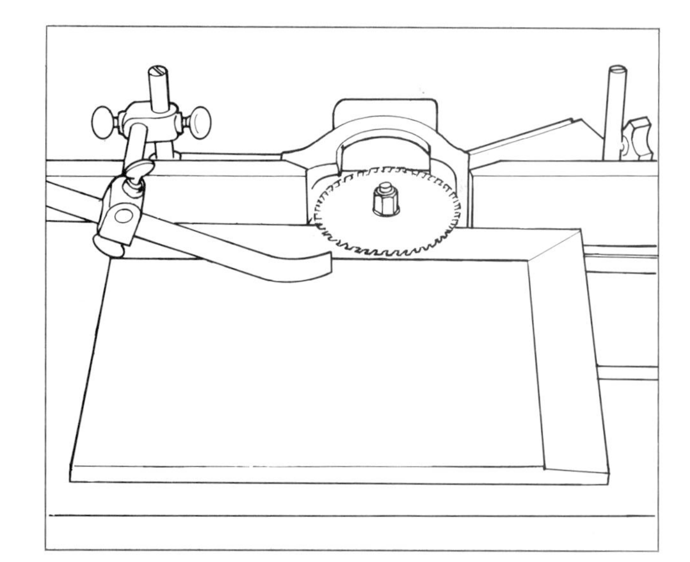

Grooving saws are widely used on the spindle molder to cut grooves, rabbets or (above) to cut fielded panels which requires a batten fixed along the fence to raise the panel to the appropriate angle. Wherever possible, use an auxiliary fence (right) for additional safety.

Stopped rabbets, moldings or bevels are best done with the aid of stop blocks pinned or clamped to the fences. Hold the end against the starting block and feed the work gradually against the cutters, keeping firm control to avoid "snatching".

Jointing or surface planing is straightforward using hold downs to hold the work firmly against the fence.

Thicknessing is more difficult, but can be achieved with great care by moving the fence over and feeding from the opposite direction. Use clamps and push sticks to avoid getting the hands near the cutters.

Shaping the edges of circular pieces is easier with the portable router. To do it on the shaper, pin two triangular blocks to a plywood auxiliary base at the right distance apart to bring the edge at the right contact with the cutters.

The special 6in circular block is similar to that used on a planing machine. It is used with two or more cutters to plane edges or to cut rabbets on the spindle molder.

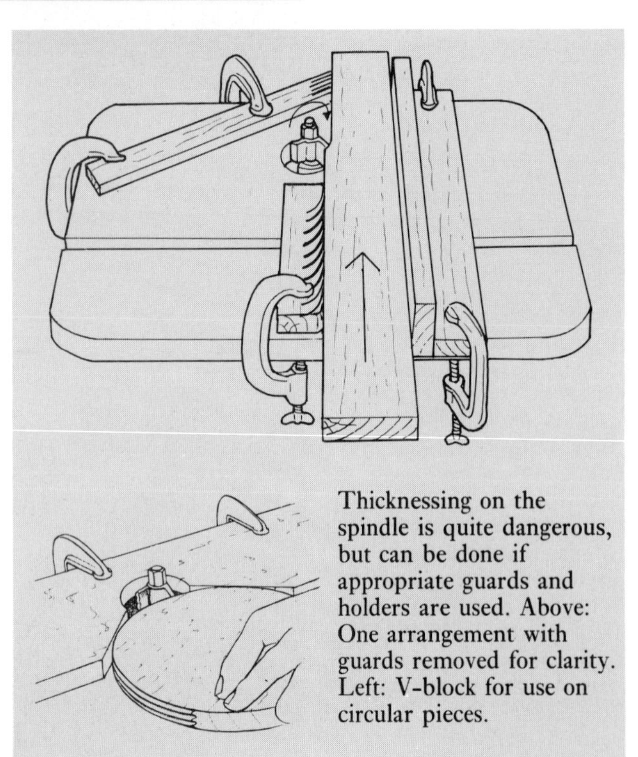

Thicknessing on the spindle is quite dangerous, but can be done if appropriate guards and holders are used. Above: One arrangement with guards removed for clarity. Left: V-block for use on circular pieces.

2 SHAPING AGAINST COLLARS While cutting straight edges is done against the fence, curved and irregular shaped work is best done using the collars as a guide.

The principle is quite simple. The collars, fitted above below the cutter or between two cutters, run against the edge of the workpiece, so that they guide as well as control the depth of cut. Various diameter collars are available to vary the depth of cut. Plain collars rotate with the speed of the cutters and therefore tend to mark and burn the edge. In production work, ball bearing collars are used to eliminate burning and make guiding easier.

With two cutters mounted on either side of a collar, the top and bottom can be cut at once.

Starting the cut is the most dangerous part of the operation. To bring the wood against it freehand would cause it to grab the piece with a jerk. To control the start, a pin or a clamped-on block is used.

The pin is fastened in a hole in the table top and it is used to steady the workpiece at the beginning of the cut. As control can be transferred to the collar, the work is swung away from the pin to continue the cut. It must be emphasized that the edge of the workpiece must be smooth and must be pre-cut to its final shape.

A pointed starting block can be used as a substitute for the pin, but it should be stressed that it must be carefully located and clamped down firmly to avoid having it give way if the workpiece is kicked back at the beginning of the cut.

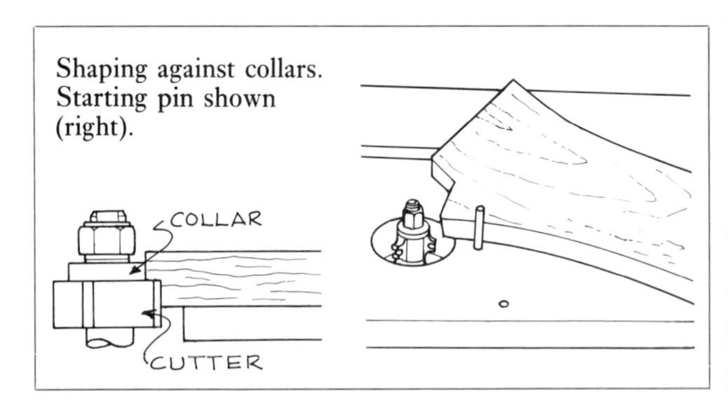

Shaping against collars. Starting pin shown (right).

A leg being cut to shape on the spindle using a template running against the collar below.

3 TEMPLATE SHAPING Template shaping is very common in production work and it is here that large savings in man hours can be made by clever application of jigs and templates.

As with all template work, the template made from plastic or good quality plywood should be accurate with edges which are regular in shape. Any irregularities will automatically be copied onto the workpiece by the cutters.

Template shaping is similar to collar shaping. The main difference is that the template and not the workpiece runs against the spindle collar. This means that the entire edge can be cut which was not possible in ordinary collar cutting. But the main advantage is that the shapes can be rough cut to approximate size on the band saw and then cut out accurately on the spindle. The amount of projection beyond the template can vary from one piece to the next without affecting the finished product.

The template, mounted over or under the workpiece, is held down with quick release clamps for industrial applications but in the small workshop, the best way is to hold it in place with anchor pins whose sharp points project at least $\frac{1}{8}$in to grab the workpiece.

The cut is started as for collar shaping by using a pin to steady the work at the beginning of the cut. On some industrial machines, a ring fence is used for the same purpose. It fits over the spindle and is shaped so that the template can be brought into contact with it outside the cutting circle. Once in contact, the template is brought around until the cutters are engaged.

Template work is as varied as there are numbers of different shapes in furniture. Typical applications include chair legs and shaped cabinet door frames.

4 SHAPING WITH A GUIDE has fewer applications. The main uses are for cutting dovetails and stair stringer housings.

The workpiece is moved over the cutters and its motion is controlled by a template. In dovetailing, the two pieces are clamped in the template jig which is moved over the dovetail cutter mounted on top of a special spindle. The shaft of the cutter runs against the sides of the template.

A few industrial machines can be fitted with an overhead guide pin, much like the overhead router The guide pin runs in the template, fixed to the top of the workpiece and the cutter mounted below copies the shape on the shape on the underside. This arrangement could quite easily be adapted with a bit of ingenuity to do many types of template work.

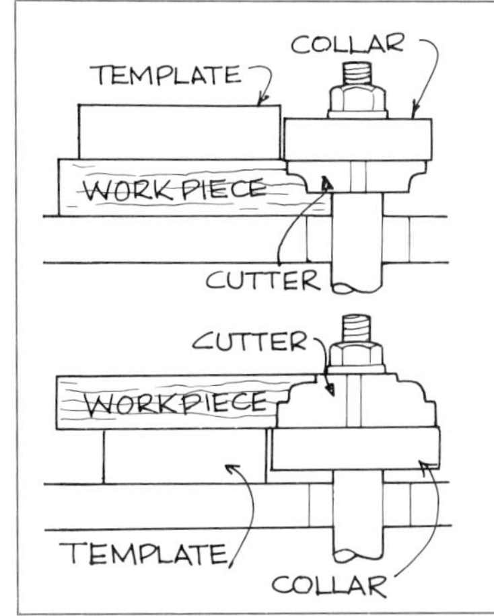

In template work, the template can be above or below the workpiece, as shown. In template shaping, the entire edge of the workpiece can be shaped.

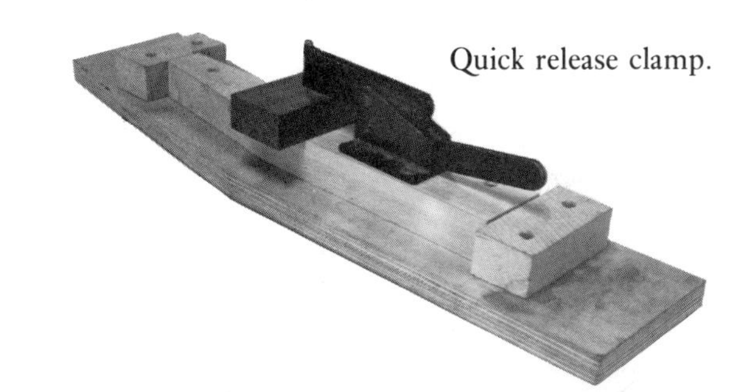

Quick release clamp.

Above: Ring fence with template below the work used as a guide. Right: Dovetail accessory. Below: Template and guide used to cut stairway trenchings.

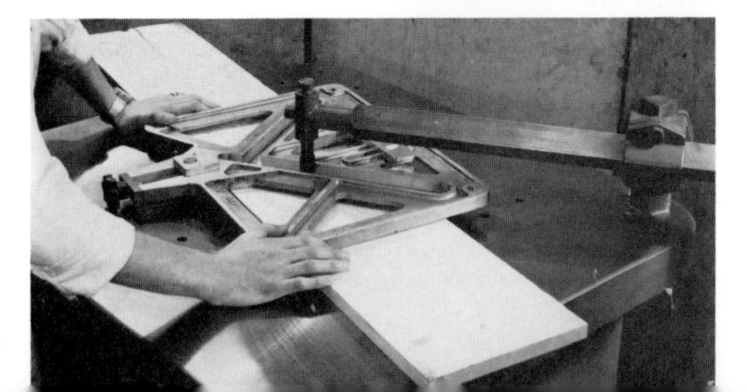

MACHINE DRILLING

The term drilling is loosely used to refer to general purpose or semi-precision hole making for screws and other fixings, whereas boring refers to precision drilling for dowel joints, for example. Drilling is done with a drill press, portable electric drill (see page 54) or with various hand drills. Boring is carried out on more sophisticated horizontal borers with one or more bits operating at once.

Accurate drilling is essential for all woodworking, and holes cleanly cut to the right depth and located exactly, is as much a sign of proper cabinetmaking as tight fitting joints and good finishing.

Drill press It's nearly impossible to drill accurately with hand-held tools. In most workshops the drill press, either a bench top or a floor standing type, is used to guide the drill accurately and to hold the work firmly. The floor standing drill press is more convenient since it has more clearance between drill bit and base to accommodate larger pieces.

The essential features are the motor with a variable speed belt and pulley system, the chuck which holds the drill bits, the table which can be tilting or stationary, and is adjustable up and down, and the base and column which support the other parts. The drill bit, which is tightened into the chuck with a matching key, is lowered into the work by operating the spring-loaded handle or feed lever. Precise configurations vary from model to model but most have a depth stop, a depth locking mechanism and a belt tension knob.

The drill press can be floor standing or bench mounted, as shown.

Prices for drill presses are usually quite competitive particularly for the lightweight models, and it is easy to find secondhand ones. These should be examined carefully for general condition, for missing parts and particularly for play in the moving parts. The small coil spring which retracts the drill is often broken on old drills and replacements may be difficult to find. In this case some form of counter-balancing pulley system can be attached to the feed lever.

As a substitute drill press suitable for lightweight work, mount the portable electric drill in a stand fitted with a lever, and depth stops as shown on page 57).

Using the drill press If possible, locate the drill press on or between long benches so that long boards can be supported horizontally. Successful drilling depends on choosing the right bits, centering accurately and supporting the work properly. For general purpose work such as screw holes, it is enough to mark and center punch the hole, then hand support the workpiece while drilling the hole. But for accurate, precision boring a machine dowel bit should be used and the workpiece should either be clamped down or held against a fence or other guide to steady the work. As with other machines, the use of the drill press can be extended considerably by the use of jigs, templates and holders to aid in positioning and guiding the work.

Drilling speeds vary from about 600 to about 4000rpm, depending on the model, and it is usually changed by moving the belt to another pulley at the top. The lower speeds are for large diameter drilling in hardwoods and for drilling in metal. The higher speeds are for smaller diameters in wood and for auxiliary functions like routing and grooving which are occasionally done on the drill press. A good general purpose speed is about 1800rpm.

For straight boring, adjust the table height and set the depth stop to the correct height to a mark on the side of the workpiece. Most tables have a clamp ledge on either side for holding down small pieces. For through drilling, hold a scrap piece underneath to prevent breakout or better yet, set the depth stop so that the brad point will just extend through, then turn the piece over to make a clean hole on the other side.

Keep in mind that counterbored holes are always drilled larger bit first, which keeps the center location for the next bit.

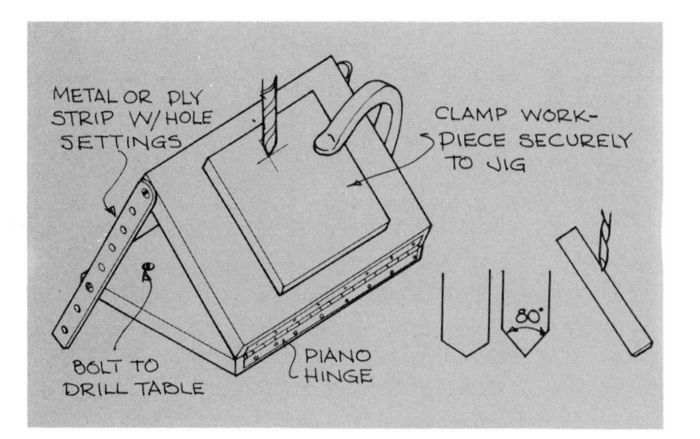

There are various ways of holding down small, awkward pieces using shop made devices but the special metalworkers' holder shown can accomodate various shapes and sizes.

Drilling at an angle requires a sharper grinding angle so that the center point enters before the wings, as shown above. The angle drilling jig made from plywood bolts to the table of the drill press and can be adjusted to any angle. Right: Forstner bit drilling at the edge of the workpiece, a job that is difficult to do with other bits.

Two shop made holders. The block on the right is used to hold small dowels. Protruding pin heads hold the dowel steady. The holder below is used to drill at 45°.

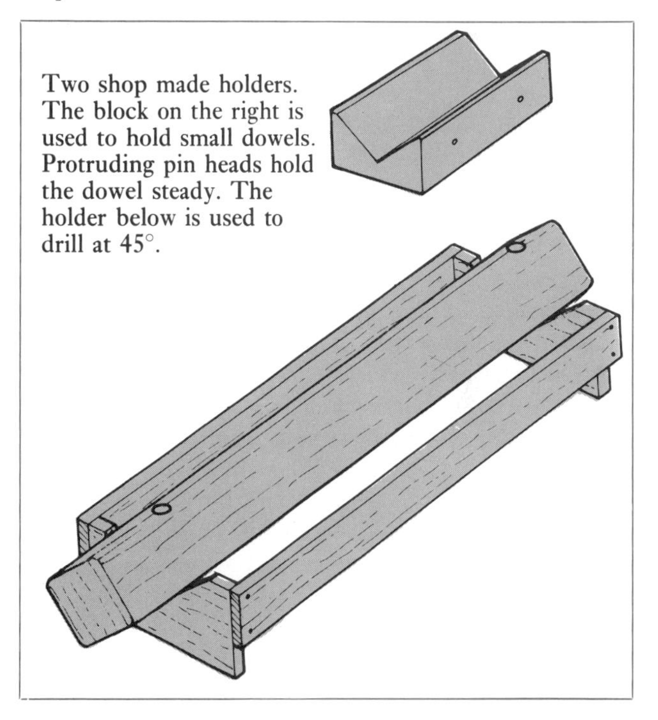

Drilling at an angle is best done by tilting the table to the appropriate angle (checked with a sliding bevel), then clamping the work to it making sure that the drill point will not come through to hit the metal table underneath. Where no tilting mechanism is available, an auxiliary support must be built to hold the work at the right angle.

ANGLE SUPPORT To build a simple support for holding the work for angle drilling, join two 10 x 12in pieces of $\frac{3}{4}$in plywood with a continuous hinge. Drill counterbored holes in the bottom piece positioned so that it can be bolted through the slots in the drill press table. Hold the top piece at an angle using a piece of plywood at either end screwed to pivot at the top with a series of holes corresponding to specific angles drilled along the length.

For acute angles, the pivots of twist drills must be ground to a sharper angle of about 80° and machine dowel bits may also have to be carefully re-ground to make sure the brad enters the wood before the spurs. But for drilling angled holes, such as fixing the rails to table tops as shown on page 226, it is better to use Forstner bits which are guided by the rim rather than the center.

To hold small pieces, either devise a simple clamp system or use a special holder, intended for use in metalwork, which can be bolted to the table. Dowels should be drilled using a homemade V-block with brad points extending slightly through the side to help positioning.

A similar holder can be made to hold squares, such as table legs, at a 45° angle, for example for drilling bolt holes to take corner brackets which support the rails.

Two devices for drilling regular spaced holes. Block and bolt (left) and template (right).

Jig with two stops to locate hole.

Fence bolted to table for lining up holes.

In production shops, various jigs and devices are used to position holes automatically. For adjustable bookcase supports, for example, one way is to clamp a block to the fence with a retractable pin which locates in the previously drilled hole. Another way is to use an accurately drilled template (also shown on page 56) which is pinned or clamped to the work and used as a guide to drilling.

A fence is frequently incorporated in an auxiliary table which is clamped to the drill press table and used with stops to drill repetitive holes in the face of boards.

Dowel boring into end grain is difficult on the drill press because of the limited clearance and the difficulty of holding work upright. It can be done by rotating the table 90° so that the top is parallel with the drill bit, but a better way is to use the radial arm saw (page 103), a horizontal borer or a hand-held electric drill with a drilling jig, (page 57).

Horizontal borer These are production machines for precision boring of dowel joints which are today universally used in industry as a substitute for the mortise and tenon, but are also convenient for the small workshop. To hold the dowel pins properly, the holes must be absolutely accurate and the drill must leave a clean hole suitable for gluing. Burned holes polish the wood and decrease the holding power of the glue, as do woolly holes drilled with blunt bits. Special tipped machine bits are therefore in frequent use to guarantee a proper joint. Horizontal borers can have one or more heads boring at once in any number of patterns which are pre-set to suit the particular job. The exact feeding mechanisms vary, but the most common uses a foot pedal which locks the work in place and brings the bits forward to drill the holes. A large number of variations are possible, particularly with more modern automatic, multiple spindle machines.

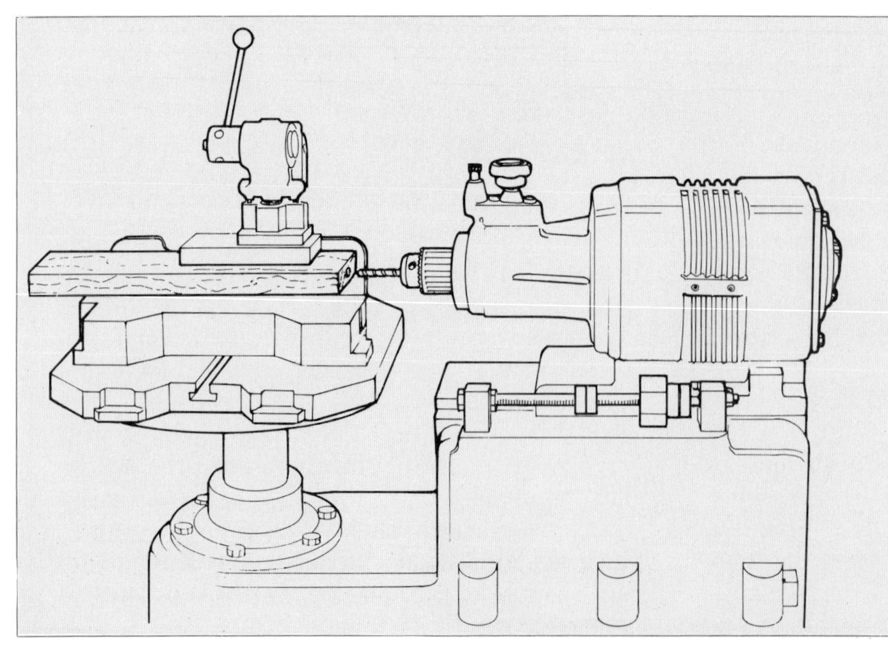

The horizontal borer can have one head as shown here or several heads which drill simultaneously. The workpiece is clamped in position and fed into the drill by a foot lever.

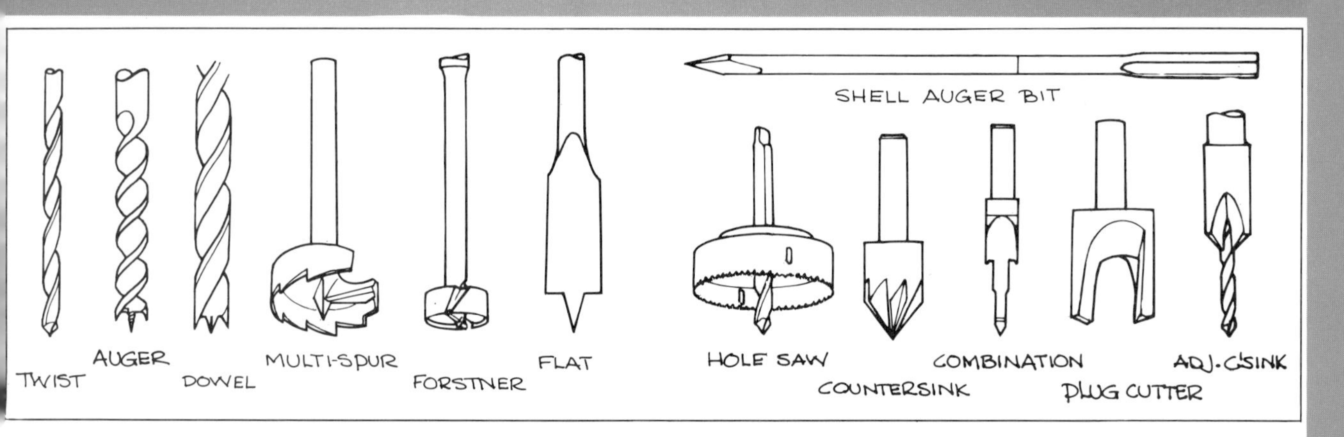

Drill bits

Twist drill bits are technically intended for metalwork but they are nonetheless widely used in woodworking for general purpose drilling of screw holes. They are not as suitable for precision boring for the bit is difficult to center accurately without a point, and without spurs the entry hole is not as crisply cut. But because of the large variety of sizes and ease of sharpening they are widely used. Sizes range from $\frac{1}{64}$in to about $\frac{3}{4}$in diameter in $\frac{1}{64}$in increments. For all-purpose work the enclosed angle of point is normally 118°, but for woodworking only it is useful to grind a sharper angle of about 80 to 85° which aids in keeping the center.

Machine dowel bits or spur machine bits Available in carbon steel, high-strength steel or tungsten tipped, these bits have a center point or brad for accurate centering, two spurs for cutting across the grain for clean entry and two cutting lips for fast cutting. The helix angle of the flute is 60° which is smaller than most bits and which assists in fast clearance of chips. These features make dowel bits perfect for drilling clean, accurate holes in wood, and accounts for their wide use in industry for dowel boring by horizontal borers or by hand with a doweling jig as shown on page 56.

Auger bits are either the Jennings pattern or solid center type. Auger bits with a square tang intended for use in the hand brace can be adapted for machine use by cutting off the tang and filing off the screw center point. With a brad point, spurs and cutting lips, auger bits are similar to machine dowel bits but not as fast cutting since their helix angle of about 45° doesn't allow fast chip clearance. The bit must be repeatedly withdrawn to clear the chips to prevent burning.

Multi-spur bit (saw tooth center bit). This bit has a brad center for accurate centering and multiple cutting teeth for fast, accurate drilling. It is particularly useful for drilling large diameter, shallow holes, and is available in diameters from $\frac{3}{8}$in to about $3\frac{1}{2}$in, though the larger sizes are difficult to find. They are particularly useful for drilling holes in thin pieces.

Forstner bit The Forstner bit is also excellent for drilling shallow holes without the danger of having a point break through. It is guided by the circular rim which makes it suitable for work where centering is difficult such as for holes in round chair legs, for drilling semi-circular holes at edges of wood or for drilling holes at angles. Diameters are from $\frac{3}{8}$in to 3in, though the most common are from $\frac{1}{2}$in to 2in.

Flat bits (speed bits) are inexpensive, easy to sharpen bits suitable only for high speed drilling. Because of their low price and a wide range of sizes from $\frac{1}{4}$in to $1\frac{1}{2}$in diameters, they are popular for general purpose drilling but unsuitable for precision dowel boring.

Hole saws There are several configurations of hole saw but the most popular has a rim of small teeth which fits on a mandrel with a $\frac{1}{4}$in diameter twist drill mounted at the center. Diameters range from about $\frac{1}{2}$in to $3\frac{1}{2}$in.

Adjustable countersinks These fit over drill bits to drill the screw clearance hole and the countersink at the same time.

Combination drill bits drill the pilot hole and clearance holes and countersink for screw holes. Each bit is made to match a particular number and length of screw.

Plug cutters are available in various sizes up to 1in diameter and are used to cut short plugs which are glued in counterbored holes of matching diameter, to cover screwheads.

MORTISER AND TENONER

Light duty mortising can be done with a drill press or drill placed in a drill stand and fitted with a mortising accessory. For continuous and heavy-duty use, a mortising machine is required. Unlike most other machines which are frequently adapted for other uses, the mortiser is used for one thing only; to cut rectangular holes, or mortises to match the male part or tenons in mortise and tenon joints.

The work is held against the fence in the table by a clamp. The table is moved backwards and forwards to line up the cut, and then sideways to make the several square holes required to make the rectangular slot. The bit and chisel mounted above the workpiece is brought down into the wood, as for a drill press, either by a long handle on the side or for more automatic machines by a foot pedal. Stops located at the appropriate points control the depth of cut and the length of the mortise.

Chisel and bits Chisel and bit pairs come in sizes from $\frac{1}{4}$in to $1\frac{1}{4}$in with $\frac{1}{4}$, $\frac{1}{2}$, $\frac{3}{4}$ and 1in the most common. The rotating bit mounted inside the chisel does most of the cutting and removes all the chips leaving basically the four corners to be trimmed by the chisel. The bit and chisel must be mounted precisely to allow the tip of the bit to clear the chisel to prevent rubbing and bluing up the cutting edge. The wings of the bits are quickly worn down and the bit is periodically replaced after many sharpenings, rather than risk overstraining the chisel which is much more expensive to replace.

The heavy-duty mortiser above has several useful features essential for quick and accurate work. The chisel can be replaced by a chain for chain mortising. The depth stop has two settings, one for the mortise and one for the haunch, where required. And large wheels and locking mechanisms make it easy to mount the work and adjust the cut. As a lightweight substitute, use a portable stand fitted with a heavy-duty electric drill.

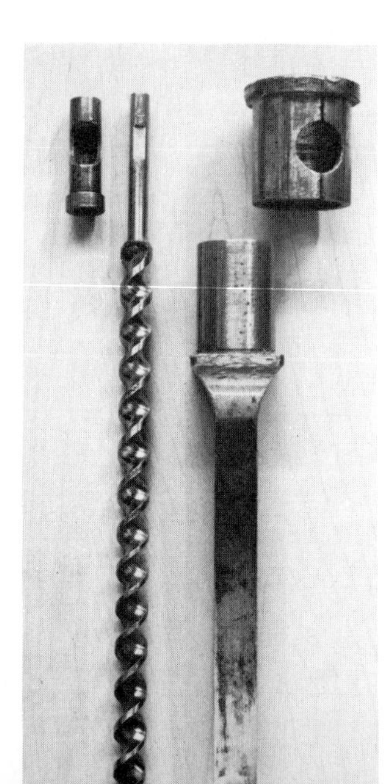

A chisel with matching bits and bushings. The bushings made to fit a particular machine compensate for the various diameters of chisels and bits and must be inspected occasionally for wear. A worn out bushing causes the bit to rub excessively against the chisel, ruining the cutting edges.

Mounting the bit and chisel Exact holding devices vary from model to model, but the chisel is always fastened separately from the bit. Select the correct size bit and chisel and, after making sure that both are sharp, place the bit inside the chisel by pushing it through from the bottom. Select the correct size chisel bushing and bit bushing and place these over the chisel and bit ends respectively. The bushings all have the same outside diameter to fit the holder but the inside diameters vary to suit the various drill or bit sizes. Push the chisel almost all the way up leaving a $\frac{1}{16}$ to $\frac{1}{8}$in gap at the top depending on the size and the hardness of the wood. A coin is a handy spacer to place at the top.

Temporarily tighten the chisel locking screw, then push the bit up until the bottom of the wings are exactly level with the corner tips of the chisel. Lock the bit tightly in this position with the set screw, then loosen the chisel and remove the coin spacer so that the chisel can be pushed all the way up leaving the appropriate clearance for the bit at the bottom. Before fixing the chisel in position, set it absolutely square to the fence, holding an engineer's square in place to check.

Aligning the cut Mortising is usually done on several identical components such as table legs or framework stiles. The mortise is marked on one piece only, and the stops are set for repeating the same cut.

Mark the face and edge on all the pieces and clamp the piece with the mortise outline on the horizontal table, good face against the fence. Move the table in or out to position the chisel to the marks. Then set the depth stop, which limits the downward travel of the head, to the right cutting depth against a mark on the end. Newer models have two depth stops, one to mark the bottom of the mortise and one for the shallower depth of the haunch where applicable.

Finally set the two stops on the sideways movement of the table so that the correct length of mortise will be cut. Also set adjust the stop so that each piece is placed in exactly the same location as the previous one. On older machines which may not have these stops, blocks clamped on the appropriate points will do the same job. Alternatively, all the mortise lengths can be marked at once with all the boards held together, ends lining up.

Place coin above chisel to get the wing tip clearance right. Below left: Tips level with chisel. Below right: Correct clearance.

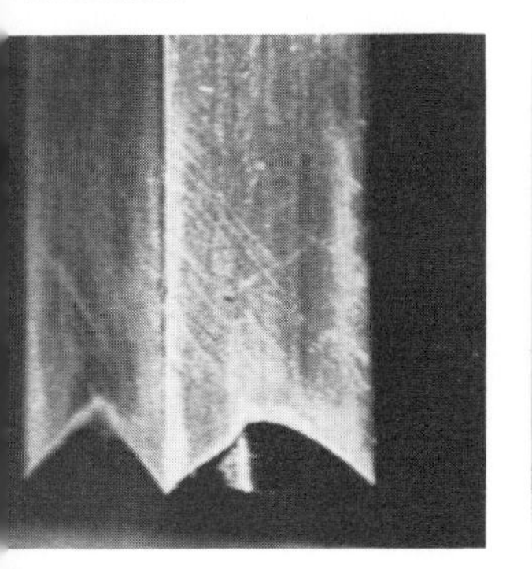

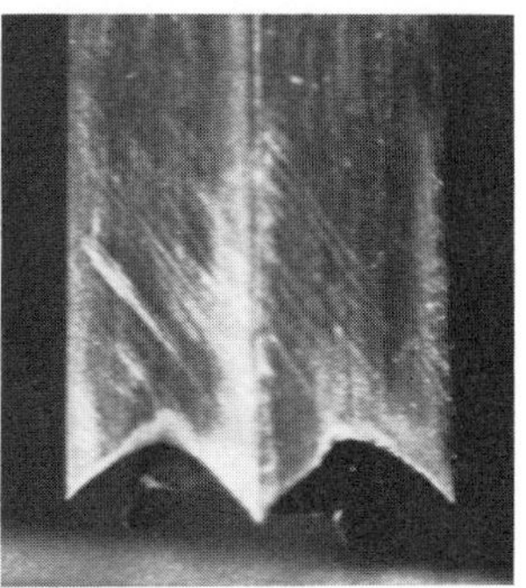

Cutting the mortise Make the right hand cut first, lowering the bit slowly to allow it to cut cleanly and to clear the chips. Depending on the wood, make the last cut next, then clear in between with successive cuts, moving the table horizontally by turning the wheel with one hand, while the other pulls down the lever. On hardwoods which may be difficult to cut, make the cuts in succession instead to make it easier to feed the chisel through the wood. Then cut mortises in the rest of the pieces before changing the setting for other cuts.

For through mortising, where the cutter must exit on the bottom side, it is difficult to cut holes with clean exit holes. The chisels must be very sharp and a close fitting support strip must back up each cut. Alternatively the mortise can be cut from both sides, but this is difficult to line up.

If the cutters burn, it is either a sign of a dull bit or of too fast a feed rate or perhaps an indication that the chips are clogging inside the chisel. Difficulty in feeding the cutter into the wood usually indicates a dull chisel. It is important to remedy these faults to avoid having the tips overheat which cause the tips to blue over and lose their hardness. A worn bit bushing will cause uneven rotation and rubbing against the chisel. The resulting noise is unmistakable and is a warning to stop the machine at once to correct the fault.

Sharpening the bit and chisel The bit is sharpened like an ordinary auger bit using the appropriate section precision file for each job. Always file from underneath and inside, maintaining the original cutting angles and removing as little steel as possible to prolong the life of the bit. Sharpen the spurs from the inside to avoid changing the cutting diameter, using a flat section file. Then sharpen the cutter with a triangular file, holding the bit diagonally downward onto the bench to expose the underside of the cutters. Try to maintain the wings and cutters to the same height to balance the cutting action.

Use a special tool to sharpen the chisel. Record Ridgway sell sets which include files and sets of countersinking sharpening points to match the various sized chisels. Each countersinking tip, has a round pilot at the end which fits inside the chisel top to keep the bit in line. To use the sharpening bit, fit the bit in the holder which is held in a brace. The bit, held upside down in the vise, is sharpened with a few turns of the brace, then the corner points are sharpened separately by hand using the triangular shaped precision file.

Chain mortising Many production mortisers can be converted into chain mortisers for faster cutting. The chains, available in various widths, come in sets consisting of a matching guide bar and driving sprocke

The width of the mortise cut with one thrust is determined by the width of the guide bar plus two thicknesses of the chain. For wider mortises, a wider guide bar is used or more than one cut is made.

Chain mortising leaves a rounded bottom to the mortise which may be squared off by hand or allowed for in the length of the mortise.

Chain mortising is the only mortising method which leaves a clean exit hole, so it is ideal for through mortises.

Sharpening the spurs from the inside with a flat section file.

Sharpening the cutters on the underside with triangular file. Right: Chisel sharpener.

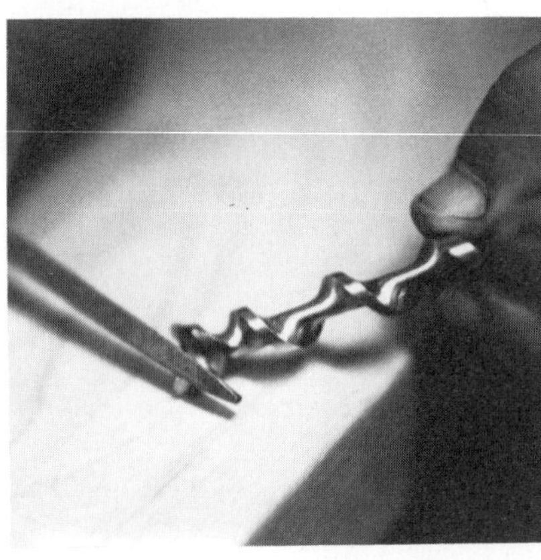

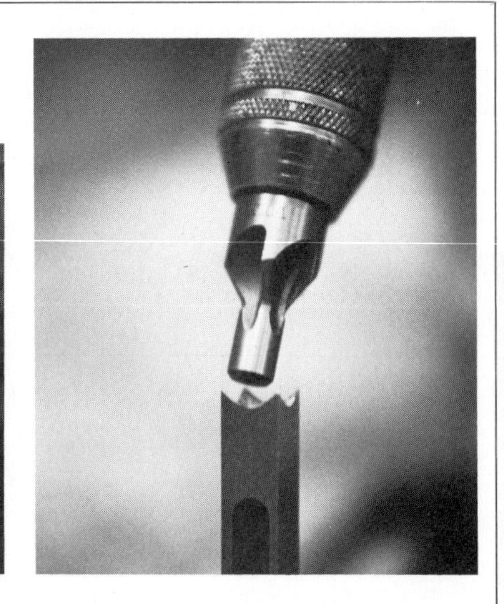

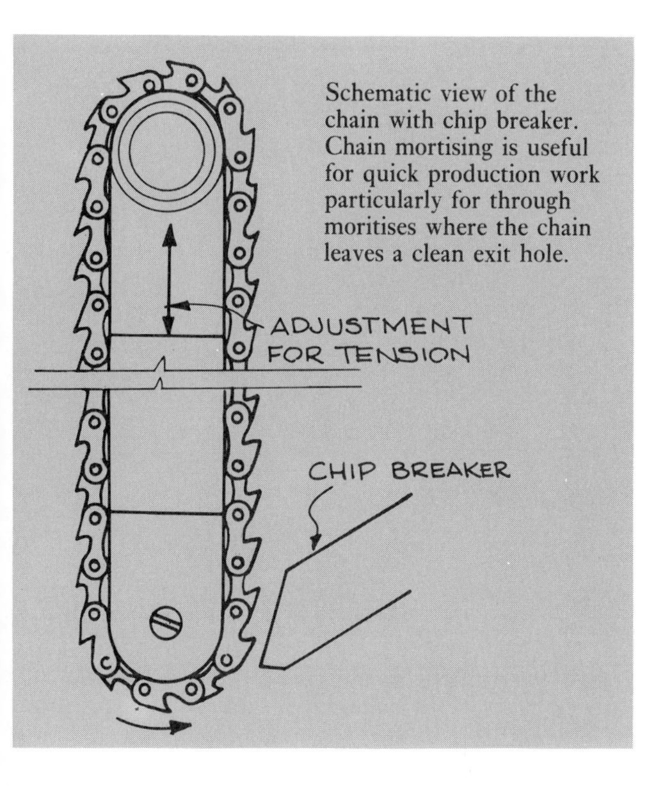

Schematic view of the chain with chip breaker. Chain mortising is useful for quick production work particularly for through moritises where the chain leaves a clean exit hole.

Maintaining the chains requires grinding of the cutters, repairs to the links and proper lubrication. Many machines have a built-in attachment for grinding the chains.

Storage of the chain is also important requiring setting up a special rack to allow the chain to sit in an oil bath.

Refer to the manufacturer's instructions for grinding angles and other maintenance details.

Tenoners The mortise is always cut before the tenons which are then cut to match. Tenons can be cut on the table saw, radial arm saw or spindle molder, but in production shops a tenoner is used. There are many sizes and styles of tenoners from the standard single ended model to the automatic, double ended types which do everything but assemble the pieces.

The single ended tenoner, normally too expensive for small furniture shops, is widely used by joinery shops for window and door work. Two tenoning heads fitted with special rounded cutters which cut with a shearing action, cut the top and bottom parts of the tenon. A scribing head, fitted behind the main heads, shapes the shoulders to match the molded shape on the stile. In addition, a cut-off saw may be fitted behind these to cut the tenon to length automatically. More advanced machines include additional heads, and double ended tenons have an identical set up on the other side for cutting both ends of the rail at the same time.

Left: Single ended tenoner with sliding table pushed forward for clarity. The cut-off saw is mounted in front of the two main cutters and the scribing heads behind.
Below: Details of heads.

GRINDERS

A sharp tool is basic to good woodworking. Dull tools not only do bad work, but are also dangerous since more force with resulting imbalance is required. Large production shops often have a special sharpening department which deals with circular saw blades, band saw blades, planer blades, etc., but for the small shop these are best sent out for attention by a specialist. Most other tools including hand saws, chisels and turning tools, are easily cared for in the shops.

The basic sharpening kit should include a high-speed grinder, one or two oilstones, a few slipstones plus a leather strop for a fine edge on carving tools.

It's a good idea to set aside a small, relatively dust free corner of the shop for a sharpening area, keeping it well-lit and supplied with a bucket of water for cooling off the edges during grinding.

Grinding wheels are used to re-grind edges which have become damaged or badly shaped. Edges ground on round wheels are hollow, that is they have a slight circular shape from the circumference of the wheel. The hollow edge is most often honed on an oil or slipstone to make it flat and to bring up a sharper and burr-free edge. On some tools such as turning tools (page 145), the ground edge is sufficient.

The grinder is also used to shape and sharpen French head, slotted or Whitehill cutters widely used on the spindle molder (page 124), and it is used for restoring other metal tools such as screwdriver or bradawl points, and cutting edges on knives.

The old-fashioned, water lubricated grindstone cuts slowly with less danger of bluing the edges, but these have largely been replaced by high-speed electric or hand driven grinders.

There are several variations on the standard grindstone, using continuous abrasive strips instead of wheels. These offer several advantages for they are easily replaced and don't reduce in diameter or gloss over with wear.

Principles of grinding As with sandpapers, the grinding wheel works by having hard-faceted particles strike or cut the steel to remove material. These particles or grits, made from various hard materials, cut until they wear loose or until they break, in which case new sharp cutting edges are produced. In this sense grinding wheels are to some degree self-sharpening.

There are a number of components which go into making grinding wheels and these can be varied to suit particular applications.

The particles or *grits* are labeled with letters and are made from aluminum oxide or aloxite (A), silicon carbide or carborundum (C), or diamonds (D), in a number of mesh or grit sizes from 36 (rough) to 220 (very smooth). The range for woodworking tools is about 36 to 80. Diamond grit is the hardest and limited to tungsten carbide tools.

Aluminum oxide is available in standard or white. The former is used on ordinary carbon steel tools, the latter on high strength steel cutters. Silicon carbide also has two forms. The standard is not used for woodworking tools but the green form is used for the rough grinding of tungsten carbide. The standard grit for all-purpose shop grinding is SA 60 grit (standard aloxite, medium size).

The *bonding material,* as with sandpapers, must hold the grits in place and the hardness of the bond is referred to as the grade of the wheel. Various materials are used and graded according to hardness from D for soft to K for hard. The general rule in wheel design is to provide a softer bond for grinding hard materials and vice versa.

The coding on a wheel relates to the grit and bond and, as an example may be SA 80 KV, standing for 80 size grit made from standard aluminum oxide with a vitrified (V) bond of K hardness.

These factors and the wheel speed (about 3000rpm) which are built into and stamped on each machine, determine the cut of a grinding wheel.

Sizes and shapes Wheel diameters vary from 5 to 10in with 6 to 8in the most common. Thicknesses range from $\frac{1}{4}$in to about 1in and shapes vary from the regular straight types used for normal work to cup wheels used for long, straight cutters and dish wheels for use on the side of the rim.

Using the grinder Plane irons, chisels and other straightedges can be hand-held but are often clamped in a guide at the correct angle to give a 20 to 30° beveled edge. The tool is offered up to the wheel and moved back and forth until the sparks just come over the top, indicating that the bevel has reached the tip. The hollow shape and the burr are then removed by rubbing on an oilstone. Gouges are sharpened in the same way (see sharpening turning tools, page 146), but are rolled back and forth to take account of the shape. These are finished off with slipstones.

Grinding molding cutters for the spindle molder is done carefully freehand using the edge and side of the wheel (see page 126). A $\frac{1}{4}$in thick wheel is useful for cutting fine French head shapes.

Keep in mind that the edges must not be overheated or "blued" which would cause annealing, resulting in a softer edge. To prevent this, dip the tip often into a bucket of cold water placed nearby.

All grinding wheels tend to glaze over as they heat up and this glazed surface must be removed or "dressed" using either a diamond tipped or a star wheel dresser. This is held against the wheel and moved back and forth to restore the sharp cutting quality of the wheel.

The circular wheel leaves a hollow edge which must be honed flat.

A narrow, $\frac{1}{4}$in grinding wheel especially useful for grinding French head cutters (page 124).

Dressing the wheel with a star wheel dresser to remove the glazed surface.

LATHE

Above left: Various holders. Above right: Access to belts through headstock cap.

The basic components of the woodturning lathe are shown on the photograph. The size of the lathe determines the largest pieces which can be turned on it. For spindle turning, a length of 30 to 36in is standard, allowing for turning table legs which are usually between 26 and 30in long.

Bowls and other faceplate work can be turned on the inside (right) or the outside (left) of the headstock. The latter is always preferred, since the bed doesn't get in the way allowing larger diameter work to be turned, but most important of all, allowing greater freedom of movement which is essential for turning. Since some models do not have the facility for out turning, the size of the largest diameter work that can be fitted onto the faceplate over the bed (called the "swing") is the second criterion of the size of the lathe. Most heavy-duty lathes have a gap bed which allows larger diameter work to be turned on the bed side of the headstock.

The most important criteria to consider when buying a lathe, besides the size, are the sturdiness of construction and the availability of changing speeds.

The freestanding models should either be bolted to the floor or alternatively be weighed down with a pile of bricks or bags of sand placed safely on the shelf underneath. The bench top type should be bolted to a very sturdy bench (which can also be weighted down) or to a homemade stand preferably of 3 × 3in or 4 × 4in legs with something like 2 × 6in stretchers to carry the top.

Variable speed is indispensable and all but the cheapest models have three or four speeds say, 425, 800, 1400 and 2300rpm, with the former used for the larger pieces to reduce peripheral speed, and the latter for fine cutting between centers on small spindles and for finishing work. The table opposite gives a general guide to the various turning speeds for particular diameters.

The headstock or driving end of the lathe, houses the gears or pulleys and carries the main spindle or mandrel. The ends of the mandrel are threaded to accept chucks and faceplates of various types. Most mandrels are also taper drilled to take morse taper centers.

The basic set of holders normally supplied with the lathe should include a 6in faceplate for bowl turning, plus a pronged driving center with a dead center for turning between centers. Other holders available as accessories include the screw chuck for small work such as egg cups, a coil grip or Myford three-in-one chuck for small or long pieces, a standard three jaw chuck for drilling work, and a ball bearing "live" center to replace the dead center when burning or overheating from friction of the dead center becomes a problem.

TABLE OF LATHE SPEEDS

	Type of cutting		
Diameter	Roughing rpm	General cutting rpm	Scraping/ sanding
0–2in.	1000–1200	2000–2600	2800–3500
2–4	700–1000	1800–2200	2200–2600
4–6	500– 800	1300–1800	1800–2200
6–8	400– 600	900–1300	1300–1800
Over 8	300– 400	400– 500	500– 800

Note: on most lathes, only four speeds are available limiting the range of speeds possible. Select the speed nearest to the one in the table.

Tools Turning tools divide into two groups, the cutters and the scrapers. Cutting is done by chisels and gouges.

GOUGES The gouge is U-shaped or deeply fluted in section which helps break up the shavings where "dig in" is a dangerous possibility, as for the end grain section on bowl turning.

Gouges for bowl turning where there is more danger of digging in should therefore be deep fluted. A complete set of deep fluted bowl gouges should include the $\frac{3}{4}$, $\frac{1}{2}$ and $\frac{1}{4}$in widths, the last two being the most versatile.

Gouges for turning between centers or spindle work are divided into two types, roughing-down gouges and coving gouges.

Roughing-down gouges are between $\frac{3}{4}$ and $1\frac{1}{2}$in wide and are semicircular in section. The wider ones are used primarily for fast removal of stock of square sections turned between centers.

Coving gouges which have a shallow flute usually come in the same widths as bowl gouges, from $\frac{1}{4}$ to $\frac{3}{4}$in widths, with the smaller sizes the most useful.

CHISELS Chisels, like gouges, cut rather than scrape. Two chisels are required in turning; one ground straight across and 1 to 2in in width, the other ground at an angle or skew, and about $1\frac{1}{4}$in wide. Both are used for spindle turning only.

The parting tool, about $\frac{1}{4}$in wide, is ground so that it is slightly wider at the tip to prevent rubbing on the side as it cuts straight into wood. It is used to sever wood from the lathe and for other delicate cuts.

SCRAPERS Scrapers are used primarily at the finishing stages in bowl work to produce a more even finish and also to level surfaces which have been worked with the gouge. Scrapers can also be used for spindle work, where the gouge can't do the cutting because of its shape.

Scrapers are available in various sizes and shapes. The complete set should include the straight across, domed, rounded, half round and the left and right skew shapers. These can be bought in sets or alternatively can be made from old files whose teeth have been ground away or from other sections of steel such as that used for French head cutters (page 125).

MISCELLANEOUS TOOLS The other tools required are rulers and dividers for measuring, inside and outside calipers for checking diameters, a smoothing or try plane for producing a flat surface for mounting bowls, and if possible a band saw for rough cutting bowls.

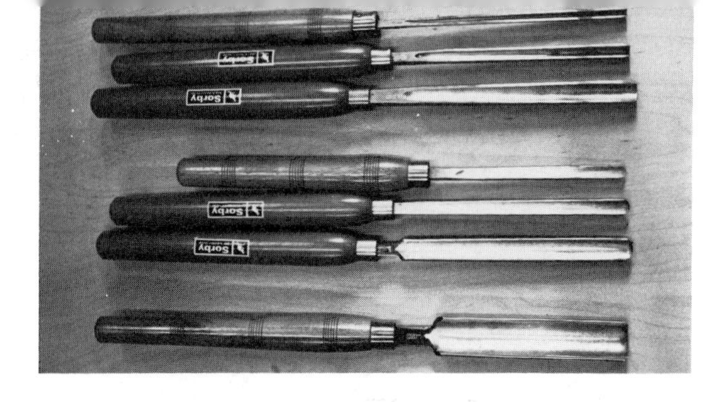

Above: Set of long and strong Sorby gouges.
Below: Plan showing profile of three types.

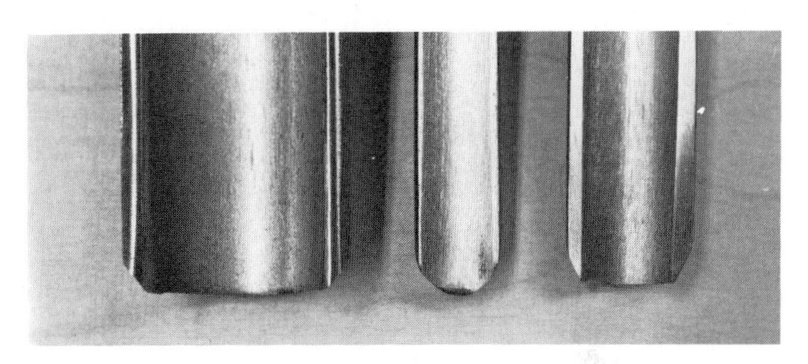

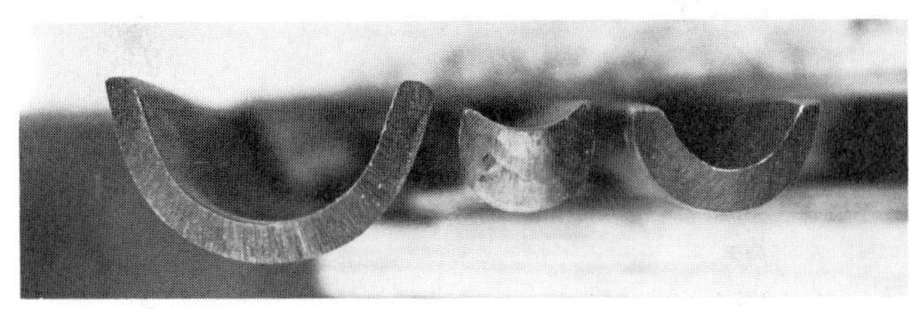

Above: Roughing-down (left), spindle gouge (center), bowl gouge (right). Below: Profiles of gouges.

Below: Sorby chisels: parting tool (left), three skew chisels and straight chisel (right).

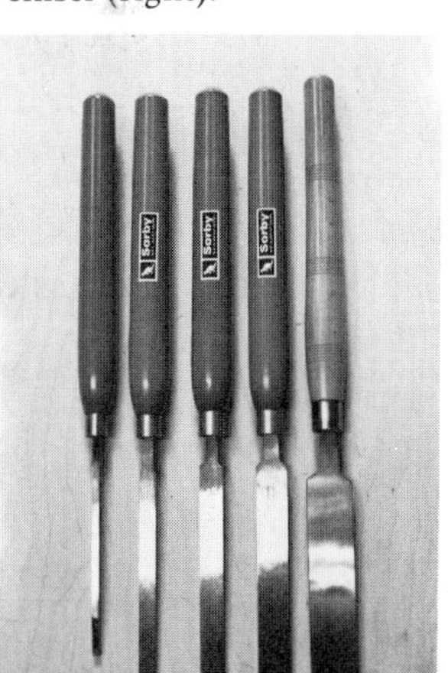

Below: Three Sorby scrapers. A scraper made from file (left) and French head steel (right).

Sharpening As in all woodwork, sharp tools are indispensable to good results in turning. The grinding stone should be located close to the lathe so that sharpening doesn't unduly interrupt the work. The old whetstone method is still the best, since the slow and cool cutting is capable of a sharper edge without the danger of softening it, but most workers use an ordinary grinder fitted with two $\frac{3}{4}$ or 1in wide, dry silicone carbide, carborundum wheels 5 or 6in in diameter, one coarse or medium grade, the other fine. The wheel should be dressed frequently using either a star wheel or a diamond type dresser to remove the glazed surface caused by the heat generated during sharpening (page 143). And while sharpening, place a bucket of cold water beside the bench to periodically cool off the iron.

DEEP FLUTED BOWL GOUGES Depending on the turning method used, deep fluted bowl turning gouges are ground either straight across or with a slight nose, both with a bevel of approximately 45°. (see bowl turning page 156). To grind the bevel, hold the gouge against the rest, positioned approximately so that the bevel will be cut at 45°. Bring the center back or heel of the bevel in contact with the wheel first, then roll the gouge carefully from side to side to grind the complete edge at the same angle. Remove as little steel as possible, and try to do it in one go without lifting the gouge up to check the progress. The edge is reached when the sparks just start coming over the top when the grinding should stop. Remember not to press down too hard or the edge will blue over and loose its hardness. Grinding always produces a hollow shape from the curvature of the wheel. In some work this is removed by rubbing on an oilstone, but for turning gouges, the edge produced by grinding can be used directly. Since the gouges are sharpened frequently this saves considerable time.

Grind gouge by rolling bevel back and forth using the tool rest set at the proper angle.

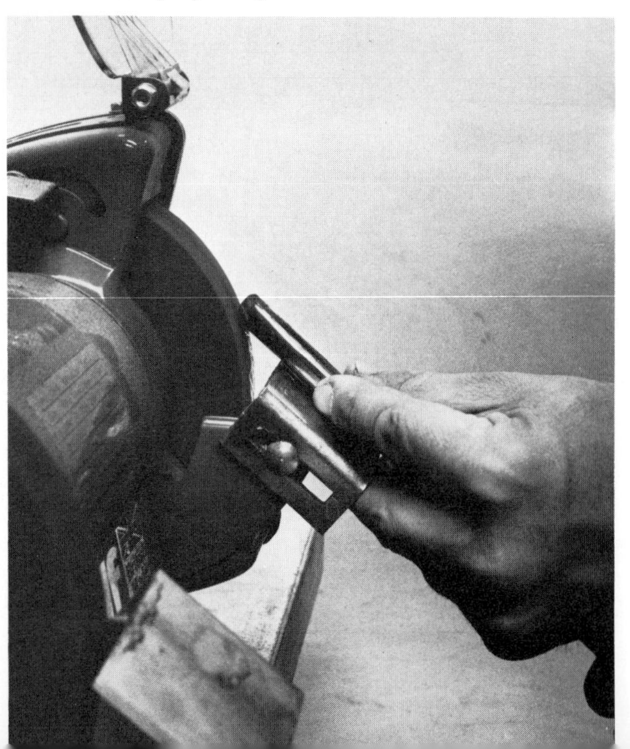

ROUGHING-DOWN GOUGES for spindle work are sharpened the same way, ground straight across with a 45° bevel.

COVERING GOUGES do more delicate work and can therefore be ground to a larger bevel, say between 30 and 35°, and should be honed with an oilstone on the outside and slipstone on the inside after grinding for fine work. Honing is best done with the gouge held stationary, under one arm pointing upwards with the oilstone held in the other hand to rub against the bevel. Finally remove the burr with the slipstone, the gouge held flat on the bench.

Hone the bevel with the gouge held stationary under the arm. Below: Using slipstone to remove burr.

CHISELS The chisel too is more of a medium use tool than a roughing tool and should therefore be honed after grinding. Chisels are beveled on both sides either in front of the grindstone like the gouge, or on top freehand where access is easier. The front method, though more awkward, is the best since the stand allows more control. A 30° bevel on each side is about right and is produced by moving the chisel straight across the grindstone for an even edge. The bevels can be left hollow ground for some work, but they are best flattened by rubbing on an oilstone, rubbing on the entire bevel with no secondary bevel as for ordinary chisels. Rubbing across a leather strop glued to a block of wood to produce an exceptionally fine edge is optional.

PARTING TOOL This is used both as a scraper and as a chisel. Used as a scraper (as it enters the wood) it loses its edge more quickly than other chisels. It is therefore best left hollow ground off the grindstone but can, of course, be honed as well if desired.

SCRAPERS Scrapers are ground to a very shallow clearance angle of about 10 to 15° to prevent the bevel from rubbing against the work when in use. Hold the scraper blade against the rest, so that the bevel is just rubbing. Grind it until the sparks just appear over the top edge, then move the edge straight across while holding this angle. For all but the roughest work, it is best to hone this edge, then bring over an even burr, just as on a hand scraper, using a burnishing tool or a ground off Swiss file. Refer to the instructions and photographs on page 126 for burnishing scraper edges.

Old files make excellent scrapers. Grind off the serrations 2 or 3in behind the tip, then bevel the edge as above to any required shape. Grind the edge to any shape required for a particular job.

Steel used to make French head cutters as shown on page 125, is available in 1, $1\frac{1}{2}$ or 2in width bars, up to $\frac{5}{16}$in thick. These make excellent scrapers since the steel is hard enough to hold the edge but not as hard as files and therefore easier to grind to shape.

Chisels are ground to an angle of approximately 30° on both sides. Set the tool rest and grind while moving the chisel back and forth until sparks just appear over the top. Quench frequently. Below: Hone the hollow ground edges on a stone and if desired, finish off on leather strop.

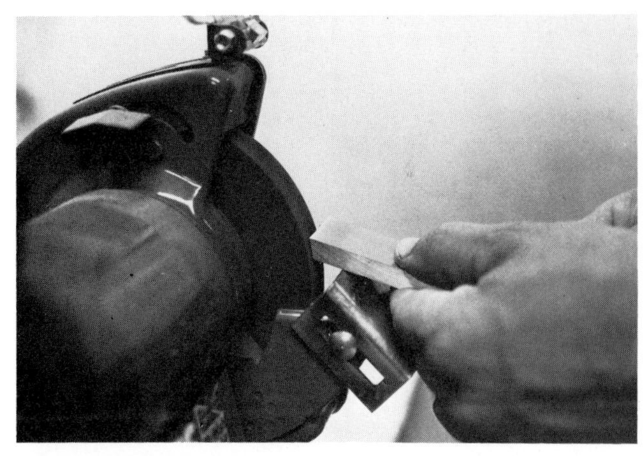

Use scraper straight from wheel or use burnisher (below) to bring over fine cutting edge.

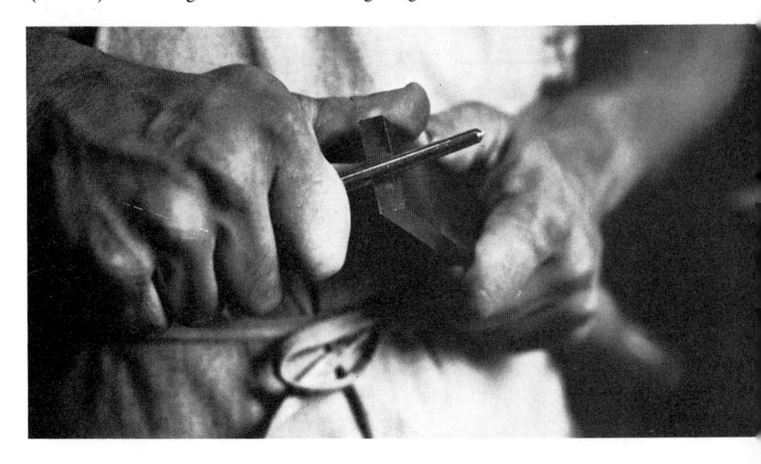

Spindle turning between centers

Spindle turning which is easier and safer than bowl turning, is used to produce a variety of articles such as tool handles, table legs, lamp bases, egg cups, salt shakers, etc. And drill bits mounted in the tailstock can be used to center bore work such as lamps or small jars with the work revolving and the drill bit stationary.

PREPARING AND MOUNTING THE SQUARE There is no need to plane down or saw off the corners of the square to be turned. Roughing down to the round will be done with the large gouge. Prepare the ends by drawing diagonals across to mark the centers. In hardwoods, drill a small center hole, otherwise use a center punch. At the headstock end, cut two shallow saw lines with a tenon saw or the band saw, to make grooves to take the two or four winged driving center which is then tapped in place with a hammer or mallet. Mount the square by pushing the driving center into its tapered hole, then sliding the dead center across to take the other end. A bit of grease or oil on the center will prevent burning and squeaking.

To mount the square, first find center point with diagonal lines, then punch indentations for driving center (left). Above: Dead center in place with a dab of grease to cut down friction.

ROUGHING DOWN The largest roughing-down gouge is used to turn the square into a cylinder. Adjust the hand rest at or just below center and make sure it clears the revolving wood by rotating with the motor off. Present the gouge very carefully to the wood, handle downwards so that the edge will cut with the hand holding the blade firmly against the rest. Bring the edge gradually into contact with the revolving corners of the square, and holding the gouge U-shape upwards, traverse the length of the workpiece several times in either direction to gradually remove more wood.

As the wood is smoothed down to a cylinder, or near cylinder, the gouge can be rolled slightly from side to side while traversing so that more of the edge does the cutting, which means that the edge has to be resharpenéd less often.

Start roughing down using only the center part of the gouge. Then use the entire edge (below).

Use the chisel to finish off the spindle with the hand rest raised to cut near the top.

Use calipers to check diameter of work, preferably with the lathe stopped.

Tracing the shape on the work from a template with the lathe in motion.

Depending on the eventual shape, reduce the cylinder to its final diameter preferably with a wide chisel, either straight or skew. Move the hand rest up slightly so that the chisel can be made to cut further up the cylinder. The nearer the top the easier it is to cut.

Traverse the wood back and forth as shown, holding the chisel at an angle to the work, with the bevel rubbing so that it cuts into the wood rather than scrapes. At all times avoid having the trailing corner touch the workpiece which would cause it to dig into the wood. As a substitute for the chisel, use one side of the large roughing down gouge to make the wood straight.

When the cylinder gets near to its final size, check it frequently and at several points along its length with outside calipers or with templates made from cardboard. It takes practice to get a straight cylindrical shape by eye. Beginners may prefer to use the widest scraper to get it straight.

An alternative method is to use the parting tool to cut several notches along the length to the final diameter, then remove the wood in between with a chisel or gouge.

TEMPLATES It's always best, particularly when producing two or more identical pieces, to draw the shape or contour on a piece of plywood or cardboard to make a template which can be used first to mark the locations of coves, beads, etc. and then to check progress along the way. I find it best to cut out the contour on one side of the template and to make semicircular cut-outs on the other side for checking the various diameters along the length.

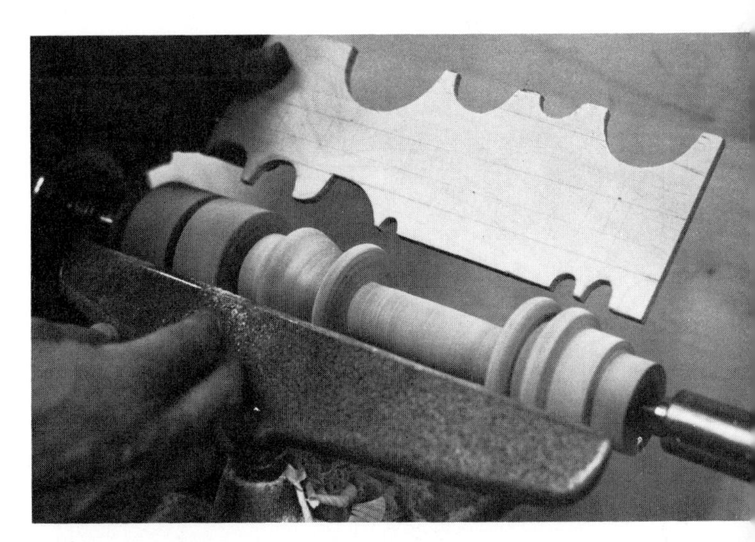

Two way template: one side has profile cut-out (above), the other side is used to check relevant diameters along the length (below).

PUMMELS Turned legs for chairs or tables often have the top and/or bottom section left square for the mortise and tenon or dowel joint connecting the rails. Start with a piece of planed up wood and mark the ends to be left square with a heavy pencil line. Also mark the face and edge marks, if applicable, for cutting the joints later (though the mortise is more often cut before the turning is done). Mount the square, adjust the rest to clear, and use the skew chisel for starting the cuts. Hold the chisel perpendicular to the work, with the handle well down so that the lower edge or point of the chisel will start to cut into the corner of the wood as the chisel is brought slowly into contact. For the square cornered pummels most often prefered, tilt the chisel so that it makes a square cut, not a V, as it enters. After the first shallow notch is made, remove the chisel and make further cuts to wider the V-shaped cut until the bottom of the V forms a cylindrical shape. Make sure the chisel edge cuts as it follows the tip into the wood. Cleaning the cylindrical shape in between is done with a roughing-down gouge. Traverse the cylinder, either using the bottom of the U-shape to do all the cutting or better still, rolling the gouge over gradually to have all the edge cutting.

Cutting pummel: The skew chisel makes initial entry, roughing-down gouge clears between pummels.

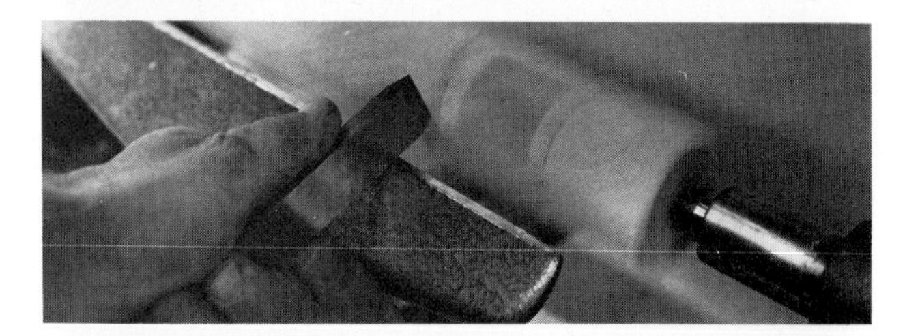

Enter the wood straight on (above), then move the handle down to cut near the top (below).

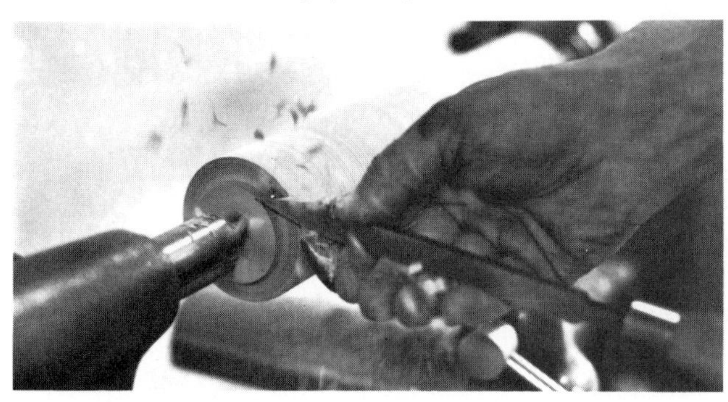

Cutting beads with a skew chisel. Keep the point in the wood throughout the cut.

BEADS A cylinder is rarely left plain. A combination of beads and flutes is most often used to decorate the turning. The parting tool used to make beads, is essentially a chisel which cuts, not scrapes, but on initial entry into the wood it is safer to start it off as a scraper at mid-height of the workpiece. Then as the edge is fed safely into the wood, the handle is lowered and the point brought up so that the bevel is rubbing against the wood and the point is cutting into the wood near the top of the cylinder. This way the edge of the parting tool cuts like a chisel and the

Above: Cutting left side of the bead. Below: Using skew chisel to clean up end grain.

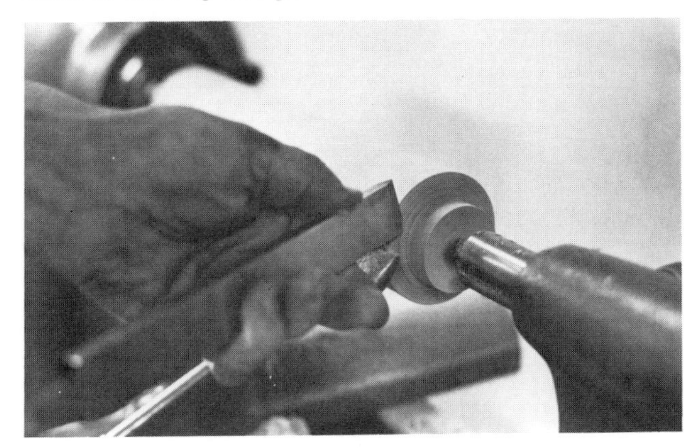

edge is kept sharp for a longer time.

Beads can be left square but are more frequently rounded off. The best method is to use either the straight chisel or the parting tool itself, to round the corner off with a clean slicing action. Start at the top, then lean the chisel over so that the leading point starts to cut with a slicing action. Continue tilting the cutting edge along the shape required, until all the wood is cut away. For a clean, even curve make sure to do it in one smooth motion with the tip not leaving the wood. The other side is rounded in the same way working, as always, from large to small diameter.

Similarly, any end grain cuts, except the final parting cut, can be cleaned up with the slicing action of the chisel. These cuts can either be curved as above or straight, square cuts where the chisel is inclined slightly to avoid cutting in a V-shape. As with cutting pummels, the point of the skew chisel enters high up on the cylinder with the handle held well down, then makes a neat slice as the handle is brought up to bring the tip nearer the center for finishing the cut. This takes a bit of experience to do perfectly, so it's well worth practicing on a piece of waste wood first.

COVES Coves are like beads, but with the curvature of the cut reversed. The parting tool is first used to cut the bottom of the cove, but the sides are cut with a small coving gouge. With the gouge on its side, flute facing toward the cove, hold the tool firmly and start the cut with the tip. The tip will tend to pull away from the cut digging into the wood. But with practice, it becomes quite easy to start the cut neatly and then to keep the bevel in contact with the wood as you continue the cut, turning the gouge over gradually to do so. Finish at the bottom point, then repeat several times until the required depth is reached. Remember to cut downhill only. The opposite cove should be cut from the other direction with the cut ending at the bottom point.

PARTING OFF The parting tool is used to cut the work to length after all the turning and finishing is complete. Make the cut as for a bead with the parting tool, but don't cut the last $\frac{1}{4}$ to $\frac{1}{8}$in before the other end has been cut down to this size.

The final $\frac{1}{4}$ to $\frac{1}{8}$in can be cut off afterwards with a tenon saw, but if a neat finish is required at one end, it can easily be cut off with the parting tool, after the dead center has been eased off slightly to hold the work more loosely. Finish the cut with one hand, and catch the cut off piece with the other.

Hold the gouge firmly, flute facing toward cut to start the cove. The cut should end up at the bottom of the curve.

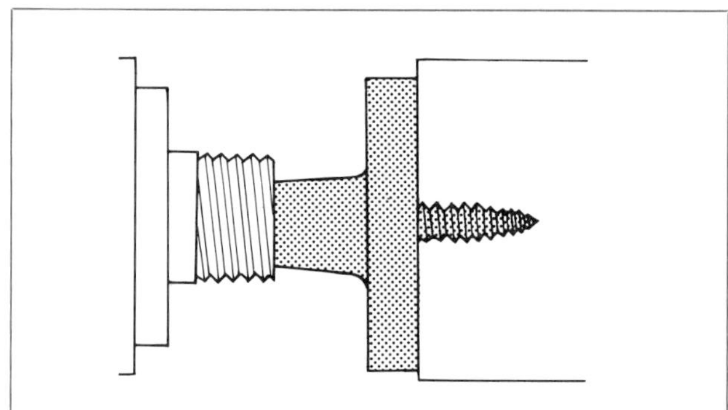

The simple screw chuck (above) can be bought as an accessory or shop made as shown below.

Turning with one center

To turn small objects such as egg cups, wooden jars and salt and pepper shakers, access to the side and one end is required just as in bowl work. Several holders and chucks are available to hold small pieces at the headstock end. The simplest is the screw chuck, where a screw protrudes from the backing plate to be attached to the center hole in the workpiece. This can be bought as an accessory or simply made by screwing a small piece of plywood to a larger one, attaching to the faceplate and turning the small piece round. Find the center, as shown, by moving the tailstock across. Then remove the faceplate and drill a countersunk hole for a large screw fixing which is adequate for most small objects.

A stronger fixing for slightly larger objects is afforded by the flanged ring holder. The Myford three-in-one chuck combines these two with an ordinary faceplate fixing for turning bowls.

Start by mounting the square blank between the dead and live center, then use the roughing-down gouge to turn the cylindrical shape. If you are using a flange ring holder, this is also the time to turn the end of the cylinder to a flange shape suitable for the holder. And make sure to make the cylinder to the right diameter for the flange, so that it's a tight fit to prevent any wobble. Test for fit by removing the tailstock and fitting the ring over the cylinder.

After the roughing out has been done, remove the blank from the two centers and mount it to the headstock using any one of the holding methods.

Making a screw chuck: Turn a plywood disk and use the tailstock to mark the center (above). Drill a countersunk hole in the disk for a no12 screw to extend about 1in into the workpiece (left). The Myford three-in-one chuck (below) incorporates a faceplate, screw chuck and a flange holder. Below right: Chuck from metal lathe adapted for turning small objects.

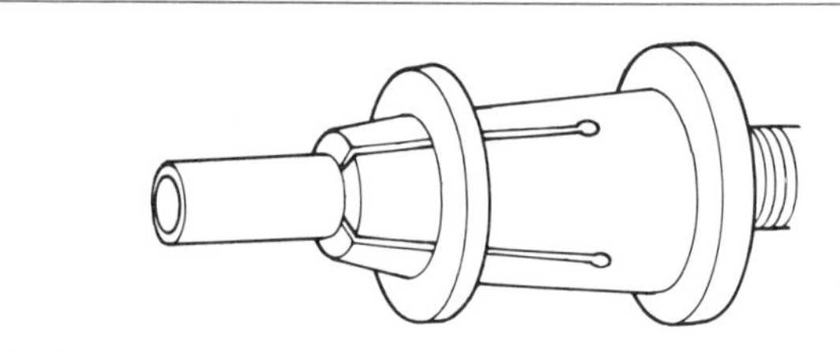

A shop made holder which uses a ring to tighten small, tapered piece onto work.

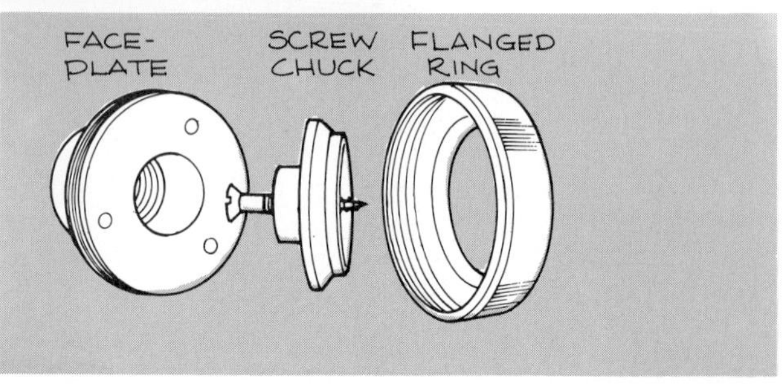

When the lathe is turned back on, it will probably be necessary to rough cut the cylinder a bit more since the original center may be lost. Also remember to make corresponding marks on the drive and the wood so that the work can be remounted in the same location if removed.

Hollow out the inside before shaping the outside. That way there is less danger of the thin walls breaking during the hollowing out.

Use a coving chisel or a small scraper ground to shape for hollowing out with the hand rest placed in front just below center. Check the shape frequently with the motor off, using a template if possible.

The inside can also be roughly hollowed out by first drilling a hole and then cleaning up with a scraper ground to shape to fit inside the hole.

After the inside cutting and scraping is done, the outside shape is turned. For making sets, it's advisable to make a template so that the shapes will be just about identical.

Start cutting the outside with a parting cut about $\frac{1}{4}$in deep to indicate the bottom of the cup. Then make the beads and coves which comprise the shape using the techniques described earlier. Finally sand the piece before any decorative rings or finish are added.

The fine black rings often found on delicate turnery can be added very simply using the back of an old hacksaw blade to burn shallow grooves.

To finish the piece, cut off the base with a parting tool either straight across or with a slight hollow to make sure that it will stand correctly.

Above: Use the parting tool to make a shallow cut at the base. Below: Sanding.

Turning the inside of egg cup using a small scraper ground to a curved shape.

Burn in decorative rings by using the back of an old hacksaw blade.

DEEP HOLES To make containers such as pepper mills where a deep hole must be hollowed out inside, either use a drill to bore the center hole or, if a cleaner bottom is required, make up the length using several shorter lengths glued and turned one at a time. Rough cut one long length into a cylinder, then divide this into several pieces, numbered so that they will go back in the same order. After hollowing out the first piece, glue the second piece onto this using pressure from the tailstock to hold them together until the glue sets. Then hollow out this piece to the same inside diameter as the first, set to inside calipers. When all the pieces are added, the outside can be decorated to hide the joints.

To make deep holes, either bore them (below) or glue successive pieces onto the spindle (above).

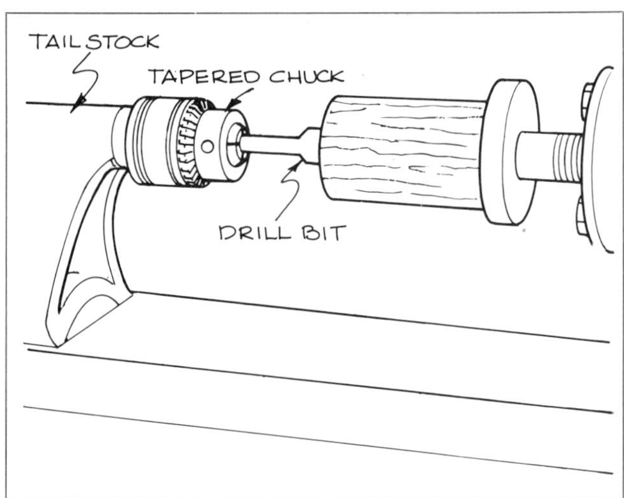

BORING END HOLES End grain boring can be done with various drill bits, using a slow lathe speed. Fit a chuck with a morse taper shank into the tapered hole of the tailstock to hold the drill. The drill is kept stationary as the workpiece, held in the headstock rotates. The best bits are saw tooth machine center bits which cut clean holes as deep as is practical. The Forstner pattern bit makes equally clean cuts but is limited to shallow holes. Long auger bits are used to drill long, small diameter holes. For long holes, the bit should be withdrawn frequently to clear the chips.

Flat bits normally used with electric drills have to be used with more care since they cut with considerable force which may split delicate work. Provide a small starting hole for the drill point to center the bit as it enters

Making copies of spindles Automatic lathes in production shops work off a master template and turn out any number of identical spindles by the hundreds. A few manufacturers sell duplicators which can be fitted onto the hand lathe. Most of these work on the same principle as key copying machines, a pointer follows the original to guide the cutter to an identical shape. This kind of set up would be particularly useful for the small workshop producing, say a limited run of a design; not enough quantity to make an automatic machine affordable, but enough to make it too laborious to do by hand.

To copy by hand, it's best to make a template as shown above or to turn one master from which the others are copied. For one or two copies it is enough to hold the master up to the spindle by hand but for more copies, attach the master to a piece of plywood hinged to the table at the base so that it can simply be leaned over to the spindle to compare shapes.

On an automatic lathe, the knives which are stationary are fed into the rotating workpiece to cut the shape almost instantly.

Faceplate turning

MOUNTING THE WORK Bowls, plates, trays, table tops, etc. are all turned on the headstock end while held to a faceplate. Sizes of faceplates vary but the most useful sizes are the 3in diameter for small work and the 6in for large work including bowls. The photograph on page 144 shows a special 15in diameter aluminum faceplate which is used to turn large pieces of flatware such as trays, table tops and chopping boards where the large diameter requires additional support.

Normally the faceplate is screwed directly to the workpiece. Three $\frac{3}{4}$in long no10 screws are enough to hold even the largest sized bowl placed so that tools will not hit the screw tips from the other side.

Attach the faceplate as near to the center of the roughly cut out workpiece as possible. A large piece of wood mounted off center will be difficult to handle when it is revolving at 400rpm. The screwholes can be filled afterwards with a matching filler or alternatively covered with green baize.

To avoid screwholes completely, use the paper joint. Screw a piece of scrap to the faceplate. The scrap should be at least $\frac{3}{4}$in thick to take the screws and about the same size as the bottom of the bowl. Mount it in the headstock, turn it to size and make the face flat with a scraper. Then glue a piece of heavy brown paper, such as wrapping paper between the scrap and the bowl. Clamp the pieces together using the tailstock for pressure if necessary. When set, the glue joint is strong enough to take any kind of work. It's best to use animal glue such as Scotch or pearl glue since it is easier to remove afterwards. When the bowl is finished, detach it by first cutting into the paper joint with a razor blade or sharp trimming knife, then by tapping or pulling the two pieces apart. Leftover paper is easily cleaned off the base with hot soapy water.

OTHER HOLDERS To hold more delicate work, especially pieces which have to be turned over to work on the base, it's possible to make simple holders by turning pieces of waste hardwood into either a hollow chuck or a mandrel chuck depending on the use. A hollow chuck is used for example to hold the rim of a bowl to turn the undersides. The recess in the hollow chuck obviously has to be carefully cut, stopping the lathe frequently to test it against the workpiece for fit. Too large a recess can be packed out with newspaper or if the fit is too close, wet with a sponge to expand it to size.

The mandrel chuck is made the same way, but to hold the work outside the rim as shown.

Before mounting the wood, draw a circle with a compass (mark the center clearly) and cut the shape to rough size on the band saw. Then plane the inside face flat and screw the faceplate in position as close to the center as possible.

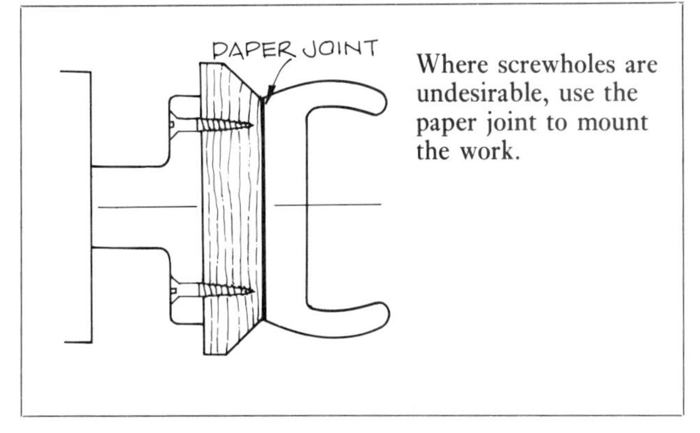

Where screwholes are undesirable, use the paper joint to mount the work.

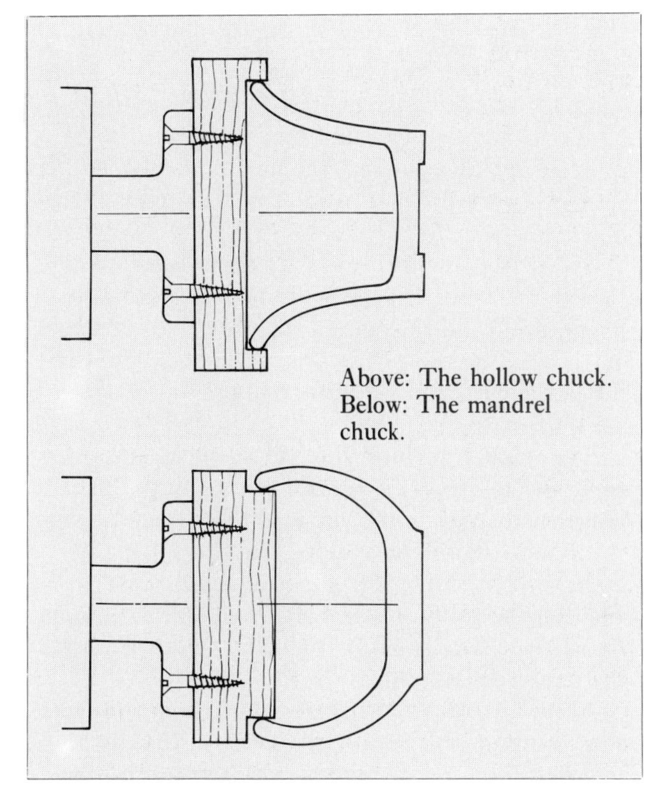

Above: The hollow chuck.
Below: The mandrel chuck.

Bowl turning

Bowl turning is more difficult and dangerous than spindle turning, partly because larger objects are involved and partly because of the orientation of the grain. In spindle turning, the grain is always along the spindle so that the fibers get sheared off safely. But in bowl turning the turning tool cuts into end grain twice on every revolution, and if the correct technique is not used, the gouge can dig into the end grain tearing off a chunk which can fly dangerously off the wheel. The emphasis is as always on sharp tools and the correct technique. The gouges should be sharpened frequently on a grindstone which is an essential tool in wood turning.

TURNING THE OUTSIDE OF THE BOWL The dried blank is faceplate mounted and the appropriate lathe speed selected according to the size of the bowl (see table). Bowls should be turned on the outside of the headstock to allow more access. Set the hand rest at about the center line of the bowl and rotate the wood by hand to make sure the rest is about $\frac{1}{8}$in clear of all points.

Even wood which has already been roughed out or cut to circular shape on the band saw will tend to rotate unevenly so the initial cutting must be firm but careful until a smooth, circular shape is achieved. Every turner seems to have his own method of bowl turning. Some use round, pointed gouges. Others grind the gouges straight across and use the whole length of the deep, fluted edge to do the cutting. I have tried several ways and though I find it more difficult in the beginning, the method explained here seems to work the best in the end. It is certainly the quickest.

To use this method, use only deep fluted gouges to turn the bowl and a few scrapers afterwards to finish up. Grind the gouges straight across and to a 45° bevel. Use the $\frac{3}{4}$, $\frac{1}{2}$ or $\frac{3}{8}$in gouge to do most of the initial cutting, then change to the $\frac{1}{4}$in gouge to get a finer cut.

The important thing is to get the hold and stance right and to keep firm control of the gouge. The longer and stronger the gouge is the better. Cheaply made tools should be avoided. All the cutting is done without sliding the gouge along the rest. Instead, the tip of the tool is moved sideways by moving the handle left or right, up or down using the hand rest as a pivot.

Always cut down the slope. Start about $\frac{3}{4}$in from the bottom of the bowl (right side), so that for the first cut, the tip only moves a short distance from left to right. Present the tool carefully to the work, flute upwards and handle down, so that the bevel is flat against the wood surface without cutting. Keep a firm grip as you lift the handle slightly to point the edge into the wood to start cutting. This is a crucial moment, for the chisel can dig in and make an awkward groove which leads the tool each time it is subsequently presented. If this happens, it is best to use the largest scraper to smooth it down for another start.

Once the edge is cutting nicely, proceed by moving the handle to the left and upwards so that the cutting edge is brought down and to the right. At the same time, roll the gouge over to the right so that the left or trailing tip does not touch the wood. This is the all important thing in bowl turning. The trailing wing tip, i.e. the left tip when moving to the right and vice versa, must *not* be allowed to touch the wood or it will dig in causing nasty marks, flying gouges and possibly flying chunks of wood.

The cut to the right started at the bottom of the flute and finished cutting with the right wing tip, using one half of the edge. After this initial cut has created a slight slope, subsequent cuts are made in the same way starting a little further to the left each time and following the same cutting path with the tip ending up a bit lower down on the bowl than at the start, in order that the bevel be kept at its proper angle to do cutting and not scraping. It's best to do a few practice runs to get the knack of it, and to adapt one's own elaborations on these basic methods. Remember to remove very little wood with each cut until more confidence allows deeper cuts to be made.

Depending on the shape, the cutting direction is reversed right to left if the rim of the bowl slopes inward, i.e. the direction of cut is always down the slope. To cut right to left, start cutting near the extreme left side, rolling the gouge to the left.

The surface left after using the $\frac{3}{4}$in or $\frac{1}{2}$in gouge will still be slightly rough, particularly at the two end grain sections. To smooth this down, use a $\frac{1}{4}$in deep fluted gouge, in exactly the same way, working downhill, rolling the gouge with the bevel rubbing for a proper cutting action, taking fine cuts only. The deep fluted shape of this narrow gouge is particularly good at cutting across end grain, so the surface should be quite smooth after using it.

Final smoothing, prior to sanding, is done with a sharp scraper which is held point downwards so that it scrapes rather than cuts. For corners or shapes which may be difficult to scrape with a standard tool, make your own scraper out of an old file or French head steel (see page 125).

Above: Starting the cut, handle well down. Left: Bevel rubbing for correct cutting angle. Below: Finish the cut with the handle up and right edge cutting. Right, top to bottom: Reverse the cutting procedure to cut the other side.
Shavings indicate correct cutting angle.
Scraper held handle up.
French head scraper, hand-held for fine finish.

Above: Dressing the face with a $\frac{1}{2}$in gouge.
Below: Starting the cut with firm grip.

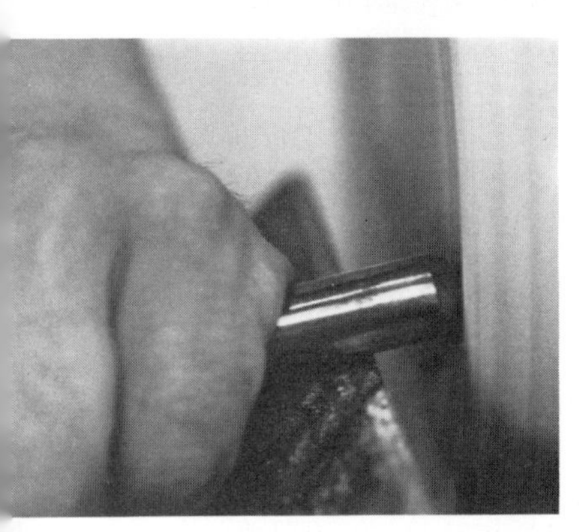

Left: Gouge entering face near rim on its side, flute toward center. The grip should be firm to keep the point in place. Below: Make successive cuts from the rim toward the center with the gouge. The cutting path is a slight arc ending up at the center with the handle horizontal. Right: Finishing rim with scraper. Far right: The completed bowl.

TURNING THE INSIDE OF THE BOWL Set the hand rest slightly below center as close to the wood as possible without its touching the wood. The face or insides of all bowls, trays, etc. are cut on the right radius only.

Before shaping the inside, dress the face, that is make the surface flat using the round pointed scraper, or with experience the $\frac{1}{2}$in deep fluted bowl turning gouge. Start in the middle of the radius and cut towards the center, starting further out on each cut depending on the size of the bowl.

Again there are many ways to turn the inside of a bowl. The method explained here uses the $\frac{1}{2}$in deep fluted gouge, ground straight across, to make cuts in one or two sweeping motions. Depending on the size of the bowl, start at the rim or for large bowls, halfway between the rim and the center.

Since all cutting is right to left, only the left part of the edge does the cutting. The right tip is never allowed to touch the wood or it will dig in causing bad marks in the wood which create a track to lead future cuts astray. Unlike outside turning where both tips were used, on inside work the leading tip does tend to dig in a little, so I find it best to grind away this tip slightly to keep it permanently out of the way. This means having a separate $\frac{1}{2}$in gouge for inside turning.

Start the cut with the gouge on its side, flute facing the center. Bring the handle down slightly, then push the edge firmly but slowly into the wood. The tip will tend to be thrown outwards but once the initial groove has been made, it will stay in line.

Continue the cut, moving the handle downwards and to the right so that the tip is moving up and to the left. Keep the bevel against the face so that the edge cuts correctly and always keep the trailing tip clear of the wood. The line of cutting goes in a shallow arc from right to left, ending up in the center. It becomes obvious when actually working on

the lathe, that to keep the bevel rubbing and to keep the right wing away, the handle has to be brought upwards and to the right towards the end of the cut. Using other methods, the tool is moved along the rest but in this method extra strength and control is required so that the left hand is kept firmly over the blade allowing it only to pivot on the rest.

Make many fine cuts in this way, until the shape is completed. Rim thicknesses vary from $\frac{1}{8}$in for fine work up to $\frac{1}{2}$in for other work.

After the gouge, the scraper is used to get a smooth finish. Move the hand rest up slightly to allow the edge of the scraper to work at near center line, tip downwards in the usual scraping action. For a really fine finish requiring a minimum of sanding, hone the scraper and bring over a fine even burr.

The widest scraper is used to get a flat bottom to the bowl, moving it back and forth with the hand rest brought as close in to the bottom as is safe.

Finishing Sanding is easy on the lathe since the revolving workpiece does all the work. The grit used depends on the surface quality left from the scrapers at the end of the turning work. The best type of paper is open coat garnet paper. Open coat papers are used for all machine sanding to prevent clogging. Start with 80 or 100 grade when a lot of rough sanding is required, 120 otherwise, and finish with 220 grit for a really smooth finish. Fold the sheet in quarters and sand the turning freehand (with the hand rest removed!), changing sides as soon as one side gets clogged up.

There are many sealers which work well on turned work. It all depends on the type of finish wanted. Some woods darken too much when sealed and are perhaps better left unsealed or perhaps lightly waxed. Other woods like teak are best oiled since the open-grain texture will not take a glossy finish well.

FILLER Whatever the finish, any minor flaws can either be left as honest reminders or filled with ordinary wood filler to match the grain.

STAIN For most decorative work the wood grain speaks for itself, but for some work such as mahogany reproduction work, some tinting or staining may be desirable. Water-based stains go deeper but tend to be blotchy, so oil-based stains are probably best. They are applied as usual with the wood stationary but the lathe can be rotated slowly afterwards to rub it in evenly.

SANDING SEALER For most woods, sanding sealers, usually cellulose based, can be applied onto the bare wood, then sanded off with 220 grit paper to give a good base for almost all the other finishes. The sealer, which is widely used in all furniture work, has an additive which prevents the paper from clogging afterwards. Special sandpaper is available for rubbing down sanding sealer.

WAX A lovely, high polish finish can be achieved with the furniture maker's standard mixture of beeswax and the much harder carnauba wax. They are melted together with genuine turpentine in an electric pot (or improvised double boiler) in the proportion desired; more carnauba wax for a harder finish. The resulting mixture is cooled to let it harden, then applied directly by hand to the wood. One or two applications give a beautiful, deep finish, made glossier by buffing with a lint-free cloth.

LACQUER To achieve a more waterproof finish, turned work can be lacquered using the polyurethane lacquer widely available in one pack cans. Dilute the lacquer with appropriate thinner for the first two coats and apply it, wood off the lathe, in thin even coats which are rubbed with 0000 steel wool on the lathe between coats. One or two thin coats are enough to give a moderately waterproof finish, ideal for use on bowls and plates to be used with food. Otherwise a coat of wax on top of the lacquer, well-burnished in gives the finish a much deeper tone.

For a high gloss finish, use any one of the plastic finishes on the market. Sometimes these are sold as clear floor finishes because of their hardness. Buy the burnishing cream that comes with it, and after several thinned down coats burnish the finish to a glass-like glossy appearance.

Bar top lacquer gives an amazing glossy finish. It, too, should be well-burnished afterwards, with the lathe moving as slowly as possible.

SANDERS

Large belt sander (above) will take 8ft lengths. Small belt sander (below) is convenient for sanding edges and ends.

Some form of mechanical sanding is essential for any shop. For the one man shop, the portable belt and orbital sanders are excellent for cutting down finishing time, which is hard work by hand. Particular attention should be paid to the portable belt sander with fine adjustment shown on page 62. This design reduces the risk of digging in which often occurs with belt sanders. It is therefore a kind of mini-stroke sander, suitable for sanding large surfaces in veneered as well as solid work.

Sanding machines come in a variety of sizes, the exact configuration depends on the application required. Disk sanders with diameters between 7 and 24in are normally fitted with a tilting table in front so that the sander can be used for accurate end trimming of square as well as bevel cuts. Used with a miter gauge which slides in a groove in the table, it is an accurate tool and ideal for the small shop. It can be used in combination with the band saw where a table saw isn't available for squaring off work for making joints. The use of disk sanders is, however, limited to end and edge grain since the small sanding area and circular motion makes it unsuitable for large, flat work.

Open coat sanding disks in a variety of grades and backings, are made to match the diameter of the metal disk. These are applied either with special adhesive sold with the disks, with rubber cement applied to both surfaces, or by using abrasive disks with adhesive coatings on the back.

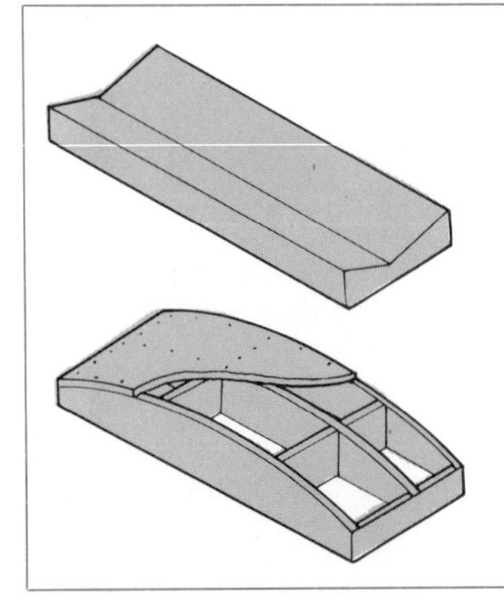

Forms such as these are placed under horizontal belts on small machines to do specific jobs such as chamfering edges (top) and sanding inside concave shapes (bottom).

Small belt sanders are more versatile than disk sanders, since all four edges of small stock can be sanded. The boards should preferably be held against a stop which should be 6 to 8in high. In addition, the open wheel ends allows the operator to sand curved surfaces freehand, which is very convenient for many types of work such as chair making.

The belt sander is often used for form sanding, where the belt runs over a form specially constructed to create the necessary shape of the finished piece. This has many applications in small production shops where the rounding of edges, beveling and so on are otherwise time-consuming operations by hand or router.

On some models, the belt can be adjusted to a vertical as well as horizontal position, convenient for sanding ends of long pieces or, when used with the stop, as a table for sanding delicate curved pieces with more control.

Disk and belt sanders are commonly mounted on the same machine, using the same motor to drive both disk and belt which together give a wide range of applications for small work.

The narrow belt sander/grinder, widely used in home workshops for grinding and sharpening tools as well as for wood sanding, may be of interest to model and instrument makers and anyone making small pieces requiring delicate sanding. Belts, available in widths from $\frac{1}{8}$ to 1in, are backed up by a steel plate to provide the solid surface required for accurate metal grinding. The plate can be removed for freehand sanding.

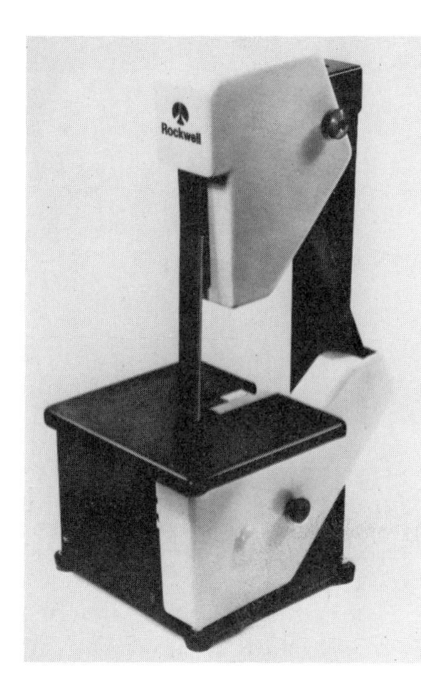

Narrow belt sanders are becoming increasingly popular. Their use in a large shop is somewhat limited, but since they are versatile enough to handle small, freehand sanding jobs as well as sharpening tools and so on, they are quite useful in the small, one man shop.
Below: A large disk sander, useful for squaring sawn ends and other small jobs. The belt and disk sander are often incorporated into one machine.

Hand stroke sander The large belt sander is undoubtedly one of the most useful machines in the workshop, but its cost and large size limits its use to larger shops. Good secondhand ones do occasionally come on the market (see page 41 for secondhand machines), but the buyer must be careful to inspect the machine in running condition since the large size, high vibration and proximity of fine dust often create problems with the moving parts.

The large belt sander consists of two pulleys. One is driven by the motor, the other is an idler, which is often left unguarded and used for freehand sanding.

The motor which drives the belts mounted on the pulleys, may also power a pump used to extract the fine dust.

Dust extraction is essential (usually required by law), for the fine dust is very harmful to the operator and may form an explosive mixture, capable of flashing over large distances given the right conditions. Where no integral dust extraction is available, a separate system must be fitted, using either a portable unit or part of the fixed system for the rest of the factory. Where the general extraction system is used, however, safety laws usually require separate pipes and bags for the fine sanding dust, to avoid forming dangerous mixtures with the other waste.

Belt tensioning devices vary from model to model, but is usually done by some form of counterweighting device which adjusts the position of the idler pulley. Some models include a third pulley above the machine for tensioning, but this creates a disadvantage since it doesn't allow the top of the belt to be used freehand for smaller pieces.

The idler pulley also incorporates a belt tracking device which adjusts to stop the belt from drifting off the pulley. Slight variations in length or variations in internal tension from one side of the belt to the other often produces a wavy or snaking roll of the belt. This, too, is eliminated by adjusting the tracking mechanism from one belt to the other if necessary.

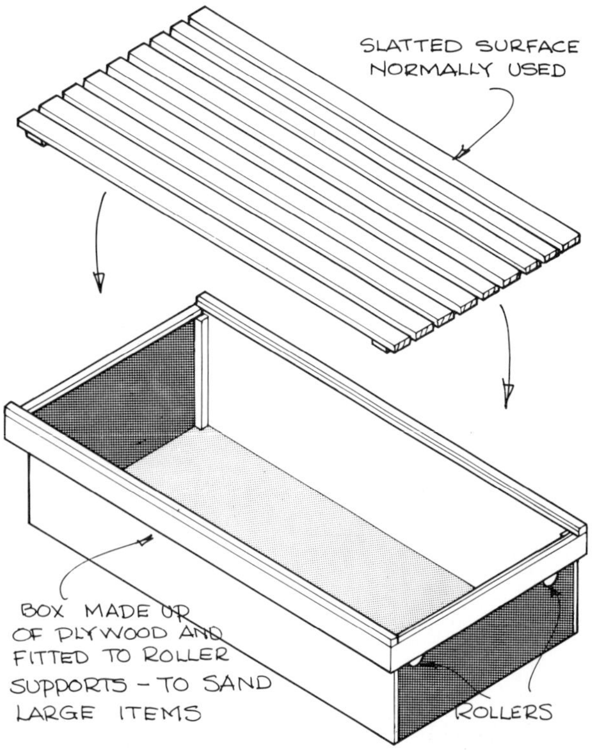

The table which holds the workpiece is adjustable up and down to allow for varying thicknesses. Tables which can be cranked all the way down have the advantage of being able to take complete cabinets or carcasses for finishing sanding. Where this is not possible, make an adaptable table as shown with a top that can be replaced with a "basket" to take longer pieces. The table runs on rollers so that it can be moved backwards and forwards during sanding.

New machines usually consist of a heavy metal body which connects the two ends of the machine rigidly so that they are fixed at the correct height. Older machines with separate left and right hand pillars are often difficult to adjust so that the rise and fall mechanism of the table works correctly. This should be kept in mind when buying a secondhand machine.

Stroke sanders can be adapted for large objects by making a box, as shown.

USING THE SANDER The size of the work is limited by the size of table or by the clearance between ends. This may be 48in for small machines and up to 12ft for very large ones, with about 6 to 7ft the most convenient size.

The work is held by a stop screwed to the top or by suction pads on new machines. To sand the work, push down onto the smooth back side of the belt with a handled block, faced with heavy felt or special modern friction-free graphite pads. The heavy felt used under saddles for horses is readily available. When pressing down it is important to move the block rapidly back and forth along the belt with one hand, while at the same time rolling the table back and forth with the other. This avoids having the belt dig in causing ridges or indentations. This would happen instantly if the block were to be held stationary. In this way the work is sanded in small overlapping areas. Work by touch to make sure the surface is even. Start at the right end and finish at the left motor end to drive all the dust from the surface. The belt sander is a powerful machine, capable of removing a lot of wood. This is an advantage when finishing a rough surface, but a distinct disadvantage when sanding veneered boards, where extra care and a very light touch should be used to avoid going through the veneer.

Most new models incorporate an automatic or hand operated pressure pad which slides back and

forth with ball bearings along a bar. The handle is levered so that much less force is required to hold down the pad. This makes it useful for solid production work but too insensitive for veneered or other delicate surfaces.

Open coated belts, which are sold in standard sizes or made up to fit a particular machine, are available in grits from 40 to about 220. The grit size required is generally larger (i.e. a smaller number) than for the equivalent hand sanding operation. Thus for hardwoods a 60 to 80 grit is used for rough sanding with a subsequent finish sanding of 100 to 120, whereas softwoods require 120 for rough and 150 to 180 for finishing. Veneers may require a 220 for final smoothing. A new belt should be used on a piece of waste first to wear down any extra large or sharp grits and produce an even wearing surface. And keep in mind that belts wear quickly. A slightly used 80 belt may behave like a 100. We keep a few well-worn 150's for a final touch to special work.

In modern factories, wide belt sanders have largely replaced the hand stroke belt sanders. The expensive wide belt sanders which take up less space and accept work of any length can be arranged with a single belt above or with a combination of belts above and below for sanding both surfaces at once. It is capable of fine adjustment making it suitable for solid work such as doors, as well as veneered panels. The modern trend is to use the wide belt sander as a thicknesser, planing and sanding one or both surfaces without using the more cumbersome planing machines. In making laminated tops such as maple butcherblock worktops, for example, the boards are face sanded, multiple ripped, then laminated with sanded faces together before being thicknessed as panels through the sander. A sanded edge glues up better than the rippled, planed edge.

Press down on the hand pad, moving it back and forth while also sliding the table back and forth to keep the belt from digging in. The hand lever (above right) is less sensitive to delicate jobs.

Below: A wide belt "thicknessing" sander.

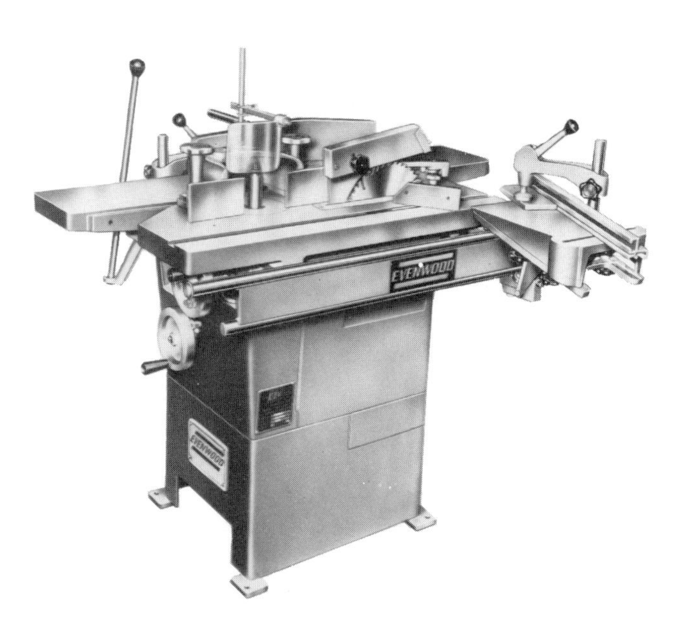

Above: Ripping. Below: Crosscutting with sliding fence. Bottom: Shaping edge with spindle.

COMBINATION MACHINE

A combination machine includes several functions working off the same motor. This saves both space and money and is ideal for the small furniture workshop or joinery shop where, unlike in the production shop, the various machines are not likely to be in use at the same time. Designs and details vary from one model to another, but most universal woodworkers include the basic machines – the saw, jointer, thicknesser, shaper plus a horizontal boring device which can be used for drilling, mortising and other operations.

Some models include a lathe, grinding wheel and sanding attachments and even a band saw, fret saw and belt sanding facilities. All of these suffer from the disadvantages of the limited size of table surface and the time it takes to convert from one fuction to another. Some machines are very badly designed, requiring cumbersome tilting, unscrewing and so on to change over, but many are very clever and require only a relocation of V belt for conversion. This is most important when choosing a machine, for the flow of work in a small shop is likely to be from the saw to the planer and then back to the saw and so on, and time-consuming change overs would nullify the advantages of the universal machine.

There are light, medium and heavy weight universal machines for various applications. If possible, choose a heavy-duty machine for they are more powerful (a minimum of 1hp is required) and often incorporate a dual motor drive by which two or more operations can be carried out at the same time.

Above left: Surface planing or jointing.
Above right: Thicknessing.

The Wadkin model shown here incorporates a saw, jointer/thicknesser, spindle molder and a horizontal borer powered by a 2hp motor. The sliding fence can be used for both the saw and the spindle.

The 10in saw is adequate for most uses giving a $3\frac{1}{4}$in depth of cut for crosscutting, ripping, bevel cuts and mitering. But, as with so many universal machines, the table is too small. To cut large sheets of plywood, supports must be added on three sides and it is a good idea to build benches nearby at just the right height to help support large pieces.

The spindle or shaper head takes $1\frac{3}{16}$in diameter spindles which are held down by a locking nut. The work is fed by hand for edging work, but for end cuts such as tenoning, the workpiece is held to the sliding table for more control.

The capacity of the jointer is 8in with a 4ft long table. The thicknessing capacity is $6\frac{1}{2} \times 8$in, which is more than adequate for most work.

Drill bits are held in a horizontally mounted chuck and the workpiece is fixed to a table which can be moved in all directions to feed the work to the bit by operating a hand lever. This arrangement resembles a horizontal boring machine and is particularly suitable for dowel jointing, where by the use of fences and stops, one or more holes can be located accurately for small production runs. The same chuck will also hold other accessories such as mortising bits and sanding disks and can be adapted to suit other uses by the ingenuity of the operator.

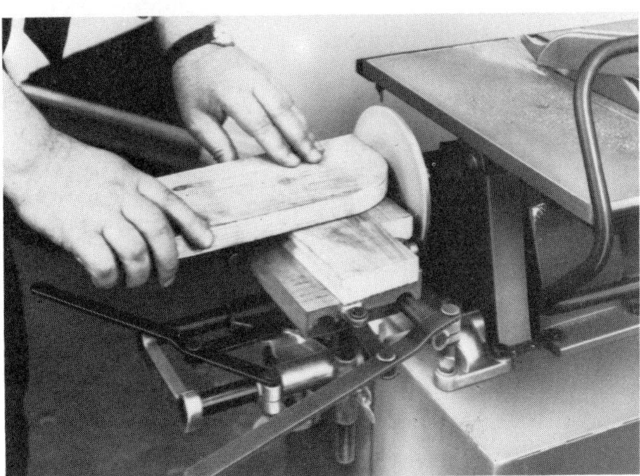

Top: Mortising with work clamped down. Above: On some models, it is possible to fit a sanding disk in the drill chuck. Below: Another accessory available with some models is a flexible drive shown with a small drum sanding attachment.

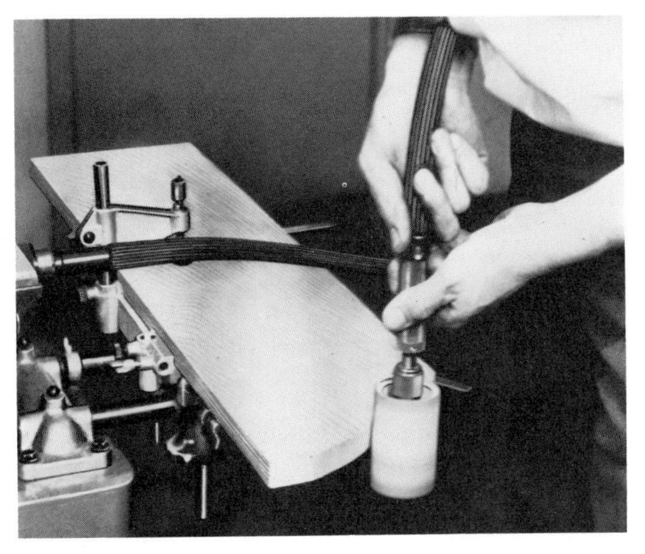

Woodworking joints

EDGE TO EDGE JOINTS

Edge to edge joints, used when gluing up several boards into a surface for table tops, cabinet parts, etc., are either plain or reinforced, but in each case rely solely on glue to hold them together.

Plain butt joints are sufficient for most work. Because of the high strength and durability of modern glues, the joints are usually stronger than the wood itself. But it is important to emphasize that all edge to edge joints, particularly butt joints, rely on good fit which means accurate planing. Badly planed boards which are forced together by clamps during gluing up will eventually release the stresses and split along points of weakness.

For thinner boards, such as those used in door panels, the glue surface should be increased by reinforcing with a tongue set in a groove on either side. Dowel reinforcement doesn't really add much strength, but it does help to align the edges when gluing up.

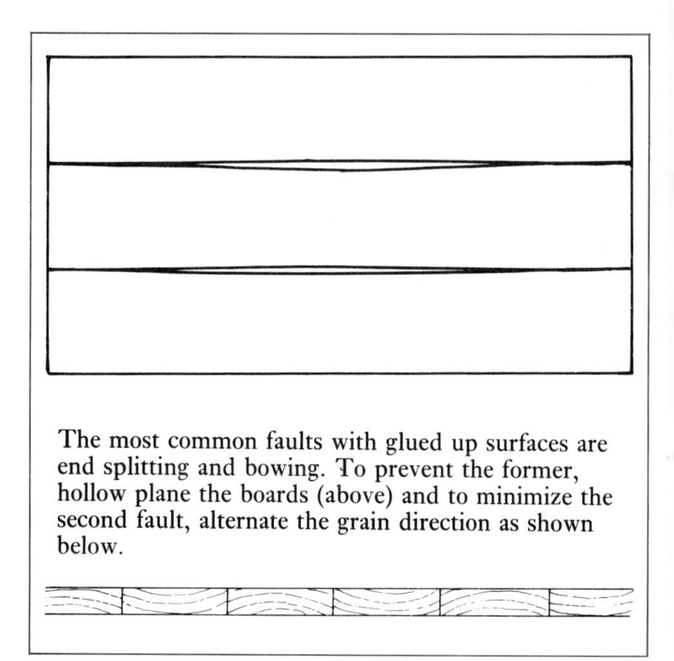

The most common faults with glued up surfaces are end splitting and bowing. To prevent the former, hollow plane the boards (above) and to minimize the second fault, alternate the grain direction as shown below.

Preparation Whether plain or reinforced, edge to edge joints require accurate planing. Machine planing on a jointer is the easiest method, particularly for long boards. The usual procedure is to plane and mark a face and an edge, then rip the second edge on the table saw. The boards can, of course, be of unequal width but in most work they are equal, which means one setting on the saw to the width of the narrowest board.

Before laying out the boards to choose the best arrangement, thickness them so that there are two sides to choose from. Lay them on battens to match them up, then mark the ends as shown in the box opposite, before planing the second edge.

Hand planing requires considerably more skill. In using a try or jointing plane, the problem is not so much getting a straight edge but keeping the edge square all along the board, to avoid leaving gaps which form unsightly points of weakness in the glued up surface. Accuracy comes with experience. The aim as in all planing, is to apply slight pressure at the front of the plane at the beginning of the stroke, and at the back of the plane at the end. And a word about the plane. It's impossible to shoot a good edge without a sharp iron, ground and honed flat and set as close to the front of the opening as possible with the cap iron about $\frac{1}{32}$ to $\frac{1}{16}$in back for fine and rough cuts respectively. The edge should be checked for square along the length and marked where it is not square for further strokes of the plane.

Very thin boards are best edge planed while held in a shooting board but this is limited to short lengths. Long, thin boards can be planed by making an improvised shooting board, placing a suitable piece of plywood on the bench as the base for the plane and adding a batten and a stop as necessary to hold the thin workpiece.

Depending on the length of the boards, they can be joined by rubbed joints or by clamped joints. See page 212 for details on gluing up. Rubbed joints, limited to short boards up to about 3ft long, require a straight edge, but edges for clamped joints are sometimes planed slightly hollow to keep the ends held together tightly. Hollow joints should be used only on boards narrow enough to deflect to fill the gap. On stouter and wider boards in a strong wood such as maple or oak, the edges are best left straight. Whether hand or machine planed, the edge is planed hollow by hand, using one or two strokes with a very fine cut. The aim is to start and end the cut gradually so that the edges come together evenly when clamped.

A useful hint for machine planing edge joints

When edge planing several boards which are to be glued up into a surface, the fence on the jointer must be absolutely square. This happens more often in books than in real life. To get around this difficulty follow this procedure:

1 Prepare the best face and edge, rip the second edge and thickness the boards.

2 Match up the boards by laying them out to select the best pattern. Alternating grain direction is optional.

3 Mark the end joints with diagonal lines, as shown. Number each joint.

4 Plane the edges on the jointer keeping the diagonal lines pointed towards the bottom of the fence for each edge. This way, if the fence is out of square, one edge will compensate for the other, ending up in a perfect fit. Of course, if the fence is too far out of square, the joint will tend to slide out of alignment during gluing up.

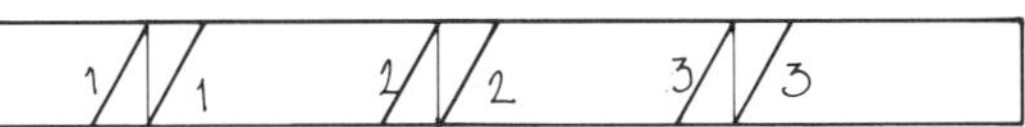

Above: Slanted lines and numbers of each joint.
Below: Planing with line pointing toward bottom of fence.

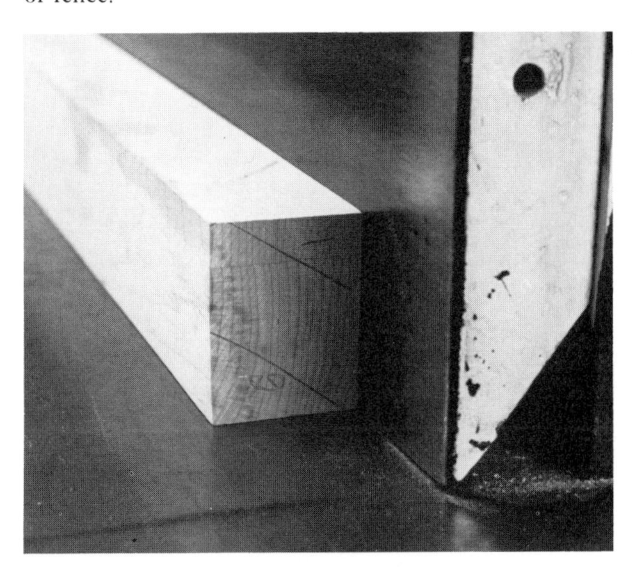

Detail of end grain table top with machined edge joint by Ashley Cartwright.

Reinforced edge joints Tongue and groove joints increase the glue area thereby strengthening edge joints and also help keep the boards in line during assembly. Two other reinforcement methods, using either doweled or slot screwed joints, have a more localized effect and do little to prevent the small splits often caused by a change in moisture content. Tongue and groove joints are not strictly necessary for thick boards, but thin boards should always be reinforced.

I find the best material is $\frac{1}{8}$in thick plywood which, unlike wood, has strong resistance to shear across the joint. To prevent the plywood tongue from showing, either stop the grooves about 3in from the ends or, on through grooves, finish the last few inches with a tongue made from the same wood as the boards. Unlike the plywood tongues which should fit in smoothly and not extend to the bottom of the grooves, these end tongues can be tight fitting so they look better, and tapped in from the ends during assembly.

Tongues can be decorative too, if made from a contrasting wood. And, if room allows, double tongues can be installed, for added decoration.

For plywood tongues, the groove must be cut just slightly wider than the plywood, wide enough to make it easy to fit (with glue) but tight enough for strength. Generally, the thickness of the tongues is determined by the tool used to cut the groove. Grooves should be slightly deeper than the tongues to hold excess glue ($\frac{1}{2}$in depth is about right).

Production shops use shapers or spindle molders fitted with a small saw blade to cut grooves. This way the workpiece can be held down against the table, and in against the fence using pressure clamps or feather boards to keep the grooves to a constant depth and at a consistent "in-set" distance. The in-set distance is all important, for it guarantees that the top edges will line up during assembly. Therefore, whatever machine is used, the best marked face should always be run against the fence or table. If no spindle molder is available, cut the grooves on the table saw, running the best face against the rip fence at all times. Clamp a feather board to the table to hold the boards firmly against the fence.

As an alternative method, fit a special slotted cutter to the portable router or use the biscuit jointer shown on page 60 to cut edge grooves.

It is necessary to make a second run after changing the setting where the kerf of the cutter or blade is not wide enough for a plywood tongue. All the boards should be run through before the fence is changed, and the second setting tested on a scrap.

If no machinery is available, cut grooves with a plough plane fitted with a grooving cutter, working with the fence against the good face.

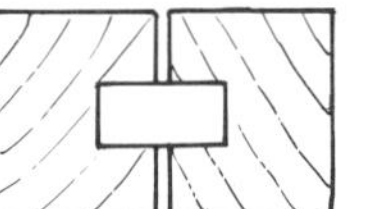

Far left: Through groove. Left: Stopped groove. Above: Faulty grooves. Below: Edge joints with decorative tongue.

Beveled or coopered joints

Applications such as making cylinders call for beveled edges where planing must be done at an angle. This is again easiest on the jointer, with the fence set at the appropriate angle and checked against a sliding bevel. Planing by hand is more difficult and is best done on a shooting board with wedges of the appropriate angle pinned to the top.

With either method, it's best to mark the angle on both ends of the board (and on the face if necessary) as a final check on planing accuracy. If reinforcement is necessary, cut a groove perpendicular to the beveled edge and insert a small tongue.

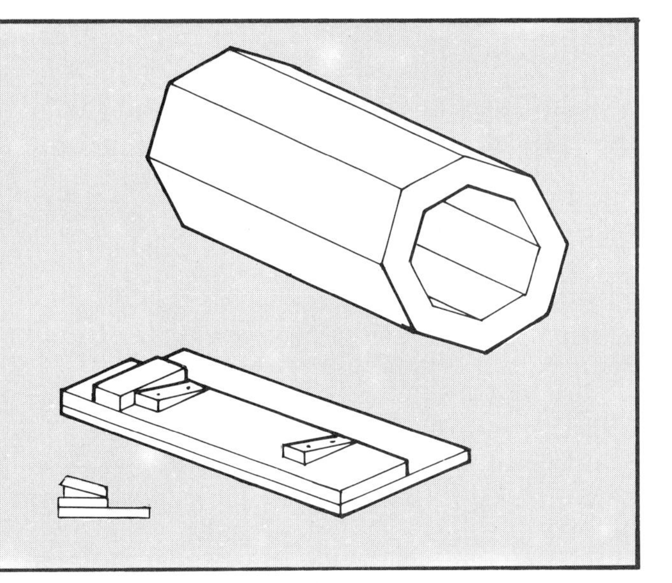

Machine joints

In mass production, the loose tongue method is too slow. Instead, special cutters, often fitted to a four cutter which planes the four side at the same time, cut various shaped tongues and matching grooves in the edges. For small runs, standard factory made cutters sold in matching pairs can be bought, or alternatively they can be made out of French head steel (see page 124) and fitted to the spindle molder.

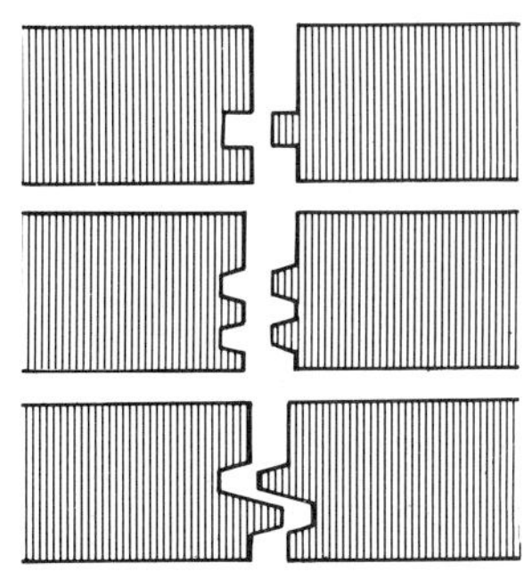

Top: Standard tongue and groove joint.
Middle: Double tongue with sloping sides for easier assembly.
Bottom: Another version of easily assembled single tongue with more gluing surface.

Grooved frameworks for panels The traditional and proper way of making cabinet doors is to mortise and tenon the sides together and to cut grooves in the sides for the solid panel which is then fixed along one edge but allowed to move along the other three.

The groove can be stopped to avoid showing at the ends, but more often it is run through and the end is filled by the haunch of the tenon.

The grooves are cut, as above, wide and deep enough to allow movement in the door panel but tight enough to hold it firmly.

Grooves for plywood panels, which can be glued all around, should be slightly tighter for the glue to hold.

A method often used in cheaper furniture is to dispense with the mortise and tenon and instead cut very small tenons in the rails which are glued into the groove cut in the stile. This saves a lot of time since the groove, which is quickly cut, serves to hold both the panel and the rail.

HOUSING JOINTS

Housings or dados, widely used in bookcase and cabinet construction, are wide, shallow grooves cut across the grain to hold the edges of shelves or dividers. The housing can take the full thickness of the shelf or it can be a narrow groove to take a tongue cut in the end of the shelf. Although the latter is preferred because the shoulder distance gives a more precise fit, either method relies on tightness of fit and to a lesser degree on glue for holding power.

The depth of housings should be about $\frac{3}{16}$ to $\frac{1}{4}$in and they should be cut tight enough in the width to require clamping or knocking with a mallet during assembly.

Marking It's best to first mark one cabinet side (cheek) then transfer the marks to the other side with the two sides laying side by side. Use a large try square to square the lines across. Mark only the bottom of the groove. If the cutting is being done by router, the box guide lines up with this line. If you are cutting by hand, the shelf itself should be stood on end, bottom edge against the line, so that the thickness can be marked on both sides with a sharp knife to form a starting line for the hand tools.

Cutting By far the easiest way to cut housings or grooves is to use a portable electric router fitted with a straight cutter, and run in a box guide (page 69) or against a simple straightedge clamped to the work. A box guide is better since it controls movement from both sides, therefore eliminating the danger of having the router stray from the guide and ruining the work. It's very convenient if the cutter diameter matches the thickness of shelf exactly, because then only one run is necessary. But this is rarely the case, so the sides of the box guide must be separated by the appropriate amount to make up the width with a smaller cutter.

For through housings, clamp or pin a waste piece across the front and back edges to back up the cut across the front and back edges, as shown in the photograph, to back up the cut and avoid splintering at the entrance and exit.

Cutting narrower housings (grooves) for tongued shelves is easier because the groove is cut first using a convenient diameter cutter roughly two-thirds the thickness of the shelf. As in mortise and tenon work, the tongues are cut afterwards to match the grooves.

Besides the router, the only affordable machine which will cut housings is the radial arm saw, but it is only practical for through housings. For stopped housings, the sloping ends left by the blade require considerable cleaning up by hand.

Standard straight through housings are the easiest housing joint to make, especially with a router. The joint relies on tightness of fit for strength.

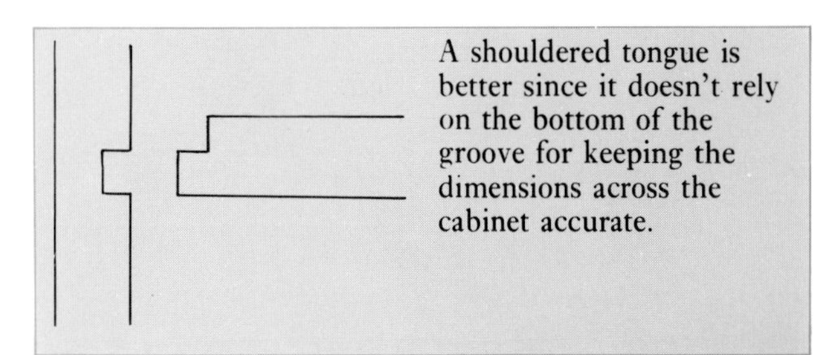

A shouldered tongue is better since it doesn't rely on the bottom of the groove for keeping the dimensions across the cabinet accurate.

Above: Cutting stopped housings with a router and box guide. Below: Pin or clamp waste strips to edges to avoid breakout when cutting through housings.

Cutting housings by hand is much more cumbersome. After the outline of the shelf is marked with a knife, cut the scored line slightly deeper with the same knife. Use a wide chisel to make a V-shaped groove for the saw, making sure to chisel the V from the waste side. Then run a short tenon saw such as a dovetail or gent's saw carefully back and forth in each groove until the right depth is reached. The rest of the groove can be cut out with a hand router, using a suitable depth guide to check level of cut, if necessary.

Stopped housings To avoid having the housing show at the front edge of the sides, the housing can be stopped about $\frac{1}{4}$ to $\frac{1}{2}$in short of the front and the shelf end notched to match. Stopped housings are quite difficult to do by hand following the methods given, but with the electric router it's easy to build a box guide with a stop at the right location. The rounded end of the groove left by the router, can be quickly squared off with a chisel. To cut stopped housings by hand, drill holes at the groove ends to the right depth and at intervals. (A Forstner bit is perfect.) Then cut out the rest of the groove with a hand router. The hand router is quite accurate, but it's a good idea to check the depth with a gauge made from a block of wood with a nail protruding the set amount.

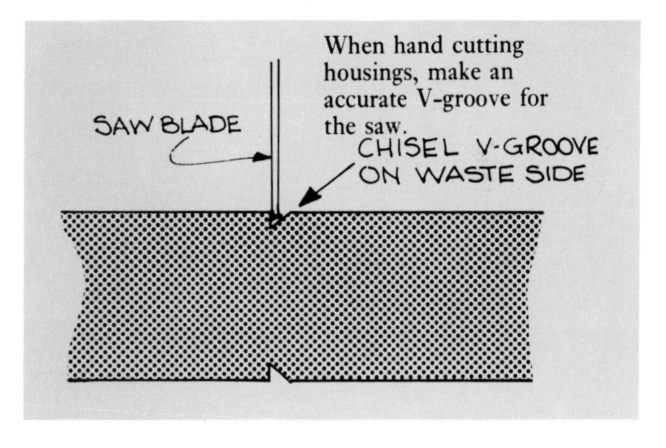

When hand cutting housings, make an accurate V-groove for the saw.

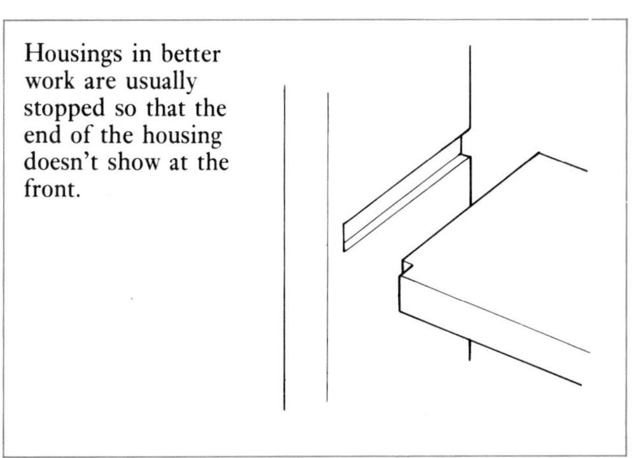
Housings in better work are usually stopped so that the end of the housing doesn't show at the front.

Tapered housings Cabinets with straight housings are awkward to assemble because all the shelves must be fitted at the same time, requiring lots of clamps and dexterity. Tapered housings, where both the housing and the matching tongue are tapered towards the front, can be slid in from the back and, if well cut, will make a tighter joint than straight housings. The taper can run the whole distance (this is easier to cut) or be limited to the final third of the width.

Whether to cut the housing or the tongue first depends on the method. Using a router, the housing is cut first. It's quite simple to taper the two sides of the box guide to get the correct cut. Remember to taper only the underside of the shelf, leaving the top edge square. By hand, the tapered tongue is cut first, then held up against the squared line on the cheek to mark the outline with a knife, as before.

Dovetail housings The dovetail housing is the best housing joint, for it relies on mechanical strength rather than friction and glue to hold the sides together. Wherever it is critical for the cabinet sides to be held firmly in place, such as for tambour cabinets, for example, (**page 250**) dovetail housings are the answer.

The dovetails can be single with the top edge left square, or double with both edges cut at an angle. They can also be tapered but this isn't really necessary for a well-cut joint. It can also, as for ordinary housings, be cut through or stopped. The through version has many decorative possibilities.

Single dovetailed housings must be cut by hand using a beveled block cut to an angle of 15 to 20° to guide the fine tenon saw. The dovetailed tenon is cut first with a saw along the shoulder line, then with a chisel at the matching angle using another guide

Tapered housings and tongues are much more difficult to cut, but allow for sliding the shelf in from behind. Make the box guide as shown with one slanted side to cut the housing which can be straight or dovetailed.

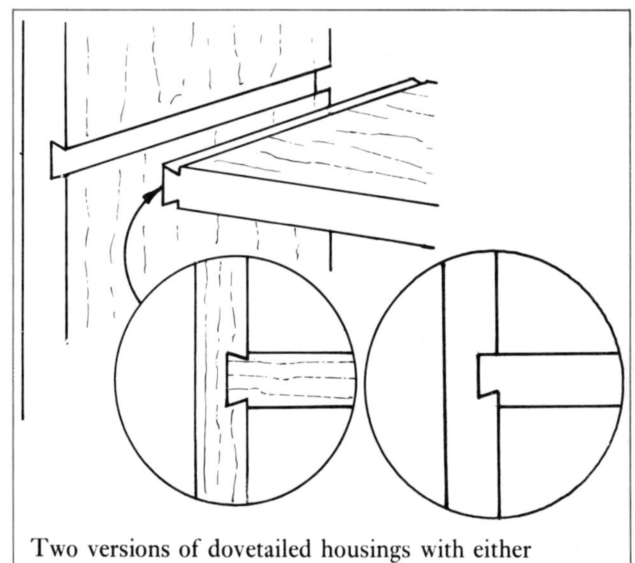

Two versions of dovetailed housings with either double or single dovetail. Single version should be dovetailed on the underside. Either version may be through or stopped.

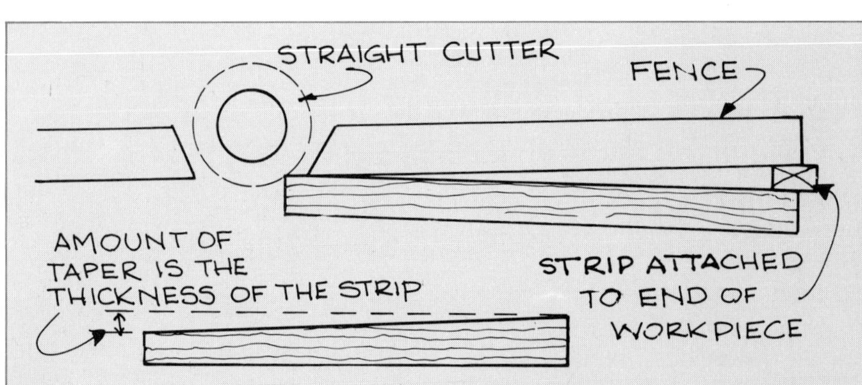

Cutting tapers
Tapers are easily cut on the jointer as shown on page 115. Another method which is more versatile and suitable for use on the spindle molder is to attach a small strip by gluing or tapering to the end of the workpiece. Work out the taper based on the length of cut (which should stop before the sliver) and the thickness of sliver.

block to cut the complement of the saw guide. Thus if the saw guide is 20°, the chisel guide is 90 minus 20 = 70°.

It is almost foolish to do the work by hand considering how easy it is to cut a clean, accurate joint by router, using a dovetail cutter. Use a box guide to cut the housing in one cut. For stopped housings the router must be backed out of the groove, and it is common to forget that and ruin the work by lifting it up at the end of the cut.

To cut the matching tongues, set up a router table (page 74) and hold the shelves vertically clamped to a holder to keep them square. Set the fence by trial and error so that after cutting both sides it leaves a tongue which fits the housing tightly but with enough room to slide it across the groove. If the shelves are too long to hold vertically, make a small jig with a block to fit on either side of the board. Clamp it to the board exactly level with the end and set the router fence the appropriate distance in to make the cuts, one from either side. As always, it is important to test the cut first, so it's wise to have a few shelf scraps handy for experimenting. Even though the tongues should be loose enough to slide in from the back, they will swell up slightly to fill the groove with the addition of glue.

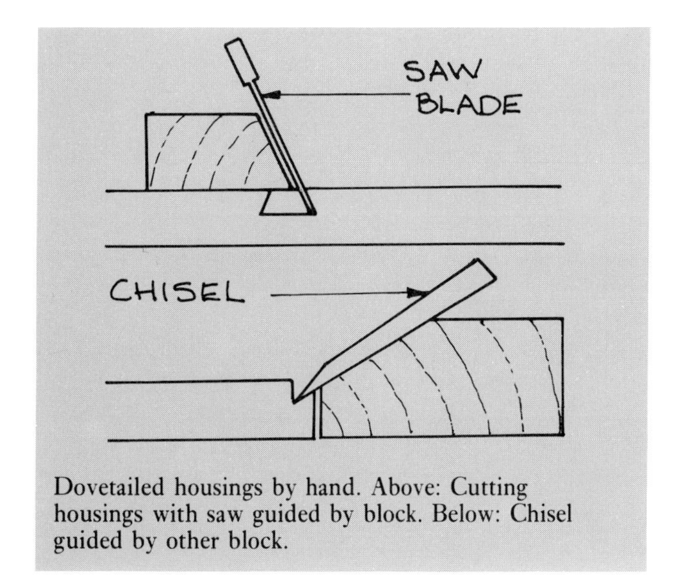

Dovetailed housings by hand. Above: Cutting housings with saw guided by block. Below: Chisel guided by other block.

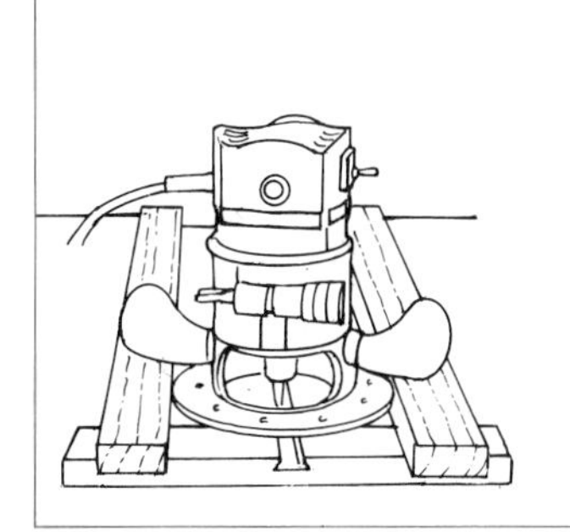

Cutting dovetailed housing with a router fitted with a dovetail cutter. Note use of box guide to keep groove accurate.

Decorative housing joint made by cutting dovetailed groove on both sides and inserting ebony butterfly strip.

Clamp identical blocks on either side and use fence with router from both sides to cut dovetailed tongue.

MORTISE AND TENONS

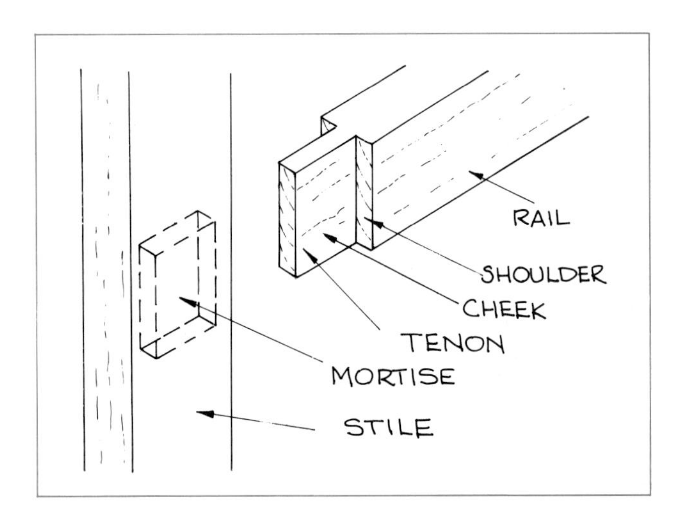

Above: Joint set out for hand cutting showing horns (waste) at end of stile. Below: Gauging in haunch depth on tenon.

Although mortise and tenons have been replaced in many applications by dowel joints, there is no doubt that a tight fitting tenon has more rigidity and reserve strength than a dowel. The main advantage of dowel joints is that they are much easier to cut.

There are numerous variations on the basic stub tenon shown in the diagram. Each has its specific purpose but the setting out and cutting of the joint is basically the same for all.

Setting out The basic procedure is to cut each female part first so that the male part is made to fit. So the mortise is cut first then the tenon, followed by any groove or rabbet for the panel door. Generally, the width of the mortise should be just over one-third the thickness of the stile or leg, but the exact width is governed by the chisel sizes available, whether for machine or hand mortising. The position of the tenon depends on the thickness of the stile. Where the stile and rail are of equal thickness, as in most cabinet door frames, the tenon and mortise are in the middle. And where there is a groove to take a panel, it is convenient if the tenon coincides with the groove. The tenon may be wider but never narrower than the groove.

Setting out can be done entirely by measurements following an accurate dimensioned drawing which should clearly show the size of opening, the size of members, the depth of groove, and the overall size of frame. The lengths can then be worked out and most cabinetmakers allow an extra inch or so in the length of the stiles for strength during cutting, to be trimmed off after assembly. An old-fashioned but very reliable method for setting out, uses a stick on which all the relevant lengths have been marked and labeled, to be transferred to the workpieces. The stick method can be used for entire cabinets and once it is accurately marked off from drawings, no further measuring is necessary (see page 244 for further details).

Whichever method is used, tick off the opening height on the stiles and the opening width on the rails, keeping the best faces up and face edges towards the inside. Then mark the width of the rails on the stiles, either by measurement or by offering the rail up. Then transfer all these marks to the inside (marked) edges. With a small try square also mark the extent of the mortise itself and the haunch, if any. Some workers prefer to leave a small shoulder (about $\frac{1}{8}$ to $\frac{1}{4}$in) at the inside of the tenon to cover any marks or rounding left from the chisel, but this requires more accuracy in cutting the tenons, for all shoulders should line up exactly for a proper fit. Cutting tenons with four shoulders is standard on tenoning machines and on the table saw.

The next step is to decide on the length of tenon, which can be a through or blind tenon, and mark this length off on the rails from the opening shoulder mark made earlier.

For through tenons, allow an extra $\frac{1}{16}$in which is trimmed off after assembly. For stub tenons, plan to make the mortise about $\frac{1}{8}$in deeper than the tenon to leave a gap for glue and also leave enough wood so that there is no danger of cutting through when making the mortise.

The method used to set out the width of the mortise slot depends on whether the mortise is to be machine or hand cut. In machine work, these setting out procedures are simplified because the various stops and fences automatically position the mortising chisels in the correct location. Nonetheless, most woodworkers prefer to mark up one frame completely showing depth of cut, mortise outline, shoulder line of tenon, etc., and use these marks to set up the mortising and tenoning machines.
With machine mortising, it is enough to mark the inside (face) edge of the mortise slot on one piece, but in handwork, the complete outline must be indicated on every piece using a mortise gauge set against a chisel. Use the same setting of the mortise gauge to mark the tenons on the rails, on the ends as well as the edges.

Cutting the mortises Machine mortising is relatively simple and if much mortise and tenon work is done, a mortising machine (page 138), or perhaps a mortising attachment to the drill is highly recommended. It makes short and accurate work of a job which is quite cumbersome to do by hand.

To set up the mortising machine, fit the appropriate chisel, and use the marked stiles to set up the depth stop and other stops which make the job nearly automatic. If there are no stops on the machine, it is possible to clamp on wood stops. Otherwise the top and bottom edges of the mortise must be marked on all the stiles as a guide for the first and last cuts.

Machine mortising between marked end lines. Make staggered cuts, then clear waste in between.

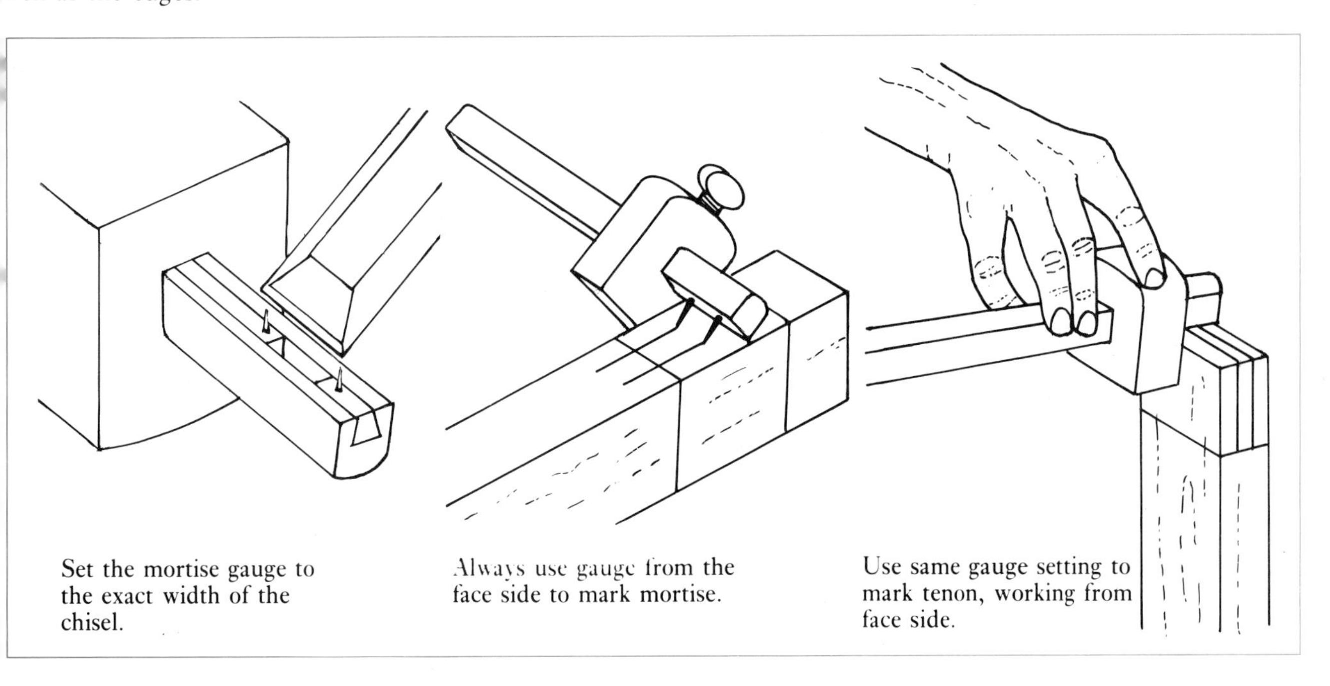

Set the mortise gauge to the exact width of the chisel.

Always use gauge from the face side to mark mortise.

Use same gauge setting to mark tenon, working from face side.

Hand mortising is harder work. It helps to drill out most of the waste with a drill press, if possible, so that the depth can be set accurately. Then, with the stile firmly clamped to the bench (use a piece of waste underneath for through mortising), chisel out the rest of the waste. Use the gauged lines to locate the chisels for the final paring cuts, which should leave a clean, straight side to the hole. Absolute accuracy and smoothness is not essential, since modern gap-filling glues will help hold badly cut joints together. But a clean exit hole is required for through mortises and the stile should be turned over and finished from the other side to prevent the sides from breaking out. The shoulders on the tenons will hide the edges of the mortise which are often splayed open slightly while using the chisel, particularly for softer woods.

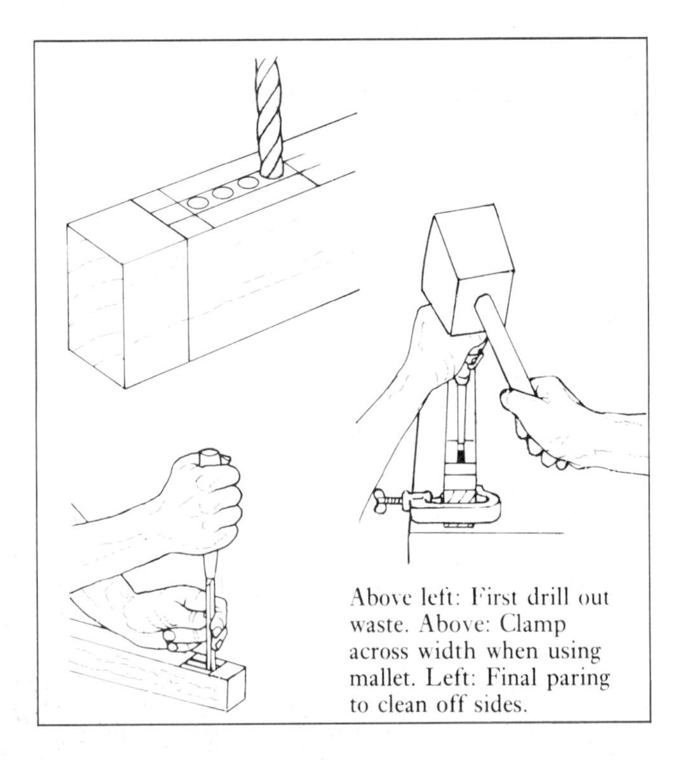

Above left: First drill out waste. Above: Clamp across width when using mallet. Left: Final paring to clean off sides.

Wedged through tenon showing faulty mortising and wedges which are too wide.

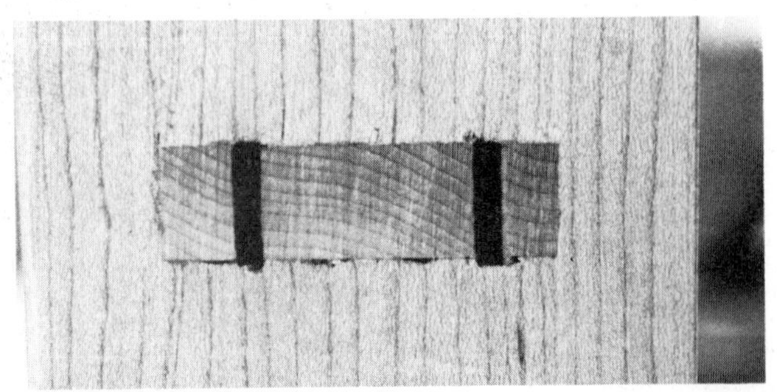

Cutting the tenons Tenoning by machine is quicker and more accurate than by hand. Since tenoning machines are expensive production machines, many shops use the table saw or radial arm saw to cut tenons. There are two basic ways to cut tenons on the table saw. In the first method, the shoulder is cut with the rail lying flat, and the cheek is cut with the rail held vertically in a special holder. Since this method is quite dangerous (the guards must be removed) and a bit cumbersome to set up, I recommend the second method in which the waste is removed by successive cuts. Use a piece of waste of the same size to set up the saw. Lower the blade to the right height by trial and error. Then set the rip fence as a stop so that the shoulders are accurately cut with the end of the rail against the fence. This method does require pre-cutting all the rails to the

Homemade tenoning jig requires extreme care.

Cutting shoulder with workpiece flat on saw.

Safer method making several passes with work flat.

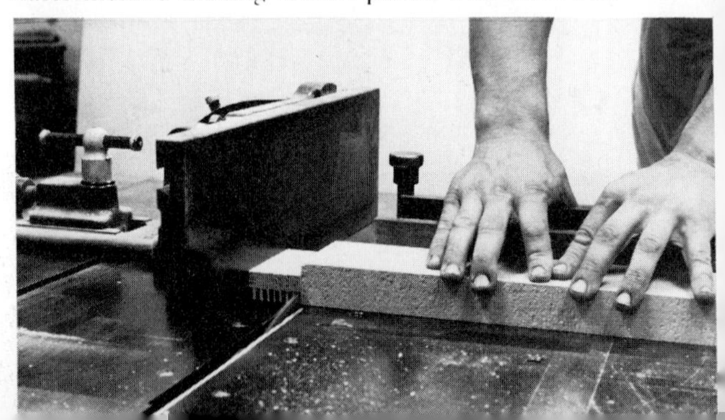

exact length, but this should be done earlier anyway. With the fence and saw blade height set (try it on the waste first) hold the rail against the miter fence and make successive runs to remove the waste. Always work with the marked good face down for the first run, then reset the height of the blade (not the fence) if necessary for cutting the shoulder and the waste on the other side. With the fence fixed, it is quite easy to cut accurate shoulders on all four sides.

In cutting tenons by hand, start by marking the shoulder lines with a knife or chisel against a try square. Holding the rail in the vise, cut the cheeks first to the lines. Then with the rail held in a bench hook cut the shoulders, using the scored lines, which should be widened with a chisel, to form a line for the tenon saw.

Left: Mark shoulder with chisel or knife, then make V-notch as guide for saw on waste side of tenon.

Right: Start sawing cheeks with wood at angle in vise.

Left: Finish cuts with wood vertical. Make sure not to cut too deep.

Right: Saw shoulder along V-notch with work held in bench hook. Again, be careful not to over-cut.

Cutting tenons with the spindle molder

Short tenons are frequently cut on the spindle molder using a saw blade as shown for bridle joints (p. 182) or by using two Whitehill blocks fitted with straight cutters. The space in between them is the thickness of the tenon, which is varied by a combination of thick plywood and thin paper washers. Several manufacturers sell sliding table jigs which clamp the workpiece in place, but it's just as easy to make a holder out of hardwood or plywood.

The portable router, fitted upside down in a homemade table (page 74) is used in the same way. Fit a large, straight cutter and use a push block to feed the workpiece through. Make several passes, adjusting the fence for each pass and turning over for each cut.

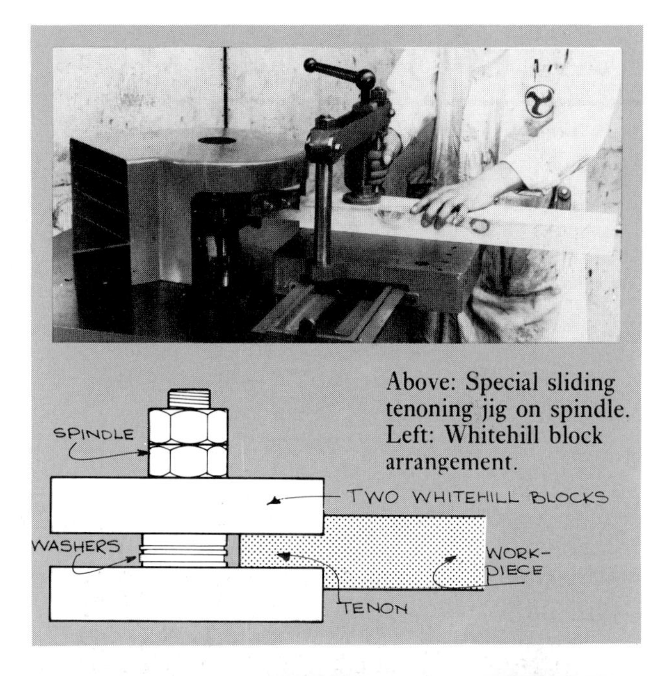

Above: Special sliding tenoning jig on spindle. Left: Whitehill block arrangement.

Use push block to cut tenons on homemade router table. Move fence between cuts.

Haunches A haunch is strictly necessary only to fill the gap where the groove for the door panel runs through to the end of the stile. But since it helps to strengthen the joint to prevent twisting of the rail, a haunch is normally included for heavy frameworks such as table frames where there are no grooves. The haunch is normally rectangular to fill the groove, and to make it easier to cut. A secret haunch is chiseled off at an angle with a corresponding slanted cut made in the stile after the mortise has been cut. Secret haunches are awkward in production work where the groove is usually run through to the end, but it's an elegant detail for handwork.

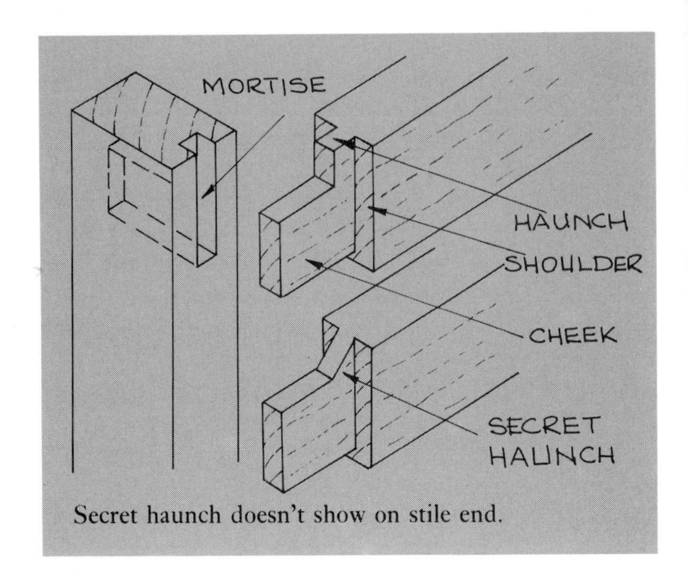

Secret haunch doesn't show on stile end.

Grooved frames In production work, grooves for holding the center panel are run through to the ends, making the grooving much easier on the spindle molder or saw. In this case, a haunch must be included to fill the end of the groove, and the location of the tenon should, if possible, correspond to the groove. In fact, in production work the tenon is often shortened so that it simply fits in the groove, doing away with the mortise completely and relying on the glued-in panel to reinforce the joint. In either case, the part of the tenon removed by the groove on the rail forms a convenient shoulder, and the end of the mortise has to be located accordingly. I find it best to cut the joint first before grooving to make it easier to locate the groove to correspond with the tenon. Where the tenon must be wider than the groove, the haunch has to be cut down to fit into the groove.

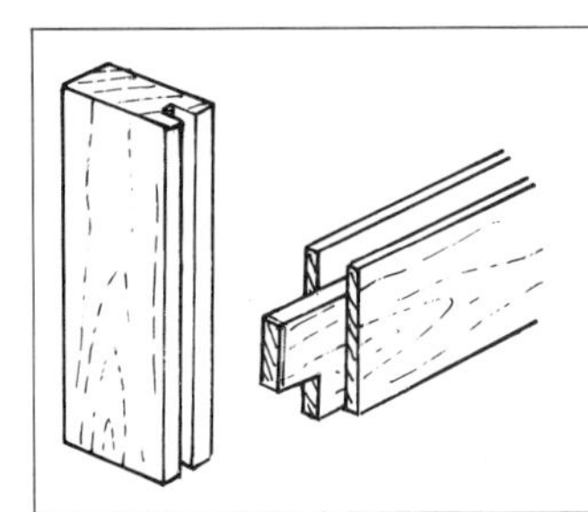
The shoulders in grooved frameworks are equal. The haunch fills the end of the grooves on the stile.

Rabbeted frames The panel can also be held in a rabbet, and since this is much easier to assemble, it is preferred by many woodworkers. For glass panels, it is of course essential to have a rabbet rather than a groove to allow replacement of the glass by removing the beading which holds it in place.

The rabbet can be cut before or after the frame is assembled. When it is pre-cut, the shoulder is left long on one side to fill the space left by the rabbet. If a haunch is required in addition, this is cut separately after cutting the mortise and rabbet.

Cutting the rabbet after assembly is much easier. The mortise and tenon are cut as normal with equal shoulders, then the frame is assembled and cleaned up before the rabbet is cut either on the spindle molder working off the head, or with a portable electric router fitted with a rabbet cutter or with a straight cutter and a special fence (page 66). The rounded corners left by the machines are easily cleaned up by hand with a chisel.

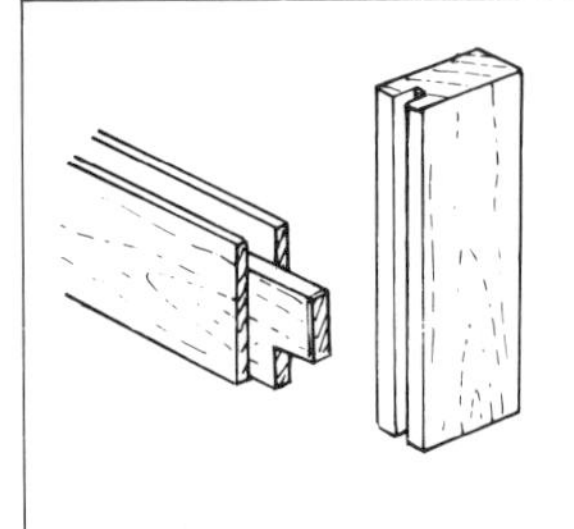
On rabbeted frames, one shoulder is set in to fill the rabbet on the stile.

Elegant wedged through tenons on bench designed and made by Alan Peters.

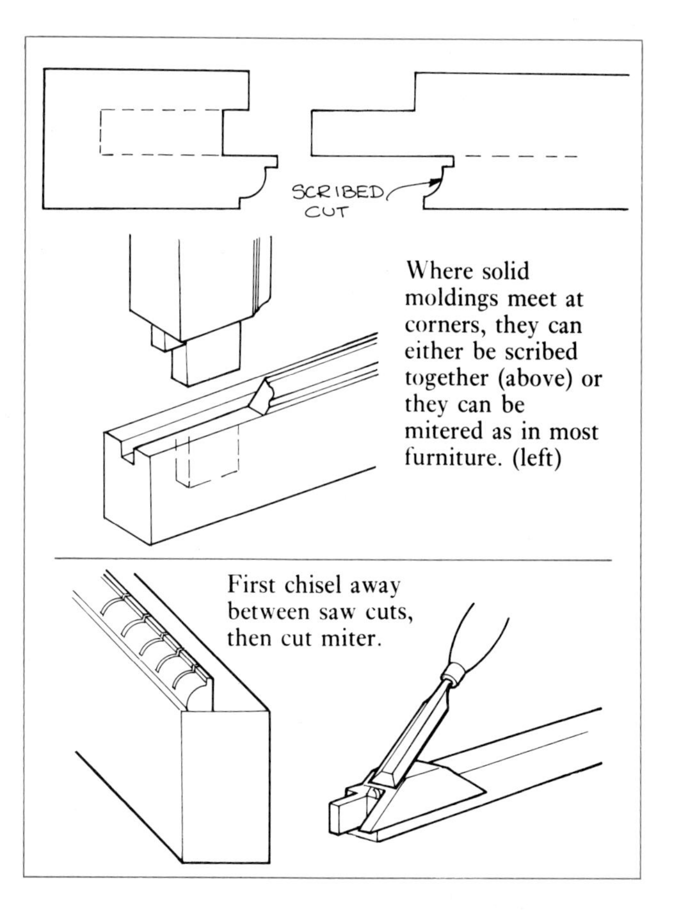

Where solid moldings meet at corners, they can either be scribed together (above) or they can be mitered as in most furniture. (left)

First chisel away between saw cuts, then cut miter.

Making a miter template

For finishing off ends of mitered panel edges or moldings, use a brass or homemade wooden miter template. Any close-grained hardwood can be used but many prefer exotic species such as padouk or rosewood for this and other accessories such as oilstone boxes. Cut a rabbet in a $1\frac{1}{2}$in square block about 12 to 18in long, long enough to make it easy to handle and to make an extra block as a spare. Then cut the opposite miters using an accurate mitering tool, finishing off on a miter shooting board.

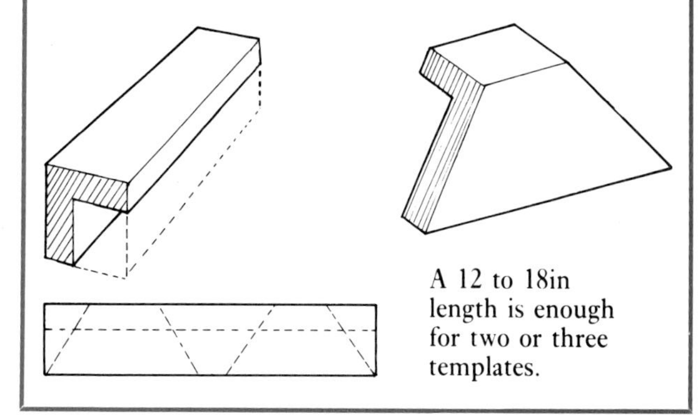

A 12 to 18in length is enough for two or three templates.

Frames with moldings Grooved or rabbeted frames with a decorative molding cut out of the solid on the front face are widely used in making window frames and in cabinetwork. In general, the rail moldings are scribed (i.e. cut to match the contour of the stile molding) for joinery work, and mitered for cabinetwork, but there are no hard and fast rules about this. Cut the joints and rabbet or groove before working the molding, which can be done by hand with molding planes or by machine with a portable router or a spindle molder.

The final step in mitered work is to cut away the corner molding to allow the joint to fit together. On the rails, most of the molding is cut away with the tenon. On the stiles, make a series of saw cuts across the molding, then chisel away the waste, leaving a flat, square surface. Cut the mitered corner carefully with a chisel against a miter template made from a close-grained hardwood such as rock maple or beech. Try the joint for fit, and trim the miters here and there if necessary. The scribed joint is worked the same way, but only the rail molding is cut using a coping saw to copy the molding outline of the stile. In a tenon machine, this is cut automatically by the scribing head.

Table leg joint in mahogany with maple splines and pegged tenons which lock the splines in place designed and made by Sam Bush.

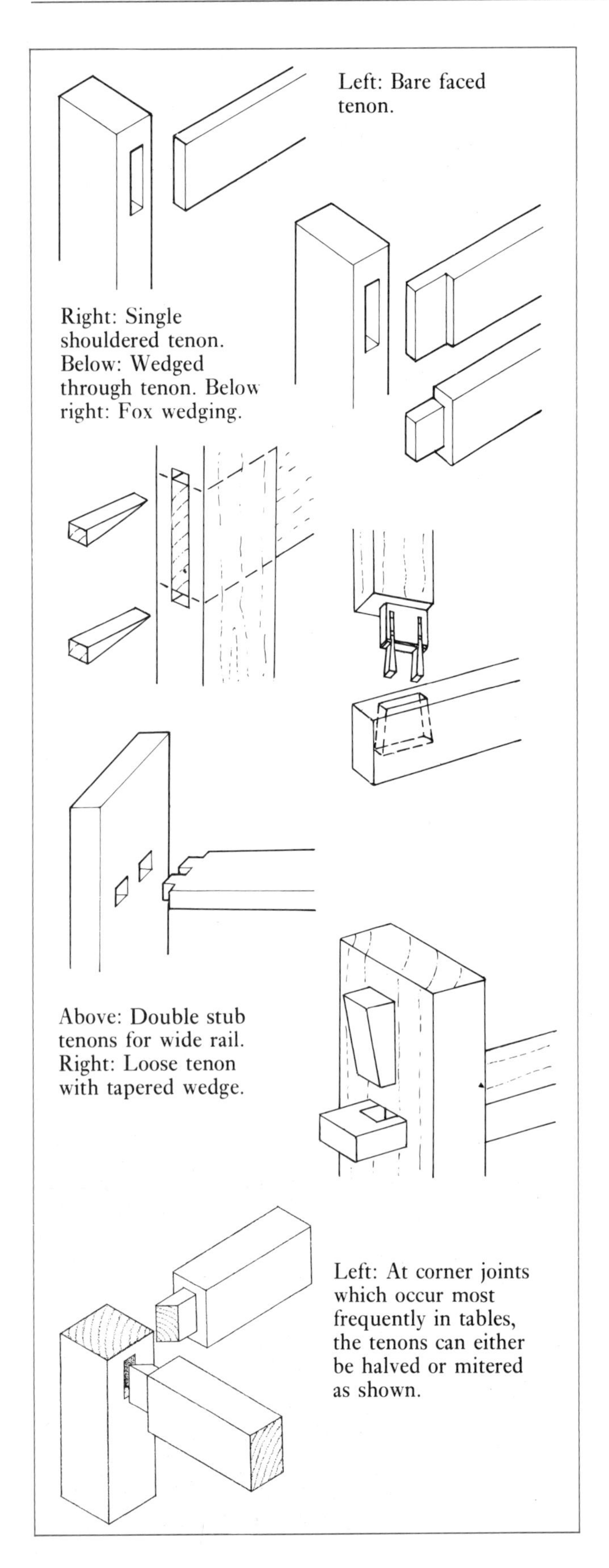

Left: Bare faced tenon.

Right: Single shouldered tenon. Below: Wedged through tenon. Below right: Fox wedging.

Above: Double stub tenons for wide rail. Right: Loose tenon with tapered wedge.

Left: At corner joints which occur most frequently in tables, the tenons can either be halved or mitered as shown.

Variations of tenon joints There are numerous variations of the basic joint and some of these are quite elegant.

Barefaced tenons are the simplest type, used where the rails are too thin for shoulders, such as for slatted shelves where it doesn't matter that the side of the mortise may show. Barefaced tenons may have one or two shoulders, top or bottom depending on the available width.

Single-shouldered tenons again are used for thin rails and are an improvement on the barefaced tenons. They are often used in leg and rail construction for coffee tables where the shoulder faces outwards to hide the edge of the mortise. In some applications, it is better to face the shoulder inwards so that if there is any tendency for the joint to move or separate, the unshouldered side would leave no unsightly gap.

Through wedges are frequently used in both joinery (windows and doors) and in cabinetwork, where they are often wedged with a contrasting wood, and used decoratively as well as structurally. The wedges can be placed in saw cuts at either end or in the middle. (In the photograph, notice the small holes to stop them from splitting.) In either case, the mortise can be angled out to make a stronger mechanical joint as the wedge is inserted.

Above: Through wedges. Below: Detail of tenon.

Fox wedged joints are even more elegant. The wedges are part-inserted into the saw kerfs before inserting into the mortise. The edges of the mortise should be undercut at an angle so that the tenon will spread out to form a mechanically strong (dovetailed-shaped) joint as the wedges are forced into the saw kerfs.

It is important to cut the mortise slightly deeper to create a clearance at the bottom of the hole to insure against the joint being too small, for once assembled it is impossible to get the joint apart. Also notice that the saw kerfs are cut slightly deep in the tenon for the same reason.

Stub tenons are used for shelves and carcass partitions. They can be single, for relatively narrow rails or double, for wide rails. If they extend through the cheek, it is best to wedge them and to use these wedges as a decorative feature.

Loose tenons are the traditional knock-down joints often used in refectory tables. The tenon is cut slightly undersized and locked in place with a tapered wedge.

Corner tenons which meet at right angles at a table or chair corner are either mitered or halved so that they won't interfere with one another.

Mitered or sloping shoulders The mortise and tenon joint can be shaped as a decorative feature by cutting away (or adding to) part of the stile. For a square shouldered joint, about $\frac{3}{16}$ to $\frac{1}{4}$in of the stile is cut away and finished with a mitered corner matched by a miter on the rail (tenon) end. The sloping shoulder joint is cut the same way but at an angle rather than square. Use a sliding bevel to mark both the mortise and the tenon. Notice that for both joints, the tenon shoulder must be set in by the necessary amount to fill the cut-out. At T-joints it may be necessary to glue a block onto the mortise to avoid weak end grain corners on the rail end. The aim, as in all work, is to avoid leaving weak corners which would eventually break off.

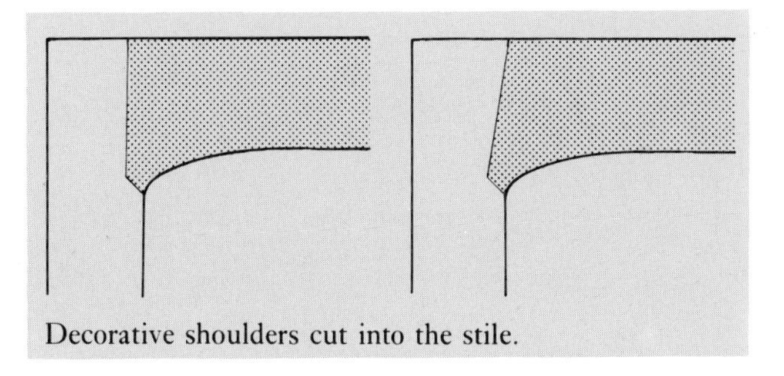

Decorative shoulders cut into the stile.

Draw boring

One way of making sure mortise and tenon joints will stay together is to lock them with one or two dowel pegs. Draw boring is used more often in joinery work but there is no reason why it shouldn't be used for cabinets, as is often seen in antique chests and cabinet doors. Before assembling the joint, drill holes straight through the mortise, fitting a waste tenon in place to avoid splinters. Then with the joint tightly assembled, put the drill bit back in the hole and turn it just enough to mark the hole center. Withdraw the tenon and drill the hole slightly nearer the shoulder than the marked hole. To assemble, use an overlength dowel peg with the end slightly tapered. The tapered end is cut off flush on the other side.

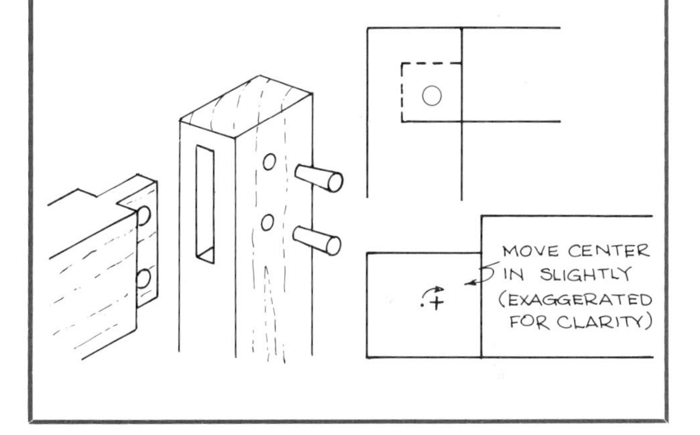

Another example of a decorative tenon joint with an arc shaped shoulder by Alan Peters.

BRIDLE JOINTS AND HALVINGS

The bridle joint is really an open variation of the mortise and tenon joint, and is similar to the plain halving joint. Each relies on glue for strength, but since the bridle has twice as much gluing area, it is considerably stronger than the halving which is usually reinforced with screws from the back.

Bridles The bridle joint is set out and cut in a similar way to the mortise and tenon. The slot should be about one-third the thickness, but as for mortises, it's best to make it the exact width of the chisel, setting the mortise gauge to the chisel and marking, from the face sides as usual.

The slot can be cut on the mortiser, leaving waste at the end which is cut off afterwards to reveal the open slot. Since hollow chisels rarely leave a clean exit hole (chain mortising is better, page 140), mortising is not recommended for fine work. A better way is to make two saw cuts one on either side to the bottom of the slot and then clean the bottom by hand, paring down with a chisel. If a spindle molder with a saw blade is available, the sawing can be done with the stile kept horizontal. First make the two outside cuts on all the pieces, then change the height of the blade to cut away the rest of the waste. If no spindle molder is available, use the radial arm saw or the table saw (though this is more dangerous) to make the two cuts with the workpiece held firmly in a special holder, available from some manufacturers.

As can be appreciated, it is quite cumbersome to make the saw cuts by machine, and unless sufficient quantity is involved to make setting up worthwhile, it may be just as easy to make the saw cuts by hand, cleaning up with a chisel afterwards after a hole has been drilled to remove most of the waste.

The tenon is identical to that for the mortise and tenon joint and therefore cut in the same way.

The bridle joint is used as a corner joint for frames, as a substitute for mortise and tenons, or as a T-joint for legs supporting continuous rails. In the mitered version used for more delicate frames, the slots are cut in the rails rather than the stiles, so that the ends are hidden in the miter, leaving a clean edge on the side.

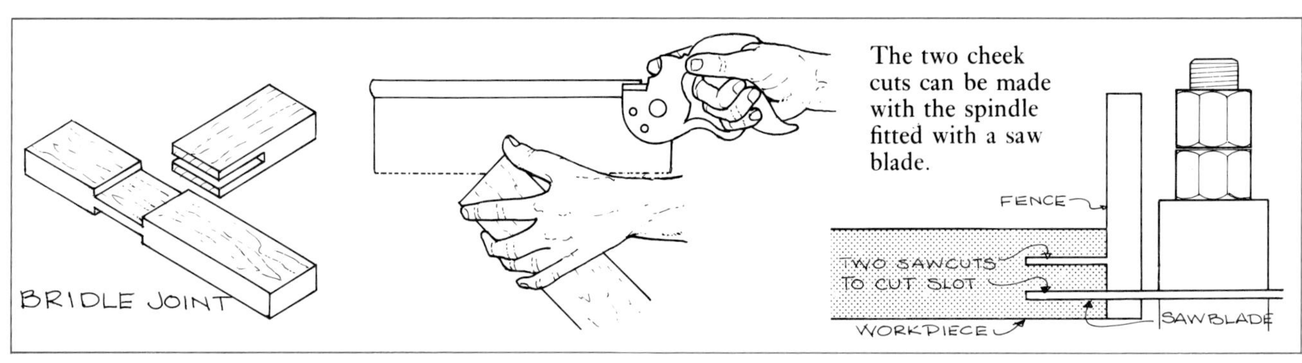

The two cheek cuts can be made with the spindle fitted with a saw blade.

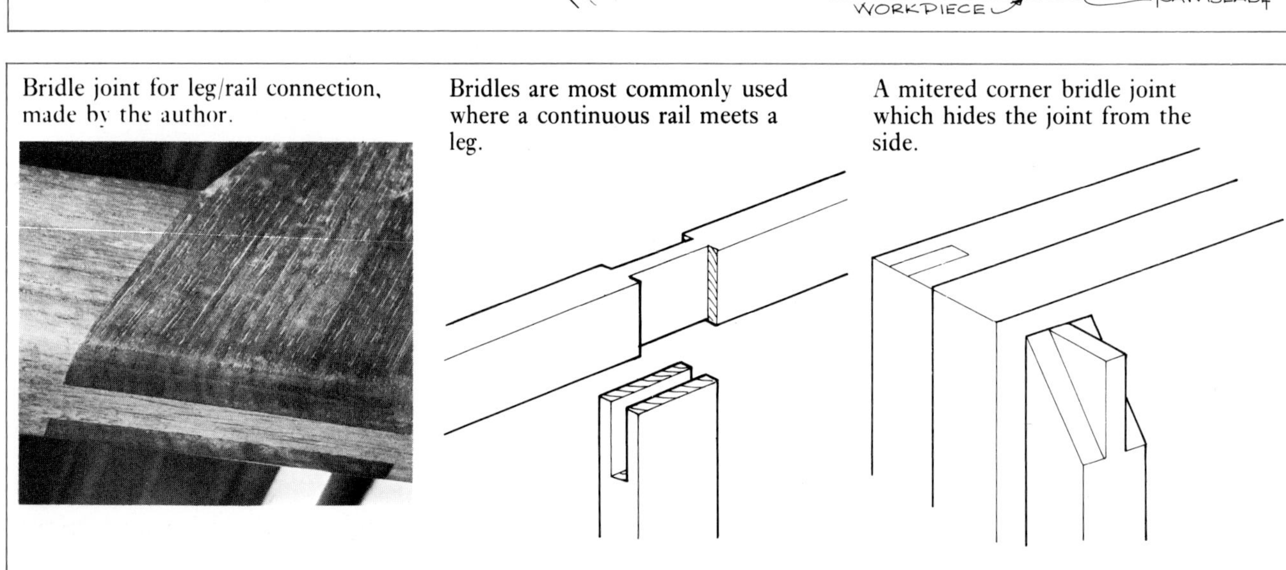

Bridle joint for leg/rail connection, made by the author.

Bridles are most commonly used where a continuous rail meets a leg.

A mitered corner bridle joint which hides the joint from the side.

Bridle joints used in paneled door frame construction made by Gil Gesser.

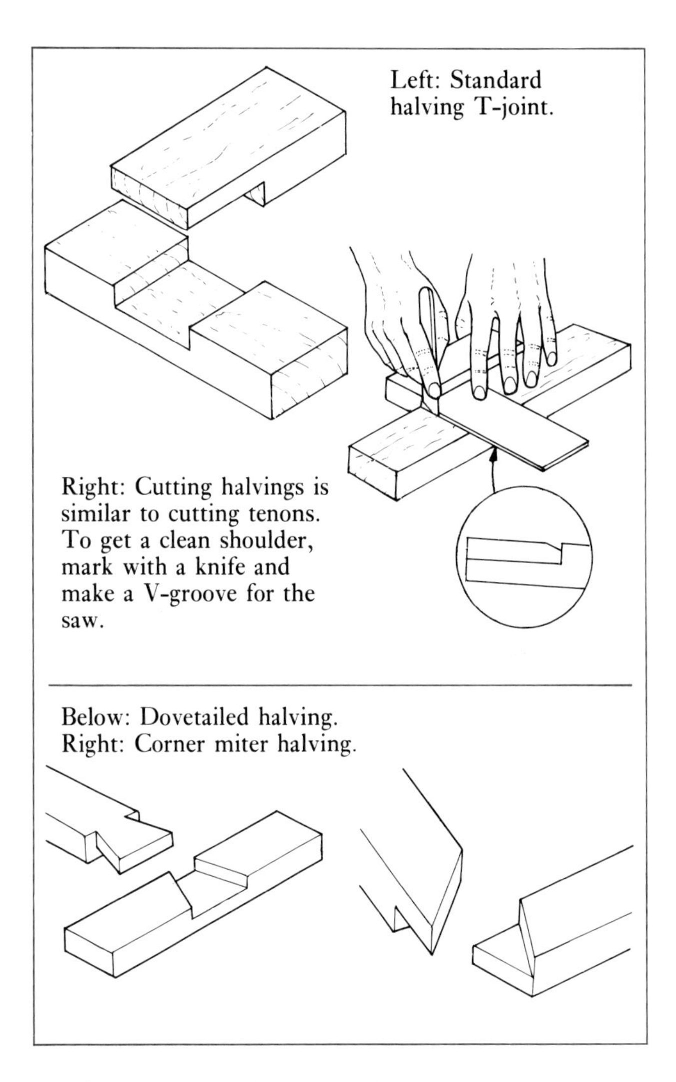

Left: Standard halving T-joint.

Right: Cutting halvings is similar to cutting tenons. To get a clean shoulder, mark with a knife and make a V-groove for the saw.

Below: Dovetailed halving.
Right: Corner miter halving.

Halving joints Halving joints, used in a variety of shapes and configurations, are easier to cut than bridles. Since both halves are open, they can easily be cut on the table saw, as for tenons, setting the rip fence to the shoulder distance and making a number of successive cuts to remove the waste. Alternatively, use the radial arm saw to cut away the waste.

By hand, they are cut as tenons, marking the cheeks with a marking gauge on the ends and sides, and the shoulders with a chisel or knife as a starting line for the tenon saw. Remove the waste with two saw cuts, one along the cheeks and one at the shoulder. Final fitting may require paring off a bit more waste with a wide chisel.

The proportions of the joint depend on the two members. If they are equal thickness, half of each member is removed, but for unequal members more waste should be removed from the thicker piece.

Corner halving joints are not normally used in frameworks for good work because the glued joint usually needs screwing from the back as reinforcement. But they are used in a variety of other applications where other joints are unsuitable.

Dovetailed halvings, used for T-joints, rely on mechanical strength as well as glue, and are therefore superior to ordinary halvings, particularly for cabinet rails which must prevent the carcass sides from separating. Dovetail halvings can be single or double and the latter can be through or stopped.

Corner mitered halvings are an elegant variation sometimes used on small frames. Because of the reduced gluing surface, it also needs a couple of screws from the back for reinforcement. Three way joints are another variation on the halving joint. See page 89 for further information.

The one-third lap joint accommodates three rather than two members for use on hexagonal tables and triangular structures. All the cuts made at 60°, require accurate setting out and cutting for a proper fit.

DOWEL JOINTS

The dowel joint is the ideal joint for anyone without much equipment, for it is quick to cut and easy to make accurately.

Tests have shown that the strength of the joint relies not just on tight-fitting dowel pegs, but also on cleanly cut holes with sides which are not woolly or burned from blunt drill bits. Therefore, it is a very good idea to buy a set of good quality machine dowel or brad point bits in $\frac{1}{4}$, $\frac{3}{8}$ and $\frac{1}{2}$in diameters, to match the three most common sizes of dowel. Another handy device is a doweling jig (see page 186), which clamps onto the workpiece and serves two functions. It lines up the holes, obviating the need for marking, and it guides the drill bit, guaranteeing accurate and square holes. It may be too slow for production work, but for handwork it is highly recommended.

Production doweling The dowel joint has almost universally replaced the mortise and tenon joint in production work because it is quick and inexpensive to cut and assemble. In production shops, two or more holes are drilled simultaneously in a horizontal borer (page 134), which is pre-set for hole spacing and depth.

Depending on the level of production, the machine may be single head and hand fed, or it may have multiple heads and automatic feed. It is possible to rig up a horizontal borer in a small shop using an old drill press and a sliding table. As a lightweight substitute, fix a portable drill stand in the Workmate, as shown on page 57. Another alternative is to use the radial arm saw fitted with a chuck and drill bit to do horizontal boring. (page 103)

Making a marking template

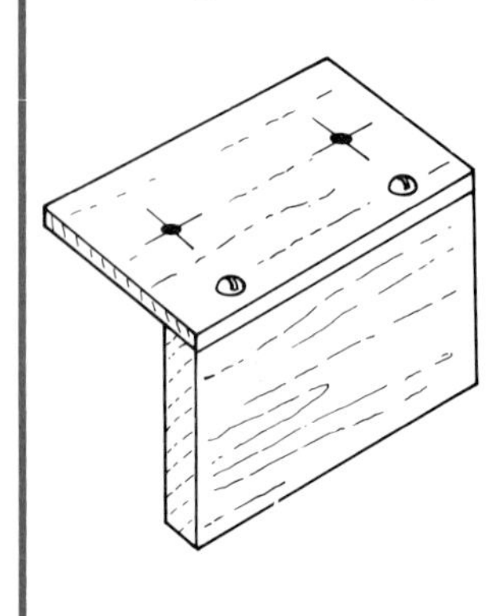

Make a template the exact size of the end of the board. Make the top piece out of $\frac{1}{8}$in plywood or metal plate and screw it to a piece of $\frac{1}{2}$in wood. Carefully mark, punch and drill hole centers to accept the top of the marking tool.

SETTING OUT No marking is necessary when using production equipment, but few small shops have access to these and must instead rely on hand methods, though the drilling itself is often done with either a portable electric drill or a drill press.

The most important part of doweling is to mark the holes accurately. First lay the frame out and number the joints, 1, 2, 3, etc.

There are various marking methods but usually each two members are carefully clamped together in turn in the vise good faces out. The center lines are squared across, then a marking gauge is used against the good faces, to find the centers, which are punched to form a starting point for the drills.

Another method involves the use of templates, each one made to fit a specific cross section (see box). Always hold the template from the good face and with the ends lining up, make center marks with a marking awl.

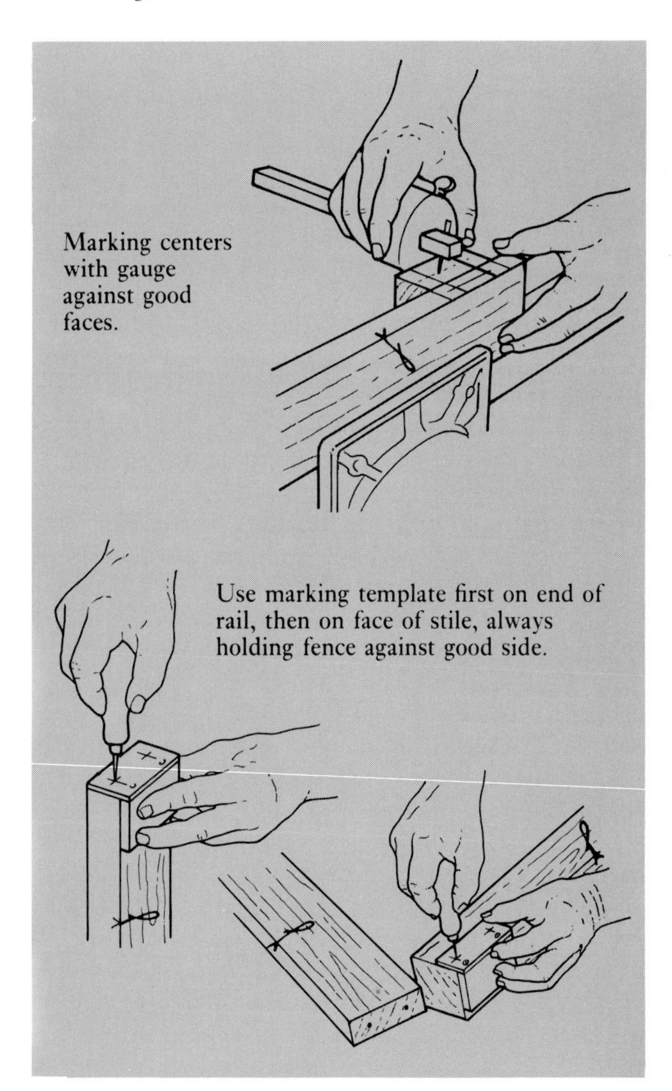

Marking centers with gauge against good faces.

Use marking template first on end of rail, then on face of stile, always holding fence against good side.

SPACING Common sense dictates the number and spacing of dowel pegs. At least two are required to prevent swiveling, though in small rails a pin or brad can be substituted for one of these to save space. In small, delicate pieces two $\frac{1}{4}$in dowel pegs may be enough, but in heavy table legs three or four $\frac{1}{2}$in dowels are required. A good compromise for most frames is two or three $\frac{3}{8}$in diameter pegs. For adequate strength, at least $\frac{1}{4}$in of wood should be left on either side at the end and the same distance between edges of holes. In general, the further apart the dowels, the more resistance they offer against racking and twisting. Keep in mind that the dowels at corner joints, as at table legs, must be staggered so as not to interfere with one another.

The length of dowels obviously varies with the application. In chipboard panels, these are often only $\frac{1}{2}$ to $\frac{5}{8}$in long, though this is inadequate for solid sections. As a general guide, dowels should protrude at least $1\frac{1}{4}$in on either side, though obviously the further the penetration the more gluing area and hence more strength.

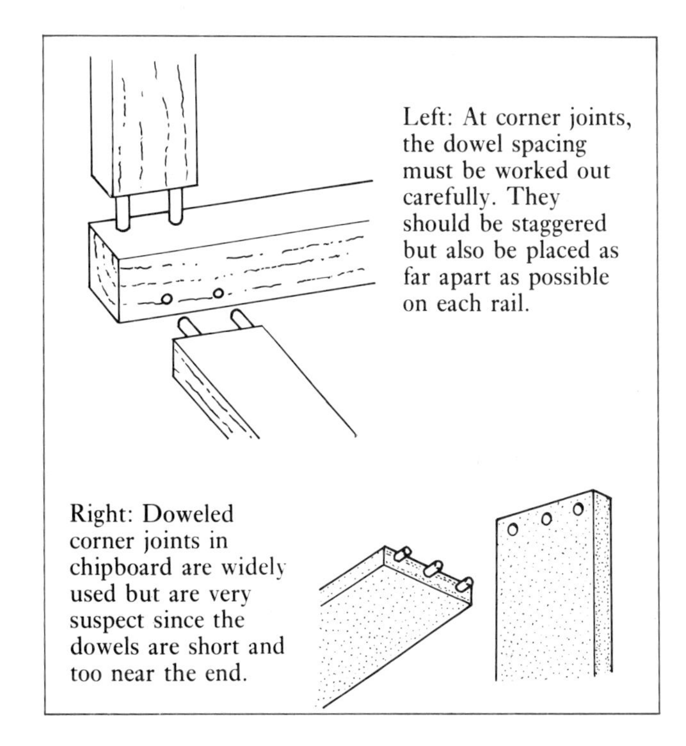

Left: At corner joints, the dowel spacing must be worked out carefully. They should be staggered but also be placed as far apart as possible on each rail.

Right: Doweled corner joints in chipboard are widely used but are very suspect since the dowels are short and too near the end.

Making dowel pegs

Dowels are available in diameters of $\frac{1}{4}$in to about $1\frac{1}{2}$in. For dowel pegs, you will need three sizes: $\frac{1}{4}$, $\frac{3}{8}$ and $\frac{1}{2}$in diameter. Be sure to use straight dowels. Keep a few lengths of each diameter in a dry place to use for pegs. As dowels may vary slightly in diameter, check the size in a drilled hole. Reject dowels which are too small, but those which are slightly oversized can be sanded down or driven through a homemade or purchased metal dowel pop after cutting each peg to length on a bench hook. Make sure to test one of the pegs for snug fit before cutting the rest. To make the grooves in the pegs, carefully run each peg back and forth over a saw clamped in a vise. Alternatively you can buy a dowel sizer with serrated teeth around the holes which will groove the peg as it is pushed through.

Then chamfer the ends slightly preferably using a belt sander which makes the job quite easy but alternatively using a rasp or a special tool like a pocket pencil sharpener called a dowel pointer. These jobs become almost automatic with a little practice and it is worthwhile to make a batch of common lengths to keep on hand in the shop.

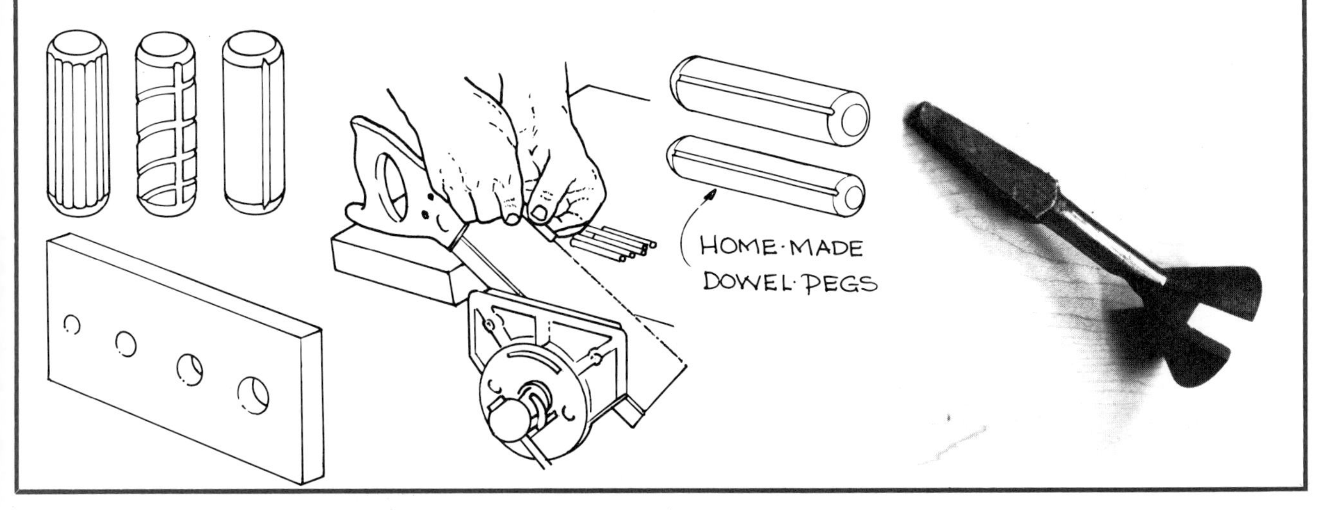

BORING Boring the dowel holes can be done by hand using a brace and Jennings pattern auger bit or by portable electric drill. The drill press can also be used but it's often difficult to accommodate the long rails to drill end holes. In this case, the radial arm saw can be set up to do small runs, but the motor must be firmly locked so that there is no play in the drill bit.

It's important to use the correct size drill bits so the pegs fit tightly. Test the bit by trying a dowel peg in a hole drilled in a waste piece, and use a dowel pop if necessary to size the dowels just right. If possible, use a doweling jig which keeps the drill bit square. There are several jigs on the market and although exact details vary, most will accommodate $\frac{1}{4}$ and $\frac{3}{8}$in bits with a charge of bushings.

The holes should be just slightly deeper than half the dowel pegs. Some electric drills come with a depth stop, but it's usually easier to drill a hole through a piece of 1 × 1in and cut off a piece of the right length which is then slipped over the drill bit.

Finally, countersink the holes. This helps locate the dowels during assembly and takes any excess glue.

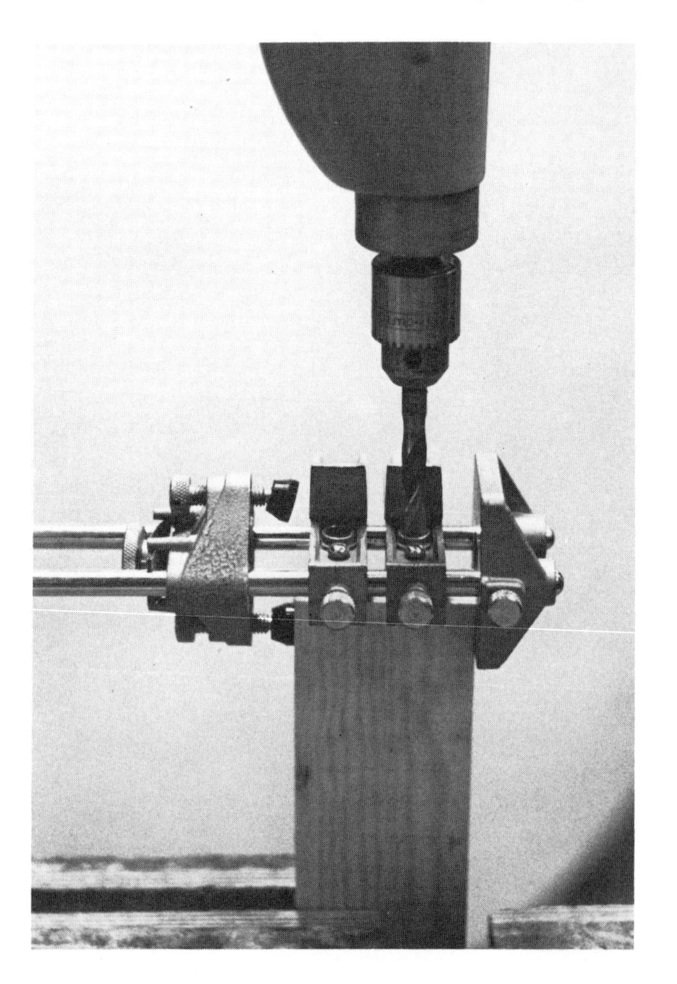

ASSEMBLY It doesn't matter which member gets doweled first, but I always tap the pegs into the rail ends before assembly. Use a gap filling glue if possible, but PVA glue if speed is important. Apply a small amount to each socket and tap in the pegs using an appropriate thickness piece of waste wood as a depth stop to make sure the dowels are tapped in the right amount. If possible, allow the glue to set before assembly, but if speed is important, assemble right away, applying glue to all sockets and pushing the joints home. Any skew holes should have been spotted earlier, filled up with a peg cut off flush and the hole redrilled.

Clamping is normally required on large frameworks to get the joints tight, but for small work, a tap with the rubber mallet should be enough to hold any well-drilled joints.

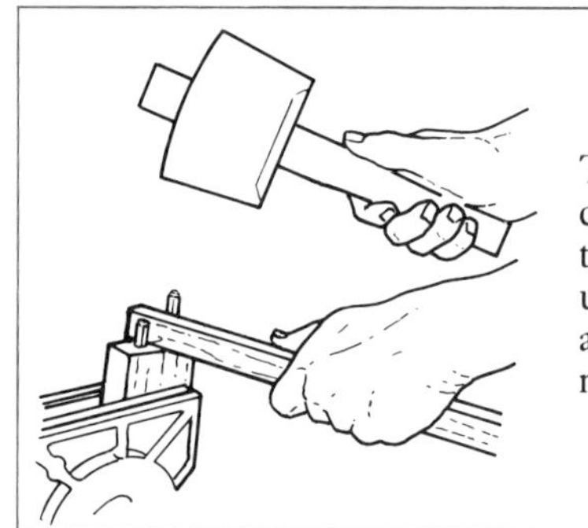

The best way to control the depth of the dowel pegs is to use a batten which acts as a stop for the mallet.

Through dowels

Through dowels are particularly useful because they can be added with the joint assembled. Hold the pieces by hand or clamped together, then drill from the outside straight through one piece into the other piece. Glue in the dowel peg, then cut it off flush and sand smooth. Adding a wedge helps to hold it tight and decorates at the same time.

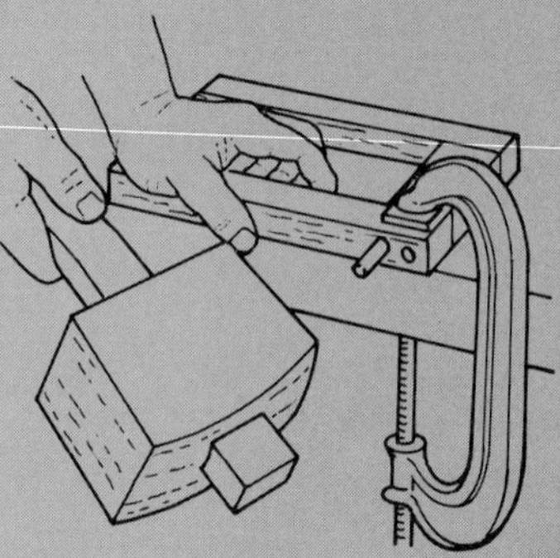

To use through dowels, first assemble the frame by clamping, etc., then with the frame firmly held, drill straight through both members before inserting the glued dowel.

Grooved or rabbeted frames Grooves or rabbets to hold door panels should always be cut after all doweling has been finished, but before the dowel pegs have been glued in place. Notice that the rail end is notched to fill the groove which holds the door panel in place.

Frames which have a molded front are done the same way, working the molding and cutting the corner miter last. I find it useful to check for fit in these cases by trial assembly of the frame using dry joints and dowel pegs which have been sanded slightly smaller for easy removal.

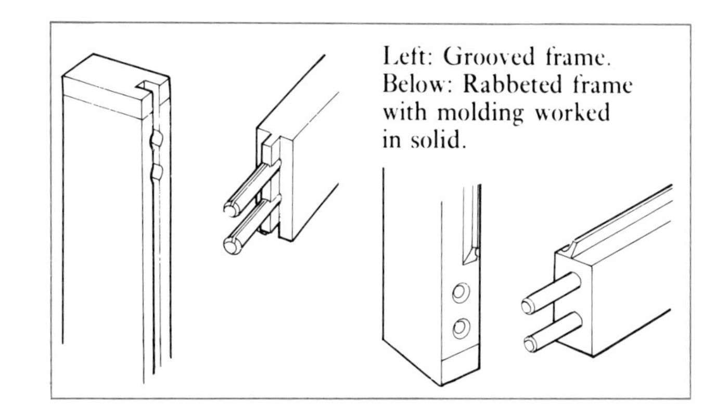

Left: Grooved frame. Below: Rabbeted frame with molding worked in solid.

APPLICATIONS

Corner joints on small frames is a common application. These can be plain or they can be rabbeted or grooved, with moldings applied or worked in the solid.

Corner joints on tables or chairs need two or three heavier $\frac{1}{2}$in diameter pegs, spaced far apart and staggered from either side.

Cross joints normally made with halvings which weaken the joints, are better made with dowels which leave the joint with full strength.

Mitered joints are often reinforced with short dowels. Drill the holes on a drill press by holding the workpiece in a jig, or use a doweling jig clamped onto the mitered end.

Sculptured joints on chairs, tables, etc., are quite straightforward to make. Set out and bore the holes before shaping the rails and legs. Where the rail meets a turned leg, the rail end is rounded with a gouge before the dowel pegs are added.

Pedestal bases on circular tables are difficult to mortise, and therefore, more often doweled together. Because of the heavy load, at least four $\frac{1}{2}$in diameter pegs of 2 to 3in length are required.

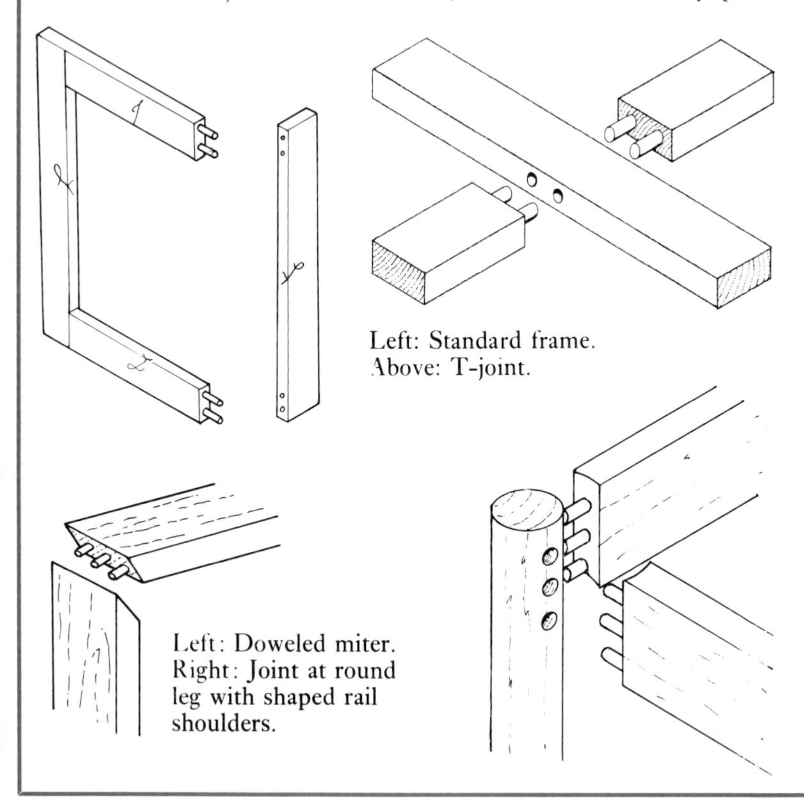

Left: Standard frame. Above: T-joint.

Left: Doweled miter. Right: Joint at round leg with shaped rail shoulders.

Joint at grooved frame is made by drilling dowel holes before cutting joint.

MITER JOINTS

Miters for members of equal section meeting at 90°, as for frames, etc., are cut at 45° and this is the standard angle for most work. However, if the members meet at any other angle or are of unequal section, the exact cutting angle for each piece must be worked out from geometry or on a drawing board.

Because the two members meet end grain to end grain, the joint is not strong enough by gluing alone, though by first sizing with thinned down glue, the glued butt joint is usually strong enough to hold until reinforcement can be added. This is the method often used by picture framers. The reinforcement is usually fine pins added from each side and hidden with matching filler.

Miter joints are more difficult to make than they appear, partly because it's difficult with some tools to make accurate 45° cuts, partly because they can be awkward to assemble, and also because any movement in the boards tends to open up the joints at the corners.

Cutting miters The best tool for cutting miters is the hand or foot operated miter trimmer which slices the wood for a very accurate and very smooth finish. The second best tool is the precision miter saw which, like the miter trimmer, can be set to any angle but locks into the common 45° position automatically. One of these two is indispensable for any shop specializing in frames or other work with a lot of decorative moldings.

They are, however, somewhat limited in that they won't cut very wide boards or plywood panels which have to be cut on a table or radial arm saw.

The traditional way of cutting miters is to mark the miters as well as possible on the section, using a miter square, then cut it with a tenon saw in a miter box. Final trimming with a sharp plane in a miter shooting board, or for wide boards in a donkey's ear shooting board, gives a smooth, accurate finish. These jigs should be accurately made and checked with a small plastic draftsman's triangle, which unlike many miter indicators, keeps its accuracy.

To cut miters on a table saw, cant the blade to 45° and test the cut on a piece of scrap before cutting the joint. Alternatively use the jig described on page 84. A good way of testing accuracy is to cut two pieces and hold them together. If the assembled joint makes a 90° angle, the miter cut is accurate. As an alternative, use the radial arm saw which in many ways is superior to the table saw for all crosscuts including miters.

Reinforcing miters Gluing the end grain of miters has little strength, so reinforcement is required. This can be added after the joint is glued up by using pins from both sides, by inserting matching veneer strips into saw cuts made across the corner in alternating directions, or by inserting a straight or dovetailed key across the joint.

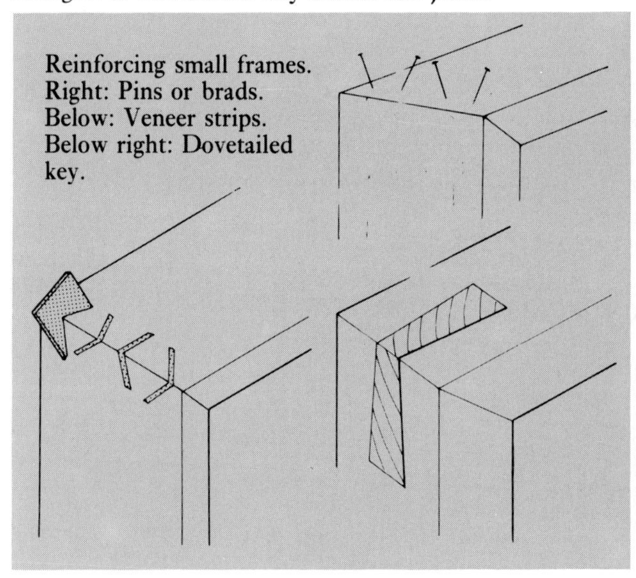
Reinforcing small frames. Right: Pins or brads. Below: Veneer strips. Below right: Dovetailed key.

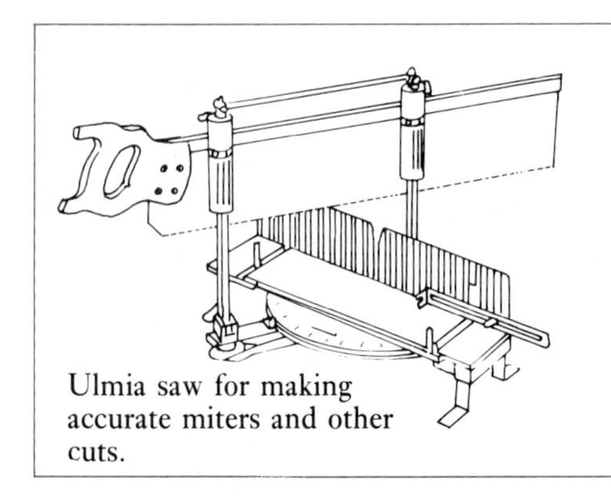
Ulmia saw for making accurate miters and other cuts.

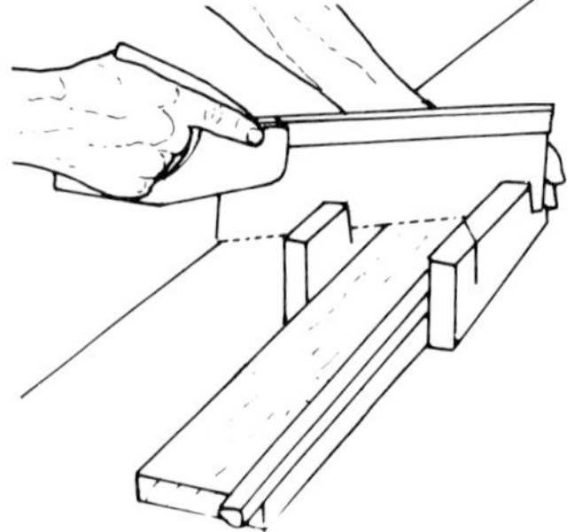
Using simple miter box to make cut with tenon saw.

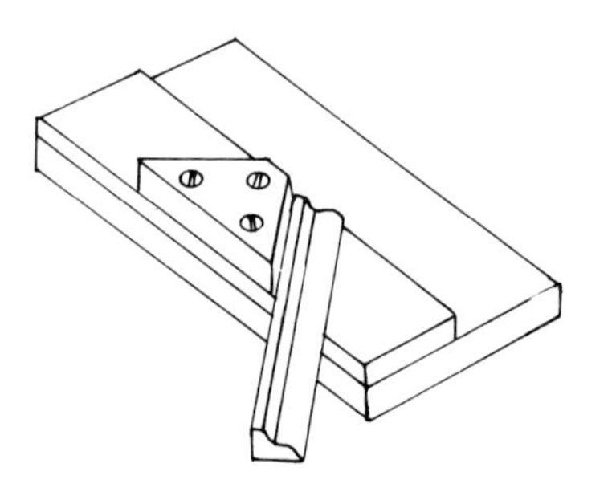
Finish off saw cut in a miter shooting board.

These reinforcements are for lightweight frames. For joints which must stand up to wear, the miter is normally reinforced with a spline or dowel. Of these the spline, because of its ease of cutting on the table saw, is preferred. It is often used for wider panels in making plywood or veneered panel carcasses.

To make the splined miter joint, first set the blade to 45° and check its accuracy. Cut all the corner miters to accurate length first. The opposite sides of any frame or box must be of identical length for the joints to close, so use a stop and cut the opposite sides in identical pairs. To cut the slots for the splines, lower the blade and use the rip fence or a block clamped to the top as an additional guide to keep the cut accurate. The position of the slot should be towards the inside to avoid leaving delicate, easily broken corners.

A spline can also be added to ordinary frame miters but for these the slot should be cut on a spindle molder, or with a portable router with the piece lying flat. A more elegant version of this, used in good quality handwork, has a spline or tenon worked in the solid to fit into a mortise.

An alternative method for reinforcement is to cut the joint so that it interlocks to reinforce itself. The simplest version is the miter and lap joint used in panel carcass construction. This doesn't strictly reinforce itself, but the keyed corners do aid in assembly. Another version which does interlock is the tongue and miter which is an improvement on the non-mitered tongue and lap joint. These are described in more detail on page 206.

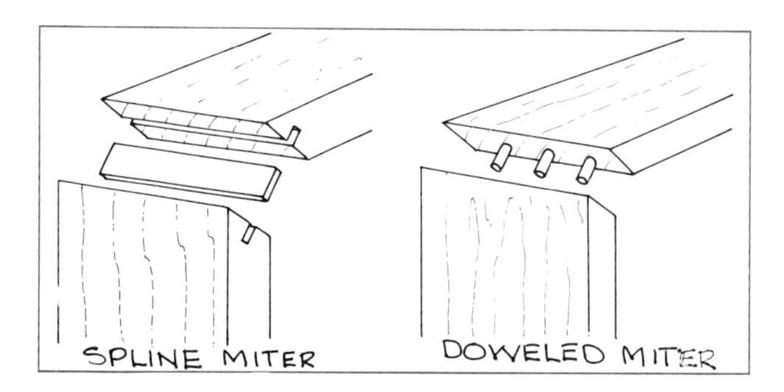

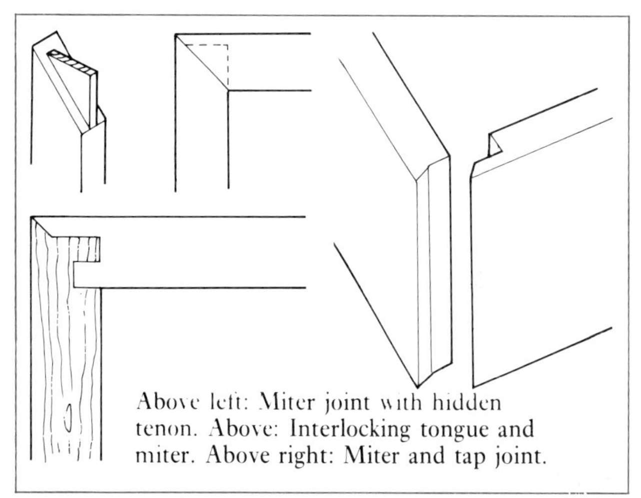

Above left: Miter joint with hidden tenon. Above: Interlocking tongue and miter. Above right: Miter and tap joint.

Butterfly keys

As a decorative and structural feature, mitered joints can be reinforced with double dovetailed "butterfly" keys. These can be made in the same or in contrasting wood, and should be tight fitting to serve as reinforcement. It's best to cut the keys first, then trace their shape to cut their sockets. Alternatively, the keys can be produced several at a time by cutting out carefully on a band saw and the sockets can be router cut by template.

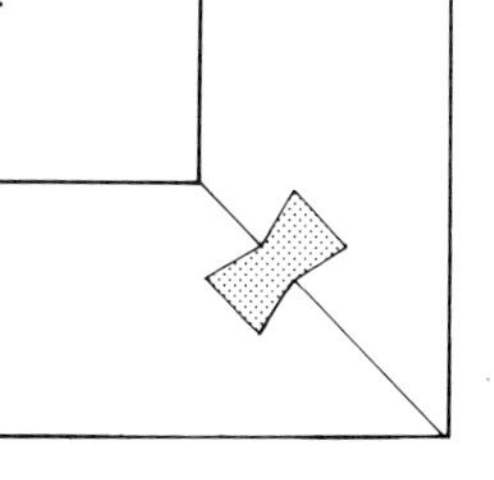

Dovetailed butterfly keys have many applications such as reinforcing miters as shown.

Top: Cutting miter on table saw. Above: Cutting groove with blade lowered. Right: Joint ready for assembly.

DOVETAIL JOINTS

Although machine dovetailing is widely used, hand dovetailing is preferred by most cabinetmakers. Hand cut dovetails are not limited to a set and even spacing which is the tell-tale of machine dovetailing. But it is undoubtedly cumbersome work, requiring accurate marking, sawing and chiseling, which has become almost symbolic of fine woodworking, a test of the cabinetmaker's skill.

Dovetails are shaped the way they are, not to be decorative, though that is an advantageous by-product, but for structural reasons. The angled shape of the tail forms a mechanical joint in one direction which can't be pulled apart. Thus for drawers and other components which are pulled a lot or which tend to separate, like cabinet sides, the dovetail is ideal. In addition to its mechanical strength, the large number of gluing surfaces guarantees that a well-cut dovetail joint will stay together for the life of the furniture.

There are numerous types of dovetail joints but most fall into one of four categories: **1** through dovetails which show on both sides, **2** lapped dovetails which show on one side only, **3** secret dovetails which are completely hidden, **4** carcass dovetails which are normally single dovetails joining a cabinet rail to the sides.

The choice depends primarily on aesthetics. Although the through dovetail is the strongest and easiest to cut, it may be unsuitable for some work such as for drawers where a lapped dovetail will hide the joint on the front side.

The choice of hand or machine cutting depends partly on the level of production, for although hand dovetails are generally more interesting, they are not stronger, and for all but the finest work, machine cutting is very convenient and accurate.

Hand cutting through-dovetails

SETTING OUT Before setting out, the sides must be cut off square to their final length allowing about $\frac{1}{32}$in at each end ($\frac{1}{16}$in to the total) for final cleaning up after assembly. Then gauge the thickness ($+\frac{1}{32}$in) of one piece on the other and vice versa, using a sharp cutting gauge (not a marking gauge which scratches the wood). Normally two pieces are of equal thickness so one setting of the gauge is enough. If they are not the same, the setting must be changed to suit each piece. Gauge all around the four sides, leaving a thin, crisp cut in the wood which becomes part of the finished joint. **(1)**

The number and hence spacing of the pins and tails is left to common sense though aesthetics, particularly in fine furniture, requires that the pins be more delicate. Whether you mark and cut the tails or pins first depends on individual taste. I prefer to cut the tails first so that I can work on the face of the board and this is the method described here.

There should always be a pin at either end. Start by deciding its size which as a general rule can be equal to the width of the other pins at the narrowest

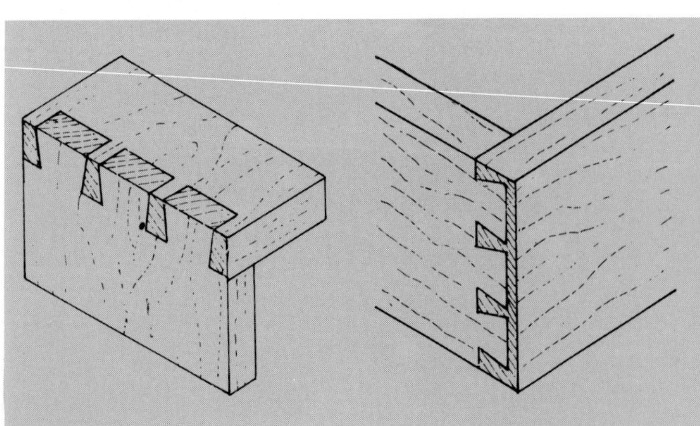

Far left: Through dovetails.
Left: Lapped dovetails.
Above: Components of secret miter dovetail.
Right: Double and single carcass dovetails.

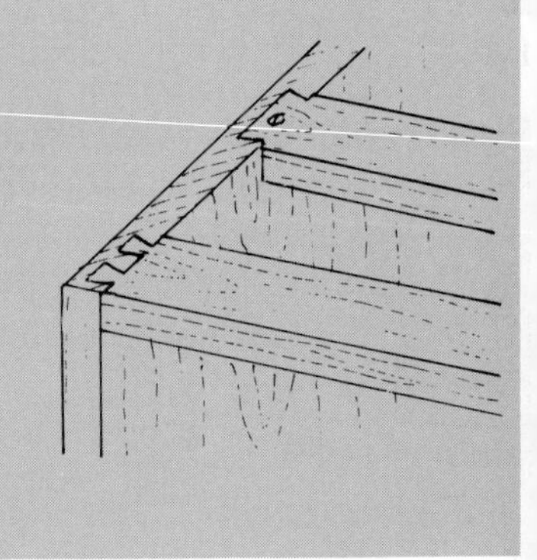

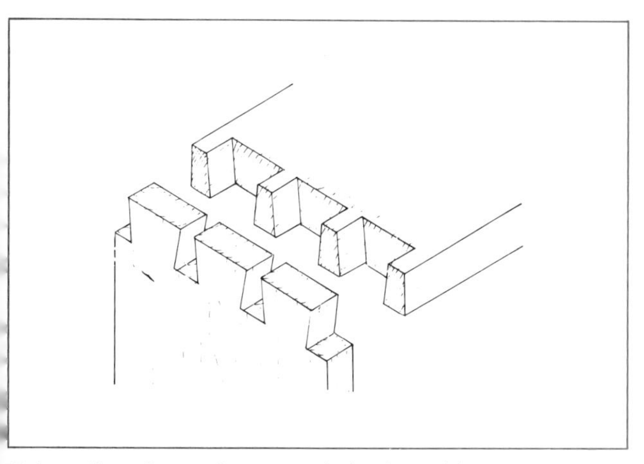

Below: Step 1, gauging around the four sides.

Below: Step 2, transferring the marks. Right: Step 3.

Below: Step 4, drawing in the rakes.

part. It shouldn't be so fine that final sanding can go through it. Mark this width on either side then mark half this width parallel to either side with a pencil (as shown). Thus, if the pins are $\frac{3}{8}$in wide at their narrow part, for example, the end pins will be the same $\frac{3}{8}$in width and two lines will be marked $\frac{3}{16}$in in from each edge. (**2**)

Divide the distance between these lines into equal spaces, depending on the size of dovetails, by holding a ruler at an angle giving convenient divisions. (**2**) Then transfer these marks by squaring lines to the end. Mark half the width of the narrow part of the pins beside each line, then using an adjustable bevel or a dovetail template draw in the rakes. (**4**)

RAKE The slope can vary from about 1 in 5 in carcass work where strength is important, to 1:7 or 1:8 for fine work. Set the sliding bevel to the required angle which can be easily marked on a piece of plywood measuring 1in over and 5, 6, 7 or 8in in from a squared line, as shown. (**3**) If you do a lot of dovetailing it's worth making one or more templates to the most popular rakes. One idea I saw recently used by an old craftsman, was a thin metal template with the dovetail outlines cut through. He simply offered up the template and tapped it with a bag full of chalk which left the dovetail outlines clearly on the wood.

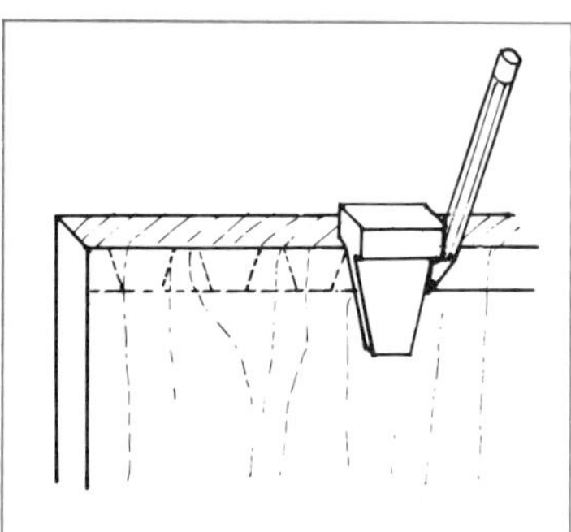

If you do a lot of dovetailing, make templates in the common rakes, 1:5, 1:6, 1:7 and 1:8 to make setting out much quicker.

After marking the tails, square the lines across the end with a sharp pencil as a guide for the saw (**5**).

A note on marking tools: many woodworkers prefer a marking knife but I've found that a sharp mechanical pencil fitted with a 2H or 3H lead and well-sharpened, is excellent for delicate marking work. Finally, pencil in the waste clearly to avoid cutting away the wrong part.

CUTTING Use a sharp, fine saw such as a dovetail saw (20 points) or a gent's saw (32 points), and with the workpiece held in a vise, cut down the raked lines, square across the top and carefully down to the gauged line (**7**). It's a good idea to tilt the workpiece over using a try square set on the bench top as a guide, so that all sawing is vertical (**6**). Before going on to clear the waste, cut away the two ends, sawing just beside the gauge marks. (**8**)

Having cut along the lines, some workers prefer to mark the pins by holding the workpiece onto the other, as shown, and scraping the tail marks through with the saw. I prefer to cut out the tails first.

In cutting out, use a sharp bevel edged chisel, slightly narrower than the waste. Hold the work on a piece of scrap wood on the bench (the sturdier the bench the better) and chisel away the waste (**9,10**). Start in from the gauged line, cutting vertically first then diagonally, in sequence, until a little over half the depth is reached, then turn over and repeat for the other side. The final cut is made on the line, paring straight down for a clean finish (**11**), although there is no harm in undercutting slightly to make sure the joint will fit tightly. It's a good idea to clean up all the corners with a newly sharpened chisel.

MARKING THE PINS With the other piece held in the vise, lay the cut piece on top and hold it in position with one hand while marking around the tails with the other (**12**). Use a marking knife or a sharp pencil. After squaring these lines down to the gauge marks on one side, make sure to mark the waste (**13**). Then saw along the waste side of the marks to cut out the pins. This is the crucial step in all dovetailing. The sawing must be just beside the line; on the line and the pins will fit too loosely, and too far over they will fit too tightly. Sawing so that the line is "split on the waste side" is exacting, but it soon comes with experience. Cutting away the waste is again done with a bevel edged chisel, after most of the waste has been removed by drilling or sawing with a fine fretsaw (**14,15**).

As with the tails, it is a good idea to undercut the waste slightly, making sure the cut starts exactly on the gauged line and slopes inwards a few degrees. The end grain has no gluing strength anyway and this way it won't interfere with a tight fit.

As a final step, examine and clean up all the joints with a sharp chisel if necessary (**16**). Some workers prefer to bevel the tails slightly on the inside to make them easier to fit.

Above: Step 5. Below: Step 6.

Above: Step 7.

Above: Step 8.

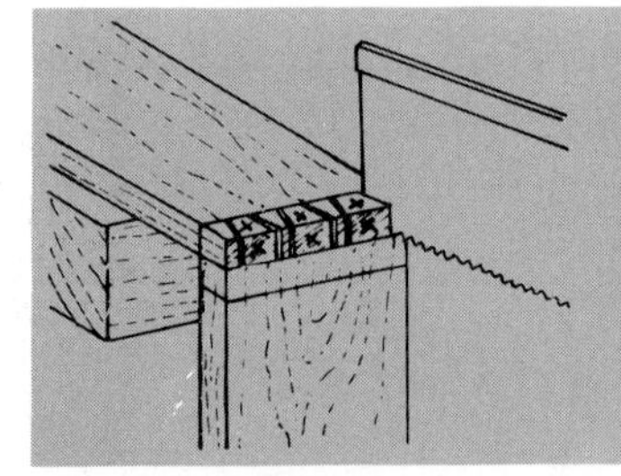

Instead of marking the pins around the cut-out tails, some woodworkers prefer to mark them through the saw kerfs before cutting the tails.

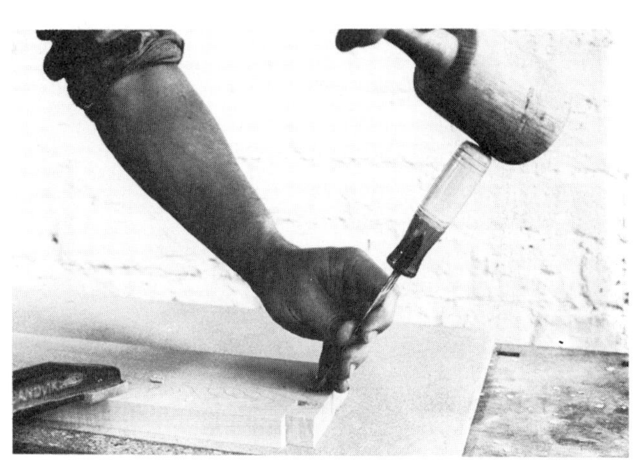

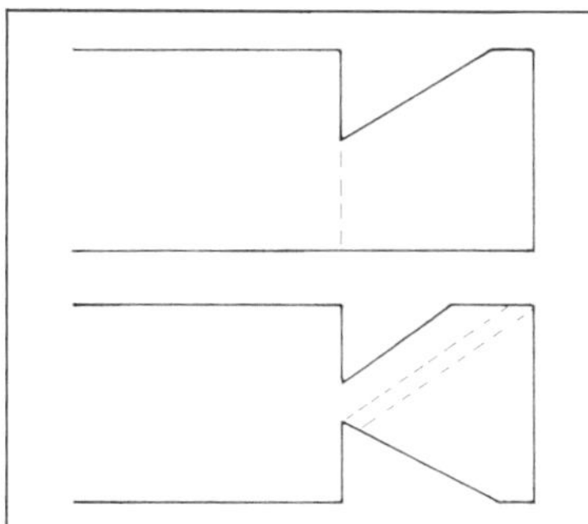

Above: Step 9.
To avoid causing unsightly breakout of the end grain, make successively bigger V-grooves from both sides. This way the waste is supported until the last cut is made.
Below: Steps 10 and 11.

The sequence of marking and cutting the pins. Above (12): Scribing the outline of tails. Left (13): The pins and waste marked. Below (14): Cutting waste with fretsaw. Bottom left (15): Final paring. Bottom right (16): Cleaning up.

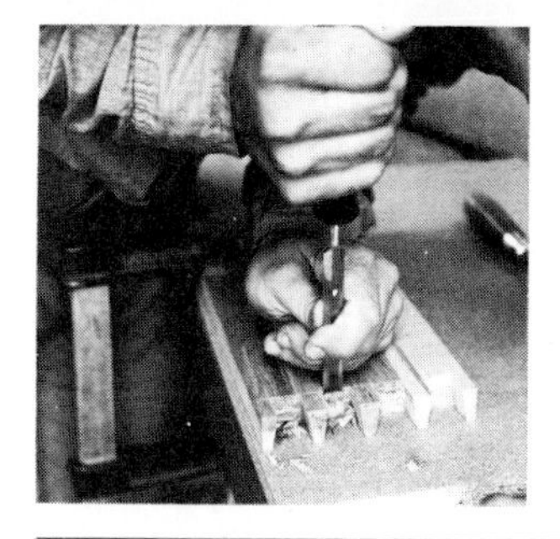

A useful dovetail template

With a metal template such as this, made from 20 gauge metal cut with shears, the lines can be squared across the end at the same time as marking the rake. For faster work, mark and saw two ends at the same time.

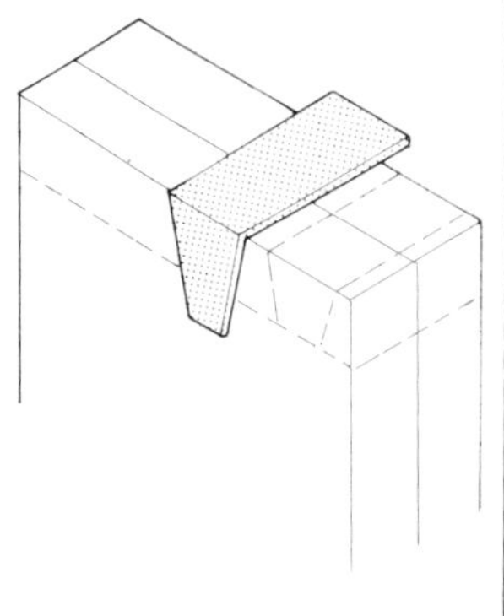

ASSEMBLY Each dovetail joint should be tried for fit, but only part-way. It should go together with a few taps from a soft-faced hammer (**17**). If it doesn't, it is too tight and may split the wood if forced. Examine the joint and trim off where necessary.

For final assembly, spread gap filling resin glue (it fills gaps and has a long enough setting time to allow plenty of time for adjustments) lightly on the sides of the pins and tails. To assemble, first tap with the soft-faced hammer, then clamp together. Since the ends will be $\frac{1}{32}$in or so proud of the faces, small softwood blocks will be necessary under the clamp heads to bring the joint together (**18**). Depending on tightness of fit, one or two clamps may be needed across each side. Be sure to have four to eight clamps on hand before the glue is mixed.

Wipe off excess glue, and after the glue has set, clean off the corners by planing or sanding (**19**).

Assembly. Top (17): Trying joint partway for fit using soft-faced hammer. Above (18): Gluing up. Notice block with small strips added. Left (19): The finished joint.

Decorative dovetails In most carcass work, the dovetail joint is set out with similar size tails, but there are no set rules regarding size and spacing. Many craftsmen prefer to design their own joints, for example alternating the large and small pins shown opposite, spacing out the large structural pins and including smaller part width pins in between. There are no rules in dovetailing, but commonsense must dictate so that enough strength is left in the joint.

Lapped dovetails Lapped or half-blind dovetails show from one side only and are, therefore, widely used for drawers to hide the joint from the front. Making this joint is essentially the same procedure as for through dovetails, only the setting out and cutting the pins differs.

SETTING OUT In most drawerwork the front is lapped and the back is a through dovetail, but if applied to carcass work, both ends may be lapped. In either case, the laps must be allowed for when trimming the pieces to final size. Notice that the exact length is used for the sides, but the front may be overlength by $\frac{1}{16}$in, i.e. $\frac{1}{32}$in at either end. This must be added to the second gauge setting when gauging the side thickness onto the front piece.

For a $\frac{3}{4}$in thick drawer front, a $\frac{3}{16}$in lap is more than adequate, leaving $\frac{9}{16}$in for the tails. Set the sharpened marking gauge to mark the lap from the inside face, in this case $\frac{9}{16}$in (line A). Mark only the ends of the tail section (front). Use the same setting to mark the tails on all four sides of the side pieces (lines B). Then reset the gauge to the thickness of the side to mark inside only of the front. (line C)

Set out and cut the tails as for through dovetails, and use them held against the gauged line (A) on the end of the front piece to mark around the tails.

CUTTING THE PINS is different from through dovetailing. The first cuts, splitting the lines on the waste side, are made diagonally so as not to cut beyond the gauged lines. However, part of the waste can be removed either by making diagonal saw cuts or by drilling to the right depth with a Forstner bit. The sawing method is preferred since you continue working with the same tool. Cut the pins with a sharp bevel edged chisel from above and the end alternatively, and clean up the corners before trying the joint for fit. As with the through dovetails, it's a good idea to undercut the end grain of the pins slightly, and I also find it useful to taper the tails very slightly towards the outside so that they will fit easily but appear tight on the outside.

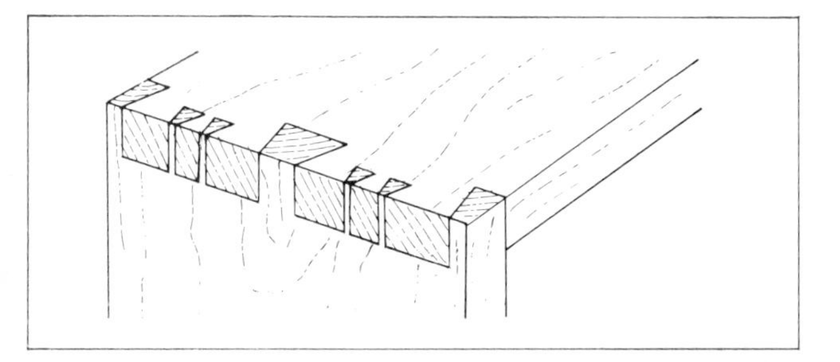

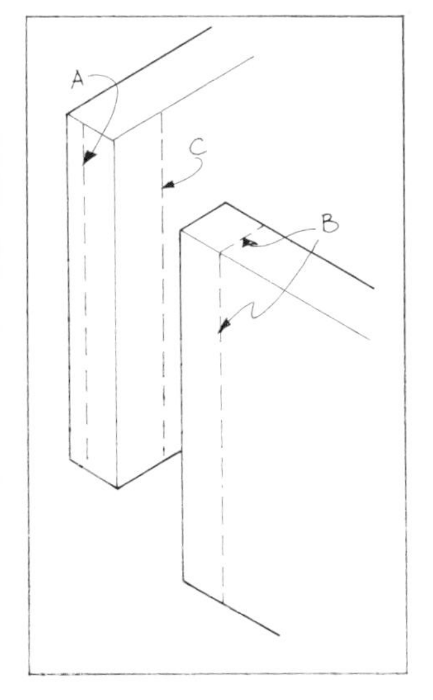

Elegant lap dovetails by Alan Peters.

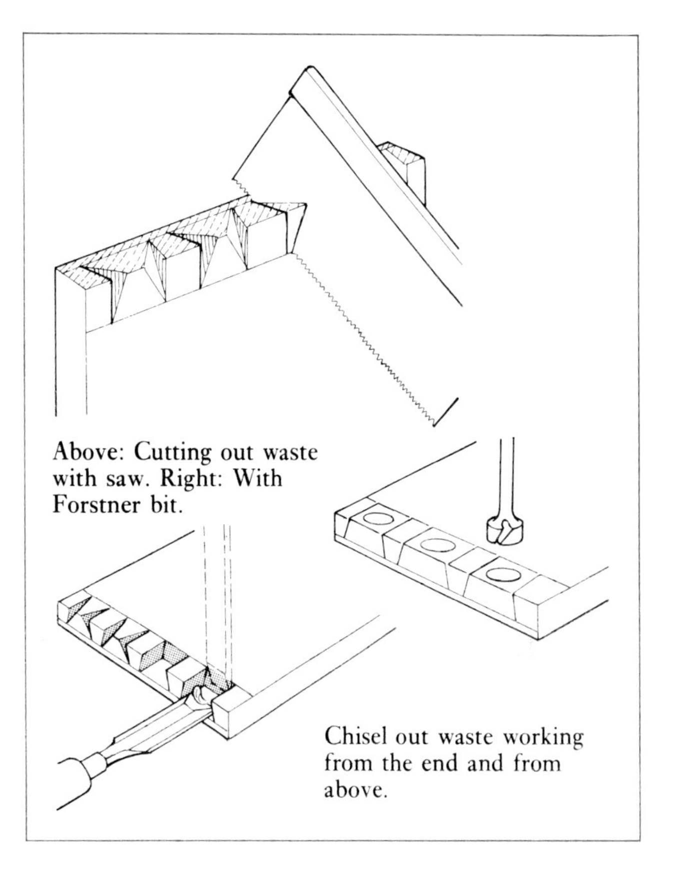

Above: Cutting out waste with saw. Right: With Forstner bit.

Chisel out waste working from the end and from above.

Secret dovetails There are two types, the double lapped dovetail where the thin lap shows on one side, and the miter dovetail which is really secret.

The double lapped dovetails are easier to set out if both pieces are the same thickness. The exact method depends on which side the lap is to show and this depends partly on aesthetics and also on which direction resistance to separation is required.

The instructions in brief are as follows:

1 gauge the thickness of X on Y
2 gauge the lap A, twice on Y, once on X
3 gauge the height of the tails on X
4 cut rabbet on Y
5 mark and cut out pins on Y
6 trace outline of pins onto X, cut out tails.

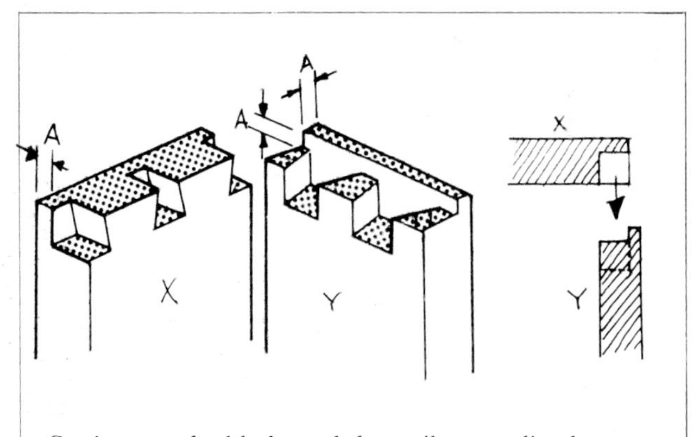

Setting out double lapped dovetails as outlined briefly in the text. Double lapped dovetails can be made in two ways as shown above and below left.

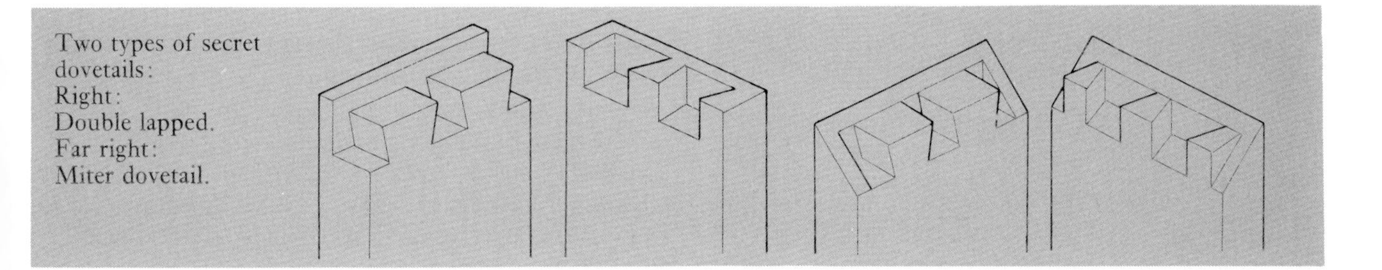

Two types of secret dovetails:
Right:
Double lapped.
Far right:
Miter dovetail.

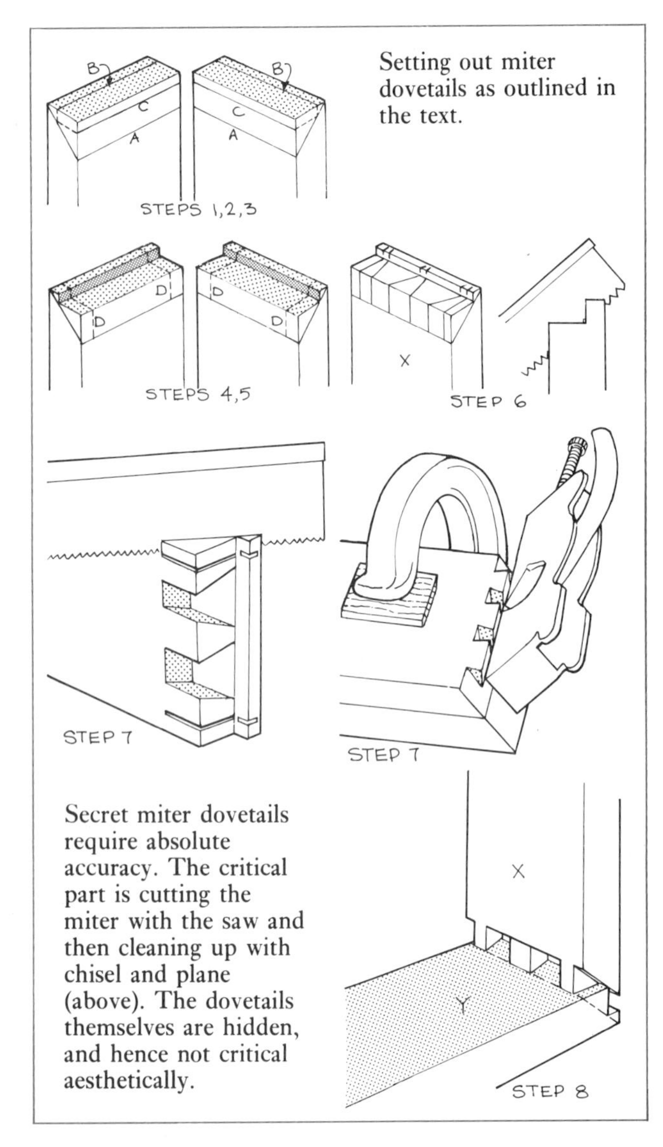

Setting out miter dovetails as outlined in the text.

Secret miter dovetails require absolute accuracy. The critical part is cutting the miter with the saw and then cleaning up with chisel and plane (above). The dovetails themselves are hidden, and hence not critical aesthetically.

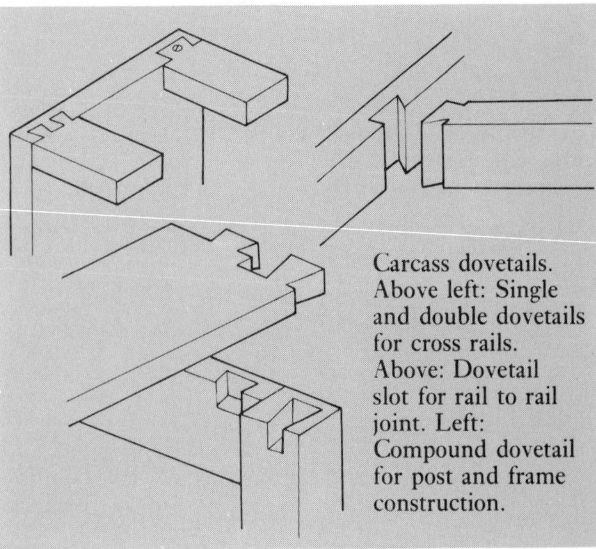

Carcass dovetails. Above left: Single and double dovetails for cross rails. Above: Dovetail slot for rail to rail joint. Left: Compound dovetail for post and frame construction.

Secret miter dovetails are more complicated because the rabbets cut on both sides must be chiseled and planed to a miter which must be accurate for the corners to fit tightly when the joint is assembled.

Keep in mind that the two pieces must be the same thickness and trimmed to the exact overall (outside) measurement of the rectangle being made.

The instructions in brief are as follows:

1 gauge the thickness on the inside faces of both sides (line A)

2 gauge the lap on both ends (lines B) and on both inside faces (lines C)

3 mark the miters on the sides using a knife or chisel

4 cut the rabbets with a fine dovetail saw (make V-shaped grooves first to give a clean finish)

5 gauge in the lap width for the side miters from both ends on each piece (lines D)

6 mark and cut the pins on piece X (1:5 rake). Note: make a special template to fit the rabbet

7 cut the side miter on piece X with a saw. Then cut the mitered end with a very sharp chisel, and trim with a shoulder plane from both sides towards the middle

8 trace the pins onto piece Y, square lines across top, cut out tails

9 cut miters on piece Y, bevel off the edges of the tails

10 clean up, and trial assemble

Carcass dovetails For large carcasses such as desks, sideboards and bookcases, the sides have a tendency to separate which is best be resisted by dovetailed cross members. In fine work the dividers or shelves can be dovetail housed (see page 172), but the standard method is to dovetail the ends of the various rails. The exact configurations vary and must be worked out for each cabinet.

Machine dovetails As with all woodworking joints, machines have been developed to cut dovetails faster to make it possible to produce quantities quickly. Machine dovetails are structurally more sound than hand cut ones because the fit is nearly perfect. But they do suffer from the disadvantage of being regularly spaced with equal pins and tails. This reduces the decorative possibilities, and the flexibility to vary sizes and spacing to suit various widths. Machines cut both the tails and pins in one operation, leaving pins with rounded ends which get hidden in the joint.

Left: Mounting numbered pieces into dovetail jig.

In production shops, semi-automatic machines are used with numerous individual cutting spindles which are capable of cutting the four joints of a drawer in one operation. These machines are beyond the means of a small workshop, but there are several dovetail cutting accessories for routers or portable drills which do a professional job and which will pay for themselves in one or two short runs.

The spindle molder can also be fitted with a dovetail accessory, but it's far too cumbersome and heavy to operate.

The dovetail template for the router is very simple to use, but demands concentration in mounting the pieces the right way. The router is limited to lap dovetails because of the rounded pins.

To set it up, follow the manufacturer's instructions which, like most manufacturer's instructions, are not as clear as they should be. Basically you screw the jig to a backing board which can be clamped to the bench or held in a vise. There are two finger templates ($\frac{1}{4}$in and $\frac{1}{2}$in) for thinner and thicker wood. Usually the kit comes with one of these and a matching template guide which is mounted to the base of the router to guide the cutter along the template.

The important thing is to prepare the pieces by trimming to accurate length and lettering them on the insides, as shown. They are then mounted inside out two at a time in the jig, as per instructions, to cut each corner at a time.

Assembly is easy since the joints fit perfectly and normally no clamping is required. Keep in mind that the drawer heights should be planned with the machine in mind in order not to leave small delicate part-pins at the ends. And the groove for the drawer bottom should be located so that a side tail will hide the groove ends.

Above: The template in position. Right: Setting cutter height with gauge supplied. Below: Cutting with router is nearly automatic. Far below: Pieces still in jig with dovetails cut. Bottom: The components and the assembled joint.

FINGER OR BOX JOINTS

This is a variation on the dovetail joint which relies on multiple glue surfaces for its strength. It is strictly a machine joint, though I have known craftsmen to try to cut it by hand, usually without much success.

The quickest and most accurate way of cutting finger joints is to use a special accessory for the spindle molder with a set of cutters which mount in order on a keyed spindle to form a spiral pattern. (see page 131) This assures that the cutters will enter the wood in order, one after the other, rather than all at the same time which would be dangerous. These sets are very expensive and do require a special spindle which is usually made to fit only industrial machines.

For those lucky enough to have a set, the drawer sides can be run off four at a time, first one end then the other, in a sequence which takes only a few seconds. It does require accurate setting up, for the last pin must match up exactly with the last space on the mating piece.

Finger joints can also be cut on the table saw using a simple jig nailed to the miter fence. Use a tipped blade, if possible, to get a clean cut. Set the blade height to just over the thickness of the boards (which should all be the same). Cut a notch in a piece of straight wood, about $\frac{3}{4} \times 4 \times 18$in long, which will become the jig. Then cut a thin sliver to fit exactly into that notch. The silver doesn't have to be as deep as the notch, but the accuracy of the joint requires that it fit tightly into the notch. Cut a short piece off the sliver, then fit the longer piece into the notch. Use the other part of the sliver to line up the jig, then clamp the jig to the miter gauge.

The sliver on the jig should be located so that it is displaced exactly one notch over from the first saw cut. To locate it, hold the sliver between the jig and the blade, then attach the jig carefully to the miter fence. It may take a little adjusting but it's crucially important to get this part right.

The cutting is simple. Make the first cut, one notch over using the sliver as a spacer, then place that cut over the sliver to make successive cuts.

The mating piece must have a notch in the end to match the space left in the first piece. Again use the scrap sliver to line up the first cut then proceed as before until all the notches are cut.

Assembly usually requires a lot of clamps because, unlike dovetails, there is no mechanical configuration to hold the joint tight in either direction.

Cutting four pieces at a time on spindle.

Detail of the four pieces.

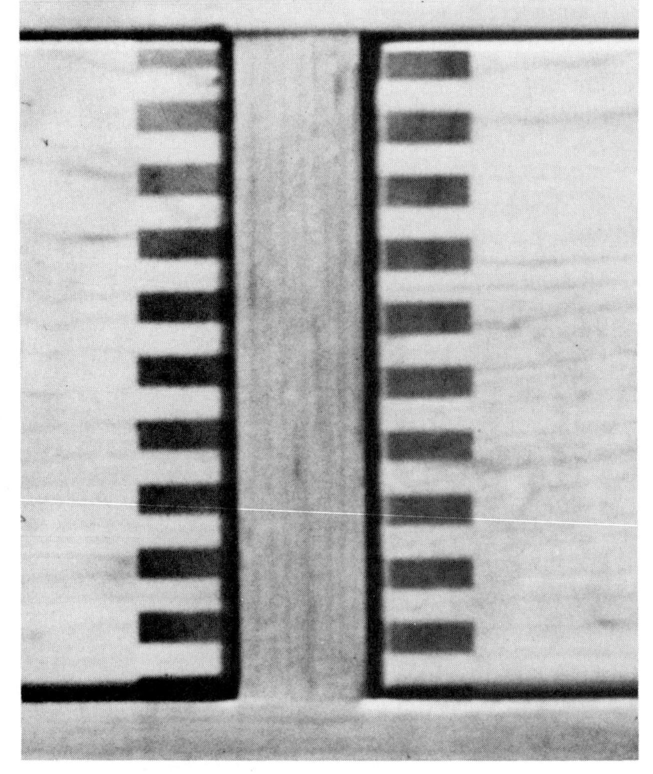

Finger joints are perfect for drawers where they are both decorative and functional.

Locating jig using two slivers.

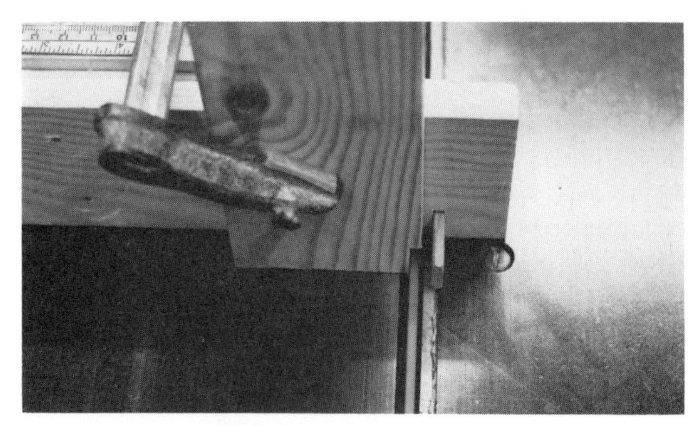

First cut of second piece.

Making first cut. Notice the clamp.

Second cut.

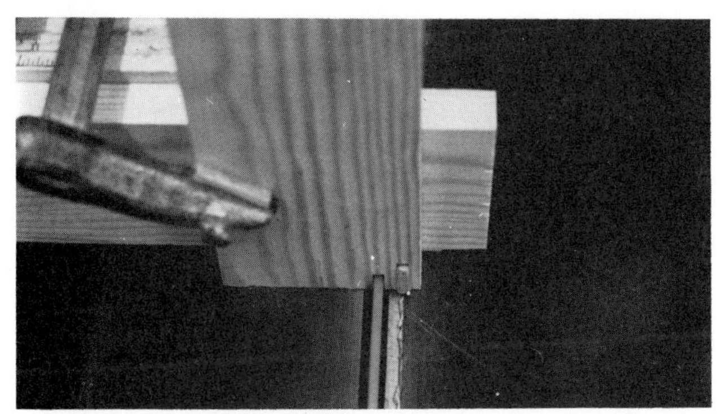

Second cut with sliver in first slot.

Last cut.

Locating workpiece for first cut in matching piece, using sliver as a spacer.

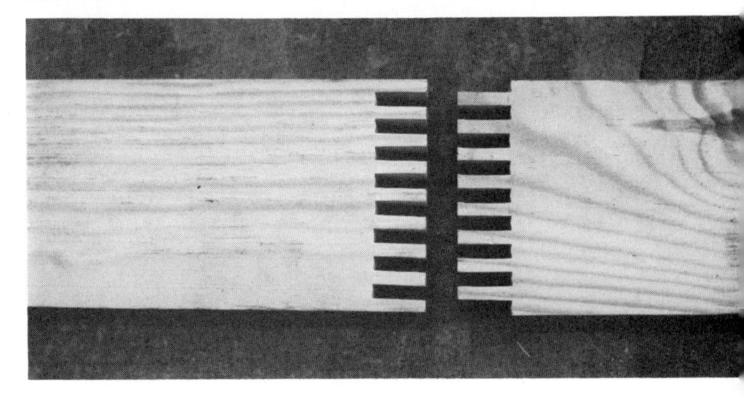

The matching sides. Their width must be worked out so that end pin fits into end slot.

KNOCK-DOWN JOINTS

Knock-down fittings have become increasingly more important in modern furniture, primarily, though not exclusively, for mass production furniture where the requirement for flat packing is all-important to cut down shipping and storage costs. Some of these fittings are very attractive and are therefore finding favor among more traditional woodworkers too.

Anyone who has had to deliver a large table up a few flights of narrow stairs will appreciate how convenient it is to have a fitting which allows the furniture to be part assembled on site in just a few minutes.

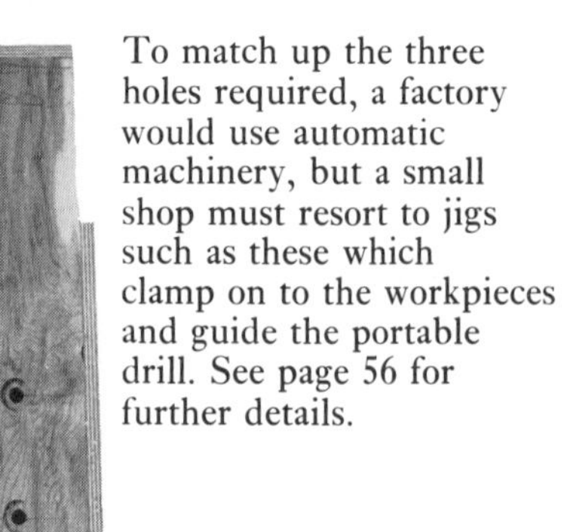

To match up the three holes required, a factory would use automatic machinery, but a small shop must resort to jigs such as these which clamp on to the workpieces and guide the portable drill. See page 56 for further details.

Bolt and cross dowel Most Scandinavian furniture today is knock-down and the joint most often used is the bolt and cross dowel (barrel bolt). This has a neat, black or brass flush round head which is tightened by a hexagonal Allen key provided with the furniture. It looks very sophisticated and it is very strong and many furniture makers use it as a general purpose joint even if the furniture is not going to be knocked-down.

To fit the bolt and cross dowel, first determine the length of bolt required to extend from the face of the first piece to at least 1in, preferably $1\frac{1}{2}$in into the second piece. The threaded bolt comes in various lengths and should be ordered to suit, but it can be cut to length with a hacksaw, if necessary.

Insert cross dowel (top hole). Then insert bolt from end, and screw on.

The bolt, cross dowel, brass head and the Allen key supplied to the customer.

Fit the post onto the bolt. Then screw on the brass head and tighten with key.

Mark the front hole and drill a shallow counterbore to make the round head fit flush, though this isn't strictly necessary. Then drill the hole for the bolt straight through. It's easiest to enlarge this enough to take the shank of the brass head and to allow room for adjustment. Since the bolts are normally used in pairs to give strength against twisting or racking, the hole spacing must match up exactly with the spacing on the matching piece. To get this right it's best to devise a drilling jig which clamps onto the workpiece (see page 55). A similar jig, drilled from the first, is used to drill the second piece. The final holes to take the cross dowels in the rails require a third jig to make sure it's centered and accurately located.

Assembly is easy. Clear the sawdust, then fit the cross dowel in its hole. The end with the screwdriver slot is fitted outwards and indicates the direction of the threaded hole. Fit the bolt from the end and turn it into the cross dowel. Then push the bolt through the other piece and add the brass head which is tightened with the Allen key. To take it apart again, remove only the brass head.

This fitting is used in a wide variety of applications. Where there is only room for one bolt, a small dowel peg or tenon should be added to the joint to prevent swiveling.

Above: A simple knock-down device: screws with washers. Left: Screw sockets. Right: Screw cups. Below: Various other K-D devices for screws.

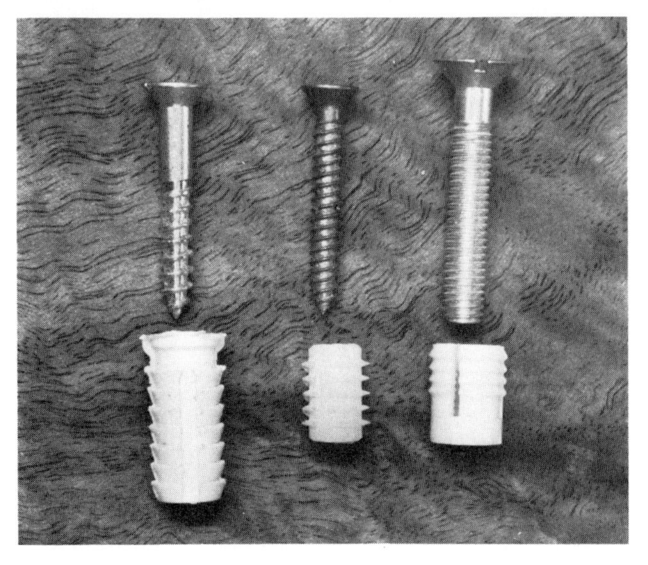

Screws and washers The simplest knock-down fitting of all is the screwed joint with washers fitted under the head. These can be surface fitting (screw cups) or flush fitting (screw sockets), but the former are much easier to fix and do not require any countersinking.

One method I have often used and found very satisfactory is to insert a wall plug into the second piece of wood. This can be the plastic or fiber type used for screwing into masonry walls. It holds by friction rather than mechanically by the threads, so it can be used in end grain and plywood, making it particularly suitable for screwing the ends to the shelves on bookcases.

There are many other knock-down fittings on the market and more are invented every day. A current hardware catalog can give much more information than a book like this. Some of these must be bought from specialists or wholesale suppliers (check the telephone book or trade journals), but many are finding their way into general hardware stores.

Through tenons (see page 175) This is the oldest knock-down joint of them all, and is still popular particularly for large refectory style tables.

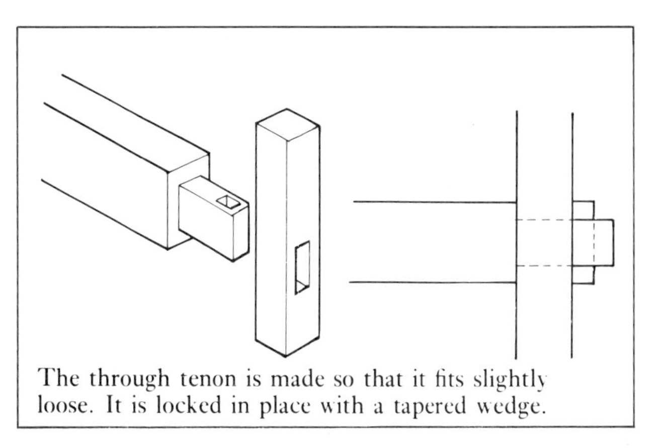

The through tenon is made so that it fits slightly loose. It is locked in place with a tapered wedge.

Above: Briefcase in pine made by Timmermännen, Sweden. Below left: Detail of hinge closed. Below right: Hinge open.

Detail of catch of the briefcase. Front knob turns plywood catch which fits over dowel pin on briefcase lid.

WOODEN HARDWARE

Making hardware out of wood adds a nice touch to furniture. For hardwearing parts a tough close-grained hardwood such as maple, padouk, rosewood or birch, should be used, and the pin should be steel rather than wood. The proportions of a hinge depends on how much load it will have to take. Small cabinet hinges can be more delicate than the heavy-duty type required for the traditional brackets to hold Pembroke table leaves, for example.

The hinges on the wooden briefcase shown in the photograph are made of pine with birch dowel pins, which breaks both these rules and testifies to the durable qualities of wooden hardware.

The matching wooden catches are made with a birch dowel as a pin, with a wood knob on the outside and a notched plywood catch on the inside. The notch is made to fit over a dowel in the lid.

Ordinary hinges are made in the same way as the traditional knuckle joint, formerly used on fold-out table legs. To cut the joints by hand, clamp them together to mark out the cutting lines, and make sure to mark the waste on each piece. Before making the saw cuts, move one piece over by a slight amount so that the first series of cuts is made on one side of the waste. Then move the same piece in the other direction to cut on the other side of the waste. It's important to cut on the waste side and not, as in dovetailing, to split the line. That way there will be enough tolerance to allow movement.

The same process can be carried out when cutting the fingers by machine (page 198). The normal cuts are made, then enlarged slightly by moving the cut slightly to one side.

Final trimming of the hinge depends on the exact configuration and how it will operate. For a 90° turn, round the ends of the fingers on both sides with the joint assembled. Then continue the rounded shapes by making V-cuts to form shoulders which will act as stops. By varying the angle of the shoulders the opening angle is changed.

Drill the pin hole with the joint assembled and immediately insert the steel pin which should be lightly lubricated with candle wax, as should the contacting sides of the fingers.

There are many other possibilities for wooden hardware such as a continuous hinge for a drop leaf table. This requires accurate setting out because the pin holes must be drilled before gluing up the surface. During gluing up, the long steel pin should be threaded through all the holes to align them.

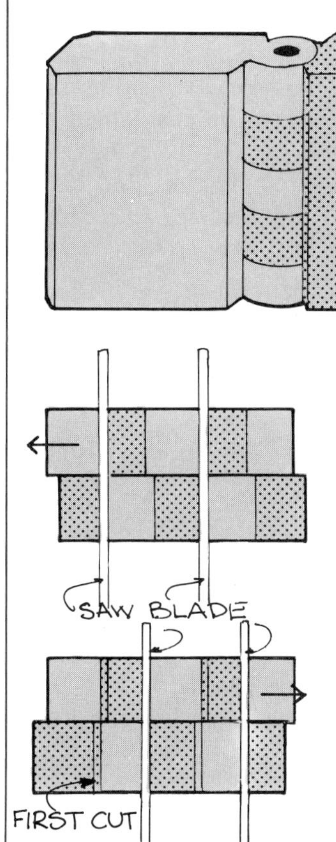

Making a wooden hinge. Set out the hinge on paper, drawing it to scale both open and closed to derive the right amount of shoulder required. To cut the fingers, mark the pieces showing the waste, and mount together in the vise. Before sawing along waste line as shown, move the top piece first one way, then the other so that the cuts are made in the waste areas.

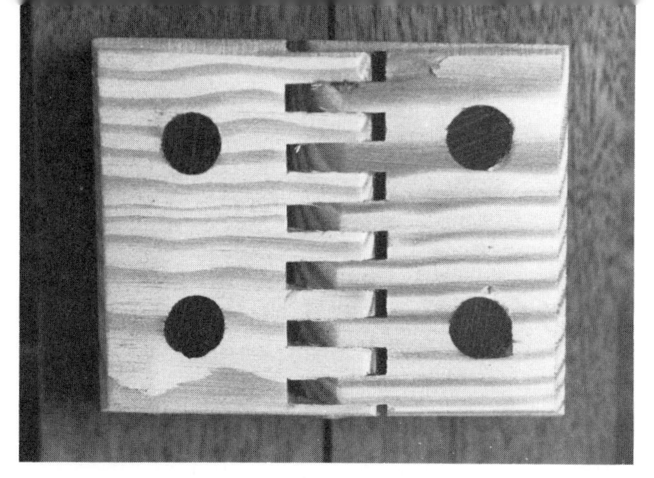

Above: Simple hinge made from machine joint. Below: Drilling hole for pin.

Above: Oak font cover by Alan Peters. Below: Door pulls by Gil Gesser.

Carved door pulls made by Sam Bush.

JOINTS FOR SHEET MATERIALS

With all these pages devoted to fancy joints for solid wood we are ignoring the fact that plywood, blockboard and chipboard, plain and veneered, are widely used in woodworking and have many advantages over solid wood. They don't expand and contract by amounts large enough to cause problems, and they cover large surfaces without the need for cumbersome gluing up. The list of advantages is long and the prejudice against them has been widely overcome especially with the fine veneer work done by many furniture makers and also by their almost universal use in production work.

Joints for sheet materials have to be designed differently than for solid wood though some joints such as splined miters are used for both. Of all the sheet materials, blockboard behaves most like wood, because of the solid wood strip cores. It can therefore be dovetailed, etc., as in ordinary work, but this is rarely done since other joints are much easier.

Screwed joints The simplest joint is a simple screwed butt joint, and many improved screws (see box) have been developed to grip better and go in faster. Special coarse threaded screws are required to grip in chipboard, but it's best to drive them into special plastic plugs inserted in the hole. Screw heads can be left exposed, decorated with cup washers, or hidden with plastic covers or with wooden plugs.

Special screws

To anyone who isn't already a devout convert, special screws such as sheet rock screws and chipboard screws will be a revelation (ask anyone who is). Sheet rock screws designed to attach sheet rock (plasterboard in UK) to metal or wood partitions, are designed to be hand or power driven with a hard point which can tap its own hole in metal. The threads are at a steeper angle for faster insertion and are coarser for a tighter grip. For all but the hardest wood, only one sized hole is required, and in many applications the point will make its own hole in the second piece particularly in the softer woods.

Doweled joints Although these are widely used in industry, they have limited holding power because they must be quite short (normally $\frac{1}{2}$ to $\frac{5}{8}$in) to fit within the thickness of the panel. I have used through dowels (page 186) successfully on kitchen cabinets particularly to connect cabinet dividers to the base where the dowels will not show. But here, too, holding power is limited and not as good as that of screws.

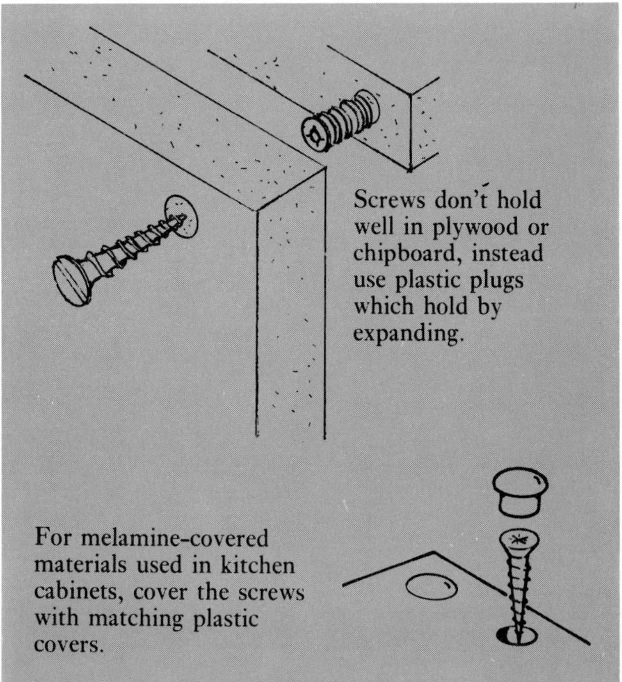

Screws don't hold well in plywood or chipboard, instead use plastic plugs which hold by expanding.

For melamine-covered materials used in kitchen cabinets, cover the screws with matching plastic covers.

Rather than using plastic covers, use plugs made from matching materials to hide screw heads.

Block joints Blocks can be glued inside the corner of a corner or T-joint to reinforce it where they will not show. Using a hide or resin glue, cover the block with glue then slide it back and forth until suction holds it firmly and the glue sets. As an alternative, reinforce the block with screws from the inside.

As a more elegant corner joint, most often used for plywood, key the two corner panels into the block with a tongue and groove joint. The tongue can be worked in the solid by cutting a rabbet top and bottom with a router to create the tongue. But it is more often loose tongued into grooves cut into the edge of the plywood and two sides of the batten. It's better, as with edge treatments, not to try to match the block, for it will always be apparent. Instead, make a feature of it by using a contrasting wood.

The corner can be left square or rounded. In either case, cut the joints carefully, making sure the corner block is left slightly proud of the plywood so that it can be sanded flush.

A block set in from the edge would require sanding of the plywood which would inevitably cut through the surface veneer.

Cut the grooves with a router fitted with a grooving blade. Work the plywood edges first, then change the setting slightly for the corner block. Bring the groove a touch further from the edge and leave the block proud.

There are many possible variations such as setting the plywood in from the edge to create a corner post or applying a decorative molding such as a bolection or ovolo, to cover the seams.

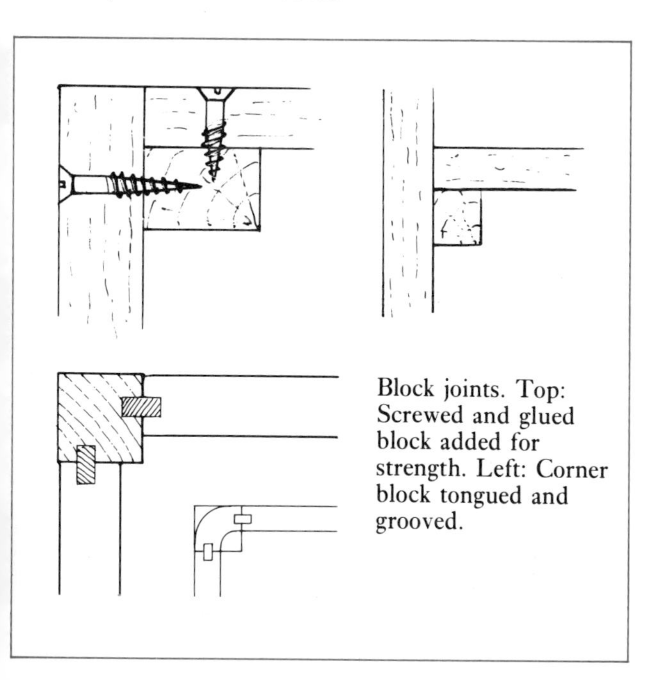

Block joints. Top: Screwed and glued block added for strength. Left: Corner block tongued and grooved.

Tongue and rabbet joint This is a good joint for simple box structures, but when used for drawers, a false front is required to hide the exposed end.

It is simple to make. First cut the groove on the table saw or radial arm saw. Its width and depth should be about one-third of the thickness, but is more often made the thickness of the saw blade. Cut the tongue to match the groove, by setting the saw blade at the appropriate height and making one or two passes. The tongue should fit very snugly and require clamping to bring it home. In addition, use a gap-filling resin glue to guarantee a strong joint.

The tongue and rabbet works equally well in solid wood. Below: Cutting groove.

Above: Cutting tongue. Below: Assembled joint for drawer with grooves for runner and bottom.

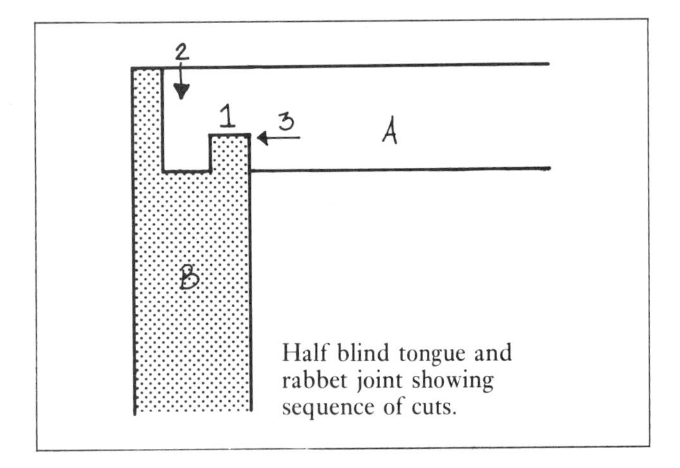

Half blind tongue and rabbet joint showing sequence of cuts.

The first step is to cut the groove in piece A.

Second step is to cut end groove.

Above: Third step is cutting off tongue.
Left: Joint ready for assembly.

HALF-BLIND TONGUE AND RABBET JOINT This is a superior variation on the simpler tongue and rabbet joint and is more complicated to cut. Three cuts are required, the first is the same as above, to cut the groove for the tongue on piece A. The second cut is more awkward because it must be made on the end of piece B. Use a tenoning jig or a homemade holding device on the table saw or set the radial arm saw horizontally. But since neither is entirely safe, it's better to make the cut on a spindle molder or with a router, with the board lying flat. Since a wide groove is required, use a dado blade or make two or more passes, resetting between each, or alternatively make this groove narrower. The third operation is a simple crosscut to cut the tongue on piece B to the right length. Try the joint for fit and make more cuts to adjust if necessary.

LOCK MITER As a further, and more complicated variation, the corner can be mitered to completely hide the joint. It is quite difficult to cut as it requires absolute accuracy for a proper fit.

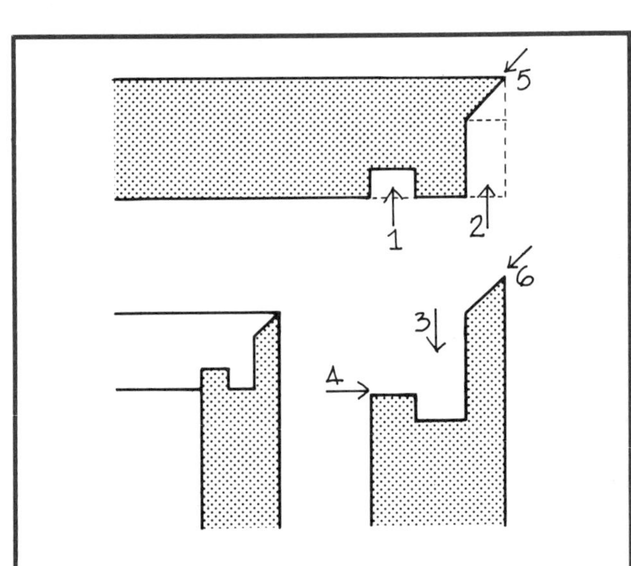

Cutting a lock miter. Cuts 1, 2 and 4 are simple crosscuts. Cut 3 is more difficult requiring end grooving with a tenon jig, radial arm saw or spindle molder. Cuts 5 and 6 are simple miter crosscuts.

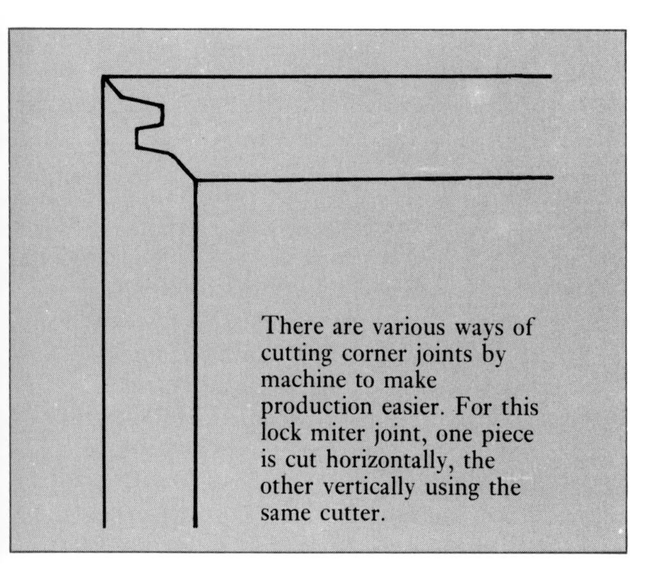
There are various ways of cutting corner joints by machine to make production easier. For this lock miter joint, one piece is cut horizontally, the other vertically using the same cutter.

Kitchen cabinets with edge of worktop showing extra thick edge grain of plywood as decorative feature. Designed by the author.

Machine joints Finger joints are often used in plywood carcasses but are limited in width by some machinery. If enough production warrants the purchase, it's best to buy solid cutters to fit on the spindle molder to cut special joints. These either have matching heads for the male and female parts or are designed to cut both components, one held horizontally, one vertically.

Edge treatments Plywood edges can be left plain as a decorative effect, particularly if good quality Finnish birch plywood is used.

The beautiful alternating bands of veneer can be put to further decorative use by cutting up strips of plywood and gluing them edge to edge.

VENEER EDGING is suitable for chipboard panels to match the surface veneer. It can be used on plywood, but because of the alternating bands of veneer, it is generally not as successful.

SOLID EDGING either of matching or contrasting wood, can be rectangular or molded in section. In cheap work it is pinned and glued. In better work it is tongued and grooved.

The tongues can be worked in the solid but it is always easier to cut grooves in both pieces (one setting on the router or saw) and fit a loose tongue. Cut the grooves and fit the edging on two sides first, then trim the ends flush and finish by cutting the grooves and applying the edging to the other two sides. Cutting grooves all around and trying to fit all four pieces at the same time is impossible to get right.

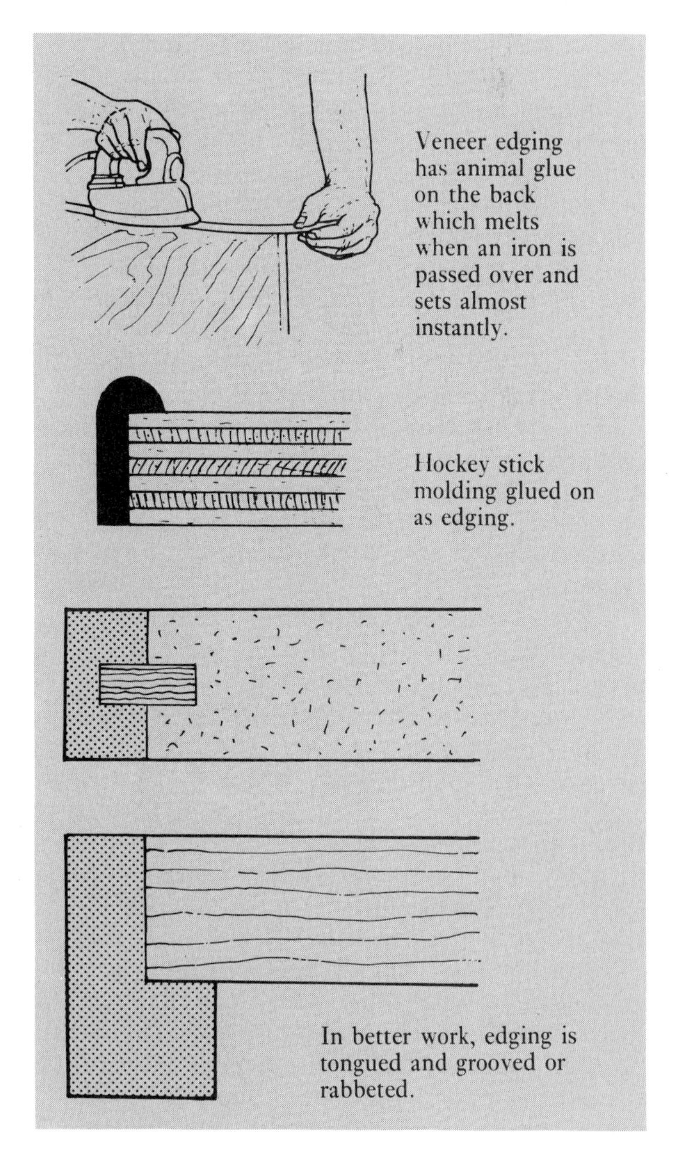
Veneer edging has animal glue on the back which melts when an iron is passed over and sets almost instantly.

Hockey stick molding glued on as edging.

In better work, edging is tongued and grooved or rabbeted.

Furniture construction

STRUCTURAL CONSIDERATIONS

Having trained as a structural engineer to study bridges and buildings, I should be able to apply the same analytical methods to furniture. But although a piece of furniture seems much more simple than a building, it is in fact a nightmarish problem to try to analyze. This is because the loads that will be imposed on it during its lifetime from people sitting or standing on it, pushing it, turning it on end, dropping it and so on, are so complicated and variable as to be beyond even computer analysis. Analysis is also difficult because all the elements of a piece of furniture work together sharing and spreading the load in quite indeterminate ways. This occurs in buildings too, but this "sharing" effect is rarely taken into account in structural analysis. Instead the building structure is simplified to make it easy to analyze in terms of components such as beams, joists and columns. This is the only approach to take in analyzing furniture.

We can analyze overall stability and identify the factors which contribute to stiffness, such as firm joints, back panels and box construction, and then study these factors in detail to see how they resist the forces. But no matter how rational our approach, we will still design furniture by intuition and tradition. Much testing has been done on joints by the furniture industry particularly on dowel joints. These results are important because the new methods developed as a result of advances in technology, such as new machinery, man-made boards and better glues have not had the testing period of traditional joints which have developed by trial and error over thousands of years.

Basic structural principles All forces are either tension, compression or shear. *Tension* tends to pull things apart. For example, when you pull on the knob to open a drawer a series of tensile forces act on the joints. By pulling the knob you try to overcome the frictional forces set up on the runners. If the drawer is heavy and the runners are not waxed the friction is greater, requiring more pull (tension) to move the drawer. The screw holding the knob must be strong enough to transfer the tensile force to the drawer front. The drawer front will pull off unless it is well enough connected to the sides, hence the need for dovetails which are perfect for resisting tension. Once the tensile forces reaches the drawer sides it equalizes with the frictional forces, acting on the base of the sides to pull the drawer open.

Compression tends to push things together. For example, when you put books on a bookshelf the weight (downward force) is equalized by the stiffness of the shelf; otherwise it would fall down. The force is transferred from the center of the shelf to its end where it is resisted by the joint connecting the shelf to the vertical support. The compressive force acts downwards on the support and if it weren't for the fact that the floor pushes up on the bottom edge it would go through the floor. The support, then, has to resist the compressive forces which push from both ends. Even the smallest member such as a 1 × 1in has high resistance to compression, but it depends on its unsupported length. The trouble is that columns such as legs, dividers and other supports tend to buckle when compressed, and to resist buckling they must either be extra large and stiff or be well-supported by shelves, etc., along the length.

Shear is less clear to demonstrate. We think of a dowel or bolt "shearing off", that is, breaking cleanly across the section. But it is more complicated than that. It is a force which tends to distort. Thus, if you pushed sideways on the top of a bookcase the sides would tend to lean over unless there was some opposing force. A plywood backing screwed to the four sides does the resisting by transferring this sideways force to the base by shear forces. We see the plywood back simply as a stiffening element, but if we exaggerate the force, it is easy to see that the back would tend to distort into a rhomboid shape as a result of the shear forces. These are transferred to the base and must be resisted by friction on the floor for the bookcase not to slide sideways. These same "distorting" forces are present inside most members that undergo "bending". Contrary to general belief, shear forces are more likely to cause breaks along the grain than across it, for even though the force on a member may act vertically, shear develops as two pairs of opposing forces acting both along and across the grain and the longitudinal stresses are usually the ones which do the damage.

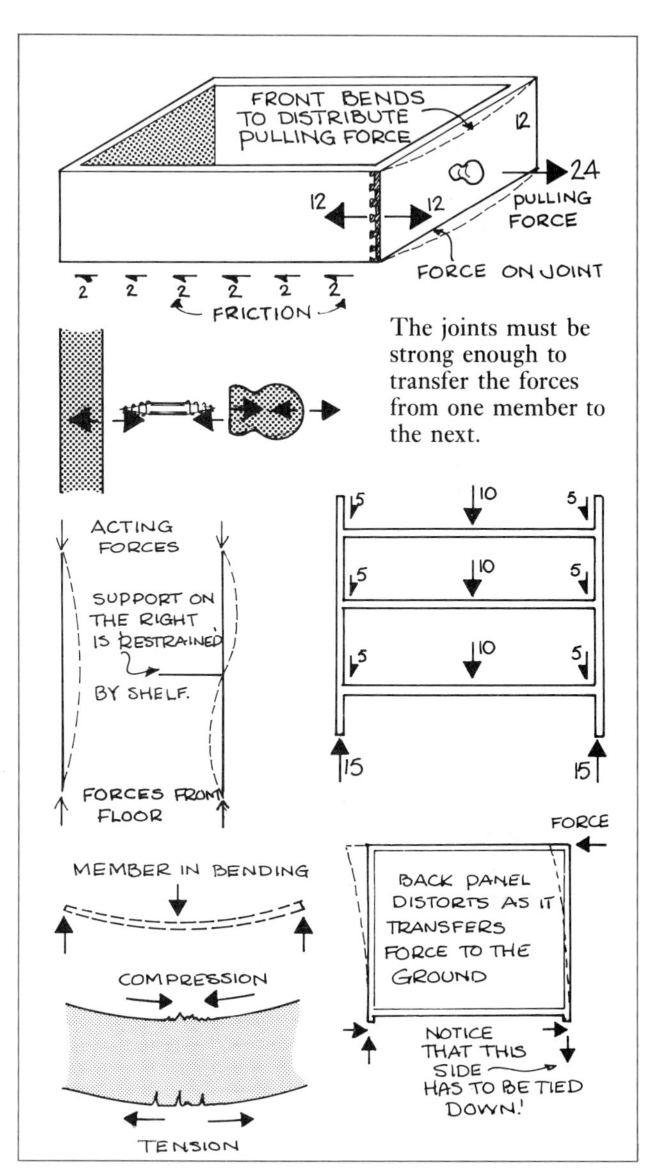

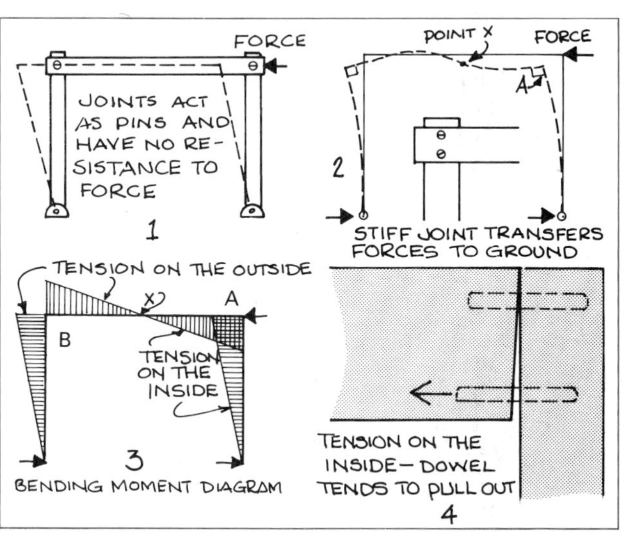

Bending Bending occurs when forces acting perpendicular to a member have to be transferred to the supports, such as when the shelf transferred the weight of the book to the support, or the drawer front transferred the pull of the knob to the corner joint. Bending is really a combination of tension and compression acting together in "couples", forces which are equal in amount but opposite in direction.

Consider the shelf. It will bend under the weight of the book. The tension set up in the bottom fibers of the shelf, acting in combination with the compression in the top fibers, forces the shelf to bend as the bottom edge elongates and the top edge gets shorter. The tensile force tends to pull the bottom fibers apart and if great enough, will break the shelf. The compressive force tends to crush the top fibers and again can cause a break. But wood is normally stronger in compression so the tensile forces tend to do the damage.

In most structures, tension, compression and the bending stresses act in combination. A member is usually, say, in compression, and bending at the same time. In that case the direct and the bending compressions add up on one side of the member and bring the fibers closer to the breaking point.

These forces act throughout a piece of furniture in an attempt to transfer imposed loads, such as weights, sideways pushes and so on, to the floor. Each element along the path must be strong enough to transfer it to the next and these points of transfer are the joints which are usually the points of weakness.

Transfer of forces Consider a simple structure such as a table rail joined to the leg with only one screw at each end. As a structure, this is unstable for if pushed, it would fall sideways. Seen as a structural model this is a mechanism with hinges or pins at each corner, with no strength to retain its shape under loading. **(1)**

If we introduce another screw at each joint **(2)**, the mechanism becomes a structure because the joints are rigid. In order to fall down (fail structurally) the screws at each corner would have to pull out or break the wood. This is another way of saying that the structure is capable of transferring the sideways force to the ground. It may bend and swing sideways a little, but this deflection is the structure's way of building up the forces to resist. In other words, force, or stress, is directly related to deflection or strain. Any structure, whether a table rail or a bridge girder, must deflect or move in order to build up resisting forces. If a skyscraper didn't lean over a little in the wind it would break.

A structural analysis of the simple rail and leg structure involves knowing (or guessing) how it will deflect or behave under load. The legs will be forced to bend sideways but the corner joints, which give the structure its strength, will tend to stay at right angles. This gives the configuration shown by the dotted lines. **(2)** In structural terms this would be represented by a bending moment diagram which is nothing more than a graph of the amount of bending forces at each part of the structure. The direct stresses, compression and tension are small compared to bending stresses. **(3)**

But it isn't necessary to understand this in order to get an intuitive feeling for the structure. It is only necessary to see that joint A resists the force by tending to open up, the 90° angle becoming more like 92°. The inside corner of joint A would crack open if the force was large enough, a sign that the tension is on the inside. You could demonstrate this to yourself by cutting out a cardboard shape, laying it flat, pinning the bottoms and pushing sideways at the top. To an engineer, the bending moment diagram shows this at a glance. The graph is always plotted on the tension side of the member, which demonstrates that the middle of the rail (point X) has no stress at all, and joint B has tension on the outside and compression on the inside of the joint.

In practical terms, this means that at corner joint A, the bottom of the rail is tending to pull away while its top is pushing against the leg. If this was a dowel joint instead of screws, the bottom dowel is in danger of being pulled out and vice versa. **(4)** The further apart the dowels are, the more bending moment resistance it has, and the stronger it is. Of course, if the pushing was left to right instead, the forces in each joint would be reversed so that the top dowel in joint A would be pulled out. In either case, only one dowel per joint would not be enough to resist sideways "racking", because it would have no moment resistance. Tables and chairs with two rigid corner joints behave much like this model, but most cabinets have at least four moderately stiff corner joints, which help add rigidity. The more joints there are in each plane, the stiffer the structure. But this analysis should be applied in the other planes as well. A cabinet can rack front to back as well as sideways, and if the cabinet top isn't rigidly attached it can rack in the horizontal plane too.

As can be seen, structural analysis of furniture gets quite complicated (this is a simple example), and in most cases it is enough to have an intuitive idea of how the members and joints will behave. Like structural engineers, if in doubt be conservative, and go for the next size.

And a well-designed structure will keep the strengths of the various elements in proportion. Luckily this happens automatically to some degree because we "feel" when a stout member is required, and design the joint accordingly. Of course, the strength of the material itself must also be taken into consideration, for some species are much stronger than others. For those who are interested, structural codes for wood available at libraries, give tables of strengths including the pull out and shear strength of nails, screws, bolts, etc., for various species.

The strength of wood The strength to weight ratio of some wood species is higher than that of steel. The tubular cell arrangement is an excellent structural configuration but suffers from the disadvantage of being stronger in one direction than the other. The strength of wood along the grain is about 50% higher than across the grain, which must always be considered when designing furniture. Wood is used correctly as long as the tensile and compressive stresses, whether direct or bending, are directed along the grain. Thus a square panel, glued up from boards, must be supported along the weak (i.e. across the grain) edge where the resistance to bending is small.

Statistical analysis of wood strengths has enabled engineers to develop a correlation between stress and strain, so that like steel it is possible to use ordinary elastic theories in designing with wood. A certain grade of oak, for example, has been given a safe compressive strength of 1500lb/in^2 along the grain, but only 550lb/in^2 across the grain. The "safe load" takes into account a factor of safety. The stress required to actually break the wood is much higher.

Furniture designers determine sizes more intuitively, but certain structural considerations are very relevant. Since wood is weakest across the grain any stresses which tend to pull the fibers apart or shear them in opposite directions are likely to cause problems. These stresses are particularly high around any imperfections or notches where the direction of stresses must deviate from a straight path. Thus halving joints, as everyone knows, are vulnerable to

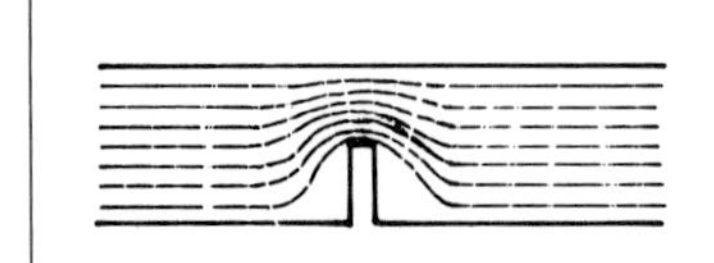

Diagram shows how stresses are increased near notches.

split, particularly where an over-sawn notch forms a start for the split.

The same phenomenon occurs in most joints where square notches leave vulnerable corners. In this sense, screwed or doweled joints are better because round holes are less vulnerable to split.

To some degree, tight-fitting joints are self reinforcing, but particular attention should be paid to areas of high stress such as in the center underside of shelves where tensile bending stresses are high, and at corners (as shown above) where stresses due to racking are high.

As structures, furniture rarely fails because a member such as a rail or leg actually breaks. The main consideration is always to design so that a carcass or cabinet will keep its shape, that is, the openings will stay square. Otherwise doors will jam, drawers will stick or, at worst, the joints will break. Many traditional methods have been developed to stiffen furniture structures. Some rely on strong joints, some strengthen the joints by adding members such as rails, pilasters or dividers, and some methods rely on panel construction such as solid backs, sides, or tops for strength.

The latter method is undoubtedly the best, for panels create a box structure which is very strong. A plywood back on a bookcase or cabinet, for example, gives it excellent rigidity. And panels serve a dual purpose. They enclose and they stiffen and are therefore perfect for chests and cabinets. For structures such as tables or chairs where panels would interfere with the function, the joints have to provide the stiffness. Chairs, for example, undergo tortuous loadings as people lean back in them, and the joints must be well-designed and made to stand up to these strains for many years. Hence, all the broken joints in cheap mass-produced chairs. In tables and chairs, unlike cabinets, a small degree of visible deflection is permissible, otherwise the components would have to be much too large.

The list below gives a few suggestions for reinforcing furniture structures:

1 Triangles make strong structures which minimize bending stresses, but diagonal members are rarely permitted functionally.

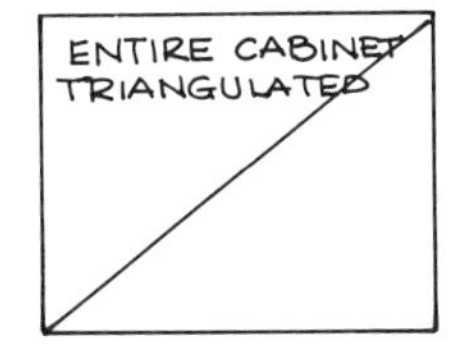

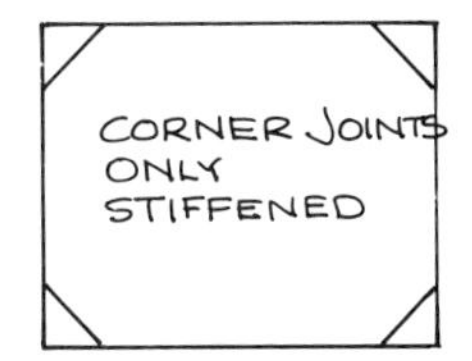

2 The corner joints could, if permissible, be reinforced with triangular gusset plates. We make them rectangular instead, extend them from top to bottom and call them pilasters.

3 An integral plinth or cornice has the same effect as side pilasters. Using both together makes it stronger still.

4 A central pilaster adds two strong joints to help stiffen the structure. Placed face outwards it gives more strength than placed edge on.

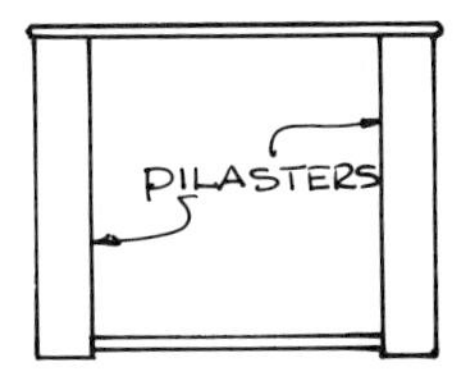

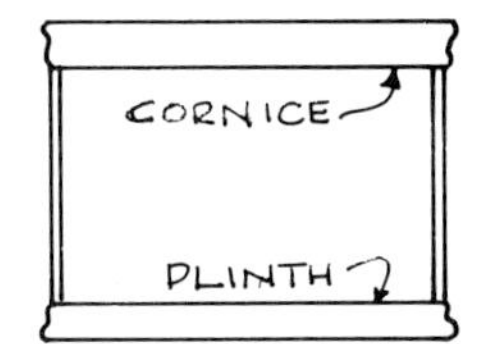

5 An open backed bookcase is a weak structure relying on the four corner joints for rigidity. Rigid shelves help, but a small rail under one or more shelves gives more strength.

6 Mortise and tenon joints have more reserve strength than dowel joints.

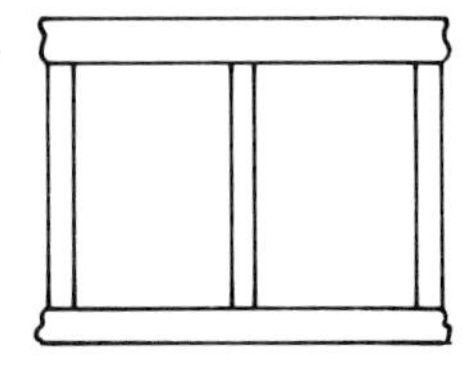

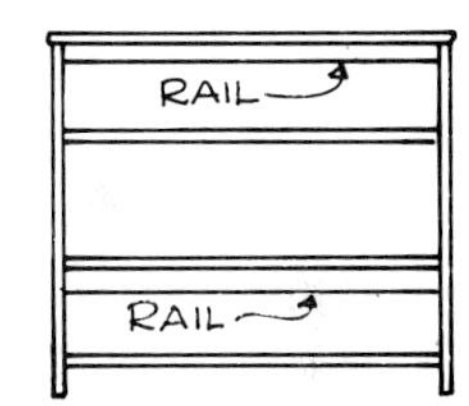

7 "Locking joints" such as dovetails or wedged or pegged tenons, should be used where there is tension or where the structure tends to separate.

8 Most joints rely on glue for strength and must therefore be tight-fitting. Wherever necessary use gap filling glue and, if in doubt, reinforce with dowel pegs or with screws from behind.

9 Doubling the amount of gluing surfaces doubles the strength. That's why finger joints have such good strength.

10 Doubling the gluing size of each surface quadruples the resistance to bendings. Mortise and tenons or bridle joints therefore should be as large as possible.

11 Shrinkage should always be kept in mind (see page 274). With time, shelves will shrink, loosening joints such as housings which rely on tightness of fit. Instead, use dovetail housings, peg tenons, or reinforce with hidden screws.

GLUING WOOD

There are many types of glues or adhesives on the market. Some, like PVA or resin glues, are widely used as general purpose wood glues. Others, like epoxy or contact cements, have special applications. In the average shop, therefore, a variety of glues must be kept on hand to deal with specific jobs, and an understanding of their properties and the development of good gluing habits are as important to the trade as accurate machining and joint cutting.

Basically, glue acts as a smoothing agent. The thin film of glue spread between the two surfaces reduces the effect of surface irregularities to bring the natural molecular forces into action. It does this by reacting chemically with the surface layers of the wood. Thus, the closer the two surfaces are brought together, the better the glue "penetrates" the wood. And the stronger the molecular structure of the glue itself, the stronger the glue joint. Some glues like PVA, for example, have low tensile strength which means that the layer of glue between the two surfaces should be as thin as possible (close-contact glues). Others, like resin glues, are gap-filling. Their structure, formed by chemical reactions, is strong enough to bridge small gaps, which makes them ideal for joints such as mortise and tenons where a tight fit is difficult to achieve.

Glue applied as a liquid sets into a hard layer, either by the absorption or evaporation of the solvent within the glue, such as with animal glues or PVA, or by a chemical reaction between the components in the glue such as with resin glues. Some glues, such as resins, set harder and are, therefore, stronger than others. Animal glues, for example, tend to move or "creep" under continuous loadings, particularly at raised temperatures. But strength is not always the important criterion. Sometimes a certain amount of flexibility is useful so an animal glue is used, and sometimes it isn't feasible to clamp the work, for example when laying plastic laminates, so a contact cement is used.

Wood adhesives, when properly applied, are usually stronger than the wood itself. Properly applied means **1** preparing the wood correctly, **2** choosing the right adhesive and mixing it well, **3** applying the glue correctly, **4** assembling correctly and in the right conditions.

Although it may sound pedantic, the importance of good gluing habits cannot be over-emphasized, especially when you consider how much work goes into the finishing of the glued-up boards.

1 Preparation of the wood

The surfaces to be glued should be as flat and smooth as possible. Contrary to traditional belief, it is better not to score or rough the surface to provide a key. The bond is chemical rather than mechanical. The only practical way of providing straight, smooth surfaces is to plane them. With hand planing it is difficult to achieve a straight, square surface, but if accomplished, it is better than a machined surface which leaves minute circular indentations. The pitch of these should be within the acceptable limits set out on page 107. Large, modern factories often thickness wood by fine drum sanding rather than planing which leaves a more even surface, but for most shops, gluing directly from the planer is the only possibility. Gluing sawn edges is not recommended, for no matter how sharp the blade, it is difficult not to leave slight indentations when ripping.

Moisture content is also an important consideration. At over 20% mc, the viscosity of the glue can be reduced, resulting in a starved joint. Under 5%, the absorption of solvents from PVA or hide glues may be too rapid which would result in a weak bond.

Generally, as manufacturer's instructions usually suggest, the surfaces should be free from dirt, dust and grease. It is always better to glue soon after machining, not only because the surfaces are cleaner, but also because they will have less time to move. And with greasy woods such as teak, the natural oils will not have had a chance to rise to the surface to interfere with the gluing action.

2 Choosing the right glue

The table opposite lists the properties of the various glues which are readily available. Wholesale suppliers often stock more sophisticated glues for specific applications, and for anyone with production uses, it is well worth seeking their advice. Check the telephone book, the large chemical companies or furniture factories for the names of local suppliers.

Generally a glue is chosen for its strength, its resistance to moisture and temperature, and to a lesser degree, for its ease of application and price. The two most popular glues are worthwhile pointing out. PVA glue is the most popular standard glue and is widely used for all sorts of woodworking applications. Its advantages are its ready-to-use containers, its low cost, relatively high strength, long shelf life, and most of all, its quick setting or

Glossary of terms

Shelf life	amount of time after package is opened that the glue remains usable. Can usually be extended by storing in cool, dry conditions in tight container.
Pot life	amount of time mixed glue remains usable (also called working life). Varies with temperature.
Clamping time	amount of time required to hold work under clamping pressure until glue develops sufficient strength.
Hardener	(catalyst or accelerator) supplied separately to be mixed with resin in form of powder or liquid for separate application glues, or in powder form mixed together with powdered resins for mixed application glues.
Gap-filling	glues such as hide glues or preferably urea formaldehyde glues are capable of supporting a thicker glue line to fill gaps in joints.
Fillers	materials added to resin glues to extend glue and give them better gap filling properties.
Creep	continued displacement under constant load, undesirable property of PVA glues.
Absorption	and (*Evaporation*), action by which PVA, hide glues, etc., solidify by losing their solvent.
Starved joint	condition in glue joint when too much glue has been squeezed out from too high clamping pressure.
Assembly time	amount of time glue remains workable after it has been applied to wood.

clamping time. Its disadvantage is its tendency to creep under loading. And it is only moisture resistant and not moisture proof, limiting it to indoor use. The new modified yellow PVA glues (aliphatic resin glues) have better, though not complete, moisture resistance and somewhat improved creep resistance under continuous loading.

Urea formaldehyde is the other popular glue widely favored by craftsmen because of its high strength and water resistance, its gap-filling properties and, in the powder form, its simplicity in mixing.

3. Applying the glue

Before applying the glue, it is important to test the surfaces by dry clamping. For joints such as mortise and tenons, part assembly is usually enough to make sure the pieces fit properly, but for edge to edge joints, the boards should be laid out and dry clamped to check the tightness of the joints.

Mixing glue

If you use synthetic resin glue regularly, you can speed up the mixing by making a beater to fit your drill by bending a rod as shown. Put the correct proportions of glue powder and cold water in a plastic container. Use the drill at slow speed to mix the glue. Clean the beater after each use.

Keep a selection of different sized containers for mixing glues in the workshop. For resin glues, use plastic containers with fairly high sides. You can then let the remaining glue set and simply flex the sides of the container to remove the glue once it has hardened.

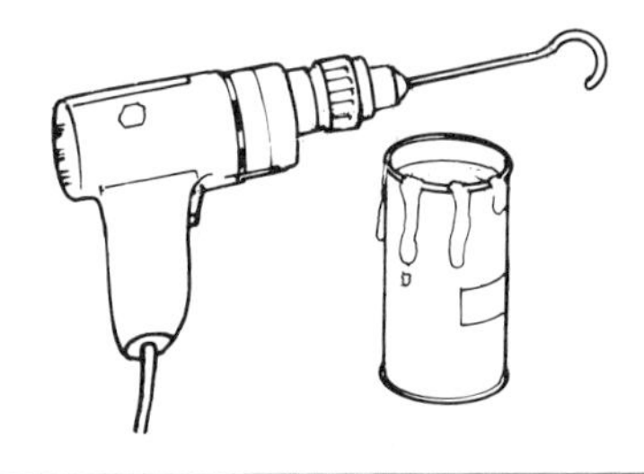

No amount of clamping will close up gaps in badly planed boards. Higher clamping pressures will only build in stresses which will be released in time to create splits at the points of weakness. Dry clamping also serves another purpose, for it guarantees that the boards are properly laid out and the clamps are ready so that no valuable time is wasted after the glue is mixed.

Spreading the glue can be done directly from plastic bottles in the case of PVA glue, but must be done with flat sticks, brushes or applicators with resin glues. The film should be thin and even. Generally more glue is applied than necessary, but it is better to err on the safe side until the right amount is reached by trial and error. The glue should just squeeze out of the joint with moderate pressure. For some resin glues, the resins and the catalyst can be spread on the two mating surfaces separately, so that the reaction doesn't begin until they are brought together.

Production shops use various sophisticated spreading devices to speed up the process and to control the amount of glue (the spread is calculated in pounds of glue per 1000sqft). For small shops the hand-held glue spreader is a useful alternative to get professional results.

Glue spreader

The glue spreader is a very convenient but expensive device for applying glue to even surfaces such as boards being glued up. If you do a lot of work of this kind it is a good investment. Pour the glue into the container. The rubber roller spreads it evenly over the surface. The only difficulty is that it must be cleaned immediately after use or the glue will be very difficult to remove. Take the spreader apart and soak all the pieces in warm, soapy water.

4 Assembling

Clamping is necessary to bring the surfaces into close contact and to keep them there undisturbed until the glue has gained enough strength to hold on its own.

Clamping pressure should be moderate and even, just enough to squeeze the glue out of the joints, but not so much that the wood is unduly compressed or the surface distorted. On large surfaces, the clamps are always placed alternatively above and below the boards, and the pressure is equalized. This prevents the surface from arching which would result in the joints opening on one side or the other. Always check at several points with a straightedge and ease off and tighten as required.

To spread the load, particularly on narrow boards, use heavy battens under the clamps, and remember that open end joints cannot be pulled together by more pressure. They will simply open up in time.

Clamping time depends on the glue (refer to the table). But it usually takes 24 to 48 hours for any glue to reach full strength, so it's wise to leave the work clamped up as long as possible (such as overnight) to allow the strength to develop before the assembly is disturbed.

When clamping large assemblies, it is critical that the clamps are ready to allow maximum time for adjustment. Notice the softening under the clamp heads and how the clamps (right) are alternated on either side of the panel to prevent bowing.

Glues and their properties

1 ANIMAL GLUES

These were formerly widely used but have now been replaced by resin and PVA glues.

Hide glue: available in sheets, pearls or cake form to be soaked in water, or in liquid form. All hide glues require heating in glue pot to about 50°F, and assembly should take place in a warm room. They have good strength but low moisture and heat resistance. Although they have general application, hide glues are more often used for veneering work today where adjustment is possible by reheating. Requires clamping for several hours.

Casein glue: made from milk derivatives with the addition of chemicals, is a strong general purpose glue with good strength and moderate resistance to moisture. It comes as a powder to be mixed with water and must be clamped for 6 to 8 hours.

2 PVA RESIN GLUES

These are very widely used as a general purpose woodworking glue. They come as a ready-to-use liquid in plastic bottles, convenient for direct application. Their high strength and short clamping time (½ to 1 hr) make them ideal for indoor use, but moderate water resistance makes them unsuitable for outdoor use. New modified PVA glues (yellow) have higher moisture and heat resistance and are less liable to creep than the standard white variety.

3 SYNTHETIC RESIN GLUES

There are several types of these very strong and very water resistant glues which are all excellent for both indoor and outdoor uses, some of which are easier to use than others.

Urea formaldehyde (UF) is the most important woodworking glue because of its high strength and durability in adverse conditions. It comes either as two separate components with the liquid or powder hardener to be mixed with the powder resin or with the powder resin and hardener already mixed to be activated by the addition of cold water. Shelf life, when stored in a cool dry place, is over a year, assembly time is about 20 minutes, but varies with temperature, down to 5 to 10 minutes on a hot day. By varying the formula, clamping times of urea formaldehydes can be brought down from the normal 4 to 6 hours to about 1 to 1½ hours, convenient for short runs. Transparent glue lines can be dyed for decorative effect using water-based poster paints, though this should be trial tested for strength.

Resorcinol formaldehyde (RF) is stronger and tougher than UF and comes in a dark liquid resin with powdered hardener which, when mixed, sets to a dark glue line, with clamping times of 6 to 10 hours. Excellent for outdoor uses.

Phenol formaldehyde (PF) also comes in two parts but requires heat for setting which limits its use to large shops, for making plywood and for making laminated structures.

4 MISCELLANEOUS GLUES

Epoxy adhesive is the strongest of all synthetic adhesives and the most expensive. It comes in two liquid parts, and is used to glue all types of materials such as glass, metal, etc. Used in the shop primarily to fix tools, to glue on glass or plastic handles, but there are indications that they may eventually be available as general purpose adhesives.

Cellulose glues are the common quick-setting glues used in model making, but they have applications in the shop, because their quick setting time requires only momentary hand pressure.

Contact cements are rubber-based, latex glues which, although they have only moderate strength and are liable to creep, are nonetheless widely used to lay down plastic laminates because no clamping is required. Glue is applied to the two surfaces and let stand until touch dry before being brought together for a rigid bond. Also convenient for spot gluing templates to workpieces.

CARBON FIBER REINFORCED WOOD

Carbon fiber technology has been around for 15 years during which time its uses and applications have expanded continuously, especially in the field of aeronautics, the field for which it was created.

The use of carbon fiber as a reinforcing material has also found its way into the production of sports equipment as a substitute for ebony as a strengthening element. It is also being used in modern engineering for helicopter blades and improved gear boxes, for example, and it will no doubt soon find its way into other industries such as the furniture industry.

Its technology owes much to that of fiber glass—today it is possible to produce carbon fibers six times stiffer than glass. At its best, carbon fiber can be two to three times stiffer than steel, but with only a quarter of the weight.

At the London College of Furniture, Sergio Mora, who feels that this technology has definite applications in the furniture industry, has been experimenting with carbon fiber reinforced wood and designing prototype chairs to demonstrate its advantages.

It is possible to obtain carbon fiber in several ways such as continuous filament tow in lengths up to 6000ft, as unidirectional woven cloth or bidirectional woven cloth. These cloths can be made as hybrids of carbon and glass fiber in different proportions of each, for example 60% carbon and 40% glass.

When used, the carbon fibers are imbedded in resins such as epoxy, polyesters and vyilesters, similar to glass fibers to give the following characteristics: increased stiffness over other materials, greater dimensional stability, higher heat distortion temperature, improved friction and wear characteristics, reduced thermal expansion and improved thermal and electrical conductivity.

When using carbon fiber as a reinforcing material in wood, it should be done in such a way that the minimum quantity of material is used in the right places so as to reinforce the structure where needed. The chair shown, for example, is a cantilever which means that the bending stresses in the bottom bend are very high. This is an ideal application for carbon fiber reinforcement since it allows the section to remain slim throughout the chair.

The frame is laminated out of $\frac{1}{32}$in beech veneer with a reinforcement of a 100% carbon unidirectional cloth. The center core veneers are laminated with urea formaldehyde and those which contain the carbon fiber with epoxy resin.

The chair frame consists of three sections, one comprising the back and seat, another the back and rear section of the leg and finally one which makes up the bottom part of the seat and the front part of the leg. The sections are bonded together using a solid beech insert where the three join together.

The laminating technique, called the fireman's hose technique, was used to glue the components together into the appropriate shape. Basically, two plywood pieces with holes for pairs of wedges surround the shaped jig. The 1 × $1\frac{1}{2}$in holes follow the shape of the jig and allow the wedges to be driven through. These apply pressure to the fireman's hose which when inflated will distribute the pressure throughout the jig.

The hose presses down onto a sandwich of veneers which include the carbon fiber cloth and a low voltage heating element made of tin plate. The lamination is then cured at a temperature of 176°F for about five hours, but this curing time can be accelerated with additives to the epoxy resin mix.

After the resin has cured and the lamination has cooled down, it is removed from the jig and trimmed with a tungsten carbide saw and the three sections are then bonded together.

The cross members are joined by an overlapped dovetail at the ends of the lamination for greater structural stability.

The seat of the chair is made of slats using the same jig as the seat section of the frame. The resultant chair is both elegant and strong.

Sergio Mora placing lengths of carbon fiber fabric on top of veneer layers.

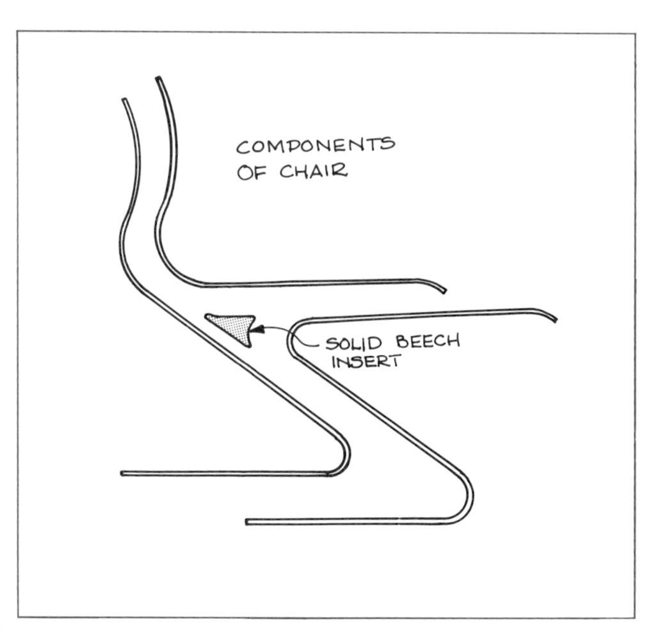

Gluing jig showing battens inserted into holes to restrain fireman's hose as it is pumped up to supply pressure to laminations below.

Left: Close-up of bend showing laminations.

Below:
The finished chair. Notice the lower cross member which is dovetail-halved into the frame.

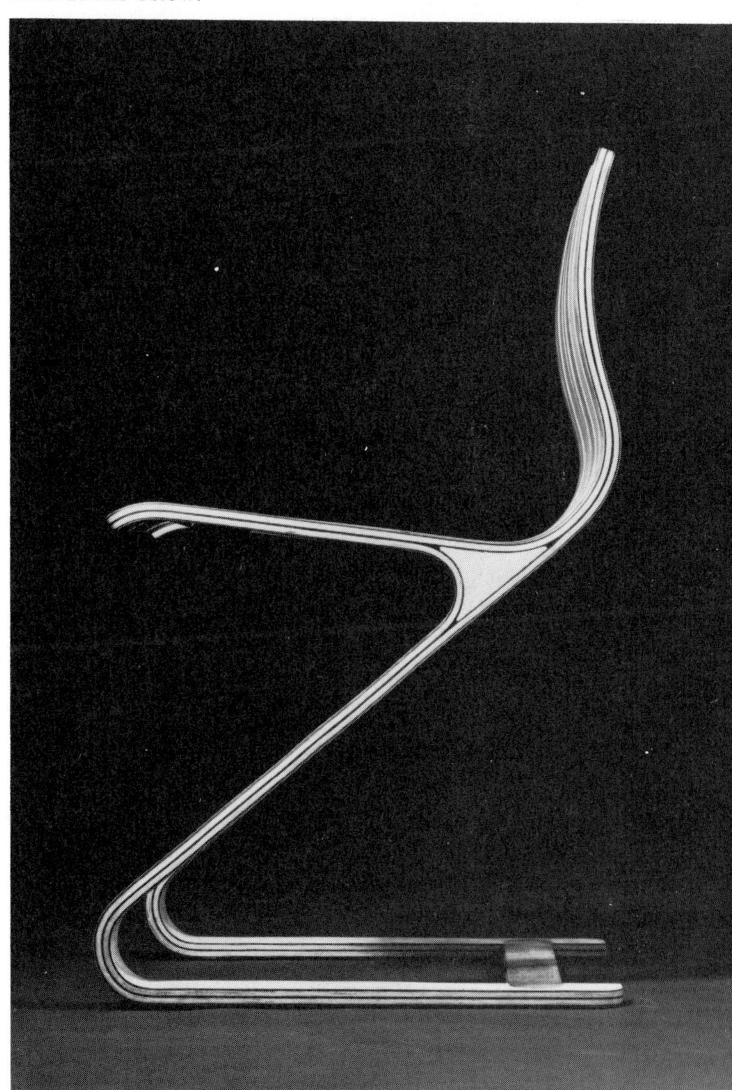

Above: Side view of chair demonstrates the advantage of carbon fiber reinforcement. It is strong enough to take high loads yet very slender and elegant.

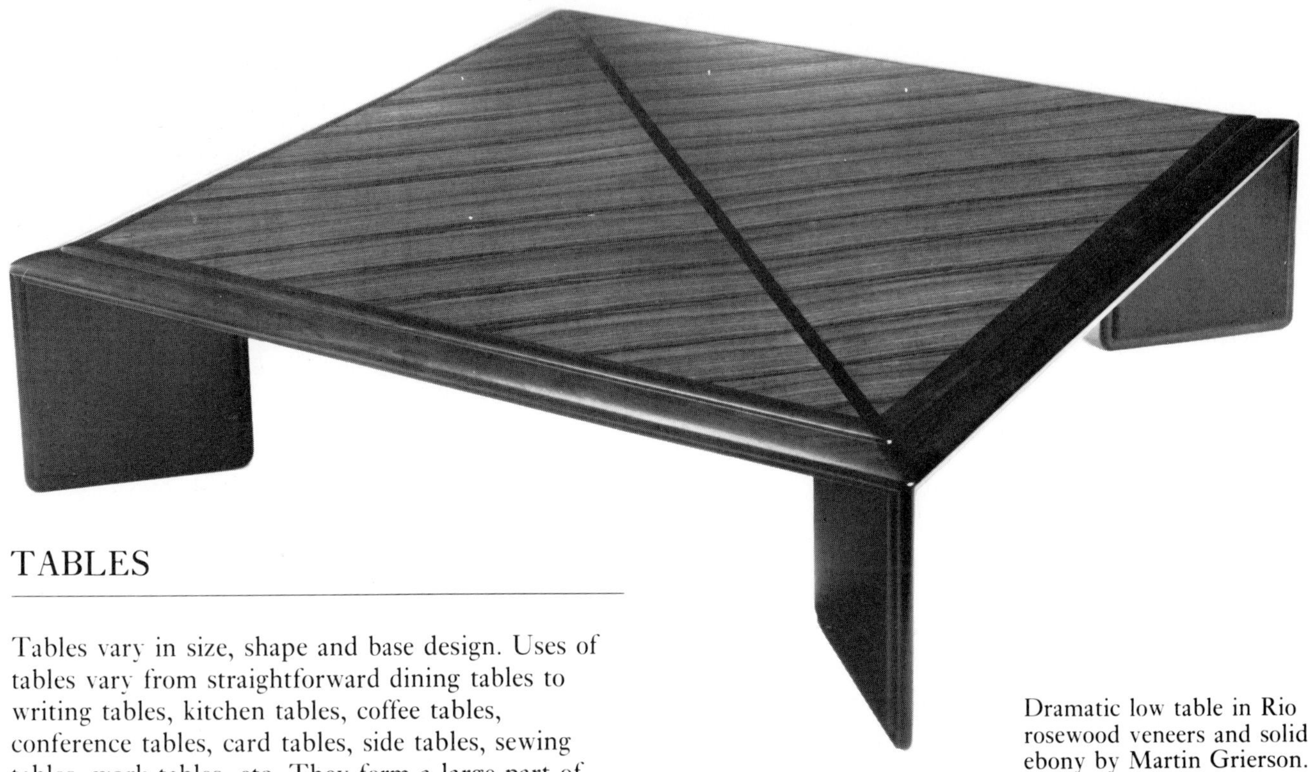

Dramatic low table in Rio rosewood veneers and solid ebony by Martin Grierson.

TABLES

Tables vary in size, shape and base design. Uses of tables vary from straightforward dining tables to writing tables, kitchen tables, coffee tables, conference tables, card tables, side tables, sewing tables, work tables, etc. They form a large part of most craftsmens' work and are relatively easy to make since no cabinetmaking drawer and door work is involved.

Shapes of tables may be rectangular, round, oval or some combination of these. Table height varies with use, but is a most important design criterion. Dining table heights are normally 29 to 30in, but this may have to be varied to suit particular people. Coffee tables are often too low. Continental tables, which are as high as 24in are more convenient.

Table tops Solid tops move with the seasonal changes in humidity, expanding as humidity goes up in summer and contracting as heating brings indoor humidity down in winter. This fact imposes certain constraints in designing table tops out of solid wood. The boards to be glued up should be of similar moisture content and the edges should be planed correctly so that no stresses will be built into the top. Any in-built stresses will find points of weakness such as bad joints or surface shakes and cause splits as they are relieved.

At the same time, any attempts to restrain seasonal movement will cause splits. A board, tongued and grooved across the ends, is an excellent but problematic detail. It is excellent because it stiffens the ends, resisting any tendency to warp, and it also covers the end grain, preventing it from drying out too fast. It is problematic because if glued, it will restrain movement which will result in splits. The solution is to glue the tongue for only a short length in the center allowing either side to move. A better solution is a dovetailed tongue, but this must not be glued at all since it has to slide on. Similarly, any batten (including rails) must be slot screwed and never glued across.

Similarly a table with a drop-in panel can cause problems. The solid wood panel can expand to push against the constraining sides, opening the joints. The amount of possible movement should be carefully considered (see chart, page 275) and allowed for in the details.

Because of the difficulty of controlling solid wood, the furniture industry today uses mainly veneered boards for table tops and other surfaces. The man-made boards will not move which makes them easier to store, to work, to assemble and to ship, and gives fewer problems in centrally heated homes. In practice this means that edgings can be glued on, rails can be screwed on, and veneers can be carefully selected and matched for the best effect. Although veneers are only a fraction of an inch in thickness and therefore more vulnerable to damage, they do have many advantages, and quality veneering is an art form with many design possibilities.

Left: Detail of rosewood/ebony table by Martin Grierson. Above: Table in solid yew by Alan Peters.

Table bases The challenge in table design is to make the base complement the top. If we think of the base as a practical sculpture, we need to consider the configuration, the lightness, the proportions, the material, and the decoration as well as the strength of the joints and the stiffness of the rails. There seem to be as many variations in bases as there are furniture designers.

Whatever the configuration, the base must fulfill certain practical considerations. Unless another method such as edging is used, the base rails should stiffen the top and hold it down to prevent warping. This is obviously more critical for solid wood tops than veneered ones, but even solid tops can be supported at points provided the top is thick enough.

The base should also be strong enough so that it will resist sideways forces in either direction and to do so without too much movement or vibration. Thus the joints which provide all the racking strength (page 209) must be carefully designed and well-made to stand up to the lifetime of wear. Of course, some tables get more abuse than others. Dining tables, for example, experience heavier loadings than coffee or side tables.

In addition, the base, particularly on dining tables, should interfere with use as little as possible. Rails which are too deep limit leg movement. The distance from floor to bottom of rail should be a minimum of 24in, but preferably more. Keep in mind that table components such as legs and low rails can get badly scuffed from shoes. It is the designer's responsibility to anticipate and design around these problems.

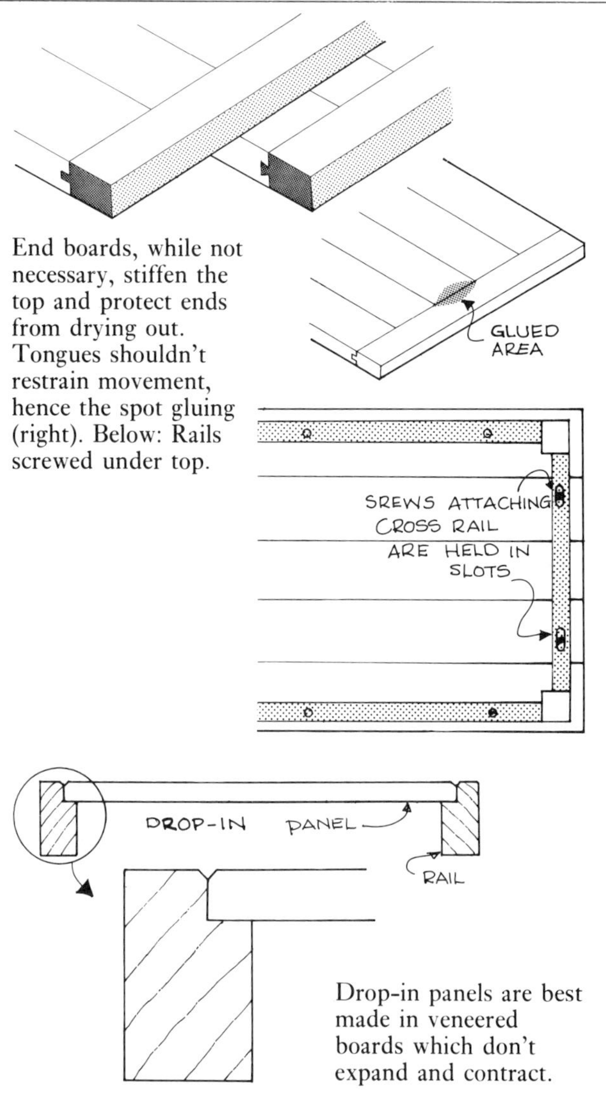

End boards, while not necessary, stiffen the top and protect ends from drying out. Tongues shouldn't restrain movement, hence the spot gluing (right). Below: Rails screwed under top.

Drop-in panels are best made in veneered boards which don't expand and contract.

Four leg base construction The standard four leg base has the legs at the corners and the rails at the perimeter of the top. For convenience, the top can overhang the base to make any lack of fit less apparent. Where the rails are flush with the top edge, it's best to shape the rail top by beveling or rounding to form a more pleasing detail and also hide discrepancies. If the rails are flush with the outside face of leg, their ends are often rounded or beveled to conceal any lack of fit.

The corner joints are either mortise and tenons, dowels, bridles, brace plates in wood or metal, or a combination of these. Tenons meeting at the corners must be mitered or half lapped, and should be shouldered to make assembly easier. The joint can be square or sculptured, as shown on page 181. Dowel joints must be very well bored with at least two, but preferably three, tight-fitting $\frac{1}{2}$in diameter dowel pegs per joint, staggered at corners.

Or the legs can be inset from the top to give a lighter appearance to the table. For dining tables they should be set far enough in from the ends to allow for a chair and leg room at either end, but not so far in from the sides as to topple over if someone sits on the edge. Similarly, the two side rails can be replaced by one center rail which must be strong enough to resist racking.

In this case the rails can be located either at the top or nearer mid-height of the legs. End rails should be slightly heavier to enable the racking load from the long center rail to be transferred to the legs. This configuration is useful for it is easy to make it knock-down, connecting the center rail with pegged tenons or bolt and cross dowel fittings (page 200). But since the top sits on top of the four legs, no stiffening is provided unless an additional crossrail is provided at the top.

Bridle joints at corners are not common but they do afford the opportunity for a decorative joint, particularly for glass topped tables. The rails from both directions are bridled, requiring a leg with a fairly large section. Where the bridles overlap inside the joint they are halved, providing decoration and stiffness at the same time.

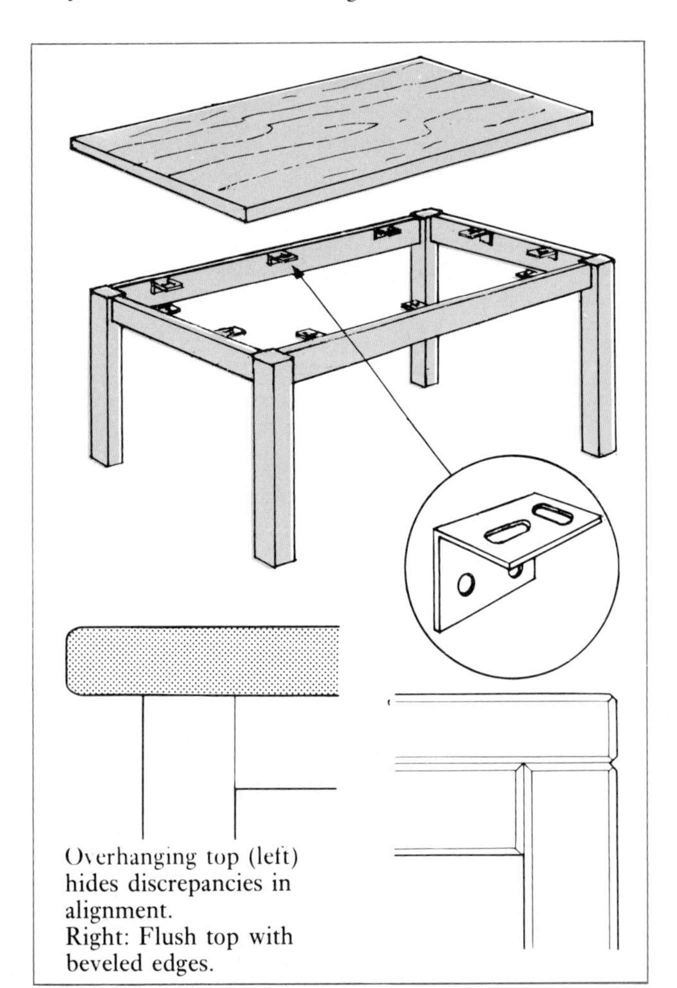

Overhanging top (left) hides discrepancies in alignment.
Right: Flush top with beveled edges.

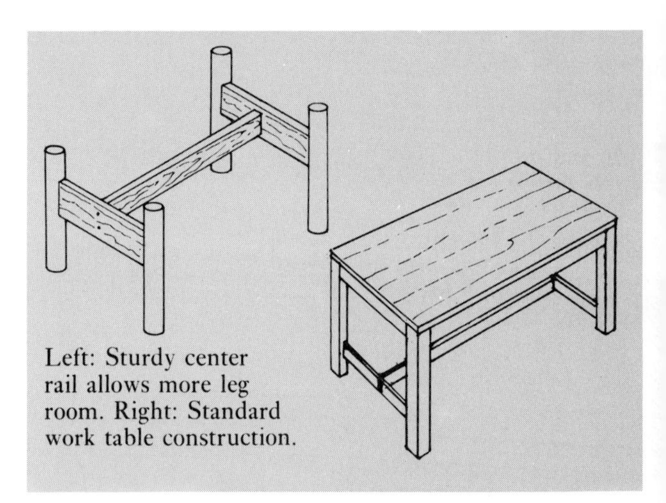

Left: Sturdy center rail allows more leg room. Right: Standard work table construction.

Below: Simple corner joint. The two rails are halved together and slot onto leg.

Corner plates

Corner plates can be wood or metal, but the metal ones are vastly superior and easier to fit. They make a superb corner joint which enables the legs to be removed for easy transport, yet are easy enough for the customer to refit.

To fit the metal version, which is available in several heavy or light duty sections, first drill a hole for the stud at a 45° angle inside the leg. Use a jig to hold the leg under the drill press as shown on page 135. The studs have a wooden thread which goes into the leg, at one end, and a metal thread which takes a wing nut, on the other end. Insert the stud into the leg by locking two nuts against each other on the metal thread. The rail ends can be stub tenoned, but this means the legs can't be knock-down. It is, however, perfectly sound to leave the ends as butt joints. Cut a notch across each rail end to take the flange of the bracket. The bracket is held in the slot by two screws to each rail. The leg is attached by pushing the stud through the bracket hole and securing with a wing nut.

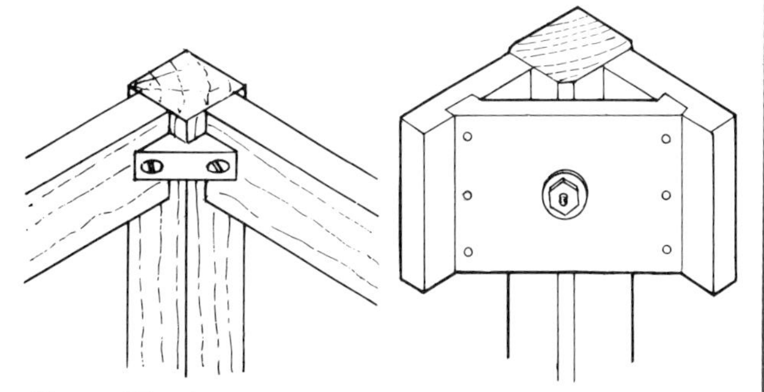

Above: Wood corner plates. Below left: Inserting stud with two nuts for metal bracket (below right).

Below: Low table with sculptured joints by Ashley Cartwright.

Left: Close-up of joint showing bridle leg joint.

Above: Variations on Caro's Red Splash. A low table in cherry and padouk by Dan Valenza.

Below: Classic side table by Martin Grierson.

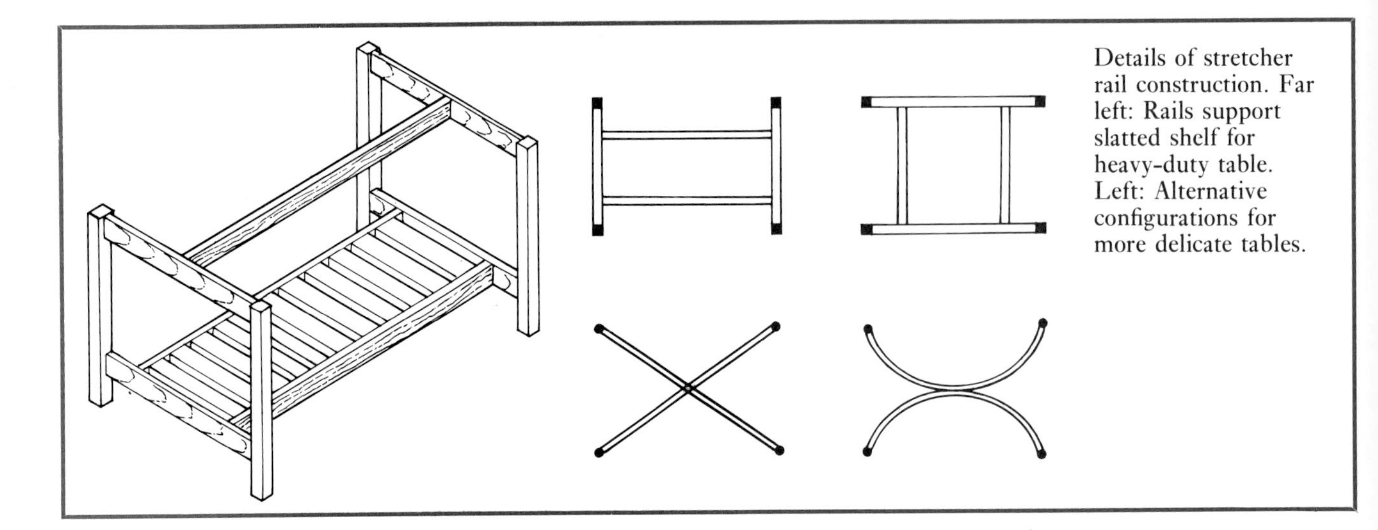

Details of stretcher rail construction. Far left: Rails support slatted shelf for heavy-duty table. Left: Alternative configurations for more delicate tables.

Stretcher rails Side tables and heavy tables such as workbenches, often require extra reinforcement with the use of stretcher rails. Their proportions depend on the design but they are usually lighter in section than the top rail. The normal configuration is to have the long rails join two side rails which connect the legs, but this may be reversed when the long rails support a bottom shelf. The shelf is either supported on an inside rabbet, if it is a solid shelf, or in a groove cut on the inside of the rail to match the grooved or rabbeted slatted shelf. For lightweight tables such as coffee tables or side stands, the bottom rails are sometimes placed diagonally or steam bent to make various decorative configurations.

Legs Legs can be rectangular, round, tapered, shaped, L-section, curved and decorated with flutes, recesses, ball feet, twists, inlays, moldings, etc. The modern inclination is for simple, undecorated, straight or tapered legs, but this will undoubtedly change with fashion.

Traditional shapes such as cabriole legs and other heavily decorated shapes will never go completely out of fashion, but they do require a great deal of skillful handwork and attention to detail, ideal for the craftsman but not feasible in modern production furniture.

Pedestal bases These have a central leg attached to two horizontal rails in an I-shape. They are popular in both traditional and modern designs. They are often designed to knock-down so that the pieces pack flat for easy shipping and storage. The end frames need to be well-jointed with good sized mortise and tenons (wedged if possible) to withstand the cantilevered load which is transferred from the side of the table to the central column. The top rail of the I keeps the top straight and the bottom should have small feet at either end, either cut from the solid or added on as hardware. The cross rail should, unlike some traditional examples, be arranged vertically, to provide stiffness against racking.

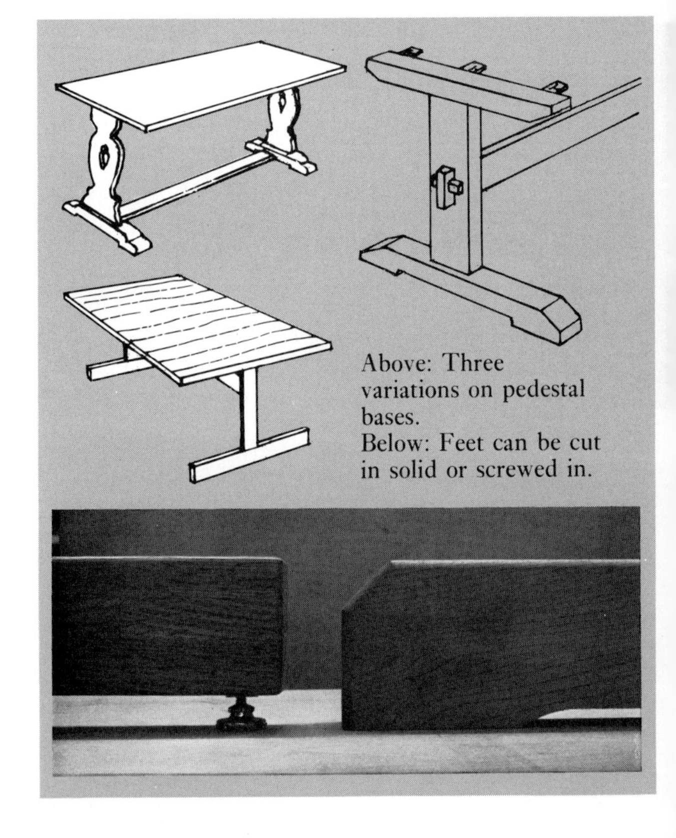

Above: Three variations on pedestal bases.
Below: Feet can be cut in solid or screwed in.

It can be connected to the ends with wedged tenons, or dowels, but a better solution is to make it detachable either with modern hardware (page 200) or with the traditional passing tenon.

Below: Side table in ebony and sycamore with sculptural base by Alan Peters.

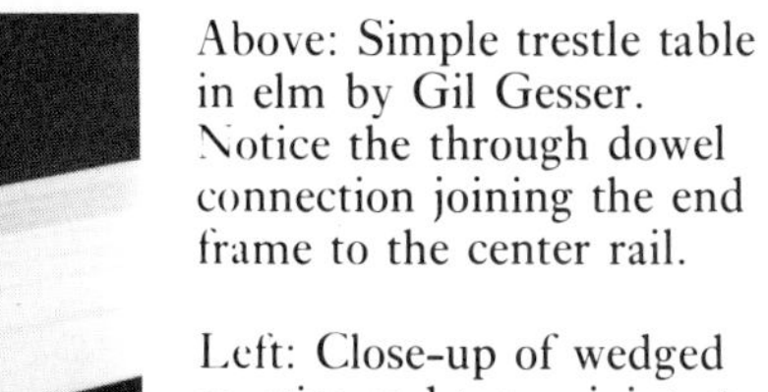

Above: Simple trestle table in elm by Gil Gesser. Notice the through dowel connection joining the end frame to the center rail.

Left: Close-up of wedged mortise and tenon joint at top end of frame.

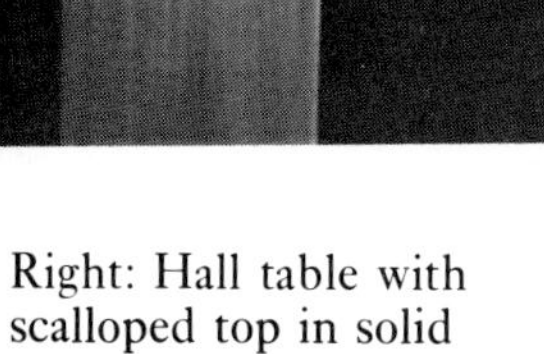

Right: Hall table with scalloped top in solid Indian rosewood by Alan Peters.

Dining table in solid and laminated Macassar ebony, designed by John Makepeace and made in his workshops by Andrew Whateley.

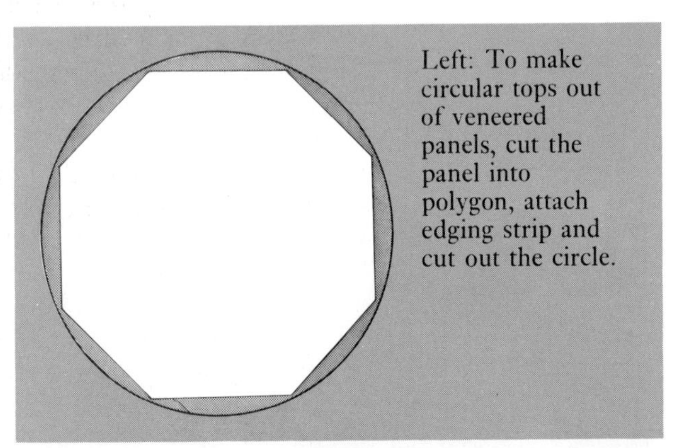

Left: To make circular tops out of veneered panels, cut the panel into polygon, attach edging strip and cut out the circle.

Detail of round table by Martin Grierson demonstrating this technique.

Round tables Round tops, whether solid or veneered, are best cut with a router, as shown on page 70. The difficulty with round veneered tops is obviously the edging, for it is very difficult to make solid edging round to fit. The usual method is to rely on veneered edge strip, but another possibility is to make the table top into a polygon, then to apply the straight edge strips with tongue and groove joints and to round the top afterwards.

Tops made from solid boards can be plain butt jointed, stop tongued or, for a more interesting effect, contrasting tongues can be exposed so that they become wider or narrower depending on their position.

Bases for round tables can be standard rail and leg construction with three, four or more legs, or they can be the pedestal type. This is a more demanding design problem because of the high loads imposed on the central column from the cantilevered top. This can cause the joints to separate with time, as evidenced by almost all antique pedestal tables which tend to rock. Leg and rail bases need at least three legs, and since tripods are very steady, this is a convenient arrangement.

Curved rails which follow the shape of the top can be cut from a thicker solid piece if the curve is slight or laminated from several layers of veneer or plywood faced with veneer if the curve is more severe.

Below: Dining table and chairs in English oak designed by John Makepeace for Liberty & Co. centenary.

Left: Detail of table top.

Traditional pedestal designs with three or four legs attached to a central column require strong joints and particular attention to grain direction when cutting out the legs. Although dowel joints are widely used (at least four ½in dowels are required), a better detail is the stopped dovetail tenon which better resists the tension in the bottom of the joint.

The column should be as heavy as possible to resist twisting and should be connected to sturdy top rails. The joints should be reinforced with steel plates or angles if necessary, to prevent or forestall the loosening which eventually occurs on most pedestal tables.

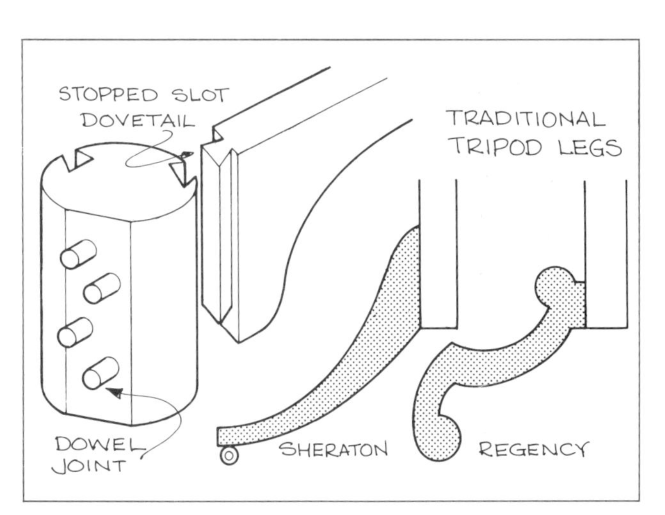

Two simple bases for circular tables, four leg base (above) and pedestal base (below).

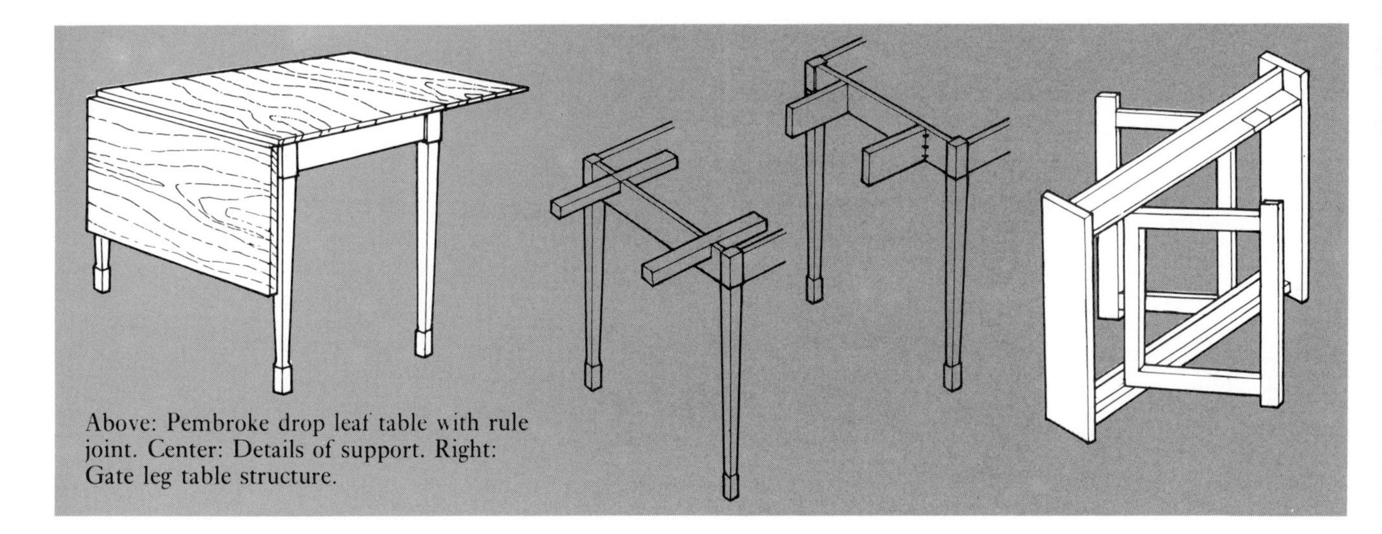

Above: Pembroke drop leaf table with rule joint. Center: Details of support. Right: Gate leg table structure.

Drop-leaf tables These are particularly popular because they take up very little space when both leaves are down. The leaves can be supported by wood or metal brackets, by sliding battens or, in the case of gate leg tables, by a frame hinged to the main frame to swing out for support. The joint between the main table and the leaves is often left as a square butt joint in modern work, but the traditional rule joint is a more elegant solution, since any pressure on the leaf is transferred to the main table.

Above: Modern gate leg table in beech with cork inset made by the author. Below: Table stored flat against wall.

Attaching table tops

In veneered work, the top can be screwed and glued as there is no movement in the top. The screws can go directly through the rails, counterbored, or be placed in pockets, drilled with a Forstner bit on a slant. Solid wood tops can't be rigidly attached. The traditional wooden brackets with a tongue which fits a groove in the rail are screwed to the top. Modern brackets are stronger. The angle bracket has slots in either direction, whereas the shrinkage plates, which are recessed into the rail come in two types and are used according to the orientation of the boards.

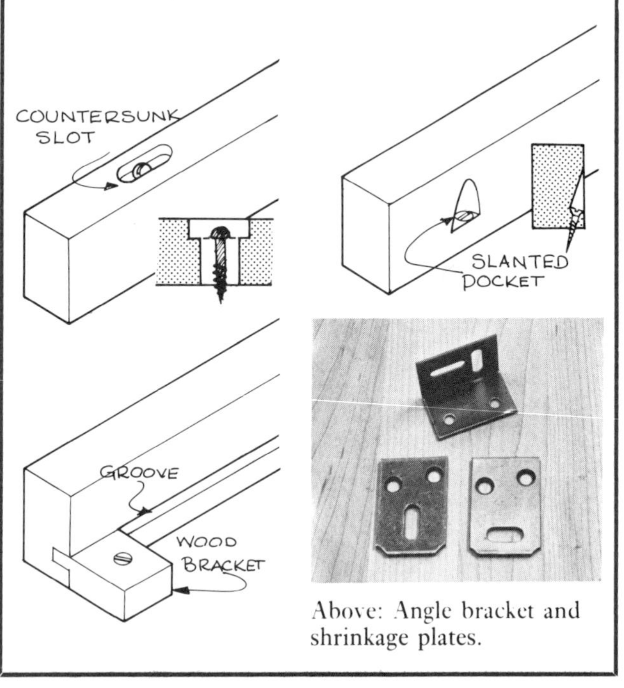

Above: Angle bracket and shrinkage plates.

Making a rule joint

The rule joint, used on fold-down tables, requires a special rule joint hinge. Set out the joint by first deciding the inset distance X (about $\frac{3}{16}$in for 1in nominal boards), then gauging this and the hinge center distance Y from each face. The distance between the two, Z locates the hinge center from the edge. Mark the radius on each end with a circle template or compass. Use a router fitted with a rounding over bit to cut one side, and a cove bit to the same radius to cut the matching edge. To work it by hand, cut rabbets then work with opposite molding planes. Recess the hinge with a deeper groove for the knuckle. Move it out a fraction from the theoretical center to give the joint clearance.

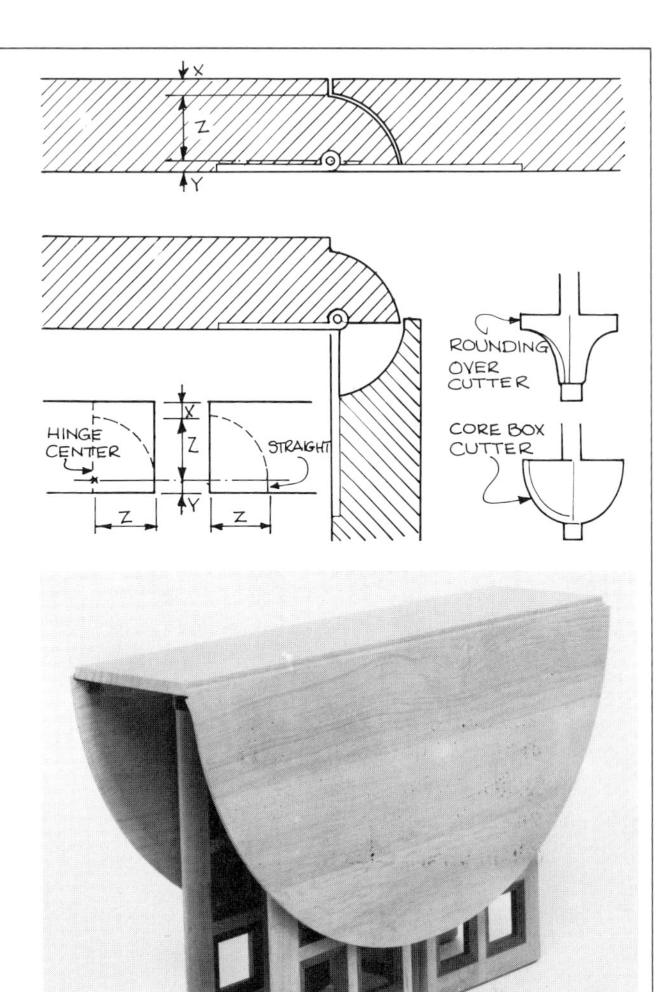

Drop leaf table with rule joint in oak and Indian laurel by Martin Grierson.

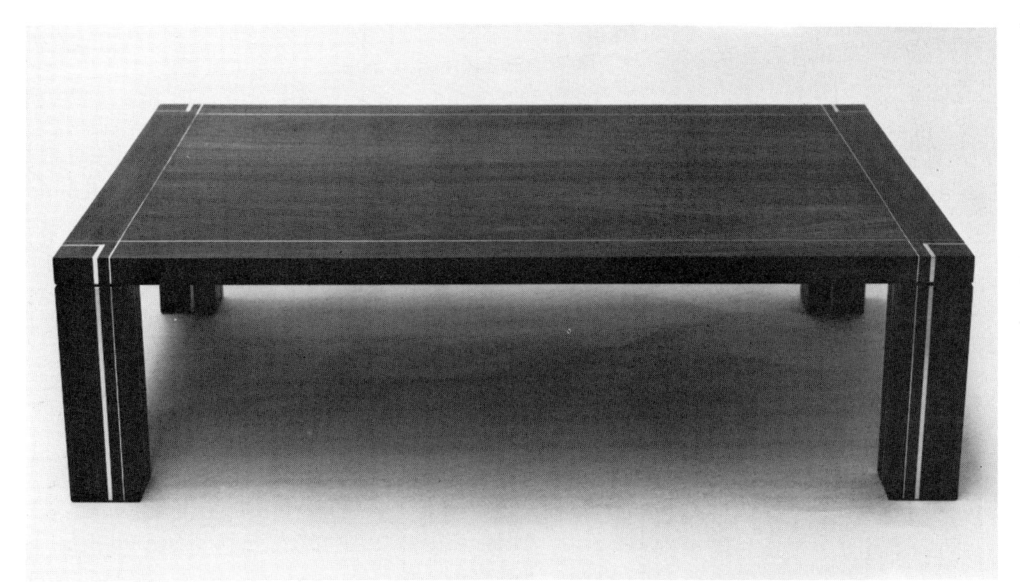

Left: Low table in Bombay rosewood and brass by David Field.
Above: Corner detail.

Above: Traditional chair with cabriole legs by Sam Bush.
Below: Detail of carving.

CHAIRS

Although there are exceptions to the rule, chair configurations are usually variations on the same basic four leg and rail structure. The basic kit includes two short front legs and two tall back legs with four rails connecting them, forming a rectangular support for the seat. The backrest is supported by the extension of the back legs. For armchairs, the front leg is taller to support the armrests which span from back to front leg.

These components are evident in most chair designs and it is a credit to designers' ingenuity that so many exceptions and variations have been found.

In most work, the legs are rectangular, round or a combination of the two. They are usually tapered towards the bottom for appearance. The back legs which extend to support the backrest, can be straight, in which case the rest is usually set back between them. It is more usually curved backwards to give the necessary angle for comfortable back support. Rails are usually shaped for appearance and for comfort since a straight front rail is not as comfortable as a curved one. Most components in mass-produced chairs are rough cut by template on the band saw, then shaped using a pattern on the spindle molder or overhead router. But in handwork, shaping identical components is more cumbersome, requiring skillful use of traditional tools like drawknives and spokeshaves, augmented by whatever machinery is available. Turned components are made on automatic lathes in industry but by hand turning in the small shop. These are relatively easier to make and also easier to join since they fit into easily bored, round holes.

The dimension and exact configuration of chairs is most important for comfort and proper support. A great deal of research has been done in recent years on the ergonomics of chairs, trying to find the right configurations. This research can be found in various books and journals, but failing that, the best approach for the craftsman is to keep an eye out for comfortable designs. I constantly try out chairs and carry a tape measure to note down the main dimensions of particularly good designs. Before producing a chair, even in small numbers, it's wise to build a full-sized prototype in pine so that final adjustments can be made by trial and error.

Chair components should be made from full-sized drawings, but complicated three-dimensional shapes are impossible to draw and must be worked from a prototype by eye.

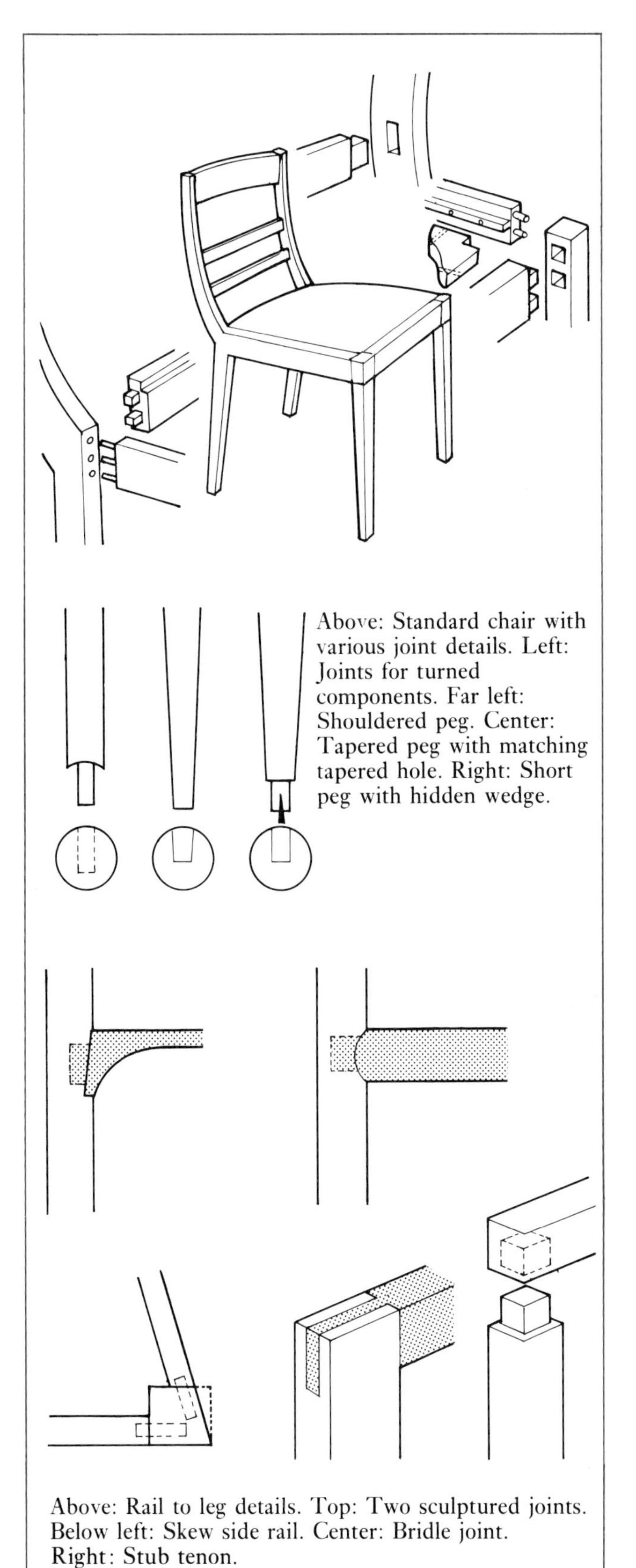

Above: Standard chair with various joint details. Left: Joints for turned components. Far left: Shouldered peg. Center: Tapered peg with matching tapered hole. Right: Short peg with hidden wedge.

Above: Rail to leg details. Top: Two sculptured joints. Below left: Skew side rail. Center: Bridle joint. Right: Stub tenon.

To minimize the weight and improve aesthetics of chairs, the sections are usually kept quite small, which makes joint design even more demanding. The mortise and tenon is the traditional corner joint and probably the best because of its reserve strength, but it has been almost universally replaced by the dowel joint in chair production, Tenons can be single or, space allowing, double stub tenons, which must be mitered or halved where they intersect in the legs. Most chairs are wider at the front which means the side rails must meet the legs at an angle. Since this is awkward to do with mortise and tenons, the side rails are frequently doweled. In most cases, the corners are reinforced with triangular-shaped blocks which are screwed and glued to the rails and notched around the leg, if necessary.

Dowel joints should have as many dowels as space allows without losing strength. Two $\frac{3}{8}$in diameter pegs is the minimum, but three is safer. The spacing can be staggered at the corners, or if space doesn't permit, the front frame can be assembled and the holes for the side dowels can be bored afterwards.

With round or shaped legs, the ends of the rails must be scribed to fit over the rounded shape. The alternative is to provide a short, flat section to provide a tight seating for square-cut rails. In many craftsman-made chairs, the joints are sculptured with either hidden or open tenons (bridles) to provide additional decoration. The easiest joint of all for turned rails is the bored hole used in many traditional designs such as Windsor chairs. The hole should be tapered (using a taper bit) with a matching taper end to the spindle, but failing that, an ordinary straight spindle should have a slight shoulder and be reinforced with a hidden wedge driven home during assembly.

Armrests can be mortised to take a peg tenon of

Detail of chair in solid and laminated sycamore designed by John Makepeace and made by Derek Christison.

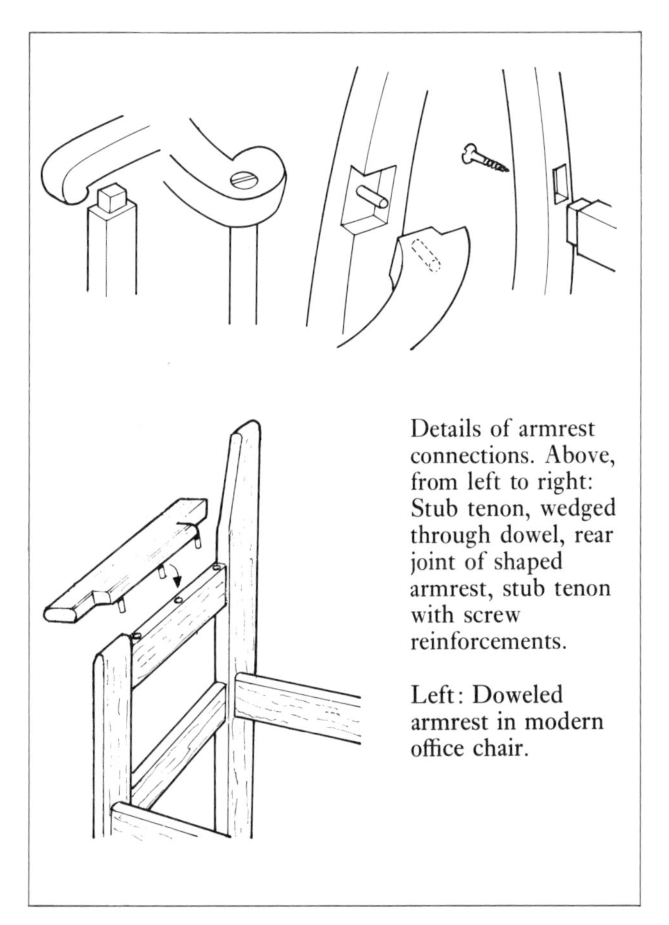

Details of armrest connections. Above, from left to right: Stub tenon, wedged through dowel, rear joint of shaped armrest, stub tenon with screw reinforcements.

Left: Doweled armrest in modern office chair.

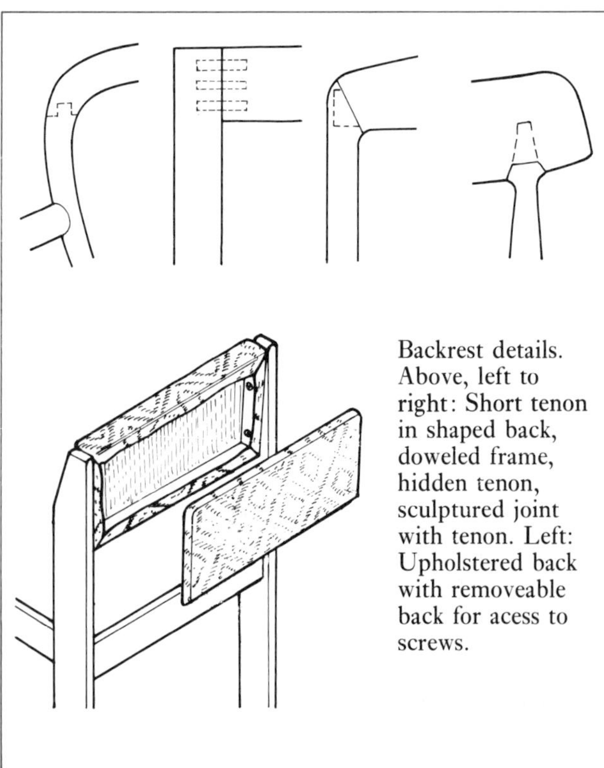

Backrest details. Above, left to right: Short tenon in shaped back, doweled frame, hidden tenon, sculptured joint with tenon. Left: Upholstered back with removeable back for acess to screws.

the leg, or the tenon can extend through and be wedged, to provide a decorative touch. The back of the arm can be halved into the leg, or dowel jointed, but in much modern work it is housed and screwed from the back, and the screw is hidden with a plug or cap.

The backrest, like the main frame, was traditionally mortise and tenoned, but is more commonly dowel jointed today. In modern work where the frame is usually rectangular, joinery is much easier than in traditional work where the backrest was often elaborately shaped, requiring ingenious joints which would be completely disguised in the finished chair.

The backs in modern chairs can be tenoned or doweled, but are usually screwed from the inside so that the fixings will be covered with upholstery. Alternatively, knock-down fittings and other decorative hardware, frequently used in production furniture, can be applied from the outside to make the fixing even easier allowing the back to be finished before assembly.

Seats Seat construction varies with the material. Upholstered seats can be designed to drop in onto rabbets or battens and screwed in place from underneath. The front edge may overlap the rail for comfort, and in the case of rush and twine seats, the resilient seat is actually wound around the four rails. Cane is normally applied to a separate drop-in frame in modern work. Cheap, modern canework, available in rolls, is pressed into a groove around the frame, but the high tension of a seat breaks the bond in a year or two.

Traditional canework, properly threaded and pegged, has a longer life, but because there are fewer craftsmen able or willing to repair cane, these are not as popular as they used to be.

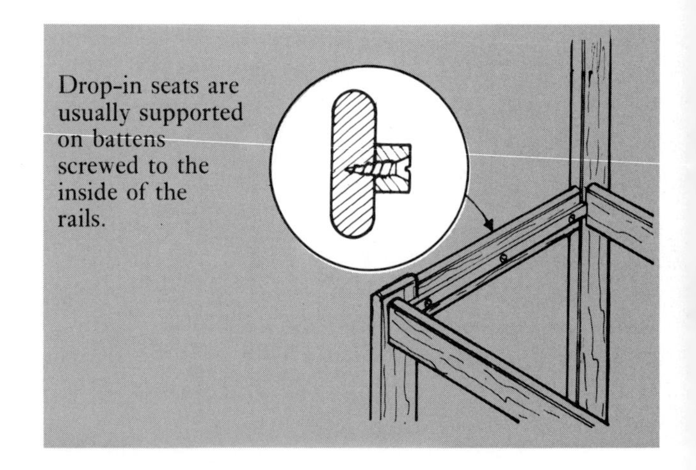

Drop-in seats are usually supported on battens screwed to the inside of the rails.

Chair in maple designed by Hans Wegner and made by Ejnar Peterson.

Chair in ebony with woven nickel silver seat and back, designed by John Makepeace and made in his workshops by Andrew Whateley.

Sycamore stool with decorative pegged through joint by Alan Peters.

Left: Chair designed by Sam Bush and made by Stuart Emmons.
Above: Detail of chair.

Left: Armchair designed by Hans Wegner and made by Poul Hansen. Above: Detail of Wegner chair.

Below left: Chapel chair in ash by Alan Peters. Below: Mass-produced "Twin" chair showing mechanical joint, designed by Tupu Kiveliö.

Above: Adjustable chair by Jim Warren.

Right: Chair designed by John Makepeace in English oak for Liberty & Co. centenary.

Below: Folding chair in fruiting cherry designed by John Makepeace and made in his workshops.

CABINET CONSTRUCTION

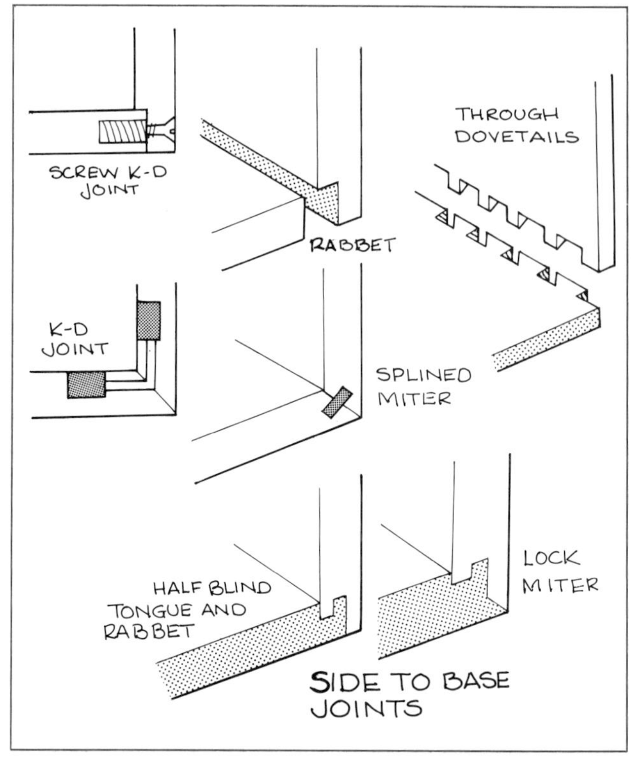

Most cabinets are boxes with one or more open sides and as such there is a lot of similarity between various types such as bookcases, chests of drawers and desks. Although aesthetic considerations and functional considerations such as drawer heights, shelf space and so on always determine the exact configurations, most designs nonetheless resemble each other because traditional jointing techniques have an important influence.

Although joints are often used as decoration as well, their basic function is to keep the cabinet square and rigid so that all close-fitting components such as doors and drawers will continue to fit (for discussion of structure see page 208).

When designing a cabinet, the components and joints should be chosen to be suitable to the function they will serve. Cabinet sides and backs must enclose and brace, rails must keep the sides from separating, and so on. Added to these considerations is the problem of movement. In tables and other relatively simple pieces it is quite easy to see how the components interrelate, but in a complicated cabinet the possible restraining effect of each member on another must be considered more carefully to avoid causing splits. The problem is simplified somewhat because only relatively wide glued up panels (12in and over) will split if restrained by a cross member. Total movement is proportional to width, so the expansion of relatively narrow pieces can be ignored. For further information on construction details to allow movement, see page 274.

Basic box construction

SIDES AND SHELVES In solid work, the panels are glued up from narrow boards, and stiffening to avoid out of plane movement (bowing or warping) is provided by adjoining members. In most cabinets, the sides and base are solid, joined by either dovetails, housings or dowel pegs. Dovetailing can be continuous across the entire width, but where only the corners are dovetailed, the joint must be reinforced by screws along the length. These can be rigidly attached without slots, because with the board direction the same, the movements of the two panels will be identical. When using dowel pegs, the base can butt into the side, but since this puts all the weight on the pegs, the base is usually set into a small rabbet in the side. Doweling must be accurate, and it is best to devise a drilling jig with a stop for the edges.

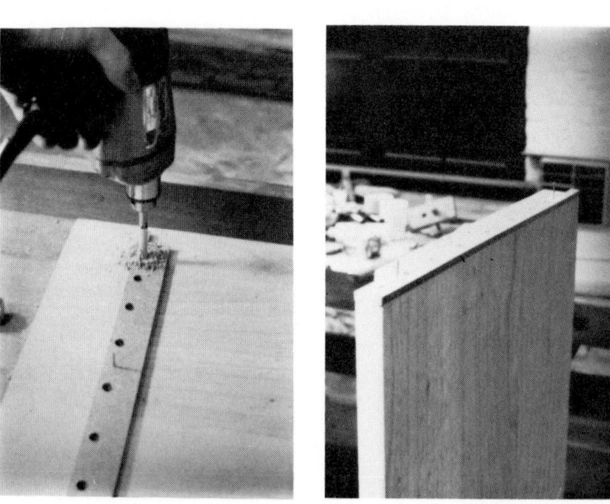

Using drilling jig to make dowel joint for side to top connection for cabinet by Gil Gesser.

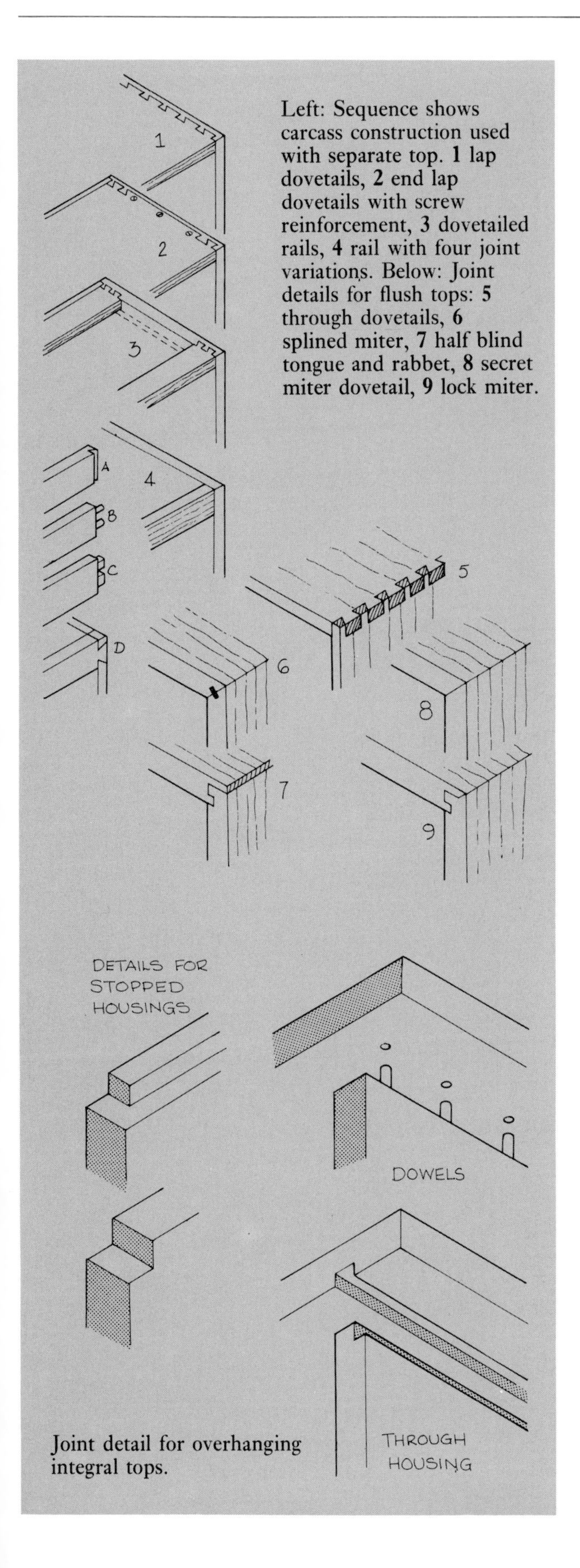

Left: Sequence shows carcass construction used with separate top. **1** lap dovetails, **2** end lap dovetails with screw reinforcement, **3** dovetailed rails, **4** rail with four joint variations. Below: Joint details for flush tops: **5** through dovetails, **6** splined miter, **7** half blind tongue and rabbet, **8** secret miter dovetail, **9** lock miter.

Joint detail for overhanging integral tops.

TOPS Carcass tops in solid work can be flush, as for most chests and cabinets, or they can overhang the sides as for most desks. Overhanging tops can be an integral structural part or can be laid onto rails which provide the strength.

The joints for flush integral tops in solid wood can be secret dovetails (though for wide tops these are usually too cumbersome), through dovetails, or finger joints. The half blind tongue and rabbet and the lock miter are useful substitutes because they are easier to machine cut. Dowel pegs should not be used. They are structurally suspect this close to the corner and since dowels don't provide a continuous fixing, gaps would be likely to develop between the pegs. Splined miters and similar reinforced joints are for man-made panels only since the tongue, glued across the grain, would restrict movement of solid wood.

Overhanging integral tops can't be dovetailed or finger jointed and are usually joined to the sides with a housing joint cut into the underside. A full housing is not as easy to assemble as a tongued housing where the shoulder on the side gives a precise dimension and an edge to clamp against. Dowels with close spacing can also be used because any small gaps would be hidden by the overhang. These joints for overhanging tops provide little sideways rigidity which is normally provided by the back panel. For open backed cabinets, it is better to provide separate rail framing and then attach the overhanging top to these allowing the frame to provide the stiffness.

Dovetailing is critical here because cabinet sides will always tend to come apart, and the only joint which is adequate in tension is the dovetail. (The cross-pegged tenon can't be pulled out either, but this joint is not suitable here.) The back rail is usually set in by the thickness of the back panel which can be screwed directly to it since the top will hide the edge. The overhanging top can be screwed rigidly to the front rail and slot screwed at the back rail.

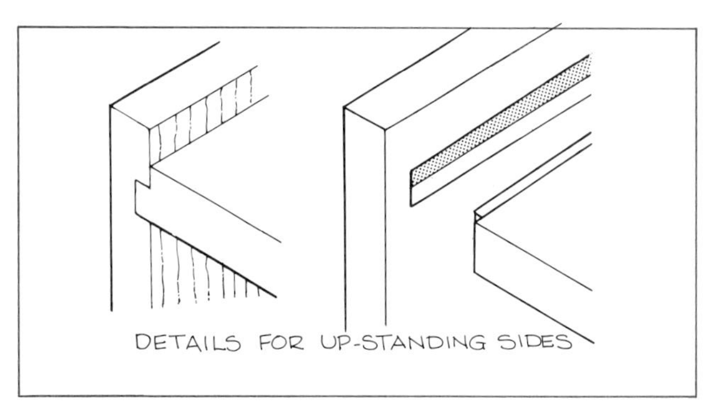

Making holes for slot screwing

Slots should be arranged across the grain to allow the round headed screw to slide sideways with movement in the wood. Slots can be cut by hand by cleaning out the waste between two drilled holes, but it is far easier done with a router using a homemade template. Make the template for the slot out of $\frac{1}{4}$in plywood, allowing for the necessary set back for the template, as shown on page 71.

Make two template slots, one for the head which is cut first, and one for the through slot for the screw shank. Use a $\frac{3}{16}$ or $\frac{1}{4}$in cutter and make as many cuts as necessary to cut through the wood.

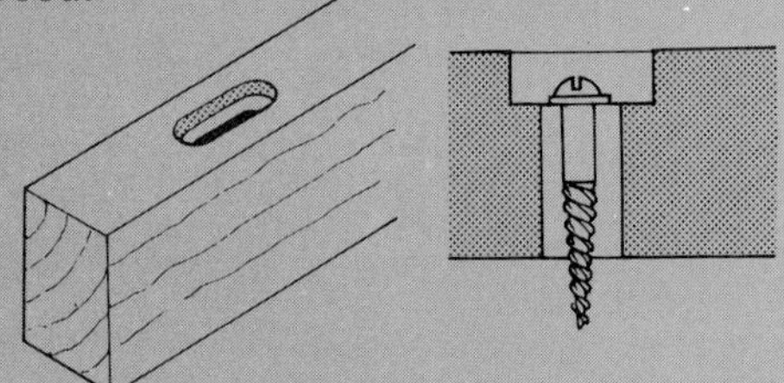

Below: Cabinet in rosewood with boxwood inlay by David Field. Left: Same cabinets separated.

Frame and panel carcass construction In box construction, such as for desks, chests and closets, the cladding and structural functions are often separated. A frame serves as the basic structural skeleton with infill panels in plywood or in solid wood to clad and stiffen the framework.

The end frame may be lightweight with relatively thin sections for cabinets, or it may be made up of heavier leg sections for desks, etc. The rails are either mortise and tenoned, doweled or, more usually, tongued into the leg groove which also holds the panel.

Intermediate rails are used, if necessary, to divide up the panels and to provide further stiffness. The panels can be set in grooves as for normal door work. Plywood panels don't expand and contract and can therefore be glued in place, but for solid panels, movement must be allowed for by gluing only one long edge and making sure any finish or stain is

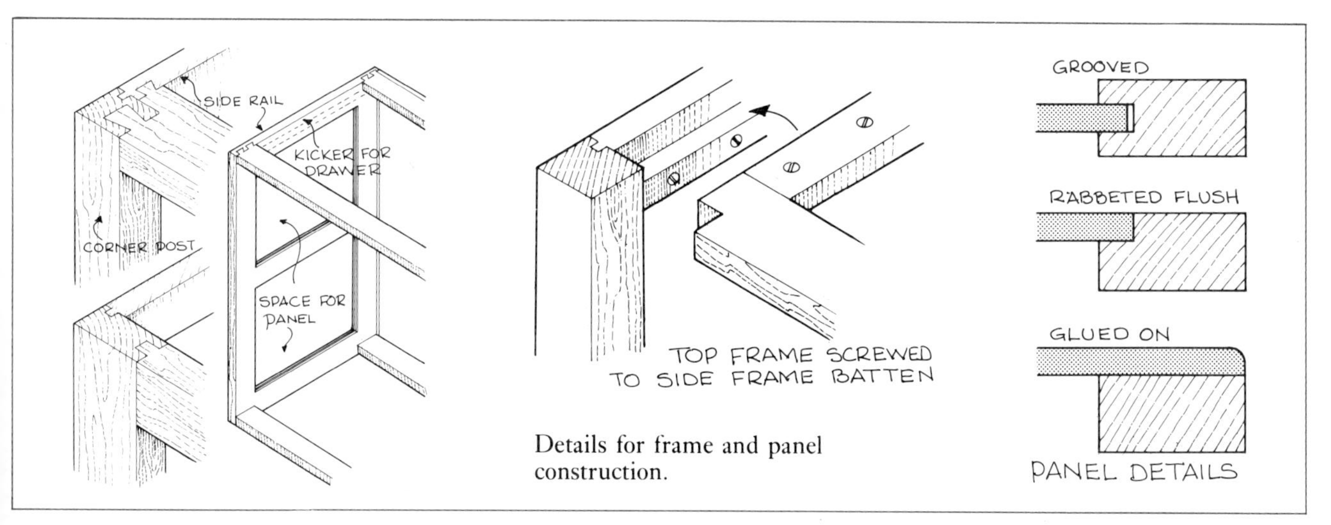

Details for frame and panel construction.

applied before assembly. Where a flush panel is required, it is better to use a plywood panel fitted to an outside rabbet. Alternatively, rabbet a plywood or solid panel to form a tongue to fit into the groove. With the solid panel, a slight bevel or groove around the perimeter will tend to mask any movement. Cross rails are the same as for solid carcasses, dovetailed to the top at the front and back.

The base can be made up of two similar cross rails dovetailed in place, with a plywood infill panel held in rabbets. But more often, the base is one panel either solid or of plywood, edged at the front and attached to the lower side rail with joints such as dowels or housings similar to the normal cabinet construction.

Horizontal dividers Shelves or drawer supports must be supported firmly and, if possible, designed to add to the overall stiffness of the cabinet. Particular attention should be paid when mixing the materials. For example, plywood or veneered shelves should never be rigidly fixed to a solid wood side to restrain movement.

Shelves are normally stop housed into the sides (4), either with a straight or tapered joint cut with a router (page 170), although dowel joints can also be fitted. For a housing, it is better to provide a shoulder to the shelf by cutting a rabbet to create a tongue on the end, for the shoulder provides a positive fit and a fixed dimension which doesn't rely on the depth of the groove for accuracy. The single or double dovetailed housing (2) is far superior to the straight sided slots, particularly where resistance against pulling apart is desired. Provided that the shelf boards are glued the long way as they should be, the movements of the side and the shelf are the

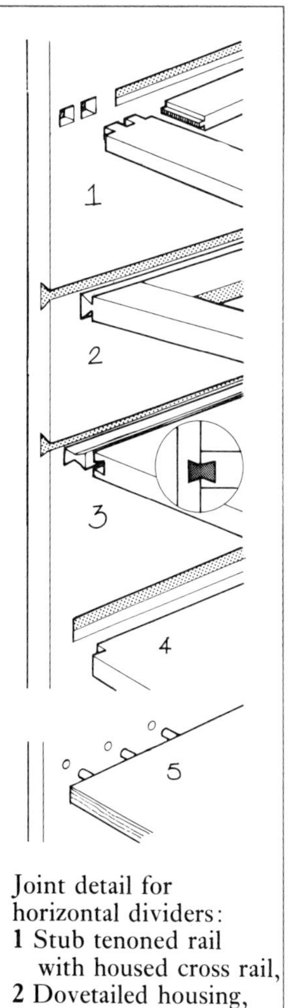

Joint detail for horizontal dividers:
1 Stub tenoned rail with housed cross rail,
2 Dovetailed housing,
3 Dovetailed butterfly joint,
4 Standard stopped housing,
5 Doweled panel.

Display bookcase in Macassar ebony by Martin Grierson.

Showcase construction

Left: Showcase in yew designed by John Makepeace and made by Alan Amey. Below: Detail of showcase joint.

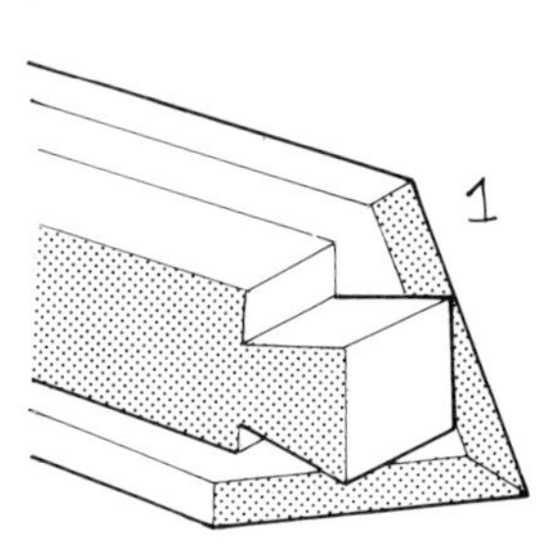

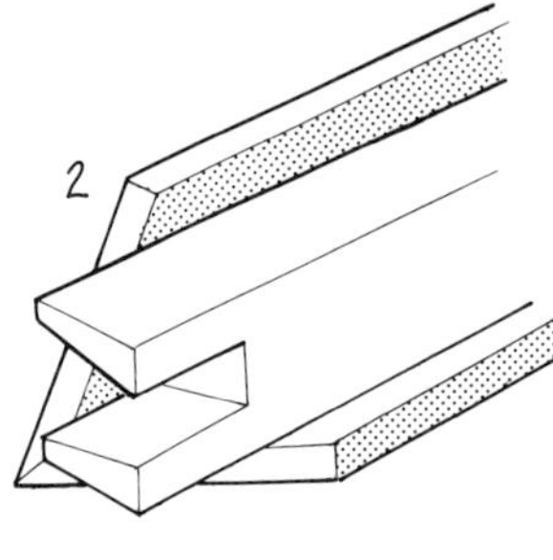

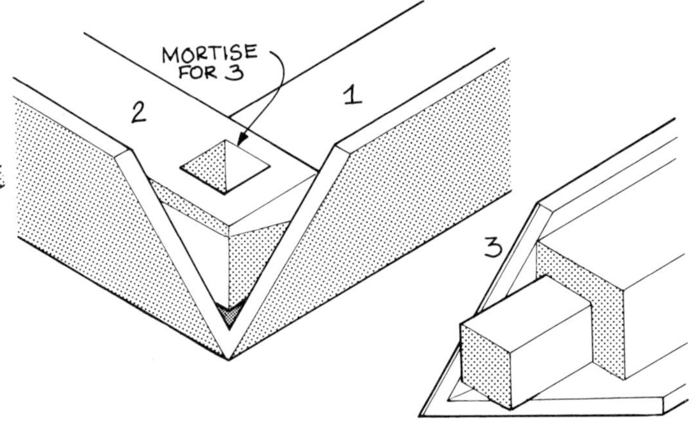

Above: Two sides ready to be dovetailed together. Right: Mortise is cut after assembly and the third side is glued in.

same. This allows the shelf to be glued into its housing and reinforced from underneath with a screw or two, if necessary.

Drawer supports are more difficult to join to the sides becuase the side runners can't be attached rigidly. It's best to stub tenon or dowel the front and back rails to the sides, and fit the side runners into housings cut in the sides (1). The side runners are further joined to the front and back rail by a tongue and groove joint which can be glued at the front only. A central runner, if required, can be rigidly joined to the rails by dowels, but is usually tongued like the side runners and glued front and back.

A superior fixing is provided by a double dovetail housing (3). Cut a matching dovetail tongue in the side runner and a dovetail groove in the end of the front and back rails before assembly.

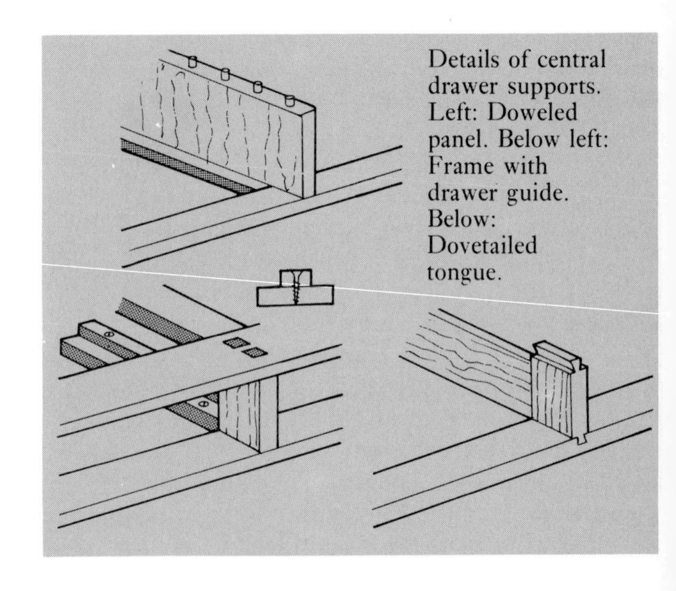

Details of central drawer supports. Left: Doweled panel. Below left: Frame with drawer guide. Below: Dovetailed tongue.

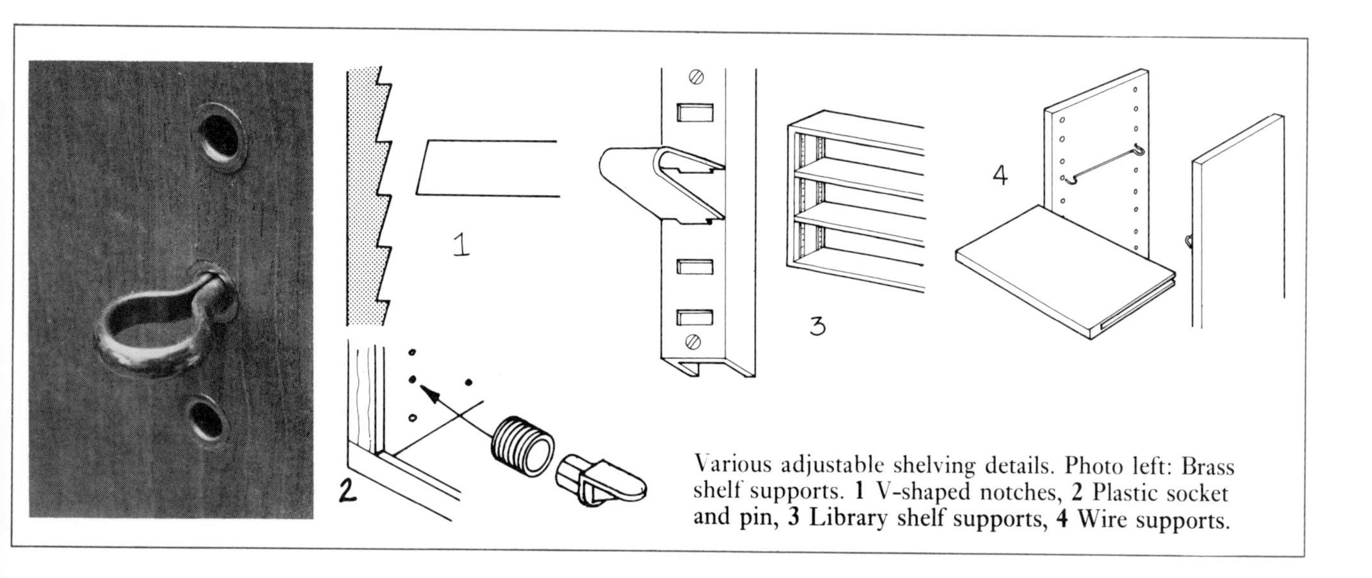

Various adjustable shelving details. Photo left: Brass shelf supports. **1** V-shaped notches, **2** Plastic socket and pin, **3** Library shelf supports, **4** Wire supports.

Adjustable shelves

In bookcases and other storage cabinets, it is often better to make the shelf spacing flexible by providing adjustable supports. A traditional adjustable fitting is provided by V-shaped notches (1) in the uprights with a matching shelf end. But this detail has been largely replaced by various hardware such as brass or plastic socket and pin fittings (2) in which sockets are inserted in regularly spaced holes with a pin used to support each shelf.

Adjustable library shelf supports (3) are easier to fit since they screw onto the surface or in grooves of the uprights. A more recent design involves wire supports which fit into pairs of small holes on the upright and in hidden grooves cut in the shelf ends (4).

There are various homemade alternatives as well, such as simple dowel pegs with one flattened side which are placed in matching holes drilled at regular intervals on the upright. Drilling holes at regular intervals requires the use of a drilling jig, as shown on page 55.

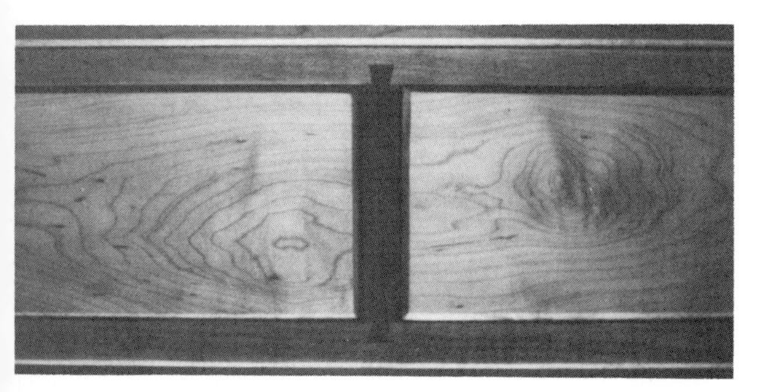

Above: Central drawer support upright connected with dovetailed housing made by the author.

Cabinet in Rio rosewood and sycamore by David Field.

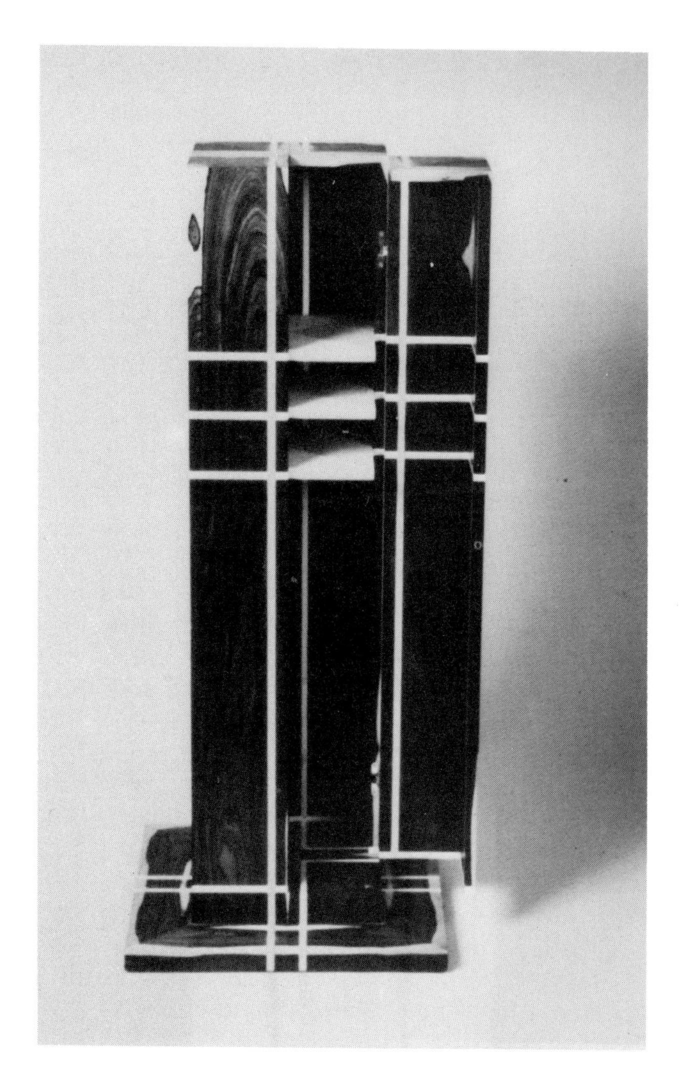

Cabinet with paneled back by Gil Gesser.

Cabinet backs Planning the framework should always allow for setting in the back panel where required. The back can be cut of tongue and groove boards, but $\frac{1}{4}$in or $\frac{3}{8}$in plywood panels, veneered to match, are superior in providing stiffness and considerably easier to make up and apply than one made up in the solid.

The back can be held in a groove and glued all around, or it can be screwed into a rabbet and covered with a fillet or simply left exposed as shown in the drawing above.

In good quality work, the back should be made out of a lightweight frame, say $\frac{1}{2}$ to $\frac{5}{8}$in thick boards, with $\frac{3}{16}$ or $\frac{1}{4}$in plywood infill, divided up into smaller panels if necessary.

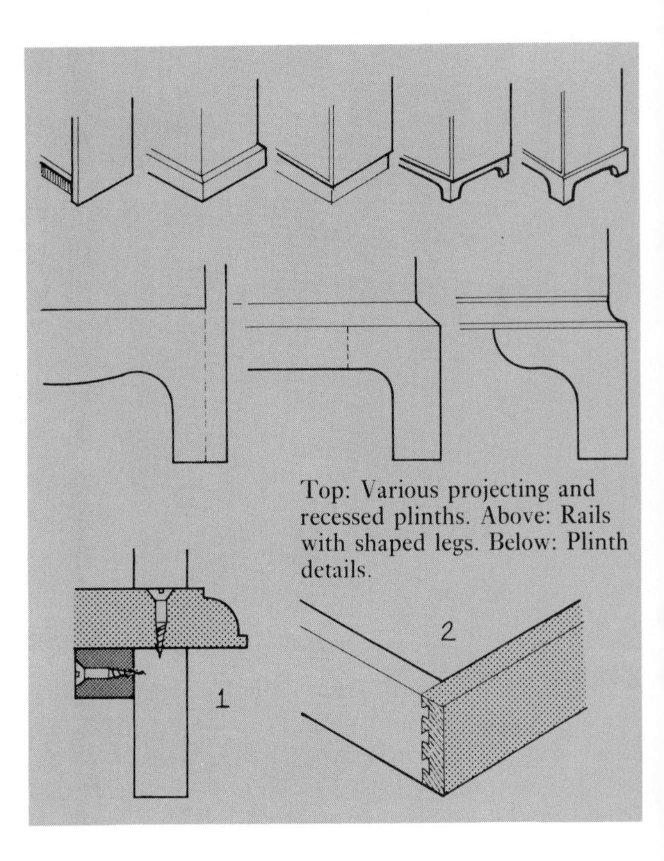

Top: Various projecting and recessed plinths. Above: Rails with shaped legs. Below: Plinth details.

Plinths There are several ways of holding a cabinet off the floor, such as using legs integrated into the frame and panel structure, or by allowing the sides to extend down for support, but the most elgant solution is to place it on a plinth, which can be projecting or recessed depending on the style of the cabinet.

The plinth is set out from the finished carcass to make sure of a good fit. The sides of the plinth can be solid and dovetailed (2) at the front and tongued and rabbeted at the back with glue blocks for extra stiffness. Alternatively they can be made up into rails and tenoned or doweled to glued up L-shaped corner legs which are cut to pleasing shapes, as shown in the drawings above.

The base is usually set off by a projecting molding, screwed and glued to the rails underneath and precisely mitered at the corners (1). The whole plinth is screwed, or for solid work, fixed with slotted brackets to the base, unless it is detachable in which case corner blocks underneath the cabinet base help to locate it.

In its simplest form, a recessed plinth is just a rail doweled or tongued to the extended side supports and screwed under the front edge. Alternatively, it can be a separate box structure, as above, set $\frac{1}{2}$ to $\frac{3}{4}$in in from the edges.

Above: Cabinet by Ashley Cartwright.
Below: Detail at corner.

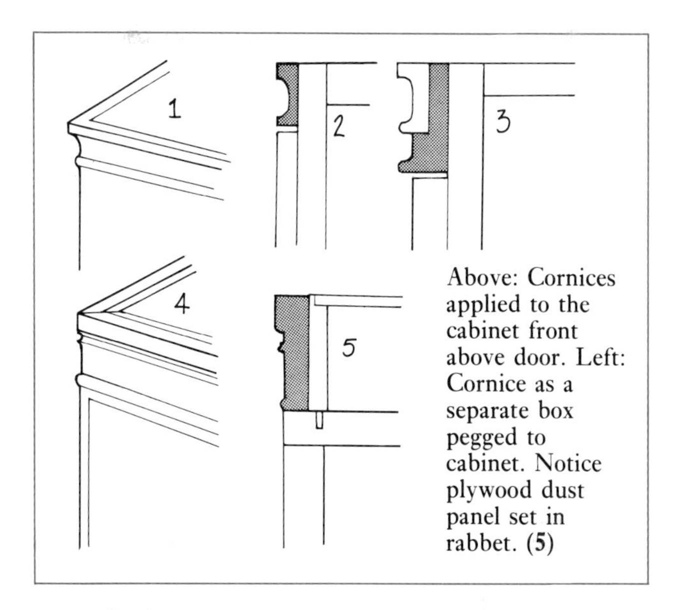

Above: Cornices applied to the cabinet front above door. Left: Cornice as a separate box pegged to cabinet. Notice plywood dust panel set in rabbet. **(5)**

Cornices It is more common today to leave the top plain, but cornices do provide a means of applying decoration to the top of a cabinet, particularly tall cupboards. On shorter cabinets where the top is exposed, the cornice can be applied to the front and side edges, flush with the top and can be mitered neatly at the corners (1, 2, 3). But in this case, the doors should be planned carefully to fit underneath and set out far enough so as not to leave a gap above.

On taller cupboards, the cornice can be a separate box simply laid on top around locating blocks, or alternatively, screwed to the carcass top (4, 5). The top can be open, but in better quality work it is closed off with a thin plywood panel screwed into rabbets, as shown.

Above: Corner cabinet in Indian laurel and pear by Martin Grierson. Right: Cabinet by Gil Gesser in elm with zebrano handles.

Above: Elegant cabinet with flush paneled doors by Gil Gesser.

PANELED DOORS

Cabinet doors can be hinged, sliding, folding, up and over, flop down, or tambour (see page 250). But the most popular doors are traditional paneled doors connected with double or triple brass butt hinges, depending on the size of the door. The panel, which can be made from glued up solid boards or from plywood veneered to match, is held either in a groove or rabbet. The rabbet can be cut in the solid or it can be created by pinning and gluing a molding strip along the front, as shown in the diagram.

Rabbeted panels can be added after frame assembly. All glass panels, for example are held in place with a bead pinned on (but not glued) from the back.

The door frame can be a simple rectangular surround or it can be divided up into smaller panels with horizontal rails and vertical muntins, for example as for cabinet backs. The traditional joint is the mortise and tenon which is usually stopped in cabinetwork, but through-wedged in joinery. The mortise and tenon does have more strength and better stiffness than the dowel joint, particularly if haunched. The dowel joint is widely used, however, because it is considerably easier to make. Keep in mind that a minimum of two dowels is required and for a proper fit, holes should be accurately drilled with a machine bit (page 134) to get a clean hole with maximum gluing strength.

Construction details for the panel frame construction.

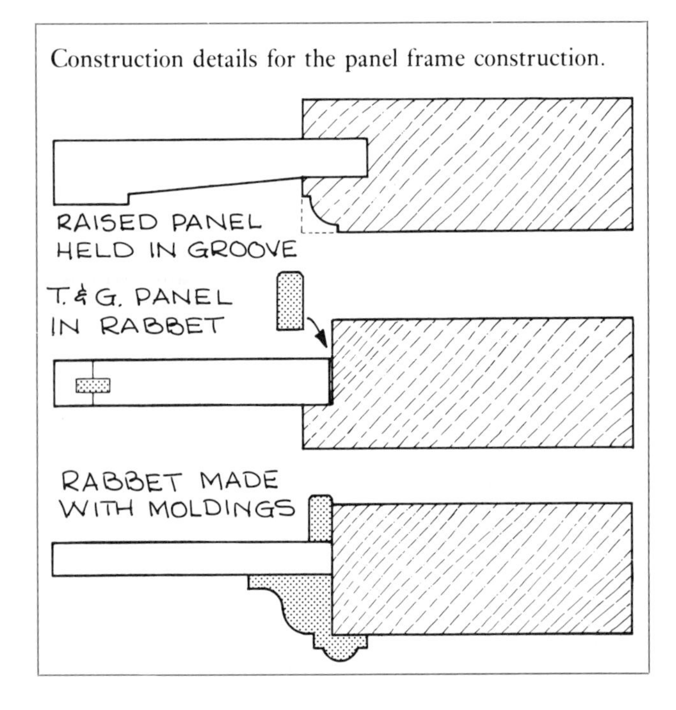

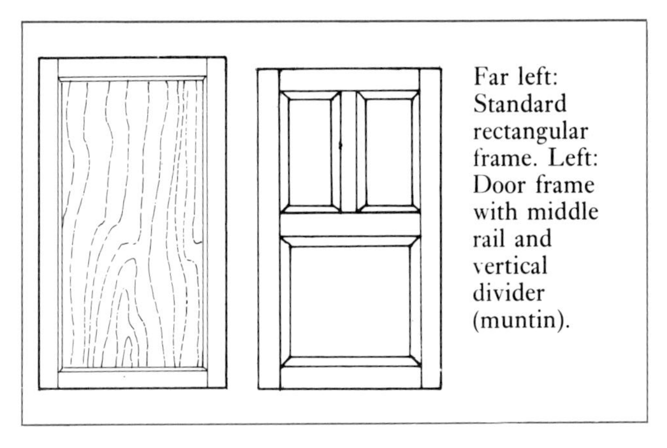

Far left: Standard rectangular frame. Left: Door frame with middle rail and vertical divider (muntin).

The tongue in raised panel doors must be stopped to prevent showing on the face.

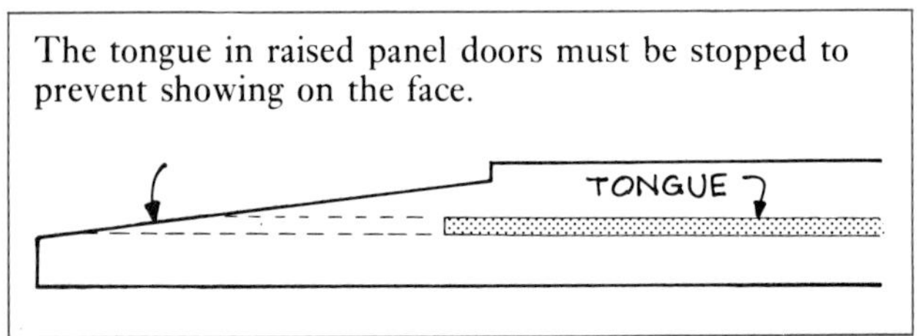

Cabinet with carved details by Sam Bush.

Door with raised panels for sideboard in maple made by the author.

Panels Veneered plywood panels are undoubtedly easier than solid and if the veneer is well-matched to the rest of the piece, they are not necessarily a compromise solution. These must of necessity be flat, however, and this is not as interesting as raised or fielded panels which can be made in solid. But veneered panels do not move, so they can be glued into position to help stiffen the door frame. Solid doors can be glued along one edge but must be free to move along the other three. They should therefore be finished before assembly so that there is no chance of an unfinished line showing along the edge as the panel shrinks.

The thin ($\frac{1}{2}$ to $\frac{3}{4}$in) glued up boards should be tongued (page 169), but the tongues should stop so that they will not show on the face when the edges are cut. Although the beveled edge on fielded panels is often cut with a saw blade, this requires laborious and inexact cleaning up. The best method is with French head cutters which leave a smooth, scraped surface.

Setting out For machine cutting the joints, it is best to work out exact dimensions of the components, and to make a cutting list with mortise depths, tenon lengths, groove depths, and so on noted. In planning rail lengths, allow $\frac{1}{8}$in at mortise bottoms for excess glue, and cut the stiles 2in overlength to allow 1in horns at either end for eventual trimming. For panels, allow about $\frac{1}{16}$in either way for tolerance, although large solid panels will require a little more to take the extra expansion into account.

Many workers prefer to make the overall size of the frame about $\frac{1}{16}$in oversize to be trimmed when the door is fitted. Once the cutting list is complete, the necessary quantities of each component can be machined accurately. But even though the measurements may be accurate, most machinists still make up a trial frame, particularly to test the fit of the tenons. Remember that mortises are always cut first, and the tenons made to fit.

Setting up for handwork is, of course, more

laborious, for each piece has to be marked since there are no repetitive cuts. In many ways this is an advantage. Each frame is marked from its opening, guaranteeing a proper fit.

The best way to mark the stiles and rails is to use the stick method, either using two separate sticks from which all the stiles and rails are marked respectively, or by holding each set of frame components up to each opening to mark the size correctly. This is the more accurate method.

With all the pieces thicknessed to accurate size, and marked face and edge, cut the stiles 2 to 3in longer than the height of the opening, but mark and cut the rails to exact width of the opening.

Set out the stiles first. Hold one up to the opening to mark the overall height, allowing an equal amount of waste each end for the horns (marks a). Offer up the rails to mark the rail widths (a–d) top and bottom, and then decide on the depth of the haunch and set off that distance (a–b) as well. Next mark the rabbet or groove depth on the edge (c–d).

Now clamp the two stiles together, good faces outwards, and square the marks across. Also gauge in the location of the mortise and the rabbet or groove. One side of the mortise should always coincide with the side of the rabbet, and for grooved frames, the mortise should fit exactly in the groove.

Separate the two pieces before marking along the back faces the depth of the groove or rabbet (equal to distance c–d). I always find it helpful at this stage to mark the rabbets or grooves on the ends, which makes it easier to see the joint developing.

Cut the mortises between lines b–c to the desired depth (see page 176). Check the depth with a depth gauge or a combination square.

Then cut the rabbets or grooves, either with a table saw (page 84), a spindle molder (page 131), or by the appropriate hand plane. At this stage the stiles for grooved frames are completed, but those for rabbeted frames must have the sockets for the haunches cut in. Mark them on the ends, then saw down either side before chiseling out the waste.

To set out the rails, first mark off the stile width, from either end, e–g. Then gauge the location of the rabbet or groove using the same setting as for the stiles. On rabbeted frames also gauge the rabbet depth on the back faces, and for clarity mark the location of the rabbet or groove on the ends. Since, on rabbeted frames, one shoulder must be set in to fill the rabbet set off that depth on the edge, f–g.

Holding the two rails together, square the lines across, then separate them to square the location of the short shoulders on the back faces, and the long

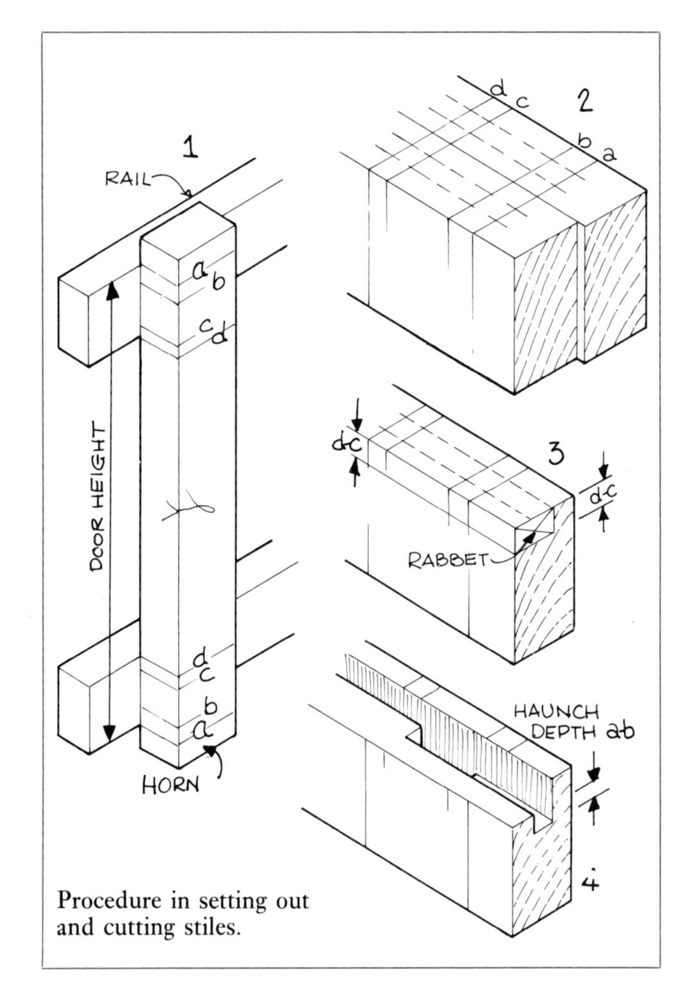

Procedure in setting out and cutting stiles.

shoulders on the front faces at each end.

Gauge the thickness of the tenons and cut them, as shown on page 177, using a knife and chisel to form a line for the saw along the shoulders. Follow by cutting the groove or rabbet, and then cut the tenon to length to fit $\frac{1}{8}$in short of the mortise bottom. Finally cut the haunch to size to fit its socket, and trial assemble the frame.

There are, of course, many possibilities for mistakes in setting out and cutting frames. The most common is ill-fitting shoulders for which there is little remedy but to cut back the shoulder and glue a thin sliver onto the face of the stile to make up the difference. Other faults, such as badly cut mortises or tenons, reveal themselves during assembly and have to be dealt with by trimming off with a chisel or by adding slivers of veneer depending whether the fault is over or under-cutting. Loose tenons can be fixed by filling in with veneer strips, or by locking them in place with cross dowels (see page 181). Either way, a good gap filling glue will help to some degree to strengthen any badly cut joint.

Molded frames

As shown on page 242, separate moldings can be pinned to the front of the frame with neat miter joints at the corners. In the best quality work, the molding is worked in the solid along the front edge before or after the frame is assembled. Working the molding after assembly is done by router, as shown on page 72. Cutting the molding on the separate pieces can be done by hand with a molding plane, or with a scratch-stock, or by machine with a spindle molder or a router. In this case the corner miters must be carefully cut using a miter template (page 179) and trimmed, if necessary, after trial assembly.

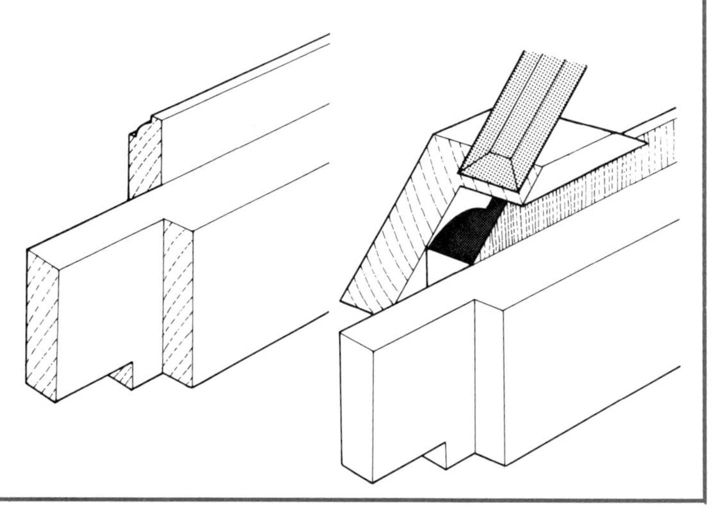

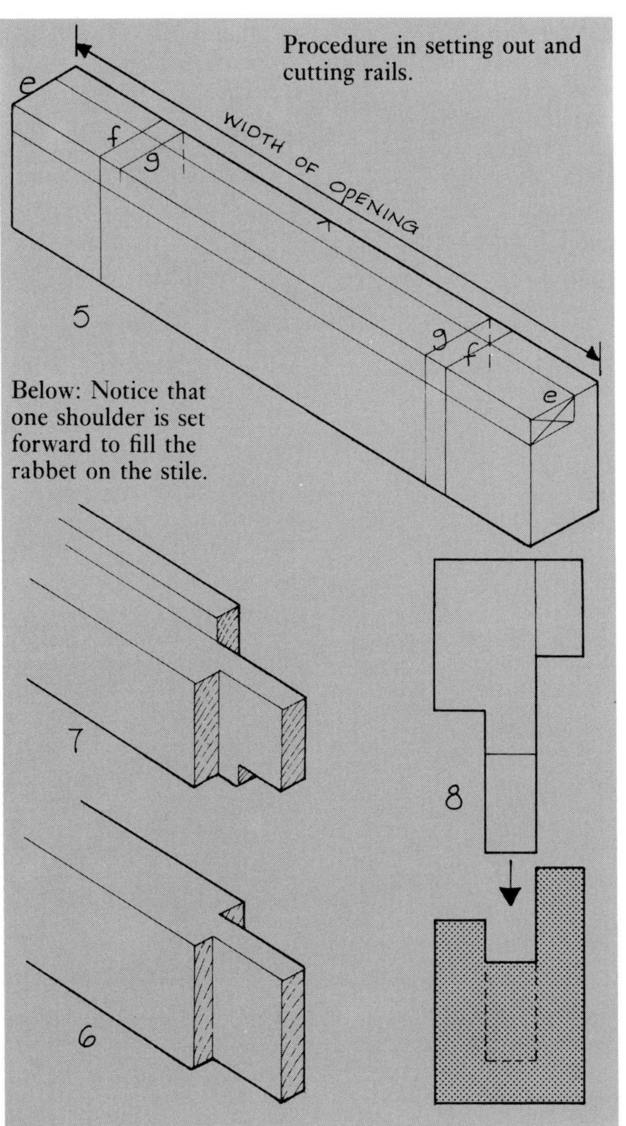

Procedure in setting out and cutting rails.

Below: Notice that one shoulder is set forward to fill the rabbet on the stile.

Making a scratch-stock

Scratch-stocks, used to cut molded shapes, are easily made by grinding the molded shape on the end of a piece of broken hacksaw blade or other suitable hard metal, then holding this tight between two pieces of wood cut to shape as shown.

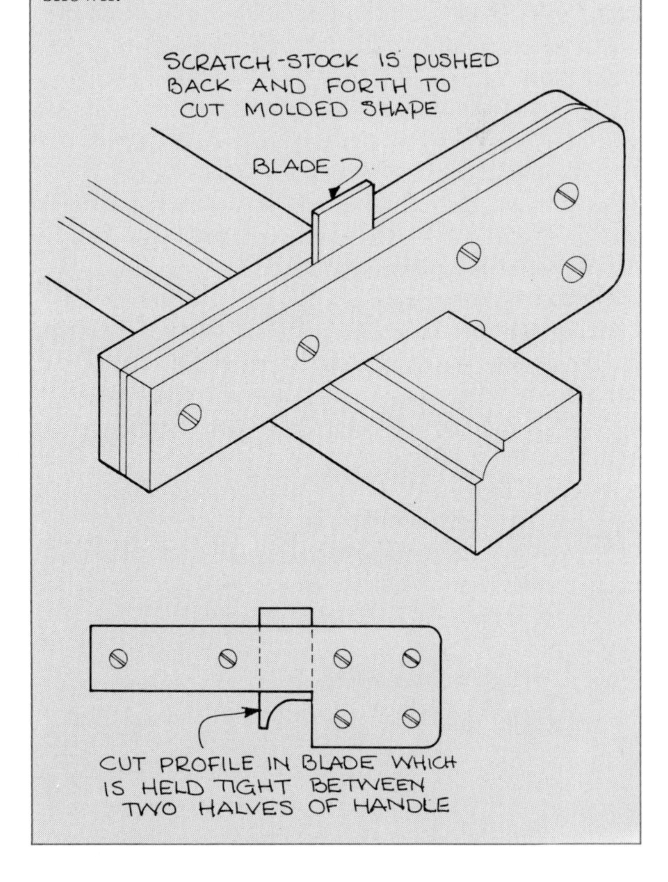

DRAWERS

Any furniture component subject to continuous handling requires joints and materials that will stand up to the loads. As tradition dictates, dovetails are best for drawers because they resist the tensions to which the joints are subjected. The normal construction is lap dovetails at the front (page 190), through dovetails at the back, with a plywood bottom set in grooves. There are, however, no set rules about the construction details, and many craftsmen prefer to make a decorative feature of delicately cut through dovetails, or use alternative joints such as finger joints, housings, tongue and rabbets or even screwed joints with the screws inserted into special plastic or fiber plugs in the end grain to increase holding power, if speed and economy are important.

Materials The drawer fronts are obviously made to match the rest of the cabinet and should therefore be chosen carefully, matching up the grain patterns by laying out the boards. The thickness of the fronts is usually $\frac{3}{4}$ to $\frac{7}{8}$in, enough to allow adequate room for the lap dovetails, but in more delicate work, the fronts can obviously be as thin as the proportion of the drawer demands, with through dovetails to hold them to the sides. The sides should be up to $\frac{1}{2}$in thick, out of any tough wood such as oak, though for side-hung drawers I prefer a close-grained wood such as maple because it runs smoother. Backs may be as thin as $\frac{1}{4}$in but it's usually easier to make them the same as the sides since all the boards can then be run through the thicknesser at the same setting.

Drawer bottoms were traditionally made of solid wood such as cedar, about $\frac{1}{4}$in thick, but $\frac{1}{8}$, $\frac{3}{16}$ or $\frac{1}{4}$in thick plywood panels are much better. Since they don't move, they can be glued into the grooves on all four sides, thereby stiffening and squaring the drawer.

Cabinet in ebony inlaid with stainless steel by Alan Peters.

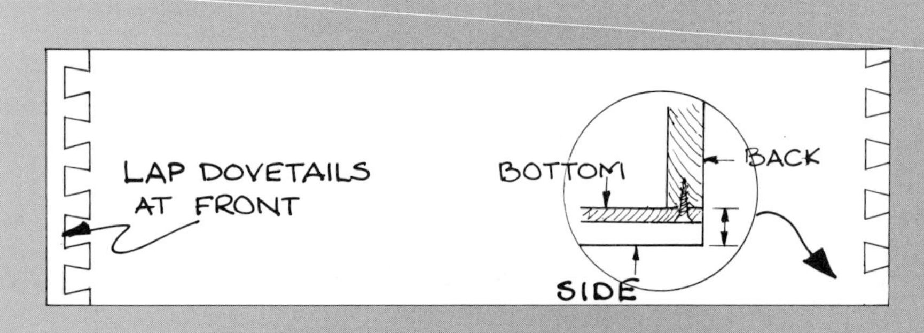

Standard drawer construction uses lap dovetails at the front and through dovetails at the back. See page 190 for instructions on hand and machine dovetailing. The back is usually cut short, as shown, to allow the bottom to be slid in after assembly.

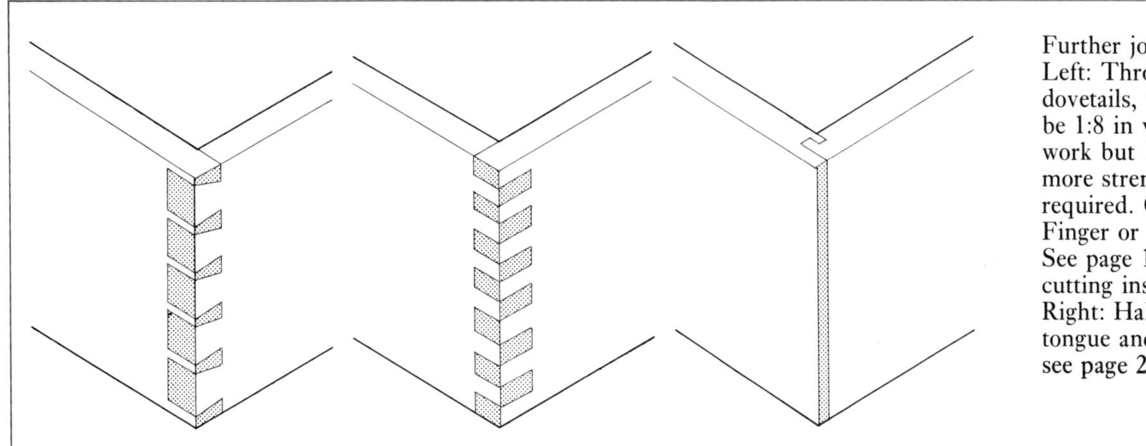

Further joint details: Left: Through dovetails, the rake can be 1:8 in very fine work but 1:5 where more strength is required. Center: Finger or box joints. See page 198 for cutting instructions. Right: Half blind tongue and rabbet, see page 205.

Flush drawers The precise details for setting out will vary according to how the drawer fits into the cabinet. Flush drawers, where the drawer fits within an opening, require very accurate setting out and fitting to get the tolerances on all four sides precise. They also require accurate cabinetwork with square openings of constant depth, so that a minimum of custom fitting is required for each drawer.

The best joint is the lap or through dovetail, but finger joints are also excellent because they are decorative and, with the right equipment (page 198), amazingly easy to cut. A third alternative which is not as strong, is the half blind tongue and rabbet (page 205) where the side is hidden by the overlapping front.

It's best to set each drawer out to it's particular opening, first cutting the front about $\frac{1}{16}$in oversize before beveling the edges with a hand plane to make it fit to half thickness into the opening. Then cut the sides to size, making them about $\frac{1}{16}$in wider than the height of the opening. Plane the top edges to a slight taper so that they will fit partway into the opening. Then mark the left and right sides respectively. The lengths should, of course, allow for the joints. And for dovetails, make sure to take the groove into account, setting them out so that a pin covers the end of the grooves along the sides. It's best to set out the grooves before setting out the dovetails. The back of the drawer should then be cut to length to fit neatly inside the opening. It should be trimmed at the bottom edge so that the bottom will fit under it, allowing the sides to be assembled before the bottom is slid in place.

After joining the sides to the front and back, slide the bottom in place from the back. The grooves should be just slack enough to allow for sliding and for a minimum of glue to be applied on the three sides. Use the bottom to square up the drawer, placing a clamp diagonally across to keep it square, if necessary, until the three or four screws are added to the back edge and the glue sets. It's a good idea to clean up any excess glue with a damp cloth before it sets, particularly along the inside bottom edges. To make this easier, seal the inside faces and the bottom with a thin coat of varnish or sanding sealer before assembly. With the glue set, clean off the joints, then fit each drawer, carefully planing off a bit here and there until the fit is smooth but snug, with no gaps at the edges. It should require just a little extra push at the end for closing.

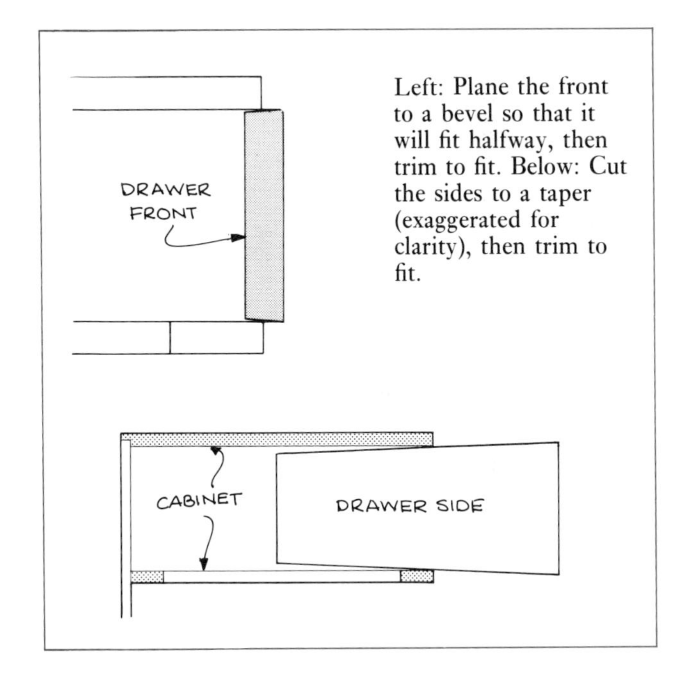

Left: Plane the front to a bevel so that it will fit halfway, then trim to fit. Below: Cut the sides to a taper (exaggerated for clarity), then trim to fit.

The final touch of rubbing carnauba wax or a hard candle along the runners, will make the drawer run even better.

Overlapping fronts These are much easier to fit because the front covers the opening, hiding any discrepancies in fit. The front extends to overlap the opening, either by using slot dovetails to hold it to the sides or by attaching a separate false front to an ordinary flush-fitting drawer by screwing from the inside. In either case, the drawer sides should be made with the same precision as flush-fitting drawers. In cheap production work they are most often made to fit quite loosely, requiring less fuss in fitting.

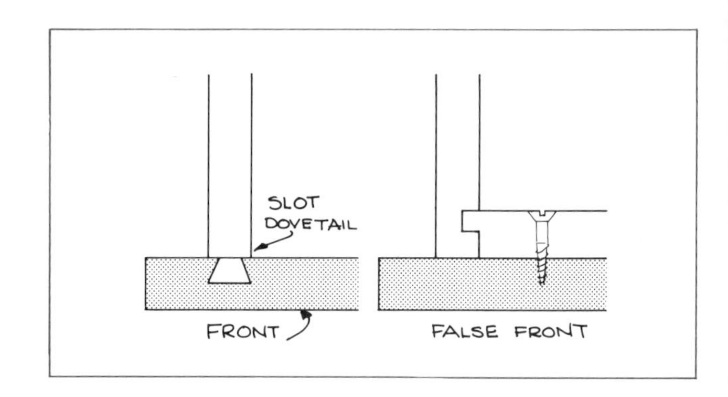

Drawer supports

Drawers are either supported from the bottom by shelves or runners fitted to the sides, or from the sides by fillets or special hardware. Shelf supports require "kickers" so that the top drawer doesn't tilt over as it is pulled out. In chests the runner for the drawer above serves as kickers for the remaining drawers.

Supporting drawers by hanging them from the sides, either by wood fillets or by metal extension slides, requires a firmly made cabinet with dovetailed rails to prevent the sides from separating. Wood fillets, about $\frac{1}{4} \times 1$in, made out of a hard, close-grained wood such as maple, should be screwed to the side (slot screwed in solid work). The corresponding slots in the drawer sides are most easily cut about $\frac{1}{16}$in wider than the fillet with a router fitted with a side fence and a straight cutter.

There are numerous varieties of metal extension slides on the market, and each comes with its own fitting instructions. These are particularly useful for large heavy drawers and for filing cabinets and all kitchen work where the drawer front can extend to cover the hardware.

Drawer stops

Flush drawers should not be allowed to hit the back of the cabinet. The standard way of stopping the drawer is to screw and glue a thin piece of plywood or hardwood to the front rail. It should be thin enough to pass under the drawer bottom. Locate it by gauging the drawer front thickness on the rail. Try the drawer in place and relocate, if necessary, before the glue sets.

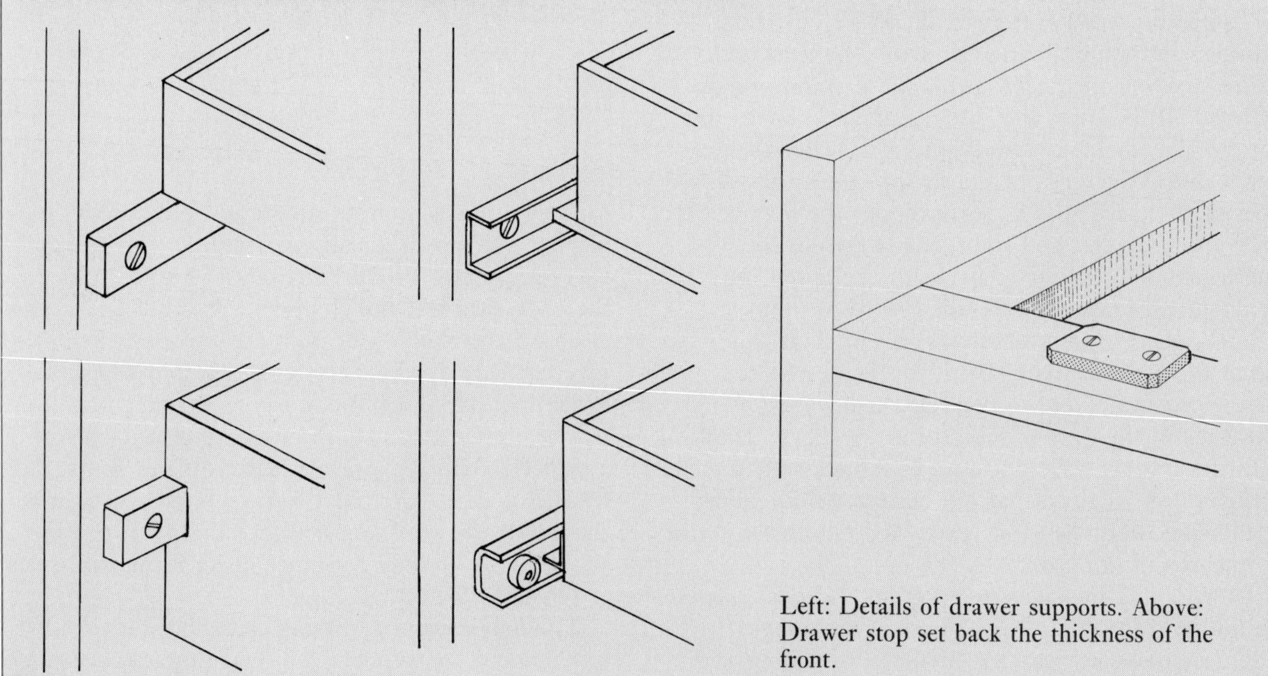

Left: Details of drawer supports. Above: Drawer stop set back the thickness of the front.

Chest of drawers in yew and Indian laurel by Martin Grierson.

Fitting cockbeads

Cockbeads are small, decorative beads or moldings about $\frac{1}{8}$in thick, fitted around a drawer front for decoration. They are often stained black but any wood can be used to contrast with the drawer front. As an alternative to the usual square or rounded bead, it is possible to use a small hockey stick molding which covers the front edges, hiding any mistakes or marks. Fit the cockbead to full thickness along the top, but only wide enough to extend to the pins along the sides and bottom. It's best to trim the top and rabbet the sides and bottom after the drawer has been fitted, gauging the thickness on the front to get the fit accurate. Make the cuts with a router or a rabbet plane.

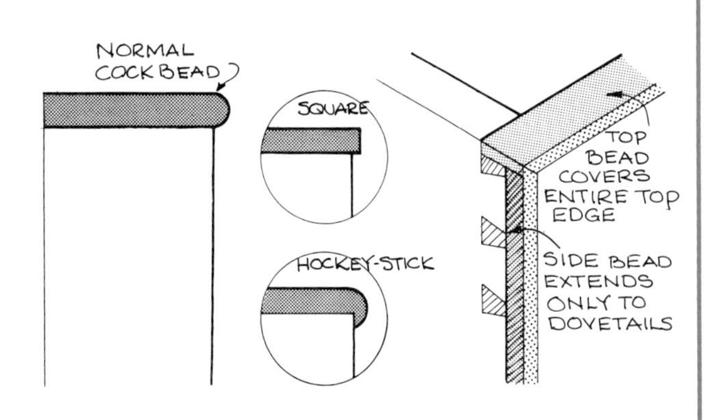

Details of cockbeads. Notice that the bead covers the top edge but extends only to the dovetails on the sides.

TAMBOURS

A sliding tambour is an elegant way of enclosing cabinet space. The technique is quite simple. Thin slats, which are glued to a backing to keep them together and give them flexibility, ride in grooves cut in the cabinet. Although the technique is simple, the execution has to be perfect for the tambour to slide smoothly and easily. The most important part is to get the cabinet absolutely square so that the two sides (or top and bottom) are parallel throughout.

Depending on the delicacy of the tambour, the slats can be from about $\frac{3}{8}$ to $\frac{3}{4}$in wide and about $\frac{5}{16}$in thick. Very small tambours can be less, as long as they are stiffened enough by the backing not to sag. For a really smooth finish, the slats can be veneered over, and cut through the back so that the closed tambour resembles a solid sheet.

Setting out The tongues cut at the ends of the slats run in grooves cut into the carcass. It is critical that the opposite grooves are identical and that the groove is smooth and at a constant depth. This is best done by making a plywood template for the router (page 71).

The groove should be $\frac{3}{16}$in deep and the width of one of the straight cutters, such as $\frac{3}{16}$in. The tongues ride in the grooves against the bottoms, not with the shoulders of the tongues against the cabinet sides.

Draw the section through the cabinet full-scale showing the groove and also the slat thickness at critical points. Corner radii should be large enough to accommodate the tongues, though the groove is usually widened slightly at corners later.

The standard construction for tamboured cabinets is to make a cabinet within to hide the interior track and to keep the tambour free from the contents.

The groove must be brought out to the edge along the back so that the tambour can be entered and the opening through which the tambour comes out must be narrow enough so that the stop or locking rail at the end of the tambour will not pass through it. To copy this outline onto a piece of $\frac{1}{4}$in plywood, either trace it, using carbon paper, or have a blueprint made and cut out the shape to be attached to the plywood.

Preparing the slats The length of the slats should be the inside clearance plus the depth of the two grooves minus about $\frac{1}{32}$ to $\frac{1}{16}$in clearance. The shoulder-to-shoulder distance should be the clear inside dimension minus a clearance which must be slightly more than the end clearance so that the shoulders don't rub against the sides.

The slats can be cut in two ways as shown. Number the slats as they come off the saw. Then sand the edge slightly and rub wax or, better yet, a thin coat of sanding sealer over the edges. This makes it easier to clean off any glue between the slats later. Reject any which are bent or badly cut.

Cutting the tongues Cut one slat to length and cut the tongue by hand and try it out in the groove to check the size. Trim it if necessary to get it right. This gives the dimensions for the rest of the slats, which can either be cut to length and tongued individually or, more conveniently, be trimmed and tongued all at once using the router, or if necessary with a rabbet plane.

Line up the slats against a batten pinned to the plywood base, clamp two battens across the top and two across the width. Put four or five strips of wide masking tape across the slats to hold them together, then remove the clamps. Trim the ends straight across with a router run against a straightedge. Then move the straightedge to cut the tongues.

Applying the backing The best backing material is leathercloth (check telephone book for supplier) which should be thin and flexible. For finer tambours use finely woven linen canvas. Cut it to size making the width about $\frac{3}{4}$in less than the shoulder distance to allow a $\frac{3}{8}$in set back on either side. The length should allow a 1in overhang at the first and last slats. Before gluing, it's a good idea to place wedges and restraining battens between the slats to make sure they are tight together. Glue the backing down, weave downwards, using PVA glue which is flexible enough to bend with the tambour. Place weights on it until the glue sets, then remove the tambour from the jig and take off the tape on the other side so that you can clean up and sand the face.

Assembly Wax the tongues and the grooves, trim the backing allowing the necessary overhang at the front, then slide the tambour in from the back, checking for any tight spots particularly at corners which usually require widening with a gouge. Since the stop or locking rail is too thick to slide in from the back it has to be sprung in at the front after it is shaped and a tongue has been cut as for the slats. Attach the locking rail with a batten screwed and glued to the back, as shown.

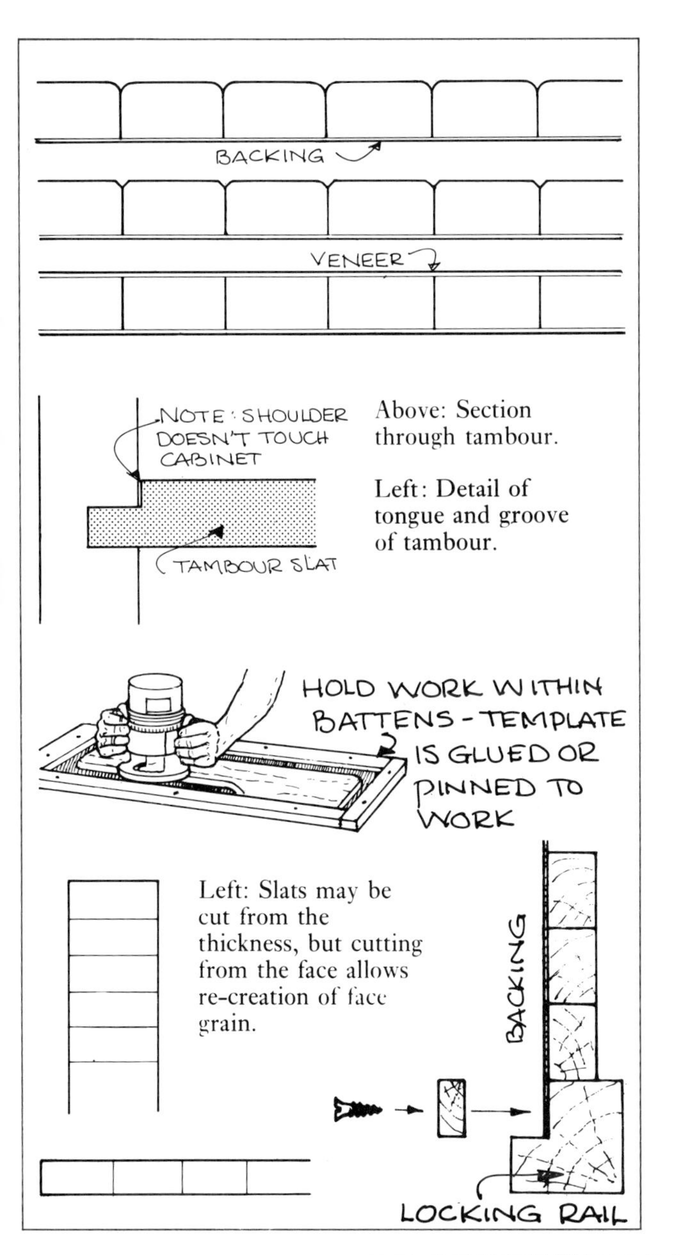

Above: Section through tambour.

Left: Detail of tongue and groove of tambour.

Left: Slats may be cut from the thickness, but cutting from the face allows re-creation of face grain.

Above: Roll top desk in muniga made by the author.

Below: The desk closed.

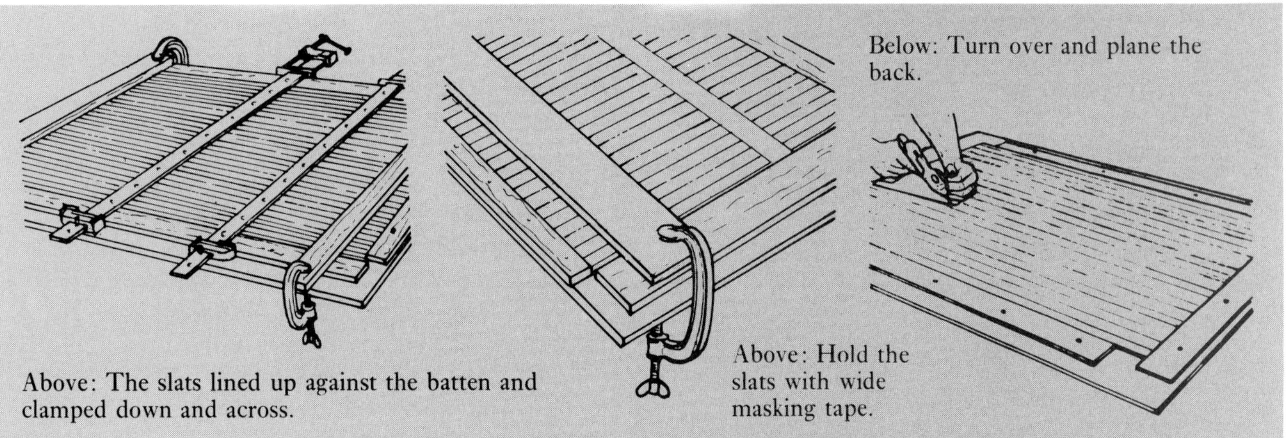

Above: The slats lined up against the batten and clamped down and across.

Above: Hold the slats with wide masking tape.

Below: Turn over and plane the back.

Wood as a material

The widespread interest in woodworking is in part due to the fact that wood is an organic material. And as an organic material, wood is unpredictable and constantly varying from species to species, depending on a variety of factors from climactic conditions to milling methods.

The entire process, from felling and conversion of the tree, to final application of protective finishes must take into account the fact that wood is a cellulose and hygroscopic material, constantly taking up and giving off moisture in response to its surroundings. It is these unpredictable qualities which present such a challenge to the craftsman who must view wood scientifically to understand and control the variations, and artistically to expose and enhance the natural beauty hidden within the cellular structure. Perhaps this is one reason why cabinetmakers are held in such high regard.

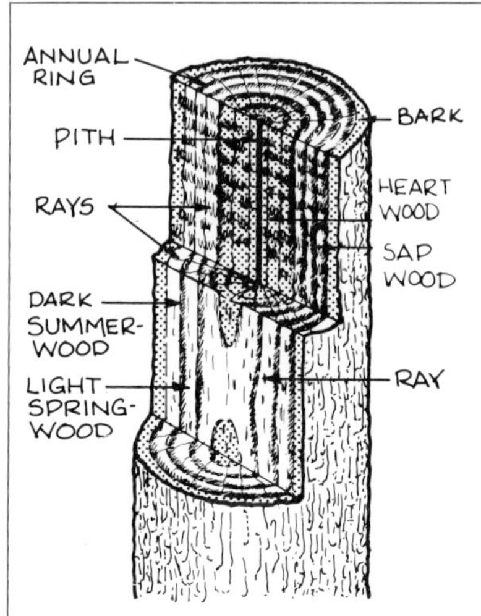

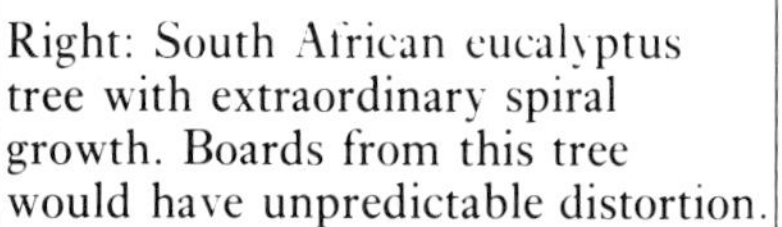

Right: South African eucalyptus tree with extraordinary spiral growth. Boards from this tree would have unpredictable distortion.

Structure and growth The basic growth pattern of trees involves a complex circulatory system of tubular cells which carry and store the various materials to sustain growth. During the growth period, sap, which is water containing dissolved minerals, is taken from the soil and drawn upwards to the leaves by pressure created by a constant evaporation of water from the surface of the leaves. Here, the minerals are converted by photosynthesis into the food required by the tree for its growth. The starches and sugars which make up the food are in turn carried slowly down the bast, which is a thin layer of cells located just under the bark. This food is distributed from here throughout the tree by horizontal radial cells or rays which also assimilate and store it to provide the cambium with continuous sustenance.

All the growth of a tree takes place in the cambium which is a thin, single cell layer located just inside the bark. The sensitive and vulnerable parts of a tree, therefore, is just inside the bark which seems a curious piece of design. Were these located in the center of a tree, the bast and cambium would have much more protection from man and insect pests, and diseases such as Dutch elm might never have happened.

Nevertheless, the cambium produces a constant layer of new sapwood cells on the inside and bast on the outside. New bark is constantly produced from a separate bark cambium to reinforce the old which splits and cracks as the circumference increases.

The layer of sapwood cells continues to grow until the oldest inside cells begin to undergo a chemical change which turns them into the heartwood cells which because they are harder and stiffer, become the structural backbone of the tree. The heartwood continues to increase in area as the tree grows, but the sapwood stays constant in thickness, changing into one layer of heartwood for each layer of new sapwood formed by the cambium.

The chemical changes that turn sapwood cells into heartwood often darken the heartwood. Sapwood is usually cream-colored but the deposition of these extractives creates many of the familiar dark woods. Because of the foodstuffs in sapwood, it attracts various wood boring pests and is therefore not generally used in furniture making.

Not only is heartwood harder and stiffer because of these chemical deposits, but it is also more immune to insect and fungal attack because the foodstuffs have been removed and replaced with toxic chemicals. Heartwood is also less permeable than sapwood, that is, water or moisture will not penetrate it as readily. This is partly because the pores of some heartwoods are blocked by small globules of gummy material called tylosis, and partly because the cell walls themselves which are capable of holding water, are filled by the chemical extractives. For these reasons, heartwood behaves differently from sapwood. A board containing both will normally have a greater moisture content in the sapwood part which can lead to problems in differential movement. Even though a prolonged drying process may have equalized moisture contents, the heartwood may shrink less than the sapwood and with a change in humidity, the sapwood will pick up moisture again more readily, with resultant swelling. Greater permeability does, however, have advantages, for sapwood will absorb preservatives, bleach and stains better than heartwood.

Cell types Wood is made up of two organic chemicals, cellulose and lignin which are arranged into four types of tubular shaped cells which carry the growth sustaining nutrients through the tree. Three of these cell types are arranged longitudinally along the length of the tree like bunches of drinking straws, and are held together by the lignin, a stiffening adhesive. The adhesive qualities of lignin are exploited in the production of hardboard which is made by compressing small wood particles causing the lignin to glue them together rigidly.

The three longitudinal cell types are *vessels*, *tracheids* and *fibers* and these differ in size and function and are present in varying proportions depending on the species. Vessels are short, large diameter cells with large central cavities which make them particularly suitable for conduction of sap. Their thin walls, however, contribute little to the strength of the tree.

Fibers are the opposite. They are long and smaller in diameter with thick walls which make them strong, but leaves little room for sap conduction.

Tracheids are less specialized than vessels and fibers in that they both conduct and provide strength adapting in proportion to suit the need.

The fourth type of cells are rays or *medullary rays* which are stacked in groups and oriented radially, perpendicular to the other three. Whereas the three longitudinal cell types provide vertical circulation and strength, rays distribute nutrients across the tree from the bast on the outside radially inwards.

Most of the gross characteristics of wood, such as grain, texture and figure, are a reflection of this cell configuration on the face of the board. Different species have different proportions and arrangements of the various types of cells, and the effect of this enables us to distinguish one wood from another.

With the exception of some tropical timbers like mahogany, growth is most active in the spring, less active in the summer and totally absent in winter. This growth pattern is reflected in the cell structure. In spring, sap carrying cells predominate. In hardwoods these are mostly vessels, but in softwoods they are thin walled tracheids with large cavities. As growth slows down in summer and there is less sap to conduct, the cells become thicker walled (fibers for hardwood, thick wall tracheids for softwoods), resulting in a denser and darker ring. This differentiation between light, porous spring wood and dark, dense summer wood results in the familiar growth or annual rings, whose number determines the exact age of the tree. The rate of growth is reflected in the width of the rings. A tree may grow fast during one part of its life but owing to changes in its environment, growth may slow down at other times. In this way it's often possible to "read" the growth rings, for a conspicuously narrow band may reflect a particularly dry summer.

Left: Magnified (62x) view of softwood, showing regular arrangement of single, tracheid cell structure. Large pore spring growth is at right. Large holes are resin canals. Below: Oak showing large spring growth vessels and small summer growth fibers in hardwood.

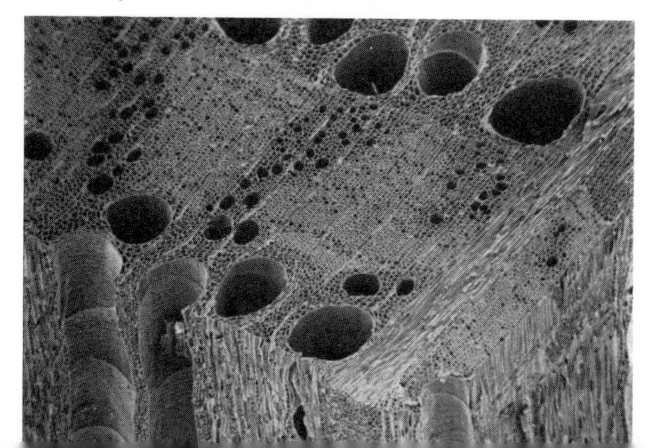

Above: Branches are cantilevers which are continually stressed. This is reflected in the growth patterns. The summerwood grows wider in the underside (bottom) to give extra support where it is needed. Left: A broken-off branch which has been covered by the cambium, results in a dead knot.

Reaction wood is found in branches or in leaning trees where the wood has been under stress from having to hold itself up. This results in "eccentric" growth rings, which are much wider on one side of the wood. This tension or compression wood is generally unsuitable for working because the in-built stresses make it difficult to machine the wood (it may spring when ripped) and to keep it flat and straight.

Ring spacing is a good indication of the strength and also the workability of wood. Evenly spaced and regular rings usually mean a mild working wood. Very defined bands of dark and light reflect a distinct variation in density which usually makes for a difficult to work wood. In hardwoods, wide rings signify a harder wood but in softwood the opposite is true.

The difference in porosity between the spring and summer wood must also be considered when applying stain. Since spring wood has larger pores it will absorb stain more readily. This produces a marked dark grain pattern on stained surfaces as the spring wood soaks up color more readily.

Hardwood and softwood Technically softwoods are distinguished from hardwoods by their seeds which are naked (covered for hardwoods), but the usual differentiation is that softwoods are conifers (needle-leaved) while hardwoods are deciduous (broad-leaved).

There are several other differences such as the presence of resins, which occur mostly in softwoods, and the marked presence of rays (those of softwoods are microscopic). But the most important difference is in the cell structure itself.

SOFTWOOD The rings in softwood consist of tracheids only. These are somewhat rectangular in section and although the center cavity is enlarged during springtime growth, they are relatively uniform in size with the marked absence of vessels which occur only in hardwoods.

HARDWOODS Hardwood rings include tracheids as well as vessels and fibers, but the presence of large pores is a certain sign of hardwoods. The size of vessels varies from species to species depending on the pattern of growth. Ring porous species such as oak and ash produce large distinct vessels in spring, then dense heavy fibers in summer. The large vessels show up on cross sections, but more important they have the effect of producing a coarse grain or texture which is difficult to finish smoothly. In diffuse porous hardwoods, however, the growth rings are less distinct because the vessels are distributed fairly evenly through the year's growth. These result in an even texture which is easy to finish to a smooth, silky finish such as for maple, birch or beech. Others have large vessels which result in a fairly coarse even texture such as mahogany, sapele or elm. A third category, semi-ring porous, falls between these two and exhibits a mixture of large vessels and smaller vessels. These species such as black walnut, have a moderately fine texture.

Another characteristic of hardwoods is the conspicuous presence of medullary rays which show up on cross sections as fine radiating lines cutting across the annual rings. The rays on some species show up on quarter sawn boards as distinct and highly decorative markings, as silver grain in quartered oak for example, or wavy or lacy patterns in sycamore. In other hardwoods the rays are fine and nearly obscure such as for ash, birch or American black walnut.

Structurally, rays have two important effects. First, because they cut across the rings, they lessen the radial shrinkage of wood as it dries out. One effect is to make the radial shrinkage differ from

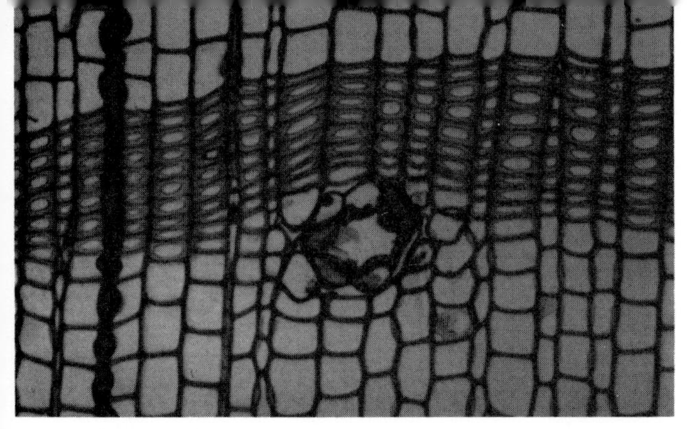

Above: Spring-summerwood variation in pine, a softwood. Below: Diffuse porous hardwood, beech.

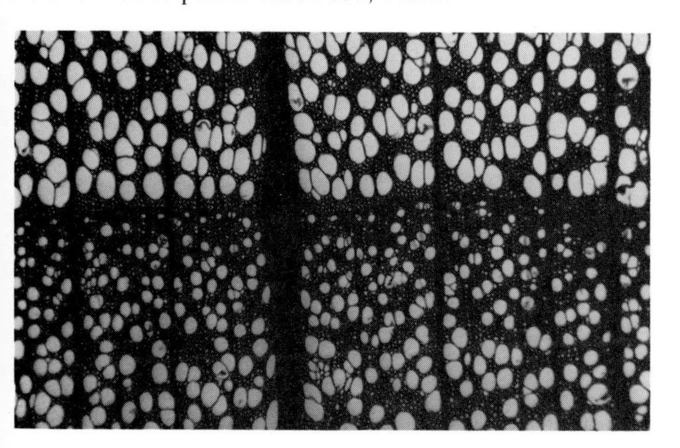

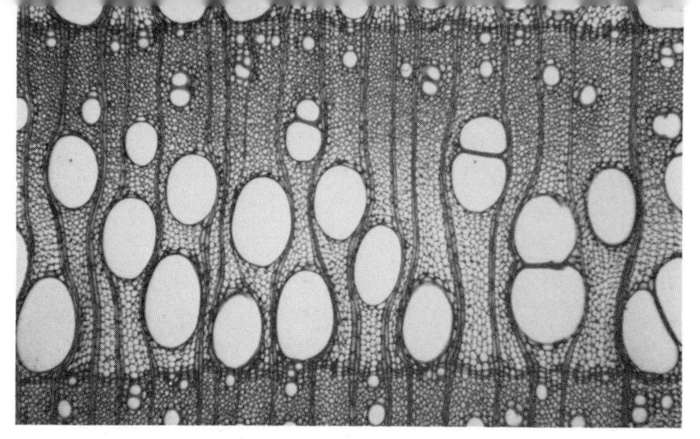

Above: Ring porous hardwood, ash. Below: Semi-ring porous hardwood, birch.

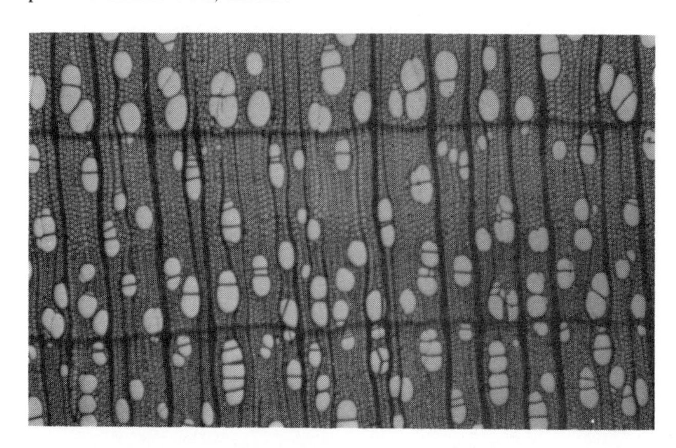

tangential shrinkage, a fact which contributes to distortion in drying wood (see movement in wood page 274). The other effect is to create points or lines of weakness along the rays. These can result in various defects such as radial shakes and end checks, which can become pronounced during kiln drying.

Wood characteristics The grain, texture and figure in wood, in other words what we see and feel on a board, relates to the growth patterns of the cells as reflected on the surface of boards. The longitudinal cells grow quickly in the spring and slowly in the summer to produce rings which form tapered cylinders or cones within the tree. Converting the tree into boards involves making straight cuts (planes) which intersect the cones in various ways depending on the orientation of the cut. It's a matter of solid geometry. Crosscutting produces a plane which shows circular rings and the radial lines of the medullary ray. If the crosscut is angled slightly, the annual rings show up as ovals, accentuated on one side. This method is frequently used to produce "oysters" from the branches of laburnum, olive or lignum vitae, which are used in decorative veneering work. But it's the longitudinal cuts which interest woodworkers, for these cuts produce the characteristic grain, texture and figure which give wood its personality.

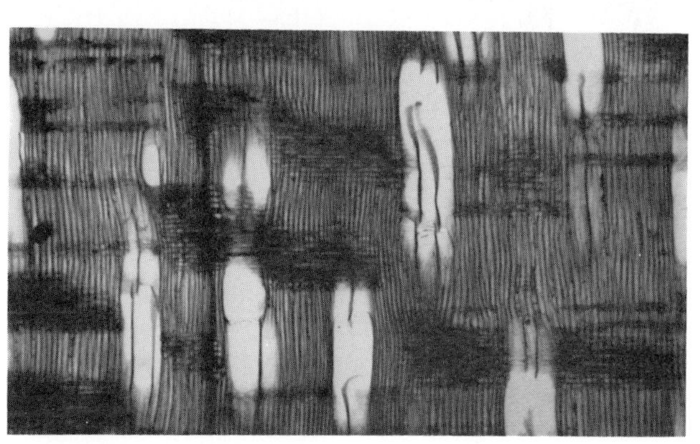

Above: Radial longitudinal section in birch. Below: Tangential longitudial section.

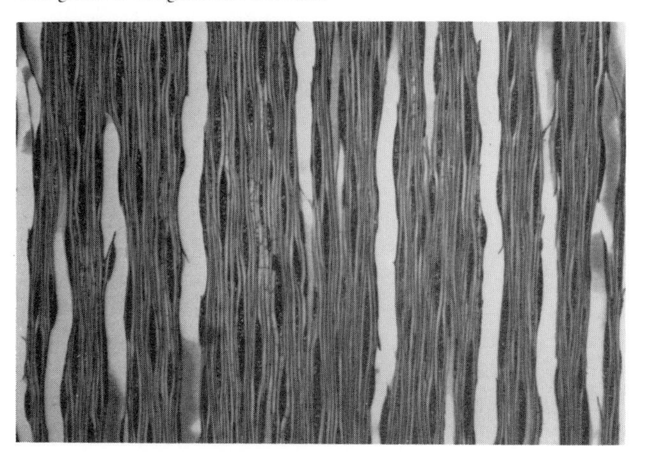

GRAIN The term grain is widely and loosely used to mean any number of things. When people refer to the "beautiful grain" they actually mean the figure, and when we talk about a close-grained wood we are referring to the texture. Strictly speaking, grain refers to the lines of the board which are the intersection of the growth rings (cone) and straight conversion cuts (planes). If these lines are parallel to the sides of the board, the grain is straight, otherwise it can be diagonal or if even, crossgrain.

Since organic growth rarely takes the regular conical form of the simplified model, grain can be wavey, spiral or stripey. Sapele, a West African mahogany, often exhibits the striped grain effect which used to be valued by furniture makers and is the result of interlocked grain, where the growth pattern changes direction at irregular intervals. As wood machinists know, this makes the solid boards difficult to work, for whichever direction it is planed it will still pick up the grain.

Face grain refers to the effect of cutting a board tangentially to the rings so that the figure or curved lines show up on the broad side, whereas straight lines show up on the edges (edge grain).

TEXTURE This refers to the effect the size of the pores has on the surface of a board. Thus ring porous hardwoods such as oak, have large pores which show up on boards as a coarse texture which requires fillers to achieve a smooth finish. Conversely, fine or smooth texture refers to diffuse porous hardwoods and some slow growing softwoods. The texture, whether coarse or fine, can be even, resulting from slow, steady growth, or conversely, it can be uneven.

FIGURE Figure is the aesthetic effect of all the structural characteristics. The conversion of logs into boards is the art of producing more interesting (and valuable) figure as well as maximum quantities of boards.

Few mills today do quarter sawing, so much of the beautiful figure of radial cuts in species like oak or chestnut is not taken advantage of. This is where the portable chain saw mill (page 259) can be so valuable to cabinetmakers, for it allows us to explore the figure of a tree as we cut into it, and to change the direction of cut, if necessary, to take advantage of a particularly interesting structural effect.

Figure can refer to any number of interesting growth phenomena, such as the silver grain which is the effect of medullary rays in quarter sawn oak. Bird's eye maple, which shows the effect of small sprouts or shoots which did not develop, is most

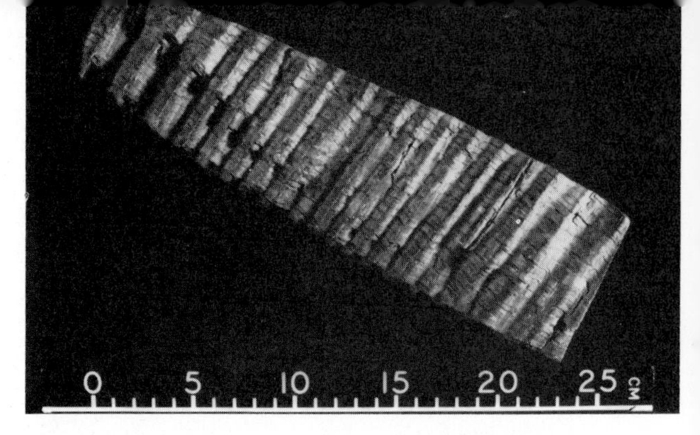

Above: Section has split along natural wavy grain.
Below: Rays reflected as silver grain in oak.

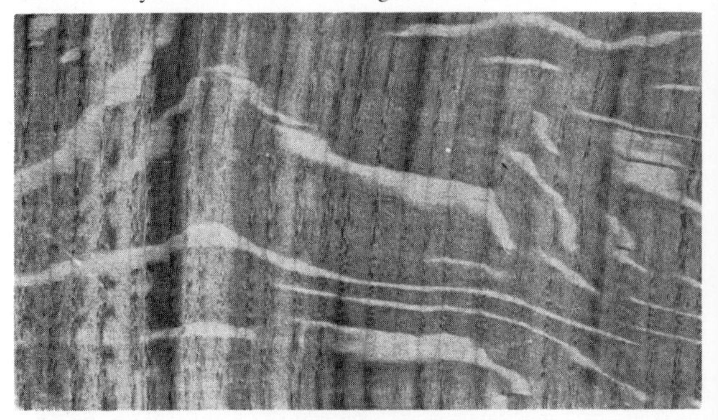

Below: Irregular growth of crotch walnut.

often present in veneer because the cuts must be made perpendicular to the rays. Burr figure is found in veneer made through areas of abnormal growth in a tree and is very striking, as are crotch or stump figures produced from cuts made where irregular tree growth takes place.

OTHER CHARACTERISTICS There are many other features of different species which makes it easy to identify them and which can be used to advantage in woodworking. The heartwood of cedar, for example, undergoes very strong chemical changes as the sapwood is converted. The resulting resins make cedar nearly impervious to insects and decay. This characteristic is exploited in making wall and roof claddings, window frames and linings for linen chests.

Teak and several related species have strong chemical properties which make them oily to the touch and difficult to glue up and finish. Teak is well-known for its resistance to decay and has therefore been widely used for boat construction and outdoor furniture, though today the high price limits its use.

Some species have a strong characteristic smell, particularly when freshly cut. The odor of Douglas fir is very familiar, but African woods such as muninga, have perfume-like smells.

Conversion Wood is one of our most valuable resources, and its widescale use to build dwellings, ships, furniture, and to provide heat and parts for tools, enables us to trace social and technological developments from times as early as man began using tools.

Most wood was used in the round and in a green state, causing tremendous problems when it dried out and twisted. Axes were used to fell and debark the trees. Later the adz was developed to hew or smooth the surface, making square sections possible.

For centuries, the only way to cut straight boards from logs was by pit saw, using one man on top to guide the saw and a man underneath to help lift it. The log was laid across a pit dug underneath to hold the "pitman" who often went blind after years of being showered with sawdust.

With the advent of the Industrial Revolution, the circular saw was developed, driven first by water power and later by steam, and advances in sawing techniques were made possible by the development of harder, tougher steels and the invention of band saws in the early nineteenth century. Today, sophisticated band saw milling machines of immense power cut large quantities of straight, parallel boards almost automatically.

This very simplified outline is just meant to give a brief idea of the historical importance of wood, and anyone interested in the subject could well spend a lifetime researching and exploring, and undoubtedly come up with a fascinating account of the role of wood in history.

SAW MILLING Depending on location, mills generally convert the log into boards in the autumn and winter months. The boards will then lose moisture gradually to avoid causing end splitting and other defects. Generally, only the butt is used for furniture grade timber but some branches or second lengths may also be converted if the reaction wood is not severe.

Above: This majestic English oak felled in 1958, lay maturing for eight years to be cut into 4in planks for the door of the Old Bailey in London. Below: Oak log stock at large mill.

Above: Saw doctor working on large milling band saw blades.

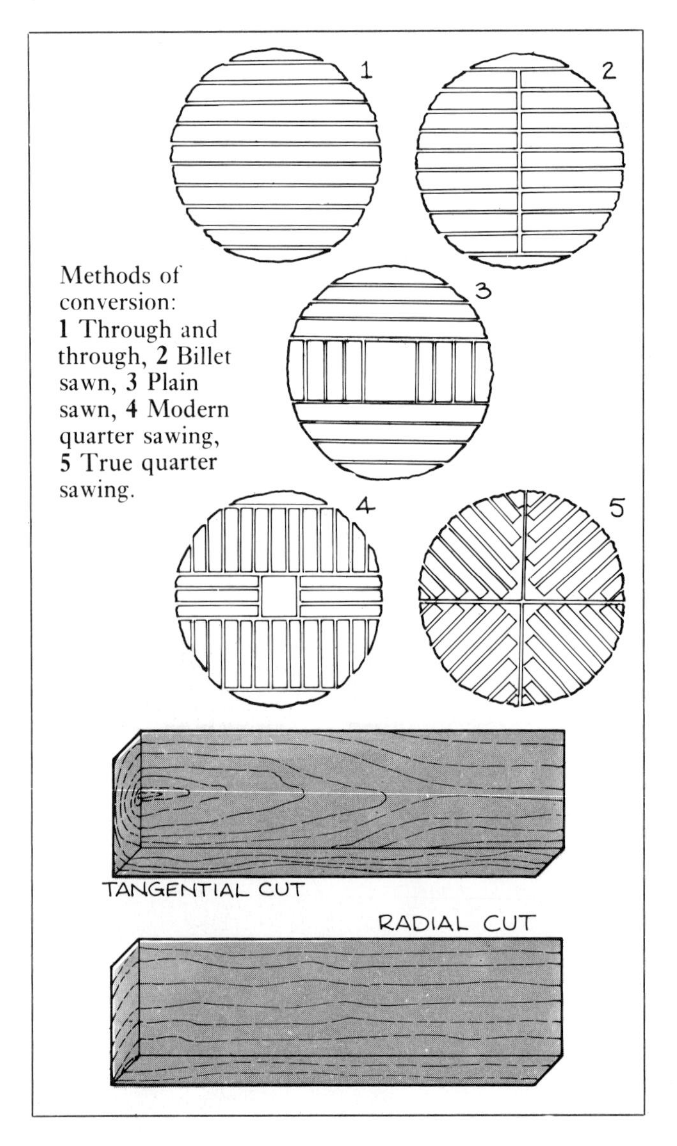

Methods of conversion: **1** Through and through, **2** Billet sawn, **3** Plain sawn, **4** Modern quarter sawing, **5** True quarter sawing.

The best quality woods from large butts usually go for veneer to be sold in sheets for veneering work or plywood manufacture.

The methods of conversion vary from mill to mill and from species to species depending on the size of the operation. Many European mills cut the logs "through and through" (also flat or plain cut) to produce waney edged boards. This produces one or two boards of quarter sawn wood but the majority of the boards show contour markings, a familiar figure in softwoods used in pine furniture, for example. Through and through boards, especially the top and bottom cuts, are more susceptible to distortion.

A variation of the flat cut is the wainscot cut for especially large diameters or where the center "star" shake is more of a problem.

True quarter sawing is rarely done by large commercial mills today because it is time-consuming and wasteful. In the modern method of quarter sawing, two through and through cuts are first made to box out the heart. Then the remaining wood is cut into narrow boards which yields a certain percentage of true quartered boards.

Wood can be bought most cheaply in log form from large merchants, but it must then be transported and converted. This can be done by small and willing local mills (most don't want to bother) or it can be done with a portable chain saw mill (page 259), which affords the craftsman more control in the choice of cut. The chain saw mill does, however, produce more waste in that the kerf is much wider ($\frac{3}{8}$in) than those of band mill saws ($\frac{1}{16}$in).

Waney edged boards are cheaper than square edged boards, but must be trimmed on either side to produce suitable planks, and this waste and time must be taken into account.

The woodworker normally has very little control over conversion and must leave it to luck and persistence to find boards which are interesting and stable. With a little prodding and at a higher price some lumberyards will allow sorting through to pick out individual planks, but most will only sell them as they rise from the pile. No one has to be reminded of the high price of hardwoods. This makes it important to select carefully and to reject boards with large defects such as long end splits, pronounced rounding, cup or ring shakes, sticker marks, sunchecking and so on. Kiln dried wood (see page 269) is usually better than air dried, but this too must be checked, for bad kilning can severely damage the wood and cause surface checks, honeycombing and in-built stresses.

Portable chain saw mill

It seems somewhat ironic that for a profession exclusively involved in exploring the qualities of wood, the most difficult part of the job is getting the wood. Most of us have, at one time or another, been turned off by the attitude of hardwood merchants who aren't interested in small orders, and if they are won't allow sorting through a pile for good boards. The perfect answer, of course, is to find or buy the logs and convert them ourselves, and this is precisely what the portable chain saw mill enables us to do. It gives us the control in choosing the type of cut, whether through and through, square edged, or quarter sawn, and without exaggeration, provides a wonderful source of inspiration which comes from seeing the grain patterns emerge.

The advantages of home milling are obvious. First, it is economical. There are many local sources of free hardwood which might otherwise be used as firewood. A good mill like the Sperber model fitted with a couple of engines, is expensive (there are smaller, less expensive models), but so is wood, and the mill will pay for itself in a few weekend's work. Second, it gives us control over the method of cutting, the thickness and length of boards and the use of extraordinary grain like crotch and burl wood. Third, it is portable, enabling one or two people to take it to the tree rather than having to arrange special equipment to take the butt to the mill.

The principle of the chain saw mill is superbly simple. The Sperber model consists of a single guide bar supporting a rip chain which is powered by two large, conventional chain saw engines, positioned at each end of the guide bar. Adjustable rollers set parallel to the chain determine the thickness of cut.

Obtaining butts Trees can often be obtained very cheaply or for free. Large trees which have to be removed from gardens might be a headache to the owner, but they are a boon to someone with the mill. The search for suitable timber will provide a new stimulus for exploring the countryside. It is often worthwhile to contact the local parks and highways departments, particularly after a period of high winds. I have found that the local road improvement plans are splendid sources of mature trees which may otherwise be thrown away.

Choosing and preparing the butt Preparation entails trimming off branches and cutting the butt into suitable lengths. It is usually necessary to use a conventional chain saw for trimming up the butt, and one of the engines can easily be taken off the mill for this purpose.

It is much easier to deal with butts which can be approached directly by a vehicle, since sharpening the chain and loading are obviously simpler in these circumstances. Unless the tree is particularly unusual or valuable, try to select straight, large diameter logs, preferably without knots, unless these will add to the interest of the board.

After the log has been felled and trimmed it must be bucked up into suitable lengths, depending on use. Remember to allow extra length for end splits and trimming to final size. The log to be milled is then positioned to get the best cutting out of it. If the log is to be cut into a beam or dimensional sizes, it is necessary to roll the log up onto ramp supports for large butts, or V blocks for smaller ones. Otherwise most of the cutting can be done with the butt lying directly on the ground, firmly wedged to prevent rolling.

Above: The Sperber two engine mill. Right: Moving the butt onto ramp support using cant hooks.

Successful milling on site does require a modicum of organization. A number of additional tools are sometimes needed, and to avoid wasted trips I find a checklist essential.

CHECKLIST OF MILLING KIT

Mill and engines	Slabbing rail
Wooden blocks	C-clamps
Wooden wedges	Two gasoline cans
Chain saw oil	Two stroke oil
Funnel with flexible end	Two cant hooks
Two ear muffs	Two pairs leather gloves
Two pairs goggles	Hand brush
Axe	Hand saw
Conventional guide bar and chain	Engine tool kit
Tape measure	Pincers
Chisel	Screwdriver
Claw hammer	Plastic box for washing out air filter
Spare chain for mill	Granberg electric sharpener
Chain saw file	Spare plugs
Starter rope	Rope for tying down boards

Milling a butt The first or "slab" cut removes the top segment of the butt. This is done by using the slabbing rail which resembles a ladder and is laid on top along the length of the butt. It is fixed by wooden blocks nailed onto the top of the butt at intervals and firmly clamped to the "rungs" with C-clamps. The sides of the slabbing rail provide straight, parallel edges which guides the rollers for the first cut which then provides the flat surface for the next cut.

If the bark is thick and the nail does not anchor firmly, it may be necessary to remove a small section of bark to allow the nail to reach solid wood. And in clamping the rail to the blocks, make sure that the clamp does not protrude above the top of the rail. In many cases it may not be desirable to have the rail sit parallel with the top surface of the log. For instance, there may be a severe taper to the log and it may be advantageous for the cut to be parallel to the pith. In this case, simply lift one end of the rail into the proper position and wedge up if necessary before tightening the clamp.

Above: The slabbing rail made of marine plywood is clamped to hardwood blocks which are nailed to the butt. Below: The slab cut.

Below: Through and through cuts are made with the rollers riding on the previously cut surface.

As long as the rail block is nailed firmly in solid wood, the rail should remain rigid.

First set the rollers to give clearance for the thickness of the slabbing rail and the deepest nail securing the blocks. Start the engines and lift the mill, placing the rollers on the end of the slabbing rail. Once the slab cut has been made, remove the slabbing rail and the slab. Adjust the rollers to the thickness of the next board required, and continue milling until the butt is sawn. It may be necessary to raise the butt to give sufficient clearance for the last two or three planks. By this stage it is light enough to be readily lifted onto a board or ramp placed underneath.

Various conversion methods

FLAT SAWING Flat,through and through sawing as described above, is the simplest and most efficient method of sawing a log. After the initial slab cut, the boards are simply run off one after the other to the required thickness. The thickness can be set between $\frac{3}{8}$in and 15in, and varied to take advantage of particular features in the log.

Make cuts until the guide bar is within about 5in of the ground. Lift the log onto wooden supports to allow clearance for the bottom side rollers. An 8in diameter log, about 7in high with a "V" cut into the top, works well as a support.

Since it is more convenient to mill with the log elevated, the log can be rolled onto ramp supports using cant hooks, as shown.

SQUARING A LOG First raise the log up on ramp supports. Draw the section of the beam to be made on the end of the log and use this as a guide for the cuts. After making the slab cut make a second cut near the bottom of the log. It may be necessary to wedge behind the saw, especially at the end of the cut. Then rotate the log 90 and repeat the process, being sure that the rail is set perpendicular to the existing surfaces.

Squaring a log. Below: Cutting the second parallel side. Center: Slab cut with butt turned 90°. Bottom: Final cut.

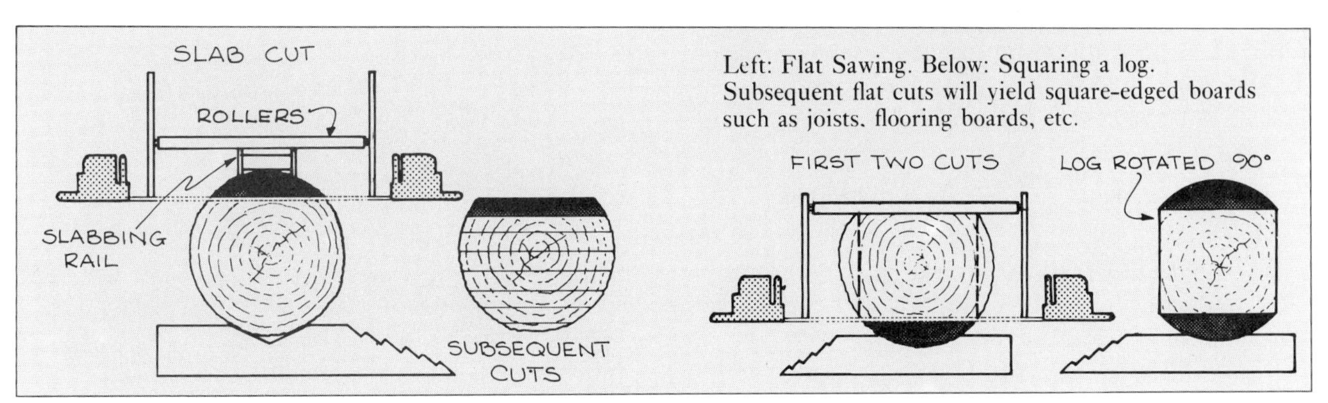

Left: Flat Sawing. Below: Squaring a log. Subsequent flat cuts will yield square-edged boards such as joists, flooring boards, etc.

CUTTING DIMENSION LUMBER The example given here is for cutting 2 × 3in but any desired size could be cut.

First raise up the log on ramp supports. Make a slab cut. Then set the guide rollers to include the lumber plus $\frac{3}{8}$in for each kerf. In this case 12in plus $1\frac{1}{8}$in. This will allow for four 3in pieces. Make the bottom cut. There is no need to remove the bottom slab at this point. Then crank down the guide rollers 3in plus $\frac{3}{8}$in. The setting should now read 9in plus $\frac{3}{4}$in. Make a cut at this setting, but this time do not cut all the way through the log. Stop about $2\frac{1}{2}$in from the end and back the mill out, being careful not to derail the saw chain – keep one engine running. It will be necessary to wedge the kerf to do this. Crank up the mill another $3\frac{3}{8}$in so the setting now reads $6\frac{3}{8}$in and make another cut. Then make the final 3in cut, again stopping $2\frac{1}{2}$in from the end of the log and backing the mill out.

Place a clamp across the slit end of the log to keep it from shifting and rotate the log 90°. Make a slab cut. Set the guide rollers at 2in and cut off groups of four 2 × 3in, cutting all the way through.

QUARTER SAWING The log can be left on the ground. Set the slabbing rail parallel to the pith, and set the guide rollers so the slab cut will pass through the pith. Make the slab cut, stopping $2\frac{1}{2}$in from the end of the log and back the mill out.

Rotate the log 90°, clamp the loose ends and set the rail parallel to the pith. Set the guide rollers so that the cut will pass through the pith. Make the cut, cutting all the way through the log. The log is now cut into two halves each half cut into two quarters that are held together at the ends.

Place one half log with the flat surface up and cut a plank. Then separate the two quarters and make alternating cuts off the two perpendicular flat surfaces. Repeat the same process on the other pieces.

Cutting dimension lumber. Left: First make flat cuts. Center: Turn butt 90° for slab cut. Right: Slice off dimensional lumber.

FOUR -3in CUTS

2×3'S

2in

CLAMP

Safety on site Proper clothing is essential, and boots, leather gloves and ear muffs should always be worn. There are hazards in trimming the butts, particularly trimming branches held under tension and removing the root. A thoughtful, deliberate approach which anticipates sudden movement is the best insurance against the unexpected – and accidents. Wet weather can add to the hazard, and milling should be avoided in sodden conditions.

Wedge the butt firmly and carefully inspect for nails. Remove any nails or metal before milling

Cutting 2 × 3's. Above: Three 3in slabs. Below: Slicing off 2 × 3's. Notice clamp.

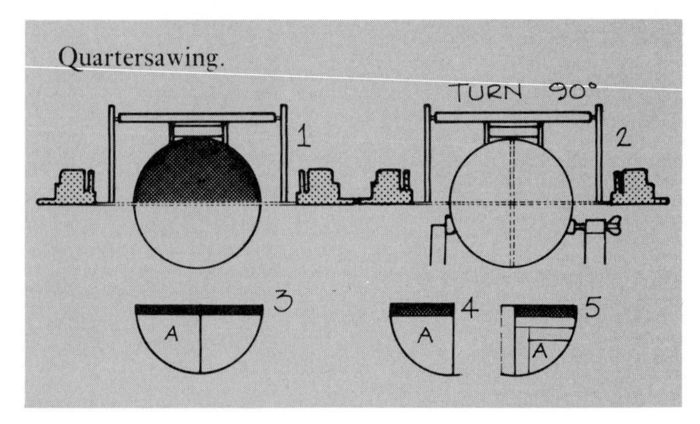

Quartersawing.

begins. Be sure that the chain tension is correct. Always begin operation with a sharp chain, a blunt one is dangerous. The actual milling is straightforward and because the chain is almost entirely within the butt, it is probably safer than conventional chain sawing. Nevertheless, vigilance, attention to detail, and keeping the engines and chain in perfect condition are major factors to avoid accidents.

Problems in milling Nails and iron, often deeply embedded, are an ever present hazard in milling. Initial inspection and removal of visible nails is obviously important. They can often be located by telltale stain marks on some species such as oak. The best safeguard is to listen and stop if the engine sound changes. You can quickly verify if the chain is blunting. By cutting down onto the point the blade has reached, thereby sacrificing the plank being cut, the iron can be removed and the chain saved. In my experience, metal detectors have not been helpful in finding hidden nails etc. – probably due to the relatively crude device used.

Maintenance

ON SITE Here the problem is mainly ensuring that the chain remains sharp, that gasoline and oil supplies are maintained and that the air filter is cleaned at intervals. The Granberg electric chain sharpener, running off a 12 volt car battery charger, is a boon for site sharpening.

Check fuel, and particularly oil levels periodically. A funnel with a flexible end is useful if the fuel runs out during milling (because the fuel cap is on the side in the milling position). And you will need a plastic box for washing out the air filter in gasoline. Site spares should include new spark plugs and starter rope.

IN THE WORKSHOP Routine care after use includes filling up with oil, fuel and cleaning the air filter. Remove the sprocket covers and brush away sawdust. Sharpen and oil the chain. This is conveniently done with a paint brush dipped in chain saw oil. I have found that problems of workshop maintenance are greatly helped by making a bench long enough to hold the mill and engines. Fasten a jig to the wall behind the bench to hold the top horizontal bar, to allow the mill to be held firmly on its side for sharpening and cleaning.

The rip chain The rip chain is the key to the success of the portable mill. The chain is designed to allow the fastest possible feed when cutting into the end grain of a log.

For fast cutting, one would logically want as many cutters in the wood as possible at all times. But the resistance of end grain is so great that too many cutters just overload the engine jamming the chain. Use a skip chain to reduce cutting resistance. Two tie straps rather than one as on ordinary chains separate each cutter. The chain itself consists of alternating pairs of scoring and raking cutters. As the chain speeds around the bar, the narrow scoring cutters first sever the wood by cutting a groove at the end of the kerf. Then the wide rakers, which are lower and shorter than the scorers, remove the bulk of the wood.

Because the chain will need to be sharpened at least twice a day when the mill is in use, it is impractical to consider anything but sharpening it yourself. Since the performance of the mill is dependent on the sharpness and accuracy of the

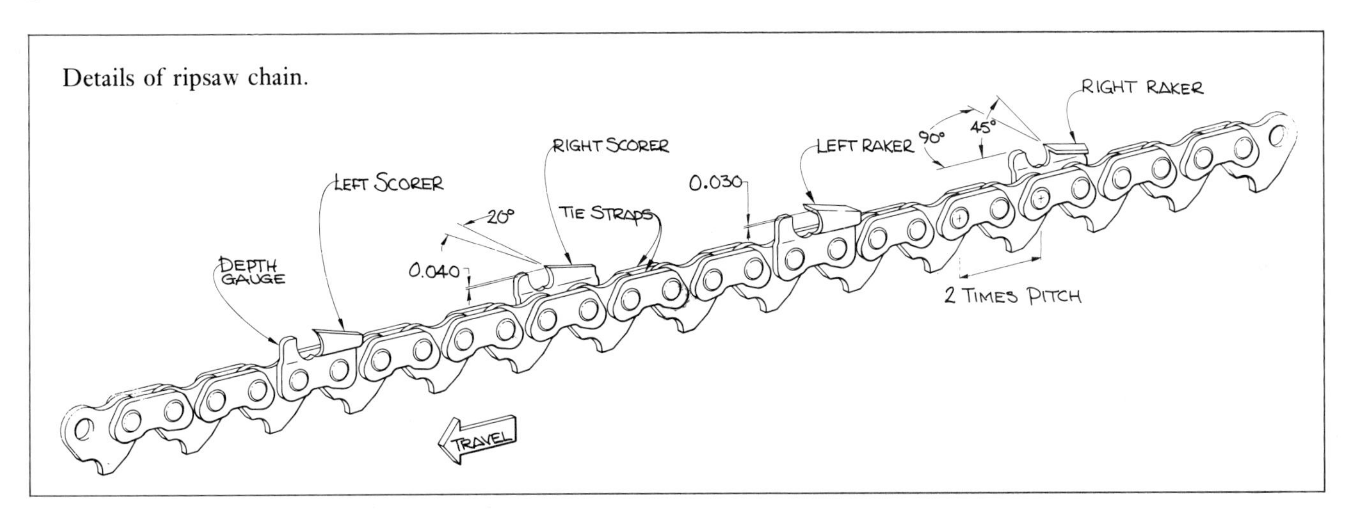

Details of ripsaw chain.

chain, it is advisable to use an electric sharpener such as the Granberg model shown. But if the chain hits a nail, rock or other debris in the log, it may be necessary to send it back to the manufacturer for regrinding.

Sharpening a rip chain Sharpen the chain at least twice a day when it is in constant use and more often if needed.

The direction sheet that comes with the electric sharpener is a good guide for setting the sharpener on the bar.

Using an AC to a 12 volt DC adaptor, the electric sharpener can be run off standard AC household current when sharpening in the workshop. Without the adaptor, the sharpener must be run off 12 volt DC current such as in most cars.

It is possible to hand file the chain while out in the field. However, it is not advisable to hand file more than twice before using the electric sharpener. Be sure that the file is the same diameter as the stone being used on the grinder.

Above: Granberg electric grinder. Below: Checking depth with auto feeler gauge and straightedge.

Moisture in wood

Wood is a hygroscopic material. It will, whether it is standing as a tree or lying in a piece of furniture, absorb and give off water or moisture in a constant attempt to achieve a state of equilibrium with its environment. The difficulty from the woodworker's point of view is that when it absorbs moisture, a board will expand in certain directions and then shrink when it subsequently dries out. This process is constant and even a heavily finished piece of 300 year old furniture will respond to changes in the environment. That's why a piece of antique furniture will often crack or split when brought into a very dry, centrally heated house.

An understanding of the relationship between wood and moisture is absolutely essential for all woodworking where solid wood is involved. The only way to get around it is to work with man-made boards and veneers which don't exhibit these characteristics and hence are much easier to manufacture, store and ship.

A growing tree can contain up to twice as much water as cells (by weight). The cells are by design made to carry water and the water is contained in them in two ways either as bound moisture or as free moisture. Bound moisture is contained within the cell structure itself, and free moisture is contained in the open pores of the cells. As soon as the tree is felled it starts to lose its free water. This is important to timber companies who must haul or float the butts out of the forest. Some heavy species like teak are actually left standing for a couple of years after having been ring barked to prevent the supply of food from reaching the bast. Evaporation from the leaves continues to reduce the water content considerably and makes the logs, once felled, light enough for transport and for floating down the rivers.

A felled log will lose much more water in the summer and this can cause splits. Butts are therefore felled in autumn and winter and the bark left on to slow down the loss of moisture.

A log continues to lose water after it is felled but because of its size, the process is very slow. Cut into planks, the wood will lose moisture faster; the thinner the board the faster the rate. At a certain point all the free moisture will have evaporated away but the bound moisture, locked in the cell walls, remains. This is called the *fiber saturation point*, and although the precise point varies from species to species it is generally taken as about 30% moisture content. Before that point the wood structure loses

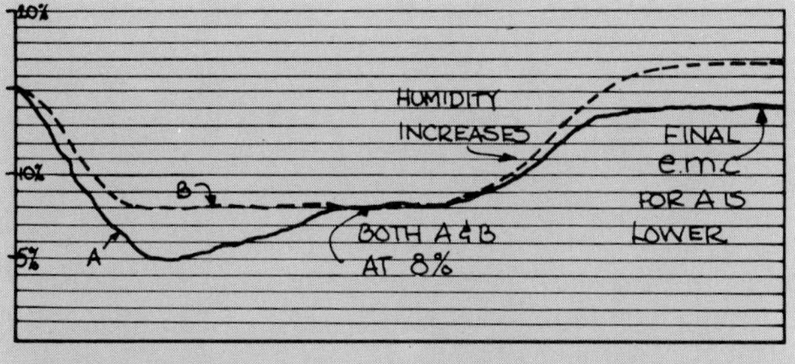

Graph demonstrates hysteresis effect. Wood A which was once drier, will not reach some emc as wood B.

water without shrinking. After this point is reached, the drying rate slows down and the cell walls start shrinking as they lose water. The movement of moisture towards the surface and evaporation from it, continues until the wood reaches equilibrium with the moisture pressure or relative humidity in the air surrounding it. This is called the equilibrium moisture content (emc) and is constantly changing as the humidity changes.

There is a time lag between the time of change in surrounding moisture and the wood reaching emc. The water particles on the surface evaporate but as the pressure is reduced near the surface, the movement of moisture through the wood is slow. This is advantageous for it means that temporary fluctuations in humidity (from a door left open for example) do not affect the wood.

"Cured" wood is a misleading term, but it does have relevance. Two pieces of wood measured to have the same moisture content may behave quite differently. This is due partly to the distribution of the moisture within the wood. A cured piece which has been allowed a lot of time to dry out will have the moisture distributed evenly throughout the thickness, so that any stresses will have had time to even out. The wood is easier to work and less likely to distort or spring during machining.

The two pieces of wood with the same moisture content may also behave differently if they have had different histories, that is, taken different paths in the drying process to reach this moisture content. As an illustration, say one of these pieces, A, has been dried from 15% to a very low moisture content (mc) of 5%, then allowed to increase naturally to an emc of 8%. The piece, B, has dried from 15% directly to 8%. If these two pieces of wood are subjected to high humidity, piece A would not reach the same emc as piece B. This "hysteresis effect" shown here in graph form can be of practical use to the woodworker, because thoroughly dried wood will not pick up as much moisture, and hence will not move as much. This is an argument for buying wood at a lower moisture content than required.

Most fungi which attack wood require moisture to grow, and wet wood is subject to attack from a number of mold or fungi such as dry rot, which feed on the carbohydrates in the cells. Below a certain moisture point (about 20% mc) the wood is relatively safe.

MEASURING MOISTURE CONTENT Moisture content is defined as the ratio of weight of moisture to the overall weight expressed as a percentage. Thus if a sample of 12oz were thoroughly dried in an oven until it no longer lost weight (i.e. had given off all its water) and were then weighed again (say 9oz), the moisture in the original piece weighed $12-9=3$oz. The moisture content of the original piece was then $\frac{3}{12}=25$, expressed as 25%. The scientific method of drying in an oven gives a very accurate overall moisture content, but doesn't give any clue as to the distribution of moisture.

As an example, consider the moisture content in wood during kiln drying. It is well known that during the kilning process, a moisture differential develops in the wood, leaving the outside drier than the core. This forms the "moisture gradient" which is the driving force in moving the moisture outwards to dry out the wood.

In many commercial steam kilns, the rate of drying is quite high with a resulting high moisture gradient. This wood must be "conditioned" at the end of the kilning process by raising the humidity in the kiln for a short time to reverse the gradient.

Badly kilned wood may retain this moisture gradient and will distort badly if it is sawn soon after kilning. An oven test will reveal an overall moisture content which is correct, but misleading because it will not reveal the distribution of moisture.

A quicker and certainly more convenient way is to measure the moisture content with a moisture meter. Two electrodes are driven into the wood and the moisture is read off the meter (see page 267).

When a moisture meter is not available, the weight of a sample of the wood being dried can be checked at regular intervals using a test piece to serve as an indicator of dryness. By weighing it periodically, it is easy to see (or to graph) when it stops losing weight which is a sign of relative dryness. By using the oven method to determine the moisture content at the end of this process it is possible to determine the moisture contents from beginning to end by adding on the weights at the various intervals.

Controlling moisture in wood

PRACTICAL EFFECTS OF MOISTURE IN WOOD Even the best cabinetmakers have had their furniture split due to changes in the moisture content. Preventing it is partly a matter of luck and partly a matter of understanding the principles involved and knowing what to do about them. There are a lot of old wives' tales connected with the curing of wood, and many of these, such as storing oak in the shade of apple trees, are based less on scientific principle than on a mysterious sense of cause and effect. The old-fashioned method of storing the wood for six months or so where the finished piece of furniture is to be located is a totally sound one, but even here, problems arise when the wood is brought back to the shop to be made into furniture.

The problem for the woodworker is to get the moisture content of the wood to a specified level and then to keep it there. There are several factors involved, each requiring careful control:

1 determining what the correct moisture content of the piece of furniture should be

2 drying the wood to that moisture content

3 monitoring the moisture content

4 maintaining the moisture content while storing the wood

5 controlling the mc while working on the wood

6 using the correct construction procedures to allow the wood to move

7 finishing the wood to form a moisture barrier.

Table A

5–8%	Very dry environments. Centrally heated offices. Radiator casings. Furniture near radiators.
8–12	Other centrally heated environments with door opening to outside. Lower range in winter, upper range in summer. Depends on geographical area.
12–14	Rooms with only occassional heating.
15	Work to go in new buildings such as churches, or other large buildings not well heated.
15–18	Outdoor furniture (depending on geographical location—in desert areas, values may be lower).
19–20	Wood at or below this moisture content is immune to rot if kept dry.
15–25	Moisture content of air dried wood. Again depends on geographical location. (17 to 23% in Britain).
40–200	Range of moisture contents for green wood. 40% is white ash, 200% is redwood. Most species have about 60 to 80% mc.

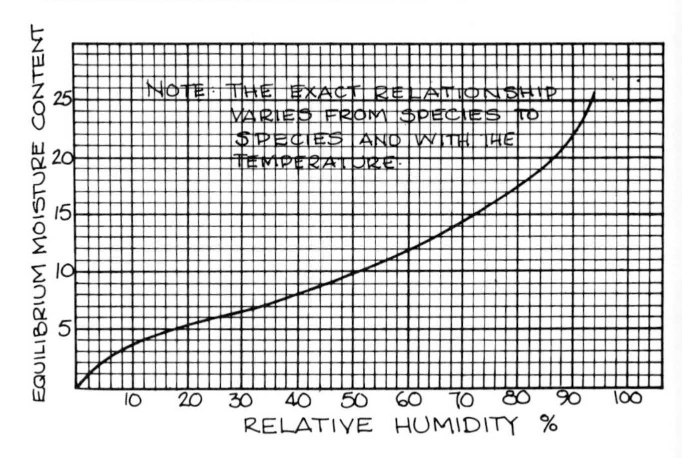

DETERMINING THE CORRECT MOISTURE CONTENT Wood absorbs and releases moisture like a sponge to keep in a state of equilibrium with its environment.

The atmosphere always contains a certain amount of invisible water vapor, the amount varying constantly with the temperature and other factors. Relative humidity (RH), measured with a hygrometer, is defined as the percentage of moisture in the air at a given temperature related to the maximum that air would hold without precipitating (the saturation point). Relative humidity doesn't actually directly measure the amount of moisture present in the air, so it would be a fairly meaningless quantity to woodworkers if it weren't for the fact that we can relate it directly to the moisture content in wood (see graph). For each geographical location, relative humidity varies seasonally in a reliable pattern. Since warm air can hold more moisture than cold, the amount of moisture in the air is greater in the summer than in the winter. But what interests woodworkers is the moisture in the air inside a house. If a house weren't heated, we could reliably take the graphs of outside relative humidity, but the pattern is influenced by the heating in the house. Thus in the summer with the heating off and open doors the RH is the same inside and out and is fairly high, but in the winter, the cool air is heated with a resulting drying out or drop in relative humidity and a lowering of the moisture content of wood as it strives to reach an equilibrium (emc) with the new conditions.

The patterns of variations of RH varies according to the region of the country. Britain is fairly wet with an average RH of between 70 and 80%. In the United States, the average RH is quite low in the dry

continued on page 268

The moisture meter

In view of the importance of moisture content in wood, it is surprising how few people have the means to measure it.

The standard oven method is actually quite difficult to apply and, of course, it is never possible to be sure that a sample will correctly represent the main stock.

The sample must be weighed (to 1 part in 1000 accuracy) and dried to constant weight, which usually means 18 to 24 hours in an oven at 105°C (221°F). The sample must then be cooled in dried air and immediately re-weighed.

This is not an easy procedure. It requires a good balance and an accurately controlled oven. Another problem is that the oven gives the average moisture content of the complete cross section of the wood and doesn't show whether the surface and core moistures are different. Electrical meters are an accurate alternative. Models such as the Protimeter, which is described here have been used by large production shops for many years and they are now becoming popular with woodworkers.

They work by measuring the electrical resistance between two points (the pins) in the wood. This is possible because although perfectly dry wood doesn't conduct electricity (hence has infinite resistance), wet wood behaves differently. When successively more water is taken up by the wood, it is held less tightly to the solid surfaces of the cells. And as soon as it is free to do so, the water begins to dissolve various minerals, and solutions are able to conduct electricity. Thus the amount of water held can be related to the amount of resistance (which is the inverse of conductivity). By repeated experiments using the oven method, the manufacturers calibrate the dial to read directly in percent of moisture.

Because different kinds of wood differ somewhat in the electrical resistance, different scales are provided on moisture meters which give the correct results for most species of wood. For the more usual woods a 0–100 scale is provided which,in conjunction with a table, gives the moisture content for over 150 varieties. The range is from about 7% up to about 30% (or "fiber saturation point" above which wood does not expand further).

The electrical meter gives instant readings so that you can make literally hundreds of checks in the time it would take to do one oven sample. Thus you eliminate the possibility of error due to the need to assume that a single sample is representative of the whole batch. Also, there's a special electrode with long pins (insulated on the shank) and a built-in hammer, which can be driven to a depth of 1in to measure the moisture at various depths. In this respect the electrical meter is much better than the oven test because it will show the distribution of moisture across the section. This is an invaluable aid when checking the accuracy and thoroughness of kiln drying either from your own kiln or a shipment from a merchant.

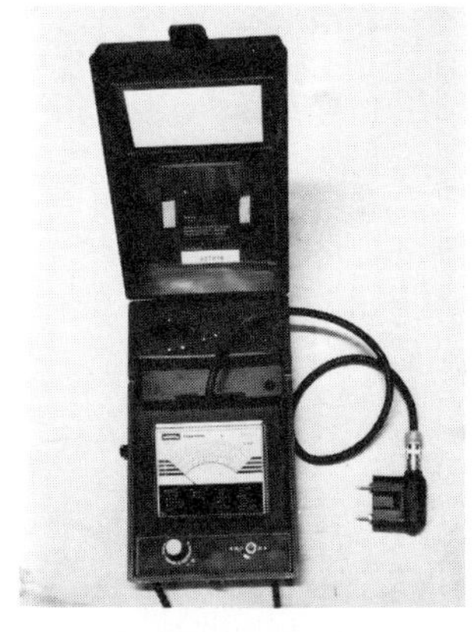

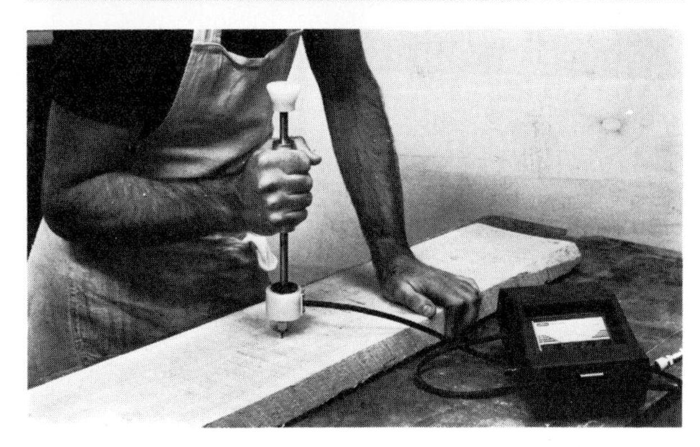

Left: The Protimeter moisture meter with standard pins. Below: Readings should be taken at end as well as on face of boards. Bottom: Using special deep probe, pins can be hammered about 1in into sample.

southwestern areas, moderate in the northeast, and high in the southeast.

To determine the correct moisture content for the wood, find out what the relative humidity will be in the eventual environment, then refer to the graph. In practice this isn't necessary, for the permissible moisture contents in various types of applications (offices, homes, etc.) are fairly well-known for each region. Table A gives an indication of average values for a moderately wet climate, but for specific applications it is worthwhile to consider installing a wet and dry hygrometer to measure RH patterns if necessary.

If these means are not available, it is best to take an intuitive approach, estimating whether the environment will be severely or moderately dry or wet, then drying the wood slightly below the point required. Keep in mind that the graph shows average values of humidity vs moisture content. As shown on the chart on page 275, the exact values vary slightly according to the species.

Drying wood One of the most exciting developments for woodworkers is the invention of chain saw mills (page 259) which makes it possible to by-pass the mills and merchants and to convert logs with sensitivity. The only problem is how to dry the boards to a practical level of between 6 and 12%, suitable for indoor furniture. Air drying, though time-consuming, is cheap, but it will only bring down the moisture content to about 16 to 25% depending on the region. To dry it further, the wood must be kiln dried. There are now several small "benevolent" kilns on the market (page 271) within an affordable range which makes an ideal partner for the chain saw mill, particularly if enough wood is available to make it pay for itself.

The alternative is either to have the wood commercially kilned or to further dry the wood by storing it in a warm room and letting it dry out slowly to the required mc.

Wood dries by evaporation of moisture from the surface, which sets up a moisture gradient causing the moisture from the inside to move towards the surface by capillary action. The aim in drying wood is to control the rate of evaporation, so that the process is gradual with moisture constantly moving outwards, replacing moisture which has already moved outwards or evaporated. If this rate is too fast the wood is likely to split or check at the surface or in kiln drying cause internal collapse (honeycombing). If the rate is too slow or the conditions are too wet, mold or fungus may develop.

AIR DRYING Even when wood is subsequently to be kilned, the normal practice is to air dry the converted log by storing it in piles with carefully placed spacers or "stickers" between each board. The rule of thumb stipulates one year of air drying for each inch thickness of board, but this is best verified by careful monitoring by weighing samples or by checking with a moisture meter.

It is best to start the air drying process in autumn or winter when temperatures are relatively low and humidity high, in order to avoid too fast an initial drying rate which can damage the wood.

Cut the log into similar lengths if possible, avoiding irregularities like crotches and burrs which often distort badly during drying (these can be dried separately). The boards should be stacked on a very firm foundation to avoid differential settlement which can distort them. Place the stack on concrete bearers

Stickered pile at John Makepeace's workshops. Ends are painted with color coded water-based paint to indicate time of stacking. Opposite page: English oak in sticks.

built up about 9in high at even spaces, or on heavy timbers such as railroad sleepers. Test for level, wedging up if necessary, then pile the boards in the order they came from the log (number them if necessary) to be able to match boards later. The stickers should be 1 × 1in or $\frac{3}{4} \times \frac{3}{4}$in made out of a neutral wood like fir or pine and well dried. To avoid leaving "stick marks" which can penetrate far into the wood, some commercial kilns use plastic stickers.

Place the stickers at about 18 to 20in centers (closer for thin boards) directly above one another, exactly in line with the foundation supports below and started near the ends. Then to prevent the ends from drying out faster than the rest of the wood, paint them thoroughly with latex or emulsion paint. At John Makepeace's workshops they color code the ends, using one color for each batch to keep track of the drying time. Finally, the stack should be weighed down to help restrain movement of the boards.

Air drying is best done under cover in open sheds which allow the free circulation of air necessary for evaporation, but failing that, the stacks should be covered with any suitable material such as corrugated roofing, to keep the rain from soaking in from the top.

To monitor drying progress, samples (small boards) placed in the piles can be removed and weighed and the weight plotted to find out when the wood stops losing weight. A better method is to go through the pile at random with a moisture meter and keep a record of the moisture contents. The deep probe sold as an accessory by Protimeter, is capable of reaching one inch into the wood to get a good idea of moisture distribution in thick boards.

After the wood has stopped losing moisture, there is no harm in leaving it a bit longer to "condition" it; that is, to make sure the moisture is evenly distributed. But for most applications, the wood has to be dried further by kilning or by air drying in a warmer and drier environment.

KILN DRYING Kiln drying speeds up the process of drying so that relatively green wood can be dried to a low moisture content in a matter of weeks. Commercial kilning is a specialist operation. The wood is stacked with stickers on small railroad cars which are rolled into large chambers. In the chambers, hot air is introduced to heat up the wood to increase the vapor pressure in the boards (i.e. higher moisture gradient) to speed up the evaporation. Since rapid evaporation would cause damage to the wood, moisture in the form of steam

is introduced at the same time and circulated through the kiln to slow down the evaporation. This balance of hot air and steam is carefully controlled and precise schedules are worked out for various species.

These schedules are not perfect, however, and often lead to high waste factors and wood which has severe surface checking and even internal honeycombing. This partly accounts for the high price of commercially kilned wood.

In buying kilned wood from a merchant, it is important to inspect it carefully for defects and to check the moisture content with a moisture meter. Kilned timber, stored in sheds (good merchants cover it with plastic sheets), begins to pick up moisture as soon as it leaves the ovens and, if the buyer isn't careful, he can end up paying kilned prices for wood which is not much drier than air dried. When checking, make sure to check moisture distribution. Incorrect conditioning at the end of the kilning cycle can lock in stresses caused by the moisture gradient.

These stresses will eventually disappear (causing movement), but if the wood is worked soon after kilning, stresses can be released during ripping leaving badly twisted boards.

Slower kilns which work on a more benevolent dehumidifying principle, are now available in many sizes to suit the craftsman wanting to kiln a few cubic feet, as well as large operators kilning hundreds of boards at a time. These are ideal for anyone with a ready supply of trees for they will bring the moisture content down from 40 to 50% to 10% or less in a month or two depending on thickness.

Large commercial kilns at William Mallinson and Sons, London.

DRYING INDOORS Since even inexpensive kilns are not available to the average woodworker, he must rely on other methods to bring air dried wood down to between 6 and 12% moisture content, suitable for indoor use. As mentioned earlier, the standard technique of leaving the wood in its eventual environment is fine, but it would take a very good client to put up with a pile of wood in his living room for six months or more. A more practical way is to bring air dried wood into a warm room, such as a storeroom, which can be insulated and slightly heated. It should again be piled and stickered with the ends sealed and carefully monitored. It's better to start out with a coolish temperature and increase it gradually. Heating the air is actually a less efficient way to dry wood than dehumidification which removes moisture from the air. So, if possible install a small dehumidifier. This is the principle of the kiln opposite.

Drying the wood in a warm room can cause problems in splitting and so on, which can be temporarily eliminated by covering the wood pile with plastic sheets to raise the humidity and slow down the evaporation. As with outdoor air drying, it is important to monitor progress by weighing a sample or by using the moisture meter. Obviously, the longer the wood stays in this environment the more cured it will be, that is, the moisture will be evenly distributed. And the better this storeroom environment matches the eventual environment of the furniture, the less likelihood of splitting and other problems.

SMALL DEHUMIDIFICATION KILNS As a further aid in controlling the quality of their raw material, craftsmen and small production workshops can buy (or make) small kilns to dry green or air dried wood to moisture contents low enough to be suitable for centrally heated environments.

When timber is being dried, the rate of evaporation is dependent on the difference between the vapor pressure exerted by the wet timber and the vapor pressure of the air. When the vapor pressures have equalized, no further drying occurs. This is the point at which the equilibrium moisture content of the wood, has been reached.

Commercial "heat and vent" kilns work on the principle of heating the wood to increase its vapor pressure. Another way of increasing the difference between the vapor pressure of the air and that of the wood, is to lower the vapor pressure of the air. This is what dehumidifier kilns do; encourage evaporation by removing moisture from the air surrounding the wood. Warm air is re-circulated through the timber stack, encouraging more evaporation. Moisture-laden hot air is not simply poured into the atmosphere as with wasteful heat and vent systems. The system isn't just a saving of money, it will also result in better quality and less waste.

Dehumidifier kilns such as the Ebac model shown here, vary in detail according to the manufacturer, but basically all need an insulated and watertight cabinet to enclose the wood plus the dehumidifier, fans, sensors and hardware such as air vents and drainage pipes.

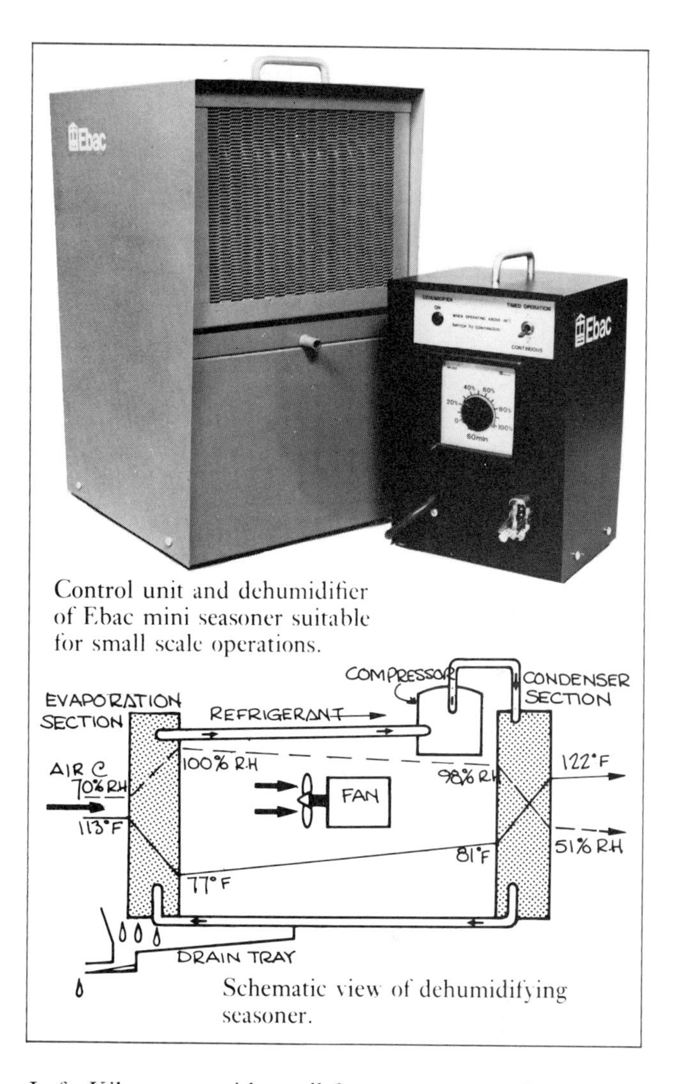

Control unit and dehumidifier of Ebac mini seasoner suitable for small scale operations.

Schematic view of dehumidifying seasoner.

Left: Kiln set up with small factory to season about 200cuft (2400 board feet) at a time. Below: Removing wood sample to check progress.

The principle on which dehumidifiers work is simple. Damp air is drawn through a refrigerated coil which rapidly cools the air to its dewpoint, which is the temperature at which 100% relative humidity occurs. Moisture thus condenses onto the coil and runs to a drain, or is collected in a suitable container. The resulting dry, cold air is then rewarmed with the heat which was extracted from it as it passed through the refrigerated coil and is recirculated into the atmosphere. The release of the latent heat which was bound up in the water vapor along with heat gained as the air passes over the dehumidifier components, means that the air leaves the machine at a higher temperature than that at which it entered. Thus not only is the absolute humidity reduced but, by the increase of temperature, the relative humidity is reduced further.

This re-use of heat ensures that, where drying is concerned, refrigeration dehumidifiers are economical. In reducing the relative humidity of a given volume of air, a refrigeration dehumidifier will consume about one fifth of the energy that direct heating methods would have required. Of course, the resultant warm air can be released in the surrounding room or workshop so that the benefit is twofold, serving as a heat source as well as a kiln.

Possible faults during timber drying Wood shrinkage is responsible for nearly all forms of seasoning faults. There is no doubt that slow, gradual air drying is less destructive and hence less wasteful than kiln drying. Correct drying techniques, particularly with regard to drying schedules, minimizes this sort of fault, though it is impossible to eliminate it entirely. The following are some of the commoner forms of kilning degrade.

SPLITTING AND CHECKING Too steep a moisture gradient is induced in the wood causing the exterior to shrink greatly onto the center, resulting in severe stresses – sometimes sufficient to tear the outer fibers apart.

Checking mainly occurs along the rays on the edges of quarter sawn material and on the faces of tangentially cut boards. This is because shrinkage along the growth rings is greater than that at right angles to them.

It should be noted that surface checks will tend to close when the timber has kiln dried to a uniform moisture content, though not in severe cases.

HONEYCOMBING This is essentially internal splitting, where the difference in stresses between inner and outer layers causes the center fibers to be torn apart.

DISTORTION This is caused by differential drying in timbers containing distorted or curved grain. Poor quality woods, and woods in which the grain is seldom straight, such as elm, are extremely prone to distortion, and these must be piled carefully with sticks at short intervals or even weighted down.

Low temperature drying schedules are recommended for woods particularly prone to distortion.

Silting up a small kiln Ebac and other manufacturers specialize in making the controls and hardware for the kilns. Some may sell entire insulated kilns, others sell only the hardware but provide instructions for making the insulated room or box required around it and for operating the controls and adjusting the schedules for various species.

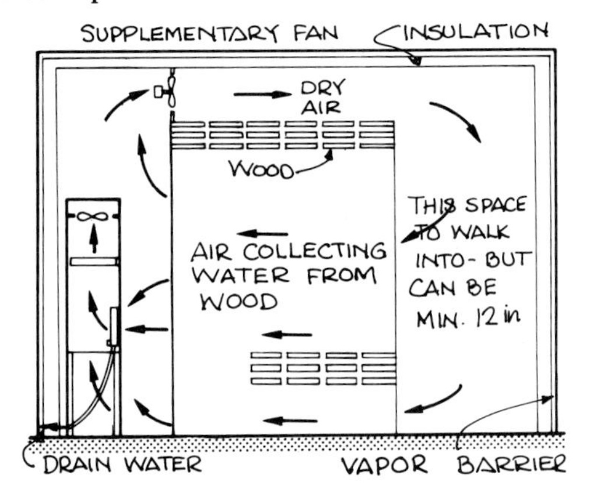

A small kiln can be large enough to walk into. For small quantities, make it out of an insulated plywood box with a sealed lid.

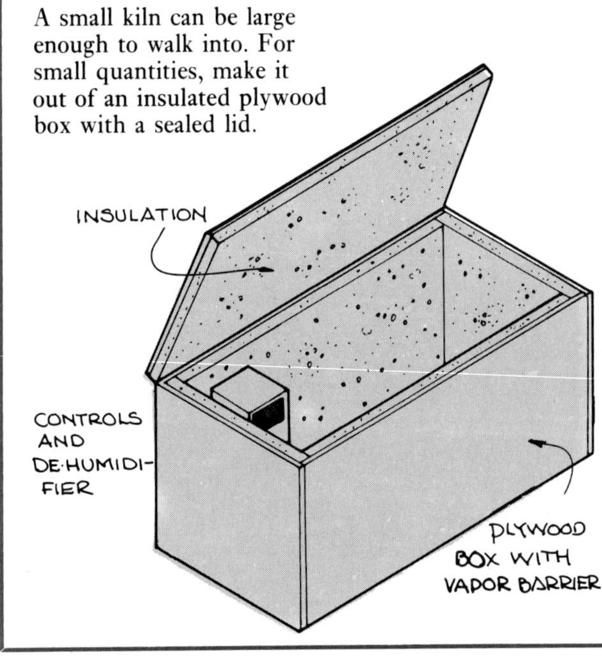

COLLAPSE Too rapid a departure of the free moisture from cells, occasioned by high temperatures, can lead the cells to collapse. Certain species of Eucalyptus are very prone to collapse, even at moderate temperatures. Collapse manifests itself as local shrinkage which, if great, may in its turn lead to severe honeycombing.

Storing and working the wood Wood which has been dried to a specified moisture content shouldn't be allowed to pick up moisture again since this spoils the entire object of the expensive exercise.

The amount of trouble and expense you go to to ensure the right storage environment depends on the level of operations. A craftsman involved in individually designed furniture or in, say, bowl turning, can well afford to keep a small closet moderately heated, but a production shop must rely on a speedy turnover in stock to minimize rises in the moisture. Many modern factories are centrally heated which is the most effective solution, for not only does it give comfortable working temperatures but it also guarantees that the work in progress, as well as stored materials, are kept dry.

The problem is less acute in summer when the humidity in the average home is raised to roughly the same level as the workshop. But in winter, furniture made with kilned wood, can be several points higher than it should be after being stored in a badly or intermittently heated shop.

One practical solution is to stack dried boards directly on top of one another and to cover them with a plastic sheet. Put plastic under it too – a lot of moisture can rise through the floor. Periodic checks on the moisture content should be made with a moisture meter. If the budget allows, install a wet and dry hygrometer which measures humidity, and refer to the graph on page 266 for suitable humidities for particular moisture contents.

If the stock is valuable enough, it may be worthwhile to install a small dehumidifier or at least a small heater to keep the humidity down. We found a good solution to our storage problems, by building a room within a larger room of boards made from hardboard sandwiched around 2in of styrofoam insulation. With a small heater we manage to keep our constant stock of about 100 to 200 cuft to acceptable levels of about 8 to 9%.

With any change in moisture content, the wood may move slightly and this movement should be allowed to take place before, and not after, machining, although this isn't always possible.

Construction procedures Beginners always have difficulty in understanding the mechanics of allowing for movement in a piece of furniture. Basically, expansion and contraction, involving tremendous forces, must be allowed but any warping or bowing has to be restrained. The methods for dealing with this are well established and involve either hardware with slotted holes for screw fixings or housing joints which allow sideways movement. A few principles and typical details are illustrated below.

1 Only that expansion which takes place across the width of boards produces problems. The amount of movement across the width of a single board is not a problem unless it is at least 8in wide.

2 Glued up surfaces behave like a solid board, but the joints, if not perfectly made, are points of weaknesses and may split if the wood is restrained or if it experiences a large change of moisture content.

3 Two wide surfaces can be joined rigidly (by gluing, dovetailing, etc) if the joint is across the ends of both (for the same species), because both will expand and contract together.

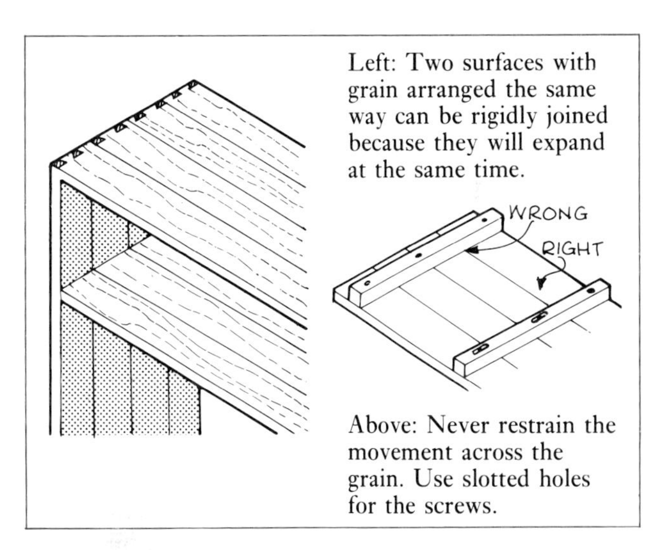

Left: Two surfaces with grain arranged the same way can be rigidly joined because they will expand at the same time.

Above: Never restrain the movement across the grain. Use slotted holes for the screws.

4 If a board such as a batten is glued or otherwise rigidly attached across a surface, it will restrain the wood from moving (wood doesn't move along its length). This will set up stresses which will find the weakest points which will split. The recommended details include slot screwing or enlarged screwholes to allow the screw to slide slightly.

In short, a board should *never* be glued across the width of a panel. It can be rigidly tied at one point by spot gluing, but the other tie points must be allowed to move.

The amount of movement can be roughly calculated from the chart opposite. This depends not only on the species, the width and the change in moisture content, but also on how the board was cut.

6 Shrinkage brackets are often used to hold table tops to the cross rails. The slots in the brackets allow for sideways movement, but the bracket holds the top down to prevent the top from warping or bowing. The rail should be rigid enough to prevent the two ends, or the center from lifting up.

7 In cabinets made from solid wood, the battens for the drawer supports must be held in housings which hold them up but allow the side to move back and forth. One point can be attached by spot gluing or screwing.

8 Blockboard, chipboard and plywood move very little, so sides made from these can be screwed and glued without any problems.

9 Edge gluing quarter sawn boards should be done with rings going in opposite ways. Tangentially sawn boards can be arranged with grain in opposite or in the same direction. Either way there will be a rippling effect if the wood moves. The narrower the boards, the less the movement.

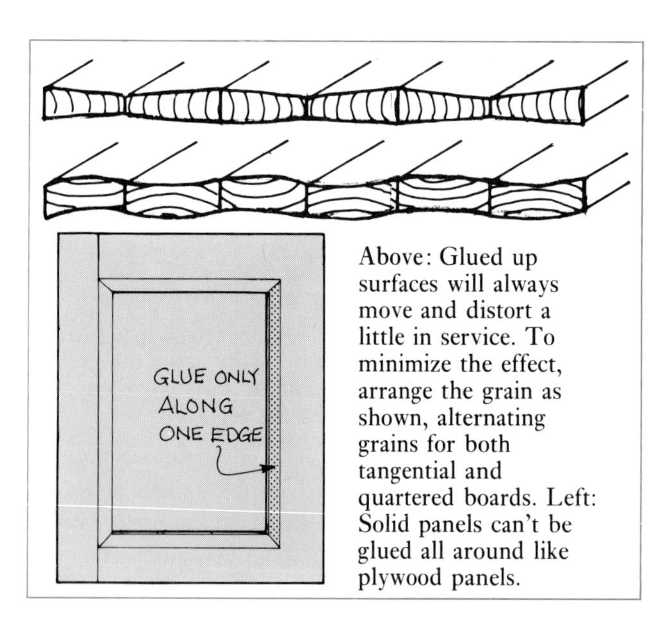

Above: Glued up surfaces will always move and distort a little in service. To minimize the effect, arrange the grain as shown, alternating grains for both tangential and quartered boards. Left: Solid panels can't be glued all around like plywood panels.

10 Panels of solid wood for doors or drawer bottoms will expand and contract, and should therefore be held in grooves. One side can be glued but the other three sides must be unfixed. Drawer bottoms are usually screwed to the back side which is less deep than the sides.

Finishing methods The aim in finishing is to protect the wood from dirt and marks and also to form a moisture barrier which helps to limit the amount of moisture absorbed when a change in humidity occurs.

Fortunately there is a considerable time delay from when a change in atmosphere occurs and the wood starts responding. Thus, day-to-day changes don't affect the moisture content which tends to average out to the prevailing conditions. An applied finish will increase the time delay but no amount of coats, even of paint or lacquer, will completely prevent moisture from getting in or out. Some finishes are better than others, as shown in the table. Linseed oil finish is not much better than raw wood, and three coats of oil based paint offers about 50% protection.

EFFECTIVENESS OF FINISHES

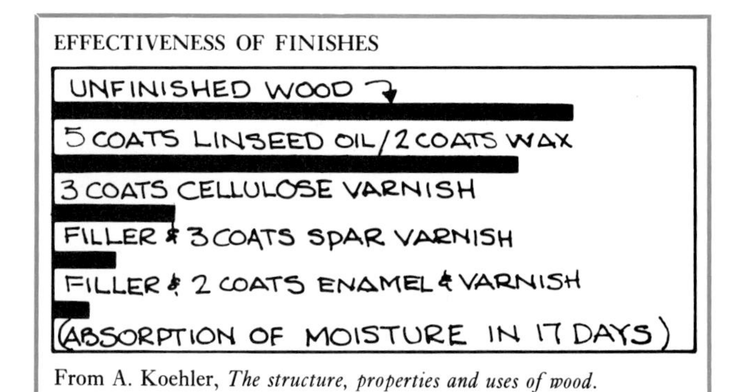

From A. Koehler, *The structure, properties and uses of wood.*

Movement in wood In the drying process, wood first loses its free water held in the center of the cells, but this does not cause any change in dimension. Only when the fiber saturation point is reached (about 30%) do the cell walls begin to lose water with the subsequent loss of volume. Thus, wood dried to the normal practical range will expand as it picks up moisture and contract when it loses it. This process is continuous, even in antiques, and seasonal since winter heating produces dry air (low humidity) and summer, high humidity (although in some parts of the world this cycle is reversed).

The amount of shrinkage (or expansion) depends on the species and on the direction of the grain.

The amount of movement is always more tangentially (along the rings) than radially (across the rings), and problems arise as a result of this differential shrinkage.

The chart opposite gives the amount of shrinkage per foot width for the tangential direction and the radial direction with a change in humidity from 90 to 60%, with the corresponding moisture contents of 20

Movement of timbers in seasoned condition due to reduced humidity

The table (right) shows the amount of expansion or contraction expected in various species with a change in humidity from 90% to 60% or vice versa. Those species with equal tangential and radial movement are likely to be the most stable causing minimum distortion.

Although the exact amount of expansion is rarely required in practice, the table is nonetheless useful in showing the relative stability of various species to aid in choosing woods for particular applications.

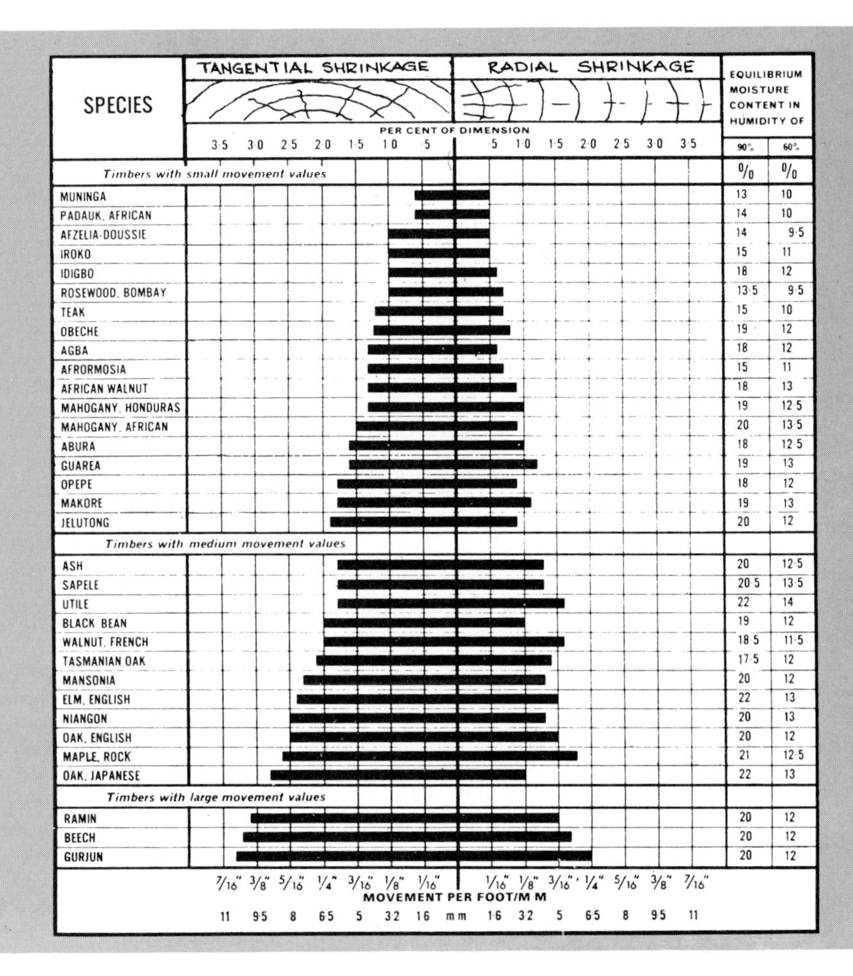

Below: Cross section of log shows how distortion depends on the board's orientation. A radially or quarter sawn board will shrink fairly evenly across the thickness and width. A tangentially cut board will cup away from the heart as the wetter, outer rings subject it to a larger pull along the outside.

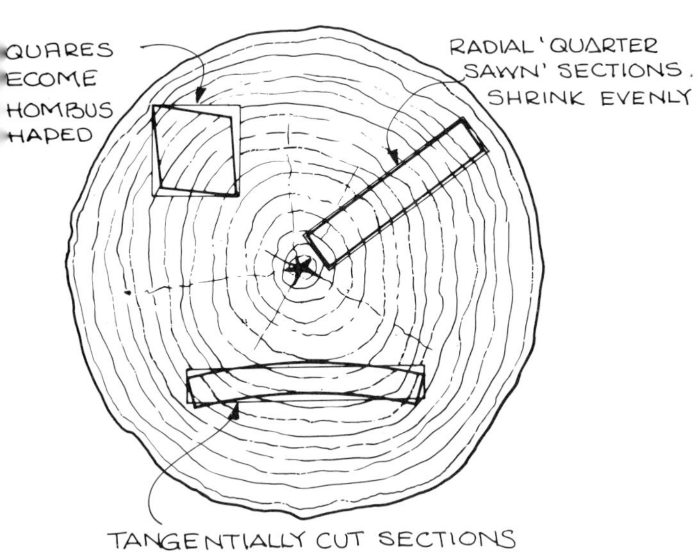

to 11% shown on the right. Species with small movements are towards the top.

The last column gives values of equilibrium moisture content for 60 and 90% humidity. Notice that the emc varies slightly for different species.

It is in this range where most movement takes place. Changes in moisture content below 11% will have less effect on movement. To work out the total shrinkage of a piece of tangentially cut African mahogany 18in wide as it is dried from 19 to 12% mc, the tangential value is read off as $\frac{3}{16}$in per 12in. Thus the total movement is $\frac{18}{12}$in $\times$ $\frac{3}{16}$in $=$ $\frac{9}{32}$in.

The effect on various boards is shown in the diagram. Tangentially cut (through and through) boards experience more shrinkage across the width (along the rings) than across it, but will also tend to cup away from the heart because the younger wood towards the outside of the tree shrinks more than the older center wood. Similarly, squares will distort and become rhombus-shaped.

A radially cut, quarter sawn board shrinks more in thickness along the rings than in width, which causes fewer problems and accounts for the popularity of quarter sawn wood among cabinetmakers. Boards with wild or uneven grain will tend to move unpredictably usually warping and twisting.

Finishing

There is a bewildering variety of wood finishes, and mastering their application takes years of experience and often requires the kind of knowledge and tricks of the trade which are only learned from old-timers who learned it from other old-timers.

Since most of us don't learn our woodworking that way, we must rely on books and miscellaneous trade advice as a starting point. The variety of finishes should be kept to a minimum and the choice will depend as much on the facilities available as on the type of finish wanted.

The facilities are important, because without proper spray equipment it is impossible to properly apply some of the modern quick-drying synthetic lacquers, widely used in production work. Most lacquers, varnishes and even French polishing require a warm and relatively dust free environment which usually means a separate room sealed off from the dusty workshop and well ventilated to remove harmful vapors and comply with safety requirements.

The success of a finish depends on good preparation and on keeping the finish simple. Good preparation requires control over the various steps in the gradual smoothing of the wood. Accurate ripping without burn or score marks requires less planing, and slower planing with sharp blades produces smaller planing marks (see page 118) which require less sanding. Following the same procedure, sanding should proceed with gradually finer papers (page 280). Of course a stroke belt sander (page 160) is a marvellous tool to give quick professional results, but it's an expensive machine and most of us have to make do with scrapers, and vibrating or belt sanders. The difference between hand and machine methods is only time and effort. The results should be the same.

The second criterion of success is keeping the finish simple. This is easy because rather than covering the wood with various layers of pigments, bleaches, paints, etc., you apply the finish for the reason it was intended, which is to protect the wood from dirt and blemishes and to seal it off from the changes in moisture content. The fact that it enhances the natural wood qualities is an added bonus. Of course fillers, stains and other specialist's methods, do have their place, but this is more relevant to the furniture industry which often wants to make one wood look like a more expensive one.

Wood characteristics The precise finishing methods vary with the texture of each wood. Ring porous woods such as oak, which show a coarse texture, usually require some form of grain filling for a smooth finish. Fine textured woods such as maple, which are diffuse porous woods, are quite smooth to begin with, making them suitable for a simple oil or wax finish as well as a lacquered finish. Semi-ring porous woods, such as walnut, fall somewhere in between so grain filling is optional.

The difference in porosity between the denser summer and the large pored spring woods is also a consideration. In some softwoods such as Douglas fir, the spring wood will soak up stain making it darker than the summer wood. The usual remedy is to apply a sealing coat of shellac, white for light woods and orange for dark woods. Alternatively, a coat of shellac mixing lacquer will seal troublesome grain. This is produced by first reducing four parts shellac with one part denatured alcohol which is then slowly mixed with an equal part of clear mixing lacquer.

Certain species such as teak, iroko and afrormosia, have a high resinous oily content which makes them ideal for outdoor use but which limits the finishing materials to teak or Danish oils or thin coats of lacquer.

Using different species of wood in the same piece of furniture also presents certain problems. In chairs, for example, there can be as many as three or four different woods, and if it is important to match these, skillful applications of stains are required.

The best way to proceed with any finishing problem is to experiment on scrap pieces, which often brings unexpected inventions to add to the list of established formulae.

Staining Stains are coloring pigments suspended in a solution of either water, oil or alcohol. There are numerous pigments some of which are natural, but most of which are chemicals such as anilines which are moderately color-fast even under constant light. As with fillers, pigments and the solvent should be chosen to be compatible with the other finishes to be applied. Check through the trade for mail order firms or dealers who supply the numerous powdered dyes which can then be mixed by formula or trial and error to produce the desired effect.

Water-based stains are the cheapest and the deepest penetrating stains, but they are difficult to apply evenly without blotching. Mix the dye in hot or boiling water until the right color is reached. Before applying, wipe clean water over the work, particularly the end grain, allow to dry and sand off. Clear the dust, then apply a light wash of water again to help spread and even the stain. This should be applied with a sponge, brush or rag and wiped off after application and left to dry before further deepening coats are added.

Oil stains which are usually a solution of dyes in boiled linseed oil and turpentine, can be mixed with wood fillers such as silica or pumice powder to make a stained filler. Oil stains don't penetrate as deeply as water stains and don't dry as quickly, but they are easier to apply to an even tone. And since they dry slowly, the laps are not likely to show. Apply a thin sealing coat of shellac over oil stains to avoid a reaction between the stain and the varnish.

Spirit or alcohol dyes are more difficult to use, but for the professional finisher the wide range of spirit soluble aniline dyes make them very valuable. They dry quickly, which means the varnish can be applied the same day. They are therefore more suitable for spraying than brush application. They do bleed or strike through most other finishes, however, which makes them difficult to use.

Traditional dyes for staining

Various chemicals and natural materials when dissolved in water, oil or alcohol, will serve as good color-fast dyes. The following are a few traditional formulae.

To stain mahogany brown, dissolve bichromate of potash in warm water. To stain it gray to imitate walnut, use sulphate of iron dissolved in water.

To stain oak brown, use a solution of bichromate of potash dissolved in hot water, or alternatively, vandyke crystals in warm water, strained through muslin. Add a tablespoonful of 880 ammonia to each pint of stain before application to get a richer color.

Aniline dyes, available in powder, or in solution with water, oil or alcohol, come in a range of wood tones such as vandyke brown, black, umber, and also bright colors such as red, blue and green.

Filling wood Open textured woods such as oak, walnut and mahogany require filling before any smooth, matt or gloss finish such as French polish or lacquer can be applied. Fillers can be paste or liquid. The former, used on open-grained wood are not as transparent as liquid filter used on close-grained wood. Basically, anything which will fill the pores and not interact with or cloud the finish can be termed a filler. Thus, several thin coats of lacquer can fill grain, but generally, a filler with more body is used to reduce the number of applications required. The traditional filling materials are milled plaster of Paris, whiting, pumice powder and fine silica (silex) or china clay, to which coloring pigments can be added to match any tone of wood. These can be mixed with turpentine, raw linseed oil and japan drier, and wiped across the grain as a thin cream. The standard formula for paste filler is to use $\frac{1}{4}$ pint of boiled linseed oil, two ounces of japan drier and $\frac{1}{2}$ pint of turpentine. Mix these thoroughly before adding silex to make a paste. This should in turn be strained through a wire screen and then finally thinned with turpentine to achieve brushing consistency.

A simpler, old fashioned formula uses fine plaster of Paris and water. Place each in a bowl and wipe a lint free cloth across the grain with a circular motion first dipping it in water and then in the plaster. Remove any surplus plaster with a coarse rag, again rubbing across the grain. Allow the filled surface to dry for about eight hours, then rub with fine paper along the grain. The final stage is to apply a thin coat of raw linseed oil to bring back the color of the wood, making sure to wipe off all surplus oil.

There are various liquid fillers on the market, but a simple method suitable for most close-grained woods is to use ordinary shellac, white or orange, depending on the color of the wood. Used as a filler, shellac should be thinner than the 4lb cut usually supplied. A 2lb cut is about right. If this cannot be bought, add one gallon of alcohol to a gallon of 4lb shellac. Apply two coats, sanding each when it is dry.

There are various other proprietary fillers on the market, in paste or liquid form. These must be chosen carefully to match the stain below and the finish to be applied above.

It was once standard to seal both the stain and filler coats, particularly in the best work. A coat of white or orange shellac is the best sealer particularly to cover an oil-based stain. But remember to keep the wash coat thin and well sanded down so that it doesn't cloud the finish.

Oil finishes Raw linseed oil used alone or by the traditional formula of $\frac{1}{8}$ part turpentine to $\frac{7}{8}$ part oil, requires numerous coats to build up sufficient patina to offer any real protection, but even then it isn't completely waterproof. (Once a day for a week, once a week for a month, once a month for a year, and once a year thereafter is the rule of thumb.) Linseed oil dries by slow oxidation and has largely been replaced by more modern penetrating oils such as teak and Danish oils with a higher resin content which react with the wood chemically to form a good protective coating in one or two coats. Oil finishes are more successful on close-grained woods such as maple or birch where no filling is required. They can be applied in a moderately dusty environment which is a tremendous advantage for small shops where a separate dust free finishing room is not available. Standard application is to soak the surface and allow to penetrate for about 1 to 2 hours before wiping off with a clean cloth and buffing with another.

Oil finishes are easy to renew as scratches or marks can easily be sanded and refinished to match the old surface. It is also possible to color tone the oil by adding small amounts of powdered stain to the oil. But it is important to try the effect on scraps first before proceeding. It is also important to mix enough for the entire job as it is almost impossible to get identical results on two different mixings.

To give oil more water resistance, warm a mixture of equal parts of boiled linseed oil, spar varnish and turpentine in a glue pot or double boiler, then rub this mixture well into the wood, rubbing with fine steel wool or wet and dry paper, if necessary, before buffing dry.

Shellac

Shellac is available in granular form to be mixed with alcohol or as a ready-to-use liquid, with the grains dissolved in industrial alcohol. It comes in orange, white or de-waxed form. The orange, which is the least pure form, is the most popular for dark woods such as mahogany, whereas the white is used for light woods.

The concentration of shellac known as the "cut", is the proportion of shellac to alcohol. Thus a 4lb cut is 4lb of shellac granules dissolved in 1 gallon of alcohol. Generally the first coat or two are 2lb cut, followed by coats of the thicker 4lb cut.

Shellac can be brushed on using a soft varnish brush or it can be sprayed but it is most often rubbed in as a specialist French polish.

It must be emphasized that whereas a well applied shellac finish is fairly resistant to normal wear, it is not suitable for table surfaces or any other surface where it will come in contact with water or alcohol. This will leave white rings on the surface.

For cabinets and other light use, shellac can be used as a simpler wood finish that leaves a nice, silky matt look by applying a 2 to 4lb cut of white shellac with a brush. Apply several coats, smoothing down with 000 steel wool between coats. As a final touch, buff with a paste wax. Two further points to remember. First, shellac has a limited shelf life so buy only small quantities from large suppliers to guarantee fresh stock. Second, because it tends to cloud over in the presence of moisture, apply shellac on dry days.

Wax polish Wax offers little protection applied onto bare wood and is, therefore, not suitable for tables and other surfaces subject to hard wear. But wax makes an excellent polish over thin protective coats of shellac, lacquer, polyurathane or sanding sealer. Use the standard recipe of 1lb of beeswax melted into $\frac{1}{2}$ pint of turpentine in a double boiler. Allow this to cool into a paste and apply liberally to the wood. Wipe off with 0000 steel wool before it is allowed to harden, and buff with a soft cloth. Adding carnauba wax to the mixture gives a harder wax with more luster. There are many other wax formulae on the market, some of which are sold to match a specific polish.

Varnish Varnishes consist of a mixture of copal gums and linseed oil mixed with turpentine. Copal gums are natural solids which have been largely replaced by synthetic resins. Varnishes are slow drying and therefore unsuitable for production work, but they are still popular with craftsmen because

Stoppings

An old fashioned formula for stopping, used to fill bruises in French polishing and other holes and indentations in general finishing, is to mix equal parts of beeswax with crushed rosin. These are melted in a tin can and pigment such as vandyke brown or burnt umber for oak and walnut are added to match the wood. Apply with a spatula or stick after heating slightly in the tin. This stopping is not heat resistant, but proprietary brands are available which are more durable.

they can be applied by brush and are relatively inexpensive. Varnishes are either oil or alcohol based. The oil-based varnishes which harden by oxidation, are the most widely used, particularly for furniture and exterior finishes. Alcohol-based varnishes which harden by evaporation, have more limited application.

There are numerous varnishes with varying proportions of oil and different types of resin which make them suitable for particular uses. Spar varnishes, for example, have a high proportion of oil to make them slow drying, elastic and tough, suitable for exterior use. Other types include floor varnish, bar top varnish and polishing varnish. For further advice, it's best to contact a wholesaler who supplies the furniture trade.

Nitro-cellulose lacquers These are the original clear lacquers which are still popular today even though the development of synthetic lacquers has brought more sophisticated and harder materials into use.

Cellulose, made mostly from cotton fiber, is mixed in solution with various chemicals such as plastics and resins to give them greater flexibility. They dry in minutes which requires spray equipment, but brushing lacquers are available for handwork. These can be rubbed down with 0000 steel wool between coats and wax finished if desired.

Before spraying, they should be thinned down by trial and error to give a smooth flow through the gun to avoid forming an "orange peel" surface. Cellulose lacquers are hard, durable and water resistant (not waterproof), but not as hard and durable as synthetic lacquers.

Synthetic lacquers There are several varieties including polyester, polyurethane, urea formaldehyde and melamine. These chemicals are also used for adhesives and for plastic extrusions. These are very hard and durable, of interest to the craftsman for brush application onto table tops, bar tops and turnery. Further applications should be applied in fairly quick succession to achieve proper adhesion between the coats. They needn't be applied in thick, glossy coats. One or two thin coats well rubbed down with wet and dry paper forms a good base for the burnishing creams often sold with the lacquers or for waxes well rubbed in with 0000 steel wool. As with all lacquers, these come in gloss, matt, semi-gloss or eggshell finish, depending on the amount of powder added to the basic lacquer.

Rubbing with pumice powder

As a final touch to varnished surfaces, rub them down with a fine abrasive to take away the glossy look and to give the surface a deep, lustrous and smooth finish.

You can buy various rubbing compounds such as special lacquer rubbing pastes, but the easiest method is to rub the surface with a mixture of pumice powder and rubbing oil such as paraffin oil or thinned motor oil. Teak oil is also suitable, or you can use a special polishing oil.

Pumice powder is available in different grades and only the finest should be used for finishing. Alternatively use fine rotten stone. Either mix the powder and oil together to make a liquid the consistency of cream, or dip a cloth first in oil and then in the pumice powder. As an alternative to cloth, use a small piece of soft, rubbing felt such as that formerly used for carpet underlay. Rub the oil in really well, working with the grain. Renew the oil and pumice finish from time to time.

The pumice acts both as a filler and an abrasive to remove any imperfections on the surface. Once the surface is well rubbed, go over it again with oil alone, buffing well to make it really shiny and smooth.

Bleaching

To lighten the tone of woods, apply special wood bleach to the boards. You can make a homemade mixture using oxalic acid, but it is safer to use a commercial bleach sold at hardware stores. Bleach is dangerous to use as it will burn skin and clothing, so wear rubber gloves and an apron and use the liquid carefully avoiding any contact with the skin.

Bleaches are usually sold in two parts. The bleach itself is applied with a fiber brush or a sponge and the neutralizer is applied to stop the action of the bleach.

Follow the manufacturer's instructions carefully and after bleaching, sand the work lightly before applying finish.

Never put bleaches into metal containers. Keep them in the plastic bottles in which they are sold. If you mix your own, use glass, or earthenware or plastic containers.

Abrasive papers

Sanding is a misnomer since no form of natural silica is used, but it is nonetheless used to mean the scraping or abrasive cutting process of smoothing down the surface prior to "finishing". The so-called "sandpaper" consists of several components. The particles or grits which have sharp crystalline edges of varying hardness and roughness, are held to a backing of paper or cloth by two layers of adhesives of varying make up.

Abrasives The relative hardness of the grits made from fractured particles determines the cutting action. The softer glass is used only in hand sanding where the cutting action is slow and the grits soon lose their sharp cutting edges. Flint is similar to glass but more rarely used today because it is incompatible with modern synthetic materials. Garnet is, like glass and flint, a natural material, though it is heat-treated for a more uniform consistency. Garnet grit is reddish-brown and tougher than most other materials which gives it a long-lasting quality suitable for hand sanding and some machine sanding. It has long been favored by woodworkers for its even and smooth cutting quality. Aluminous oxide, gray in color, is a man-made abrasive made from fusing bauxite powder with silica and ferrite in an electric furnace. It is extremely hard and tough which makes it ideal for machine sanding. Silica carbide which is the hardest grit available, is made by combining coke with silica in a furnace. The resulting particles are very hard but brittle (opposite of tough) which makes it less suitable for machine sanding. Silicon carbide is rarely used for wood finishing. It is more often used in its "wet and dry" form for polishing very hard substances such as plastics and cellulose.

Steel wool

Steel or wire wool is often used as a fine abrasive to rub down between coats. It comes in several grades from no3, which is not suitable for use in finishing to no00 and 000 and the very fine 0000, though the latter is not as readily available.

Steel wool should be used with care since it has more bite than some papers and in addition, it tends to leave small particles in the finish particularly in open-grained species such as oak. It is best to wipe off a steel wooled surface with a cloth moistened with fast drying alcohol.

GRIT SIZE The size of the particles determines the coarseness of cut. The size, stated as a number, e.g. 80, relates to the size of silk cloth screen through which that grit can pass. Thus 80 grit denotes a screen with 80 rectangular holes per inch. Grit sizes vary from 12 to 800, the finer the cut the higher the number. The practical range for hand finishing is 100 to 240 (equivalent to 2/0 and 7/0 respectively in the old grading system), whereas the normal range for machines sanding is 60 to 150, with 40 and 220 for special applications.

BONDING AND GRIT DISTRIBUTION Abrasive paper is made in a long continuous belt by passing the backing first through a coating process (bond coat) chamber. There the grits sprinkled on the pre-coated backing are ionized in the chamber causing them to stand upright which increases their cutting power and evens out the projections. The upright particles are locked in place by a "sizing" coat. The first coat is usually of hide glue to give flexibility, but the final coat of adhesive or size is usually a resin glue to give added protection against heat, which tends to soften hide glues.

The grits may completely cover the backing (closed coating) or it may cover up to 75% of the backing (open coat). Closed coats are used on tougher jobs and should, in theory, remove more wood than open coated papers. But in hand sanding generally, and for gummy or resinous woods in particular, open coated papers are more suitable since they hold the wood particles better between grits. Fine grit papers for finishing sanding are open coat only but most other grades are available in either type to suit specific applications.

BACKINGS The backing is either paper or cloth or a combination of the two. Paper backings made from jute and wood fiber are generally used for flat belt and pad sanders. Cloth backings, made of pure cotton, are more flexible and longer lasting and are used for sanding any irregular surfaces such as chair components in power sanders. Combination backings made from heavy paper reinforced with fabric is used for special purpose sanders such as drum sanding machines. Vulcanized fiber backing is tougher and best suited to disk and drum sanders.

The backings are made in a range of grades by weight. Paper is graded from A weight (75 grams/m^2) to E weight (230 grams/m^2).

"Cabinet" papers for coarse hand sanding are C or D, whereas A is for final finishing only. E weight is for machine sanding.

Spray finishing

Spray finishing with a pressure fed spray gun is quick and simple and produces the best finish. You can buy small and relatively inexpensive guns and compressors such as the one shown here for applying lacquers, water-based paint, oil paint, varnish and even stains. Its use isn't limited to furniture. Household jobs such as painting rooms, door and window frames and spray work on a car are easily done with a spray gun.

Applying paint with a gun takes a little practice so it is important to try the gun out on a sheet of hardboard or plywood so that you develop the right touch before you spray a piece of work.

The most important part of spray painting is the preparation of the material.

PREPARATION FOR SPRAYING

Most finishing materials are suitable for spray application, but if you are in doubt ask the paint supplier or the manufacturer of the spray equipment for advice. Before you fill the gun, you must follow certain steps to prepare the liquid you are spraying.

First stir the paint or lacquer thoroughly to be sure that the color is well mixed. Check the viscosity of the liquid before you fill the gun. This is an important consideration as liquids that are too thick will spray a blotchy surface and if the liquid is too thin, the paint will run.

Use a viscosity cup supplied by the spray gun manufacturer to check that the liquid is the right consistency. Fill the cup with paint and time the liquid as it runs out. Thin the liquid with the appropriate solvent if it is too thick.

In the absence of a viscosity cup add a small quantity of the recommended thinner and stir thoroughly to mix with a clean stick. Withdraw the stick and hold it at a 45° angle. When the paint runs from the end of the stick in a continuous stream, it should be roughly the right consistency for spraying.

It is also good practice to strain the paint before use and remove all impurities to minimize wear of the moving parts of the spray gun as well as helping to achieve a better surface finish. You can strain the liquid by pouring it through a single layer of nylon stocking mounted over the top of a container. As with all finishes, the surface to be sprayed should be clean, dry and free from dust.

Spraying should be done in a large, clean, well ventilated area with good lighting. Cover everything which is not being sprayed. It's also important for your own safety to wear a mask to cover your mouth

Spray equipment from the large system with industrial water wash booths to this small portable compressor suitable for the small workshop.

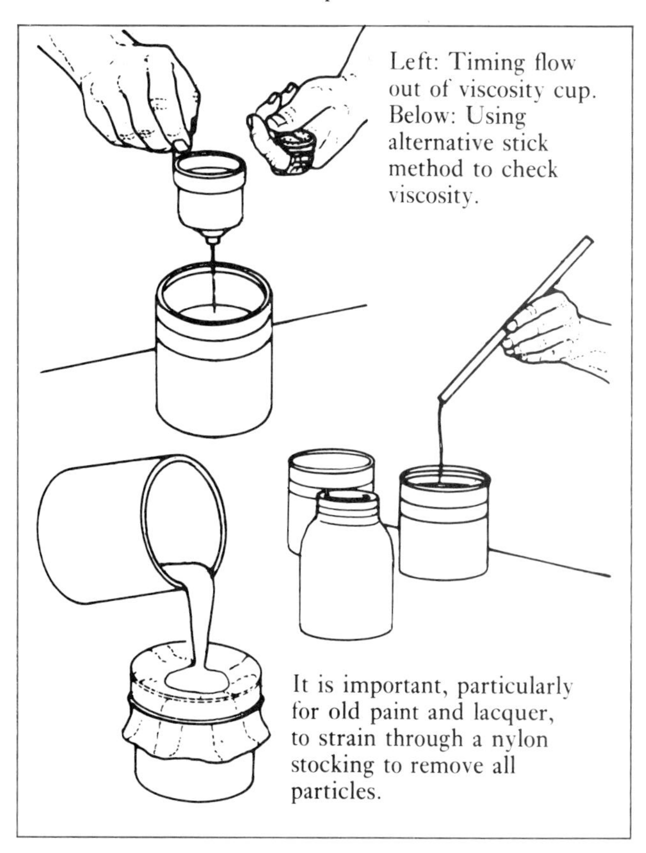

Left: Timing flow out of viscosity cup. Below: Using alternative stick method to check viscosity.

It is important, particularly for old paint and lacquer, to strain through a nylon stocking to remove all particles.

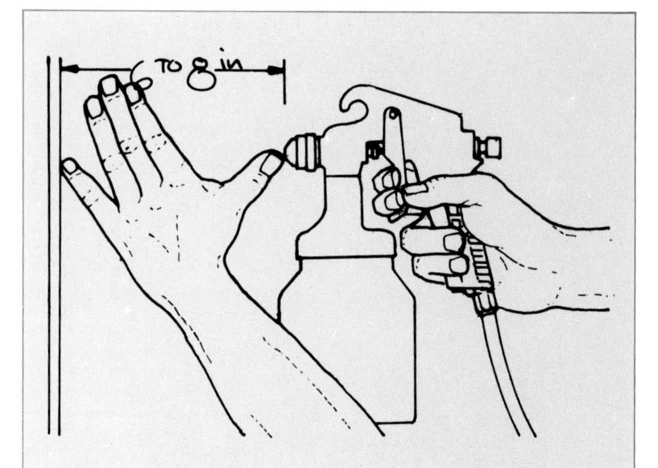

Above: The gun should be held at right angles to the work about 6 to 8in away.

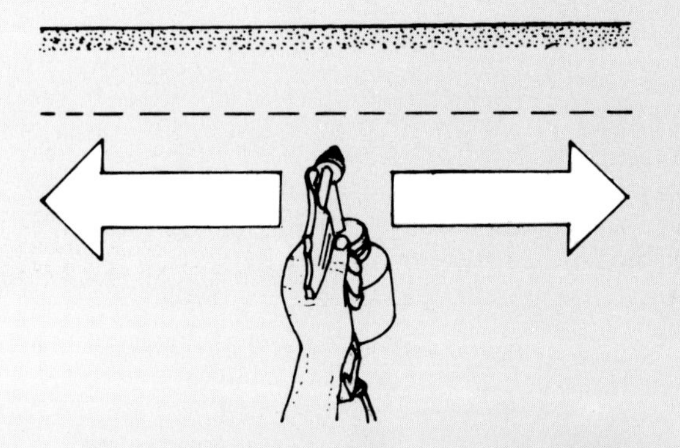

Above: Keep the strokes parallel at all times starting the movement of the gun before the trigger is pulled at the beginning of the stroke and vise versa at the end of the stroke.

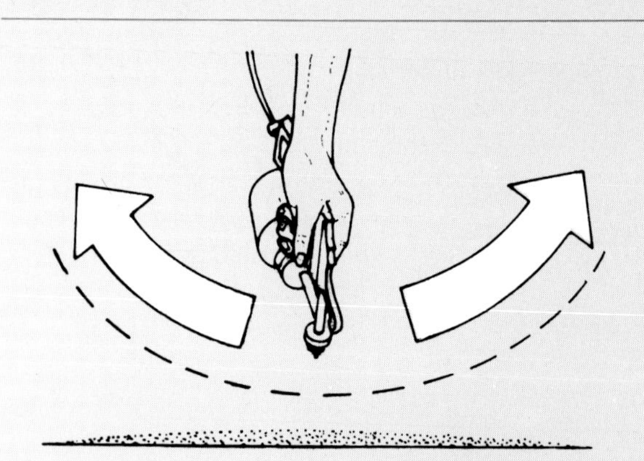

The classic beginner's mistake is to spray in a circular or fan pattern which makes it impossible to maintain a uniform thickness.

and nose when using spray equipment.

If you spray outdoors, do it on a still, dry day and do the work early in the morning to minimize insects settling on the surface. Cover the area around the work with a plastic sheet or newspapers secured with adhesive tape.

Using the spray gun Make sure that the fluid control adjustment on the gun is set correctly. While practicing, experiment with the adjustment, starting from the almost closed position to create a small pattern. Adjust the spray gradually to widen the pattern created. Notice that the more liquid that is sprayed through, the more the spray tends to break up or atomize into particles.

Practice spraying both vertically and horizontally with the adjustment set to a fine spray in a flat, fan pattern until you are satisfied with the result.

It is important when spraying to use the correct techniques. The gun should be held at right angles to the work with the distance between the surface and the face of the gun maintained at 6 to 8in. The best idea is to practice on an old piece of furniture which allows you to make mistakes and to spray in the various positions for horizontal as well as vertical planes usually required in spraying furniture.

Each stroke is made with a free arm motion across the face of the work surface with the wrist kept flexible so that the gun is kept at right angles to the surface at the correct distance from it at all times. Keep the speed on each stroke constant to maintain a uniform thickness of coating. To prevent the building up of paint at the beginning and end of each stroke, the movement of the gun should be started before you pull the trigger and the trigger should be released again before the movement of the gun is finished at the end of the stroke. Do not jerk the gun or move it abruptly.

Overlapping stokes

The edges of the spray pattern taper off slightly, so to cover the surface evenly, it is necessary to overlap the previous stroke by about 50 per cent. To do this, aim the gun at the edge of the previous stroke to get the right overlap. Then finish off the ends with one cross stroke.

To cover a large area, spray with a series of straight, overlapping strokes and also overlap the ends. Finish inside corners in one stroke. For grillwork, make sure to use a backing.

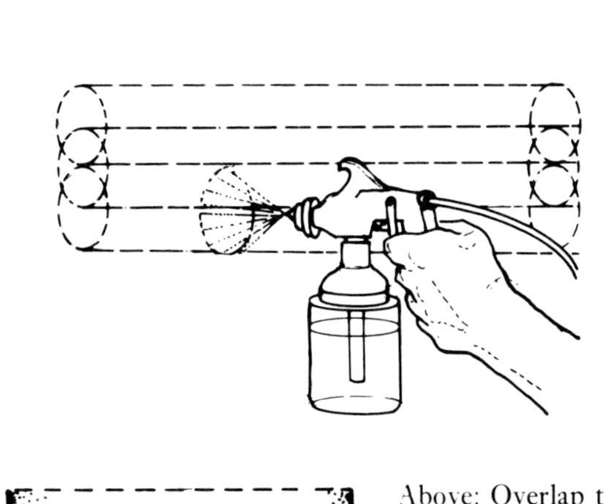

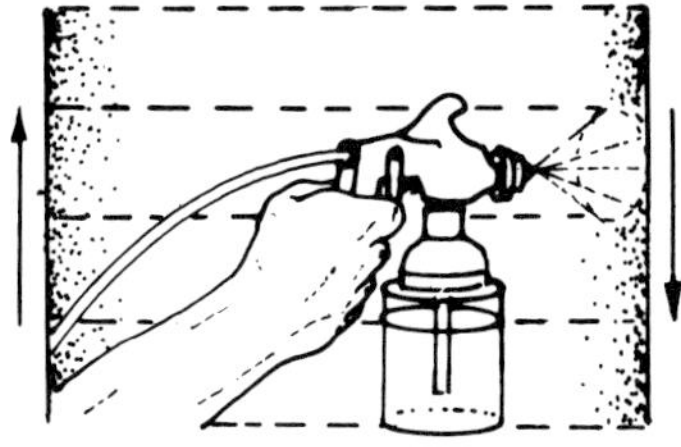

Above: Overlap the strokes by about 50%. Left: Finish off the end with one cross stroke. Below: On long surfaces, overlap each set of strokes.

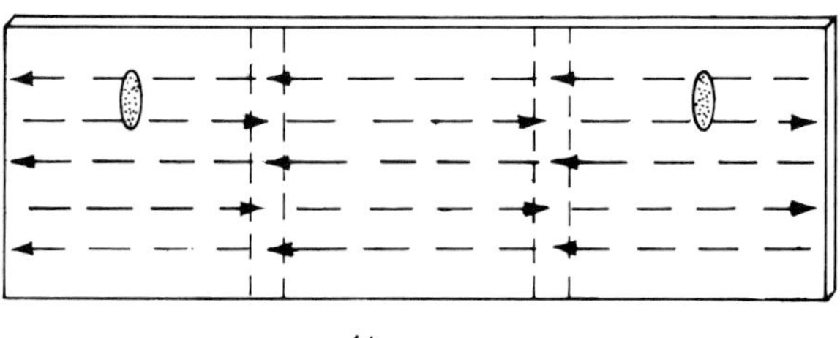

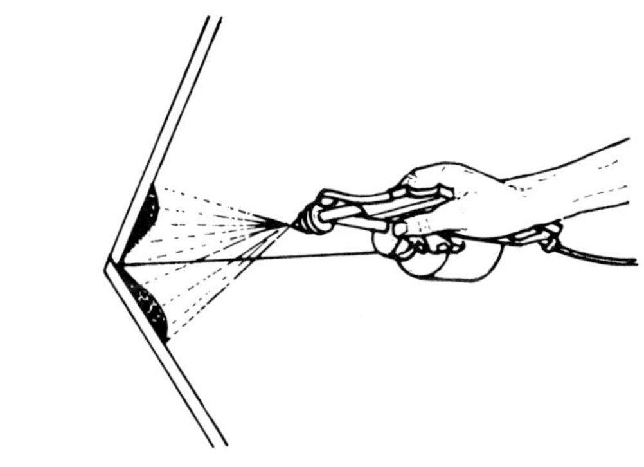

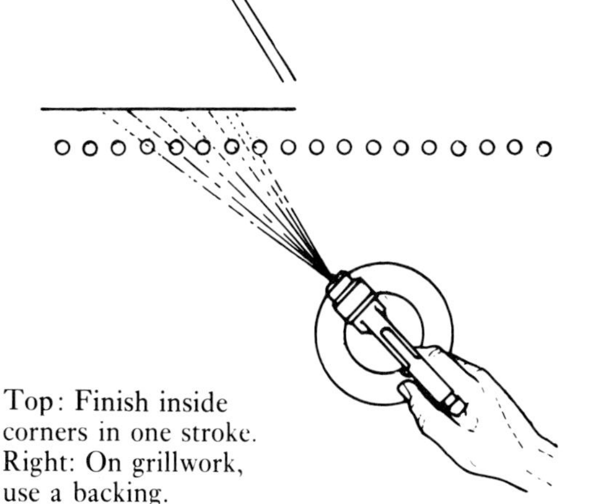

Top: Finish inside corners in one stroke. Right: On grillwork, use a backing.

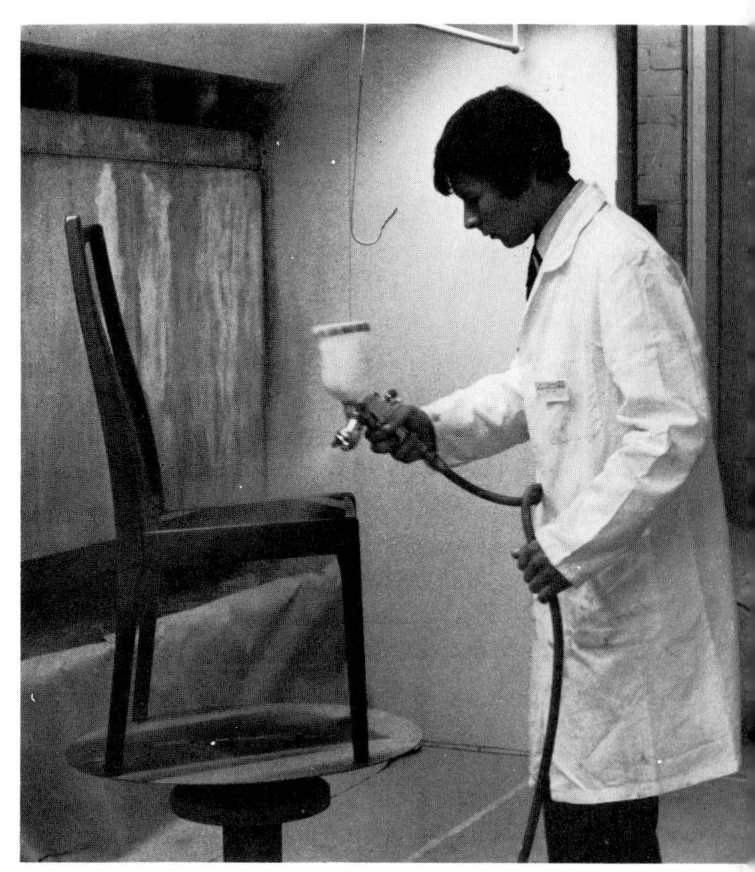

For production shops, spray booths which conform to the factory regulation standards are essential.

Cleaning the equipment

It is essential to clean the spray equipment after each use and the sooner you clean it, the easier the job.

Be sure to use the correct cleaning fluid for the finish you are spraying. When spraying is completed, stop the compressor and pull the trigger to relieve air pressure. Unscrew the paint cup and allow excess paint to drain out. Empty the paint cup and clean thoroughly with solvent.

Pour some thinner into the paint cup, reassemble the gun and spray until pure thinner is passed out. Switch off and take the paint cup off again. Dry all the pieces well. Remove the air cap and soak in thinner, cleaning the slot with a wooden toothpick.

Wipe all the pieces with a soft rag and keep in a dry, dust free place when not in use. Be sure to keep the compressor clean and clean the air inlet filters and inside the air hose each time.

Index

Craftsmen and designers whose work appears in this book:

Sam Bush
The Hill School Pottstown, Pennsylvania

Ashley Cartwright
1 Banbury Road Brackley, Northamptonshire

David Field
Barley Mow Workspace
Barley Mow Passage Chiswick, London W4

Gil Gesser
430 Rue St. Pierre Montreal, Quebec

Martin Grierson
Barley Mow Workspace
Barley Mow Passage Chiswick, London W4

Tupu Kiveliö
Lehtakatu 3 E 15 15500 Lahti 50 Finland

John Makepeace
School for Craftsmen in Wood
Parnham House Beaminster, Dorset

Alf Martensson
Woodstock
Albion Yard Balfe Street London N1

Sergio Mora
The London College of Furniture
41 Commercial Road London E1

Alan Peters
Aller Studios
Kentisbeare Cullompton, Devon

Dan Valenza
Paul Creative Arts Center
University of New Hampshire Durham,
New Hampshire

Jim Warren
128 Dunstans Road London SE22

Hans Wegner
c/o Den Permanente
Vesterport Vesterbrogade 8 DK 1620 Copenhagen V

The following companies who helped in the production of this book will supply further technical and sales information about their products.

J. Crispin and Sons (veneers)
92–96 Curtain Road, London EC2

The DeVilbiss Company Ltd (spray equipment)
Ringwood Road, Bournemouth, Hampshire, England

DeWalt/McCulloch of Europe (radial arm saws)
Clivemart Road, Cordwallis Industrial Estate,
Maidenhead, Berkshire, England

Ebac Ltd. (dehumidifier kilns)
Greenfield Industrial Estate,
Bishop Auckland, County Durham, England

Elu Machinery Ltd. (portable power tools)
Stirling Way, Stirling Corner,
Boreham Wood, Hertfordshire, England

William Mallinson and Sons (hardwoods)
130 Hackney Road, London E2

Parry and Son Ltd (tools and machinery)
329 Old Street, London EC1

Protimeter Ltd (moisture meter)
Meter House, Fieldhouse Lane,
Marlow, Buckinghamshire, England

Robin Sorby and Sons Ltd (woodturning tools)
6 Cobmar Gardens, Woodseats, Sheffield, England

Sperber Tool Works Inc (chain saw mill)
Box 1224, West Caldwell, New Jersey, USA

Philip Cole (*UK distributor of Sperber mill*)
16 Kings Lane, Flore, Northamptonshire, England

Stanley Power Tools Ltd (portable power tools)
Nelson Way, Cramlington, Northumberland,
England

Tinker Engineering Co
(Harrison graduate lathe)
253 Putney Bridge Road, London SW15

Wadkin Limited (machines and accessories)
Green Lane Works, Leicester, England

Wolf Electric Tools Ltd
(portable power tools, grinders)
P.O. Box 379, Hanger Lane, London W5